T0214332

# Lecture Notes in Computer Science 12917

More information about this subseries at http://www.springer.com/series/7412

Elisa H. Barney Smith ·
Umapada Pal (Eds.)

# Document Analysis and Recognition – ICDAR 2021 Workshops

Lausanne, Switzerland, September 5–10, 2021
Proceedings, Part II

 Springer

*Editors*
Elisa H. Barney Smith
Boise State University
Boise, ID, USA

Umapada Pal 🄳
Indian Statistical Institute
Kolkata, India

ISSN 0302-9743    ISSN 1611-3349   (electronic)
Lecture Notes in Computer Science
ISBN 978-3-030-86158-2    ISBN 978-3-030-86159-9   (eBook)
https://doi.org/10.1007/978-3-030-86159-9

LNCS Sublibrary: SL6 – Image Processing, Computer Vision, Pattern Recognition, and Graphics

This Springer imprint is published by the registered company Springer Nature Switzerland AG
The registered company address is: Gewerbestrasse 11, 6330 Cham, Switzerland

# Foreword

Our warmest welcome to the proceedings of ICDAR 2021, the 16th IAPR International Conference on Document Analysis and Recognition, which was held in Switzerland for the first time. Organizing an international conference of significant size during the COVID-19 pandemic, with the goal of welcoming at least some of the participants physically, is similar to navigating a rowboat across the ocean during a storm. Fortunately, we were able to work together with partners who have shown a tremendous amount of flexibility and patience including, in particular, our local partners, namely the Beaulieu convention center in Lausanne, EPFL, and Lausanne Tourisme, and also the international ICDAR advisory board and IAPR-TC 10/11 leadership teams who have supported us not only with excellent advice but also financially, encouraging us to setup a hybrid format for the conference.

We were not a hundred percent sure if we would see each other in Lausanne but we remained confident, together with almost half of the attendees who registered for on-site participation. We relied on the hybridization support of a motivated team from the Lule University of Technology during the pre-conference, and professional support from Imavox during the main conference, to ensure a smooth connection between the physical and the virtual world. Indeed, our welcome is extended especially to all our colleagues who were not able to travel to Switzerland this year. We hope you had an exciting virtual conference week, and look forward to seeing you in person again at another event of the active DAR community.

With ICDAR 2021, we stepped into the shoes of a longstanding conference series, which is the premier international event for scientists and practitioners involved in document analysis and recognition, a field of growing importance in the current age of digital transitions. The conference is endorsed by IAPR-TC 10/11 and celebrates its 30th anniversary this year with the 16th edition. The very first ICDAR conference was held in St. Malo, France in 1991, followed by Tsukuba, Japan (1993), Montreal, Canada (1995), Ulm, Germany (1997), Bangalore, India (1999), Seattle, USA (2001), Edinburgh, UK (2003), Seoul, South Korea (2005), Curitiba, Brazil (2007), Barcelona, Spain (2009), Beijing, China (2011), Washington DC, USA (2013), Nancy, France (2015), Kyoto, Japan (2017), and Syndey, Australia in 2019.

The attentive reader may have remarked that this list of cities includes several venues for the Olympic Games. This year the conference was hosted in Lausanne, which is the headquarters of the International Olympic Committee. Not unlike the athletes who were recently competing in Tokyo, Japan, the researchers profited from a healthy spirit of competition, aimed at advancing our knowledge on how a machine can understand written communication. Indeed, following the tradition from previous years, 13 scientific competitions were held in conjunction with ICDAR 2021 including, for the first time, three so-called "long-term" competitions, addressing wider challenges that may continue over the next few years.

Other highlights of the conference included the keynote talks given by Masaki Nakagawa, recipient of the IAPR/ICDAR Outstanding Achievements Award, and Mickaël Coustaty, recipient of the IAPR/ICDAR Young Investigator Award, as well as our distinguished keynote speakers Prem Natarajan, vice president at Amazon, who gave a talk on "OCR: A Journey through Advances in the Science, Engineering, and Productization of AI/ML", and Beta Megyesi, professor of computational linguistics at Uppsala University, who elaborated on "Cracking Ciphers with 'AI-in-the-loop': Transcription and Decryption in a Cross-Disciplinary Field".

A total of 340 publications were submitted to the main conference, which was held at the Beaulieu convention center during September 8–10, 2021. Based on the reviews, our Program Committee chairs accepted 40 papers for oral presentation and 142 papers for poster presentation. In addition, nine articles accepted for the ICDAR-IJDAR journal track were presented orally at the conference and a workshop was integrated in a poster session. Furthermore, 12 workshops, 2 tutorials, and the doctoral consortium were held during the pre-conference at EPFL during September 5–7, 2021, focusing on specific aspects of document analysis and recognition, such as graphics recognition, camera-based document analysis, and historical documents.

The conference would not have been possible without hundreds of hours of work done by volunteers in the organizing committee. First of all we would like to express our deepest gratitude to our Program Committee chairs, Joseph Lladós, Dan Lopresti, and Seiichi Uchida, who oversaw a comprehensive reviewing process and designed the intriguing technical program of the main conference. We are also very grateful for all the hours invested by the members of the Program Committee to deliver high-quality peer reviews. Furthermore, we would like to highlight the excellent contribution by our publication chairs, Liangrui Peng, Fouad Slimane, and Oussama Zayene, who negotiated a great online visibility of the conference proceedings with Springer and ensured flawless camera-ready versions of all publications. Many thanks also to our chairs and organizers of the workshops, competitions, tutorials, and the doctoral consortium for setting up such an inspiring environment around the main conference. Finally, we are thankful for the support we have received from the sponsorship chairs, from our valued sponsors, and from our local organization chairs, which together enabled us to put in the extra effort required for a hybrid conference setup.

Our main motivation for organizing ICDAR 2021 was to give practitioners in the DAR community a chance to showcase their research, both at this conference and its satellite events. Thank you to all the authors for submitting and presenting your outstanding work. We sincerely hope that you enjoyed the conference and the exchange with your colleagues, be it on-site or online.

September 2021

Andreas Fischer
Rolf Ingold
Marcus Liwicki

# Preface

Our heartiest welcome to the proceedings of the ICDAR 2021 Workshops, which were organized under the 16th International Conference on Document Analysis and Recognition (ICDAR) held in Lausanne, Switzerland during September 5–10, 2021.

We are delighted that this conference was able to include 13 workshops. The workshops were held in Lausanne during September 5–7, 2021. Some were held in a hybrid live/online format and others were held entirely online, with space at the main conference for in-person participants to attend. The workshops received over 100 papers on diverse document analysis topics, and these volumes collect the edited papers from 12 of the workshops.

We sincerely thank the ICDAR general chairs for trusting us with the responsibility for the workshops, and for assisting us with the complicated logistics in order to include remote participants. We also want to thank the workshop organizers for their involvement in this event of primary importance in our field. Finally, we thank the workshop presenters and authors without whom the workshops would not exist.

September 2021

Elisa H. Barney Smith
Umapada Pal

# Organization

## Organizing Committee

### General Chairs

Andreas Fischer     University of Applied Sciences and Arts Western Switzerland
Rolf Ingold     University of Fribourg, Switzerland
Marcus Liwicki     Luleå University of Technology, Sweden

### Program Committee Chairs

Josep Lladós     Computer Vision Center, Spain
Daniel Lopresti     Lehigh University, USA
Seiichi Uchida     Kyushu University, Japan

### Workshop Chairs

Elisa H. Barney Smith     Boise State University, USA
Umapada Pal     Indian Statistical Institute, India

### Competition Chairs

Harold Mouchère     University of Nantes, France
Foteini Simistira     Luleå University of Technology, Sweden

### Tutorial Chairs

Véronique Eglin     Institut National des Sciences Appliquées, France
Alicia Fornés     Computer Vision Center, Spain

### Doctoral Consortium Chairs

Jean-Christophe Burie     La Rochelle University, France
Nibal Nayef     MyScript, France

## Publication Chairs

| | |
|---|---|
| Liangrui Peng | Tsinghua University, China |
| Fouad Slimane | University of Fribourg, Switzerland |
| Oussama Zayene | University of Applied Sciences and Arts Western Switzerland, Switzerland |

## Sponsorship Chairs

| | |
|---|---|
| David Doermann | University at Buffalo, USA |
| Koichi Kise | Osaka Prefecture University, Japan |
| Jean-Marc Ogier | University of La Rochelle, France |

## Local Organization Chairs

| | |
|---|---|
| Jean Hennebert | University of Applied Sciences and Arts Western Switzerland, Switzerland |
| Anna Scius-Bertrand | University of Applied Sciences and Arts Western Switzerland, Switzerland |
| Sabine Süsstrunk | École Polytechnique Fédérale de Lausanne, Switzerland |

## Industrial Liaison

| | |
|---|---|
| Aurélie Lemaitre | University of Rennes, France |

## Social Media Manager

| | |
|---|---|
| Linda Studer | University of Fribourg, Switzerland |

## Workshops Organizers

### W01-Graphics Recognition (GREC)

| | |
|---|---|
| Jean-Christophe Burie | La Rochelle University, France |
| Richard Zanibbi | Rochester Institute of Technology, USA |
| Motoi Iwata | Osaka Prefecture University, Japan |
| Pau Riba | Universitat Autnoma de Barcelona, Spain |

### W02-Camera-based Document Analysis and Recognition (CBDAR)

| | |
|---|---|
| Sheraz Ahmed | DFKI, Kaiserslautern, Germany |
| Muhammad Muzzamil Luqman | La Rochelle University, France |

## W03-Arabic and Derived Script Analysis and Recognition (ASAR)

Adel M. Alimi                University of Sfax, Tunisia
Bidyut Baran Chaudhur        Indian Statistical Institute, Kolkata, India
Fadoua Drira                 University of Sfax, Tunisia
Tarek M. Hamdani             University of Monastir, Tunisia
Amir Hussain                 Edinburgh Napier University, UK
Imran Razzak                 Deakin University, Australia

## W04-Computational Document Forensics (IWCDF)

Nicolas Sidère               La Rochelle University, France
Imran Ahmed Siddiqi          Bahria University, Pakistan
Jean-Marc Ogier              La Rochelle University, France
Chawki Djeddi                Larbi Tebessi University, Algeria
Haikal El Abed               Technische Universitaet Braunschweig, Germany
Xunfeng Lin                  Deakin University, Australia

## W05-Machine Learning (WML)

Umapada Pal                  Indian Statistical Institute, Kolkata, India
Yi Yang                      University of Technology Sydney, Australia
Xiao-Jun Wu                  Jiangnan University, China
Faisal Shafait               National University of Sciences and Technology,
                               Pakistan
Jianwen Jin                  South China University of Technology, China
Miguel A. Ferrer             University of Las Palmas de Gran Canaria, Spain

## W06-Open Services and Tools for Document Analysis (OST)

Fouad Slimane                University of Fribourg, Switzerland
Oussama Zayene               University of Applied Sciences and Arts Western
                               Switzerland, Switzerland
Lars Vögtlin                 University of Fribourg, Switzerland
Paul Märgner                 University of Fribourg, Switzerland
Ridha Ejbali                 National School of Engineers Gabes, Tunisia

## W07-Industrial Applications of Document Analysis and Recognition (WIADAR)

Elisa H. Barney Smith        Boise State University, USA
Vincent Poulain d'Andecy     Yooz, France
Hiroshi Tanaka               Fujitsu, Japan

## W08-Computational Paleography (IWCP)

Isabelle Marthot-Santaniello University of Basel, Switzerland
Hussein Mohammed             University of Hamburg, Germany

## W09-Document Images and Language (DIL)

| | |
|---|---|
| Andreas Dengel | DFKI and University of Kaiserslautern, Germany |
| Cheng-Lin Liu | Institute of Automation of Chinese Academy of Sciences, China |
| David Doermann | University of Buffalo, USA |
| Errui Ding | Baidu Inc., China |
| Hua Wu | Baidu Inc., China |
| Jingtuo Liu | Baidu Inc., China |

## W10-Graph Representation Learning for Scanned Document Analysis (GLESDO)

| | |
|---|---|
| Rim Hantach | Engie, France |
| Rafika Boutalbi | Trinov, France, and University of Stuttgart, Germany |
| Philippe Calvez | Engie, France |
| Balsam Ajib | Trinov, France |
| Thibault Defourneau | Trinov, France |

# Contents – Part II

**ICDAR 2021 Workshop on Open Services and Tools
for Document Analysis (OST)**

**ICDAR 2021 Workshop on Industrial Applications of Document
Analysis and Recognition (WIADAR)**

## ICDAR 2021 Workshop on Computational Paleography (IWCP)

## ICDAR 2021 Workshop on Document Images and Language (DIL)

## ICDAR 2021 Workshop on Graph Representation Learning for Scanned Document Analysis (GLESDO)

# Contents – Part I

**ICDAR 2021 Workshop on Camera-Based Document Analysis
and Recognition (CBDAR)**

**ICDAR 2021 Workshop on Arabic and Derived Script Analysis
and Recognition (ASAR)**

**ICDAR 2021 Workshop on Computational Document Forensics
(IWCDF)**

# ICDAR 2021 Workshop on Machine Learning (WML)

# WML 2021 Preface

Our heartiest welcome to the proceedings of the 3rd Workshop on Machine Learning (WML 2021) which is organized under the 16th International Conference on Document Analysis and Recognition (ICDAR) held at Lausanne, Switzerland during September 5–10, 2021.

Since 2010, the year of the initiation of the annual ImageNet Competition where research teams submit programs that classify and detect objects, machine learning has gained significant popularity. In the present age, machine learning, in particular deep learning, is incredibly powerful for making predictions based on large amounts of available data. There are many applications of machine learning in computer vision and pattern recognition, including document analysis, medical image analysis etc.

In order to facilitate innovative collaboration and engagement between the document analysis community and other research communities such as computer vision and images analysis, etc., the Workshop on Machine Learning is as held as part the ICDAR conference, this year taking place in Lausanne, Switzerland, on September 7, 2021. The workshop provides an excellent opportunity for researchers and practitioners at all levels of experience to meet colleagues and to share new ideas and knowledge about machine learning and its applications in document analysis and recognition. The workshop enjoys strong participation from researchers in both industry and academia.

In this 3rd edition of WML we received 18 submissions, coming from authors in 13 different countries. Each submission was reviewed by at least two expert reviewers (a total of 61 reviews for 18 submissions with an average 3.38 reviews per paper). The Program Committee of the workshop comprised 34 members from 16 countries around the world. Taking into account of the recommendations of the Program Committee members, we selected 12 papers for the presentation in the workshop, resulting in an acceptance rate of 66.6%.

The workshop comprised a one day program with oral presentations of the 12 papers and two keynote talks. The keynote talks were delivered by two well-known researchers: Yi Yang of the Faculty of Engineering and Information Technology, University of Technology Sydney (UTS), Australia, and Xiaojun Chang of the Department of Data Science and AI, Faculty of Information Technology, Monash University, Australia. Our sincere thanks to them for accepting our invitation to deliver the keynotes.

We wish to thank all the researchers who showed interest in this workshop by sending contributed papers, along with our Program Committee members for their time and effort in reviewing submissions and organizing the workshop program. We would also like to thank the ICDAR 2021 organizing committee for supporting our workshop. Finally, we wish to thank all the participants of the workshop.

We hope you enjoyed the workshop and look forward to meeting you at the next WML!

September 2021

Umapada Pal
Yi Yang
Xiao-jun Wu
Faisal Shafait
Jianwen Jin
Miguel A. Ferrer

# Organization

## Workshop Chairs

Umapada Pal      Indian Statistical Institute, Kolkata, India
Yi Yang      University of Technology Sydney, Australia
Xiao-jun Wu      Jiangnan University, China

## Program Chairs

Faisal Shafait      National University of Sciences
     and Technology, Pakistan
Jianwen Jin      South China University of Technology, China
Miguel A. Ferrer      Universidad de Las Palmas de Gran Canaria,
     Spain

## Program Committee

Alireza Alaei      Southern Cross University, Australia
Abdel Belaid      Université de Lorraine – Loria, France
Saumik Bhattacharya      Indian Institute of Technology, Kharagpur,
     India
Alceu Britto      Pontifícia Universidade Católica do Paraná,
     Brazil
Cristina Carmona-Duarte      Universidad de Las Palmas de Gran Canaria,
     Spain
Sukalp Chanda      Østfold University College, Norway
Chiranjoy Chattopadhyay      Indian Institute of Technology, Jodhpur, India
Joseph Chazalon      EPITA Research and Development Laboratory,
     France
Mickaël Coustaty      University de La Rochelle, France
Anjan Dutta      University of Exeter, UK
Miguel Ángel Ferrer Ballester      Universidad de Las Palmas de Gran Canaria,
     Spain
Baochuan Fu      Suzhou University of Science and Technology,
     China
Xin Geng      Southeast University, China
Ravindra Hegadi      Central University of Karnataka, India
Donato Impedovo      Università degli studi di Bari Aldo Moro, Italy
Brian Kenji Iwana      Kyushu University, Japan

| | |
|---|---|
| Lianwe Jin | South China University of Technology, China |
| Zhouhui Lian | Peking University, China |
| Brendan McCane | University of Otago, New Zealand |
| Abhoy Mondal | University of Burdwan, India |
| Muhammad Muzzamil Luqman | La Rochelle Université, France |
| Wataru Ohyama | Saitama Institute of Technology, Japan |
| Srikanta Pal | Université de Lorraine – Loria, France |
| Umapada Pal | Indian Statistical Institute, Kolkata, India |
| Shivakumara Palaiahnakote | University of Malaya, Malaysia |
| Leonard Rothacker | TU Dortmund, Germany |
| Kaushik Roy | West Bengal State University, India |
| Rajkumar Saini | Luleå University of Technology, Sweden |
| K. C. Santosh | University of South Dakota, USA |
| Faisal Shafait | National University of Sciences and Technology, Pakistan |
| Suresh Sundaram | Indian Institute of Technology, Guwahati, India |
| Szilard Vajda | Central Washington University, USA |
| Tianyang Xu | Jiangnan University, China |
| Yi Yang | University of Technology Sydney, Australia |

# Benchmarking of Shallow Learning and Deep Learning Techniques with Transfer Learning for Neurodegenerative Disease Assessment Through Handwriting

Vincenzo Dentamaro⬭, Paolo Giglio⬭, Donato Impedovo$^{(\boxtimes)}$⬭, and Giuseppe Pirlo⬭

University of Bari "Aldo Moro", Via Orabona 4, Bari, Italy
donato.impedovo@uniba.it

**Abstract.** Neurodegenerative diseases are incurable diseases where a timely diagnosis plays a key role. For this reason, various techniques of computer aided diagnosis (CAD) have been proposed. In particular handwriting is a well-established diagnosis technique. For this reason, an analysis of state-of-the-art technologies, compared to those which historically proved to be effective for diagnosis, remains of primary importance. In this paper a benchmark between shallow learning techniques and deep neural network techniques with transfer learning are provided: their performance is compared to that of classical methods in order to quantitatively estimate the possibility of performing advanced assessment of neurodegenerative disease through both offline and online handwriting. Moreover, a further analysis of their performance on the subset of a new dataset, which makes use of standardized handwriting tasks, is provided to determine the impact of the various benchmarked techniques and draw new research directions.

**Keywords:** Shallow learning · Deep learning · Neurodegenerative disease · Handwriting · Deep neural networks · Benchmark

## 1 Introduction

Several non-invasive techniques have been developed in order to assess the presence of neuro-degenerative diseases, which is characterized by a gradual decline of cognitive, functional and behavioral areas of the brain [1, 2]. Among them, behavioral biometrics, such as speech [4], have proven to be promising in terms of accuracy in binary classification (healthy/unhealthy) for neurodegenerative diseases assessment. Handwriting behavioral biometric particularly stands out for its strict relation with the level of severity of a vast class of neurodegenerative diseases, therefore its features' changes are considered an important biomarker: [1, 2] indeed handwriting involves kinesthetic, cognitive and perceptual-motor tasks [4], resulting in a very complex activity whose performance is taken into account for the evaluation of several diseases such as PD and AD [3, 5–7].

© Springer Nature Switzerland AG 2021
E. H. Barney Smith and U. Pal (Eds.): ICDAR 2021 Workshops, LNCS 12917, pp. 7–20, 2021.
https://doi.org/10.1007/978-3-030-86159-9_1

This work proposes a benchmark of traditional shallow learning techniques with deep learning techniques for neurodegenerative disease assessment though handwriting.

This work consists of handwriting acquisitions performed online via tablet: variables like x, y coordinates as well as azimuth, pressure, altitude, in air movements and timestamps of each acquisition are collected. For the specific purpose of the study, only the final handwritten trace, i.e. the whole set of x,y coordinates and the azimuth, is used as the training data set. The handwriting procedure consists of 8 different tasks which will be show in detail in Sect. 5. The paper is organized as follows. Section 2 sketches state of art review for neurodegenerative disease assessment through handwriting, Sect. 3 illustrates the use of shallow learning technique on on-line handwriting recognition by means of velocity based-features and kinematic-based features. In Sect. 4 both offline and online deep learning techniques are presented. Section 5 shows dataset description and results. Reasoning of results is provided in Sect. 6. Finally, Sect. 7 sketches conclusions and future remarks.

## 2 State of the Art Review

The aim of this work is to provide insights about the best features and techniques to adopt into a computer aided diagnosis system for supporting early diagnosis of neurodegenerative disease. It is important to not only predict the disease, but also to monitor the progression of it during time. [1, 2]. The scientific community focused the research towards predictive models that can accurately detect subtle changes in writing behavior. These techniques will be used to help neurologists, and psychologists to assess diseases as an auxiliary tool in addition to the battery of cognitive tests provided in literature [1–9].

The acquisition tool, at time of writing, is a digital tablet with a pen. This device captures spatial and temporal data and save it inside a storage memory. After data is captures, as often happens in shallow learning scenario, features are extracted. Usually, patients are asked to perform several tasks [1].

Even though important results were achieved by the community, there is not homogeneity in tasks provided by the datasets developed. That is because scientists collected databases of handwriting tasks themself resulting in datasets with different kind of tasks, usually not connected among them and merged together, which provided controversary results. To overcome this problem, authors in [34] have developed a specific acquisition protocol. This protocol includes a digitizer version of standard tests used, accepted, tested in the neurological community used as the ground truth for evaluation. The dataset used in this work is a subset of this big dataset which is currently under development. This dataset contains well-established handwriting tasks to perform kinematic analysis and handwriting experimental tasks useful for extracting novel types of features to be investigated by researchers. Literature review on handwriting recognition for neurodegenerative disease assessment can be subdivided in two main groups: online handwriting and offline. In the online handwriting, the features computed in all the tasks are then concatenated into a high dimensional vector and then used for classification. [9] Various authors used several kinds of classifiers ranging from SVM, KNN, ensemble learning with Random Forests, neural networks and so on. [1–9]. It has also been analyzed the use of an ensemble of classifiers each one built onto one single feature space of each task [1, 2].

For the online handwriting recognition for neurodegenerative disease assessment, some of the authors of this work in [1] used several features like position, button status, pressure, azimuth, altitude, displacement, velocity and acceleration over 5 different datasets namely: PaHaw [9], NewHandPH [29], ParkinsonHW [30], ISUNIBA [31], EMOTHAW[32] achieving accuracies that range from 79.4% to 93.3% depending on the dataset and tasks.

For offline handwriting recognition, authors in [33] used "enhanced" static images of handwriting generated by exploiting simultaneously the static and dynamic properties of handwriting by drawing the points of the samples and adding pen-ups for the same purpose. Authors used a Convolutional Neural Network to provide feature embedding and then a set of classifier is used in a majority voting fashion. Authors used transfer learning for coping with limited amount of training data. Their accuracies on the various tasks ranged from 50% to 65% showing some limits of this technique.

In [35] authors explored an alternative model that used one single bi-directional LSTM layers on handwriting recognition tasks, achieving better or equivalent results than stacking more LSTM layers, which decreases the complexity and allows a faster network training. In [36] authors investigated the use of bidirectional LSTM with attention mechanism for offline and online handwriting recognition achieving important results on the RIMES handwriting recognition task. The bidirectional LSTM architecture developed in this work was partially inspired by the work in [36]. Some of the authors of this work used also computer vision for assessing neurodegenerative disease through gait [37] and sit to stand tasks [38].

## 3   Shallow Learning for Online Handwriting Neurodegenerative Disease Assessment

The term shallow learning identifies all techniques that do not belong to deep learning. In the case of online handwriting recognition, where online stands for capturing time-series of movements of the pen on a digital support, shallow learning is equivalent to perform feature extraction and classification with various machine learning algorithms. To this extent, standard velocity-based features and kinematic-based features are extracted and tested with random forest classification algorithm. The set of extracted features is shown in Table 1. All the features extracted were standardized. Moreover, Random Forest [17] ensemble learning algorithm with features ordered by relevance was adopted to make a selection of the most important criteria [18]. The random forest pre-pruning parameter was of maximum tree depth of 10 and 50 trees, in order to prevent overfitting and to balance accuracies. Random Forest algorithm [17] was also used for classification purposes: its maximum depth adopted was of 10 and the number of trees was estimated dynamically with the inspection of the validation curve. The reported accuracies are based on a 10-fold cross validation, i.e. the entire procedure was repeated 10 times, where each fold was used as a test set.

**Table 1.** Features used in shallow learning.

| Feature | Description |
|---------|-------------|
| Position | Position in terms of s(x,y) |
| Button status | Movement in the air: $b(t) = 0$<br>Movement on the pad: $b(t) = 1$ |
| Pressure | Pressure of the pen on the pad |
| Azimuth | Angle between the pen and the vertical plane on the pad |
| Altitude | Angle between the pen and the pad plane |
| Displacement | $d_i = \begin{cases} \sqrt[2]{(x_{i+1} - x_i)^2 + (y_{i+1} - y_i)^2}, 1 \le i \le n-1 \\ d_n - d_{n-1}, i = n \end{cases}$ |
| Velocity | $v_i = \begin{cases} \frac{d_i}{t_{i+1}-t_i}, 1 \le i \le n-1 \\ v_n - v_{n-1}, i = n \end{cases}$ |
| Acceleration | $a_i = \begin{cases} \frac{v_i}{t_{i+1}-t_i}, 1 \le i \le n-1 \\ a_n - a_{n-1}, i = n \end{cases}$ |
| Jerk | $j_i = \begin{cases} \frac{a_i}{t_{i+1}-t_i}, 1 \le i \le n-1 \\ j_n - j_{n-1}, i = n \end{cases}$ |
| x/y displacement | Displacement in the x/y direction |
| x/y velocity | Velocity in the x/y direction |
| x/y acceleration | Acceleration in the x/y direction |
| x/y jerk | Jerk in the horizontal/vertical direction |
| NCV | Number of Changes in Velocity, NCV has been also normalized to writing duration |
| NCA | Number of Changes in Acceleration, NCA has been also normalized to writing duration |

### 3.1 Velocity-Based Features

The choice of certain velocity-based features is dictated by motor deficits particularly present in neurodegenerative diseases. Motor deficits like bradykinesia (which is characterized by slowness of movements), micrographia (time related reduction of the size of writing), akinesia (characterized by impairment of voluntary movements), tremor and muscular rigidity [2], are particularly evident when patient is asked to perform certain tasks. These tasks are often characterized by drawing stars, spirals, writing names and copying tasks [3, 8–10, 12]. In order to model other symptoms such as tremor and jerk, the patient is often asked to draw meanders, horizontal lines, straight (both forward and backward) slanted lines, circles and few predefined sentences as shown in [11–13].

Table 1 shows the features extracted for the shallow learning classification. It is important to state that every feature is a time-series, thus statistical functions such a mean, median, standard deviation 1st and 99th percentile are used to synthetize each feature in Table 1.

### 3.2  Kinematic-Based Features

For modelling online handwriting and extracting important movements patterns, authors in [14] have used the Maxwell-Boltzmann distribution. This distribution is used to extract parameters that are then used to model the velocity profile. Its formulation is shown in formula (1).

$$mb_j = v_j^2 e^{-v_j^2} \tag{1}$$

V$_j$ is the velocity at j-th position. Another kinematic feature used for describing the handwriting pattern of velocity and acceleration profile is the Discrete Fourier Transform as shown in [15]. Its formulation, shown in (2), is composed by the computation of the DFT and the computation of the Inverse DFT, which contains the spectrum of harmonics having the magnitude inversely proportional to the frequency [16]. Thanks to the logarithm present in the formulation, components with small variations tends to converge toward 0, instead repeated peaks at higher frequencies are typical of periodic patterns described by tremor and jerks.

$$rcep = IDFT\{log[|DFT(v_j)|]\} \tag{2}$$

Again, $v_j$ is the velocity at j-th position.

## 4  Deep Learning for Offline and Online Handwriting Neurodegenerative Disease Assessment

Deep Learning techniques have been developed for various tasks such as image recognition through convolutional neural networks, but also time series analysis using recurrent neural networks with stacked layers such as LSTM and bi directional LSTM. The motivation behind this work is to benchmark deep learning architectures trained by using deep transfer learning on images generated by x,y coordinates drawing, from now on referred as offline handwriting, with respect to online handwriting models trained on time series of x,y coordinates for the RNN and shallow learning as reported in Sect. 3.

### 4.1  CNN Based Networks for Offline Recognition

For the offline handwriting recognition, 224 by 224 pixels images are generated by plotting x, y coordinates of each task and saving the generated image.

Because of the limited amount of training data, it has been decided to use deep transfer learning [20]. Deep transfer learning is useful when not much training data is available, as in this case. The idea is to use a deep neural network architecture and weights trained on a big dataset and fine tune, only final layers on our dataset by freezing the former layers. This is useful, because initial layers usually generate high level representation of the underlying patterns, while the last layers are specialized in applying the proper classification. [20] All the used deep learning architectures are originally trained on Imagenet dataset [21] and then a 2D global average pooling layer has been added followed by one dense layer with 32 neurons and ReLU activation function, and finally the softmax layer for performing binary classification. New added layers are trained on the training set for 100 epochs and cross validated on the 33% of the training set.

All labels are one-hot encoded. The following architectures were chosen depending on the importance in the literature, disk size, number of parameters and the accuracy achieved on Imagenet dataset as reported on Keras website [22]. The chosen architectures are briefly reported are in Table 2.

**Table 2.** Deep learning architectures used

| Architecture name | Size | Parameters | Top-5 accuracy on imagenet |
|---|---|---|---|
| NASNetLarge | 343 MB | 88,949,818 | 0.960 |
| ResNET 50 | 98 MB | 25,636,712 | 0.921 |
| Inception V3 | 92 MB | 23,851,784 | 0.937 |
| Inception ResNet V2 | 215 MB | 55,873,736 | 0.953 |

#### 4.1.1 NASNetLarge

The architecture of NASNet Large deep neural network was not invented by a human being, but is the result of a process called Neural Architecture Search, where parameters of the network and its architecture is discovered as the output of an optimization process which uses reinforcement learning to learn and decide what is the best choice of layer type and hyperparameters given a specific dataset. In authors experiments [23], the algorithm searched for the best convolutional layer (or "cell") on the CIFAR-10 dataset and then this cell was later applied to the ImageNet dataset by iteratively stacking copies of this cells, each with their own set of hyperparameters resulting in a novel architecture (Fig. 1).

#### 4.1.2 ResNET 50

The ResNet-50 [24] model is composed by 5 so called "stages" each composed by a convolution and an Identity block. Each convolution block and each identity block have 3 convolution layers which results in over 23 million trainable parameters. ResNET is theoretically important because it introduced two major breakthroughs in computer vision:

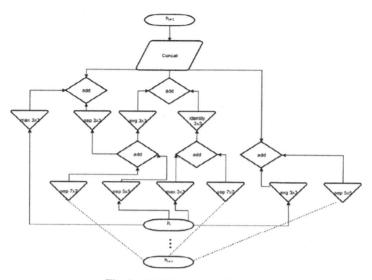

**Fig. 1.** NASNet large architecture

1. The mitigation of the gradient vanishing problem by allowing this alternate shortcut path for reinjecting information to the flow
2. The possibility to learn the identity function of the previous output, by ensuring that the later layers will perform at least as good as the previous (Fig. 2).

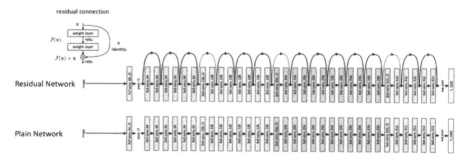

**Fig. 2.** ResNET-50

### 4.1.3 Inception V3

Inception-v3 [25] is the third release of a convolutional neural network architecture developed at Google which derived from the Inception family. This architecture makes several improvements including using Label Smoothing, factorized convolutions, batch normalization and auxiliary classifier which is used to propagate label information lower down the network (Fig. 3 and 4).

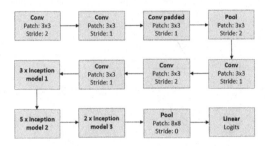

**Fig. 3.** Diagram representation of inception V3 architecture

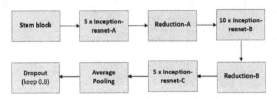

**Fig. 4.** Diagram representation of inception ResNet V2 architecture

### 4.1.4 Inception-ResNet-v2

Inception-ResNet-v2 [26] often called Inception V4 is a convolutional neural architecture that is built, as the name suggests, by fusing two major architecture families: Inception family e.g. Inception V3 and ResNet family by incorporating residual connections. This is at the moment one of the state of the art architecture used in image recognition tasks.

### 4.2 Bi-directional LSTM RNN for Online Recognition

For online recognition using recurrent neural networks, a novel Bi-Directional LSTM recurrent neural network is developed with the aim of performing online handwriting recognition. This online recognition is based solely on time series of x,y coordinates, no other information are provided. Thus, as in deep learning fashion, this architecture will automatically exploit long and short-term coherence and patterns with the aim of recognizing neurodegenerative diseases from just raw coordinates. Differently from Long-Short Term Memory RNN (briefly LSTM), bidirectional LSTM run the inputs in two ways: both from past to future and backward. This process preserves information from the future and from the past by combining the two hidden states (one for forward and one for backward) in order to preserve information from past and future. Authors in [27] have used bidirectional LSTM for modelling online handwriting recognition. The architecture developed also contains an Attention Mechanism layer. [28] The attention mechanism was invented for Natural Language Processing tasks where the encoder-decoder recurrent neural network architecture was used to learn to encode input sequences into a fixed-length internal representation, and second set of LSTMs read the internal representation and decode it into an output sequence. To overcome the problem that all input sequences are forced to be encoded into an internal vector of fixed length, a selective attention mechanism was developed with the aim to select these inputs and relate them with respect to the output sequence. [28] This attention mechanism searches for a set of positions in the input where the most relevant information is concentrated. It does so by encoding the input vector into a sequence of vectors and then it adaptively chooses a subset of vectors while producing the output. [28] The intuition here is that attention mechanism would capture very long-term relations among coordinates in such a way to increase correlations among handwriting patterns of people affected by some neurodegenerative disease versus the normative sample. The architecture developed is depicted in Fig. 5 and was trained in an end-to-end fashion. It is composed by a bidirectional LSTM layer with 32 neurons followed by a dense layer with 32 neurons and ReLU activation function, this followed by an Attention layer with 32 neurons. At the end there is a dense layer with softmax activation function that carries out the classification.

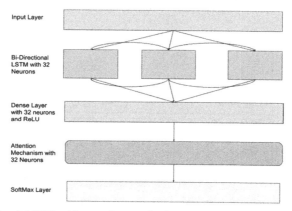

**Fig. 5.** Bidirectional LSTM with attention mechanism for online handwriting recognitionDiagram representation of Inception ResNet V2 architecture

## 5  Dataset Description and Results

### 5.1  Dataset Description

Raw data were collected by measuring x and y coordinates of the pen position and their timestamps. The pen inclination (tilt-x and tilt-y) and pressure of the pen's tip on the surface were also registered. Another important collected parameter was the "button status", i.e. a binary variable which gives 0 for pen-up state (in-air movement) and 1 for pen-down state (on-surface movement). A matrix $X = (x, y, p, t, tilt\_x, tilt\_y, b)$ where each column is a vector of length N, where N is the number of sampled points, thus can describe the whole execution process of a single task. All the tasks are listed in Table 3.

The check copying task consists of asking the user to copy a check as shown in Fig. 6.

Another task is based on asking the user to find and mark a subset of predefined numbers inside matrices, as shown in Fig. 7.

The trail test consists of completing a succession of letters or numbers inside circles by linking them with other ones generating a path of variable complexity. The example in Fig. 8 is a clear example of a user affected by a neurodegenerative disease.

**Table 3.** Taks used

| Task name | Task description |
|-----------|-----------------|
| Chk | Check copying task |
| M1 | Matrix 1 |
| M2 | Matrix 2 |
| M3 | Matrix 3 |
| Tmt1 | Trail 1 of connecting path |
| Tmt2 | Trail 2 (difficult) of connecting path |
| Tmtt1 | Trail test 1 of connecting path |
| Tmtt2 | Trail test 2 (difficult) of connecting path |

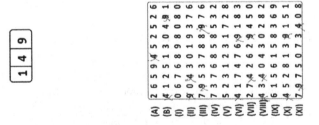

**Fig. 6.** Check copying task performed by a patient with some form of dementia

**Fig. 7.** M3 Matrix task

The user subset is composed by 42 subjects: 21 among them are affected by a neurodegenerative disease at different levels of severity which will be qualified as "mild", "assessed", "severe", "very severe". The other 21 are healthy control subjects. The dataset size is in line with sizes of other datasets mentioned in state of art review.

At this stage of the study, age and sex are not taken into account in the analysis. A deeper analysis won't be able to leave these parameters out of consideration.

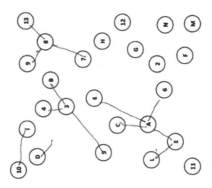

**Fig. 8.** Trail test number 2 mixing letters and numbers

## 5.2 Results

Table 4 shows the results. The accuracy is expressed as F1 score.

**Table 4.** Results of various techniques with respect to various tasks

| Task name | NASNet large accuracy | ResNet 50 accuracy | Inception V3 accuracy | Inception ResNet v2 accuracy | Shallow learning | Bi directional LSTM w/attention |
|---|---|---|---|---|---|---|
| Check copying | 0.72 | 0.72 | 0.80 | 0.78 | **0.853** | 0.72 |
| M1 | 0.68 | 0.64 | **0.76** | 0.74 | 0.677 | 0.66 |
| M2 | 0.65 | 0.58 | 0.65 | 0.69 | **0.702** | 0.69 |
| M3 | 0.67 | **0.75** | 0.69 | 0.58 | 0.704 | 0.66 |
| Tmt1 | 0.76 | 0.63 | 0.89 | 0.74 | **0.799** | 0.78 |
| Tmt2 | 0.59 | 0.59 | 0.68 | **0.74** | 0.726 | 0.71 |
| Tmtt1 | 0.84 | 0.68 | 0.64 | 0.74 | 0.774 | **0.86** |
| Tmtt2 | 0.58 | 0.71 | 0.74 | 0.63 | **0.858** | 0.70 |
| All tasks | 0.66 | 0.70 | 0.67 | 0.72 | **0.923** | 0.74 |

## 6   Results discussion

In Table 4, different CNN ("NASNET LARGE", "RESNET 50", "INCEPTION V3", "Inception Resnet V2") and RNN architectures ("Bidirectional LSTM with Attention") were tested in order to understand their performances on detecting the presence (or absence) of a neurodegenerative disease by analyzing a series of previously described tasks (CHK, M1, M2, M3, TMT1, TMT2, TMTT1, TMTT2). Moreover, further analysis was performed by running the various techniques on a dataset obtained by merging data of all the tasks. In the following analysis positive class will be represented by 1 (those affected by neurodegenerative disease) and negative class by 0 (those without neurodegenerative disease). The most promising results were obtained using predefined features and doing the analysis based on the shallow learning approach, i.e. performing feature engineering by carefully selecting features from a set of physical parameters followed by automatic feature selection to decrease dimensionality: performances were characterized by a relatively small variance between different tasks, which suggests low dependency of accuracy from the specific task dataset. The second best outcome was from "Bidirectional LSTM" with attention, this deep recurrent neural network architecture achieved the lowest variability among accuracies at different tasks. Moreover, this network was capable of successfully exploiting neurodegenerative diseases patterns capable of binary discern healthy from un-healthy subjects based solely on the raw time series of x,y coordinates. All other deep learning neural networks trained on offline (static) images show significant variations with their accuracies from a task to another. This results in high variability of accuracies between tasks and thus a decrease of confidence. This analysis suggests that online handwriting outperforms the offline one both with a preliminary features selection or letting the algorithm to learn the most efficient patterns from raw data.

## 7   Conclusions

In this work, classic features have been employed for healthy/unhealthy binary classification of subjects included in the new dataset. The main goal of this work is to provide a benchmark of accuracy of different techniques available for neurodegenerative disease detection. Indeed, the analysis was performed on a specific subset of variables acquired during the handwriting tasks performance, specifically x, y coordinates and azimuth. The shallow learning approach, with a feature preselection, outperformed all the others architectures showing small variance of accuracies between different tasks. Similar results were obtained using "Bidirectional LSTM" with attention, while other deep learning algorithms were affected by higher variability in accuracy depending on the specific task analyzed. These results suggest that online handwriting is a better approach compared to the offline one, either with features preselection and with the algorithm learning itself from raw data. This last point opens new frontiers in automatic learning specific neurodegenerative disease patterns from timeseries of raw x,y coordinates. The next evolution of this work will be to perform not only binary prediction of healthy/unhealthy subjects but also to evaluate the severity level of diseases. In this regard, as the dataset is provided with multiple sessions of acquisitions for the same patients, it will also be analyzed the inferability of increments or decrements of disease severity with time, with respect to the adoption of medical treatments.

**Ethical approval.** All procedures performed in studies involving human participants were in accordance with the ethical standards of the institutional and/or national research committee and with the 1964 Helsinki declaration and its later amendments or comparable ethical standards.

**Informed consent.** Informed consent was obtained from all individual participants included in the study.

**Conflicts of Interest.** The authors declare no conflict of interest.

# References

1. Impedovo, D., Pirlo, G.: Dynamic handwriting analysis for the assessment of neurodegenerative diseases: a pattern recognition perspective. IEEE Rev. Biomed. Eng. **12**, 209–220 (2019)
2. De Stefano, C., Fontanella, F., Impedovo, D., Pirlo, G., di Freca, A.S.: Handwriting analysis to support neurodegenerative diseases diagnosis: a review. Pattern Recogn. Lett. **121**, 37–45 (2018)
3. Rosenblum, S., Samuel, M., Zlotnik, S., Erikh, I., Schlesinger, I.: Handwriting as an objective tool for Parkinson's disease diagnosis. J. Neurol. **260**(9), 2357–2361 (2013)
4. Astrom, F., Koker, R.: A parallel neural network approach to prediction of Parkinson's Disease. Expert Syst. Appl. **38**(10), 12470–12474 (2011)
5. O'Reilly, C., Plamondon, R.: Development of a sigma–lognormal representation for on-line signatures. Pattern Recogn. **42**(12), 3324–3337 (2009)
6. Pereira, C.R., et al.: A step towards the automated diagnosis of Parkinson's disease: analyzing handwriting movements. In: IEEE 28th International Symposium on Computer Based Medical Systems (CBMS), pp. 171–176 (2015)
7. Kahindo, C., El-Yacoubi, M.A., Garcia-Salicetti, S., Rigaud, A., Cristancho-Lacroix, V.: Characterizing early-stage alzheimer through spatiotemporal dynamics of handwriting. IEEE Signal Process. Lett. **25**(8), 1136–1140 (2018)
8. Caligiuri, M.P., Teulings, H.L., Filoteo, J.V., Song, D., Lohr, J.B.: Quantitative measurement of handwriting in the assessment of drug-induced Parkinsonism. Hum. Mov. Sci. **25**(4), 510–522 (2006)
9. Drotár, P., Mekyska, J., Rektorová, I., Masarová, L., Smékal, Z., Faun-dez-Zanuy, M.: Decision support framework for Parkinson's disease based on novel handwriting markers. IEEE Trans. Neural Syst. Rehabil. Eng. **23**(3), 508–516 (2015)
10. Ponsen, M.M., Daffertshofer, A., Wolters, E.C., Beek, P.J., Berendse, H.W.: Impairment of complex upper limb motor function in de novo Parkinson's disease. Parkinsonism Related Disord. **14**(3), 199–204 (2008)
11. Smits, E.J., et al.: Standardized handwriting to assess bradykinesia, micrographia and tremor in Parkinson's disease. PLOS One **9**(5), e97614 (2014)
12. Broderick, M.P., Van Gemmert, A.W., Shill, H.A.: Hypometria and bradykinesia during drawing movements in individuals with Parkinson disease. Exp. Brain Res. **197**(3), 223–233 (2009)
13. Kotsavasiloglou, C., Kostikis, N., Hristu-Varsakelis, D., Arnaoutoglou, M.: Machine learning-based classification of simple drawing movements in Parkinson's disease. Biomed. Signal Process. Control **31**, 174–180 (2017)
14. Li, G., et al.: Temperature based restricted Boltzmann Machines. Sci. Rep. **6**, Article no. 19133 (2016)
15. Impedovo, D.: Velocity-based signal features for the assessment of Parkinsonian handwriting. IEEE Signal Process. Lett. **26**(4), 632–636 (2019)
16. Rao, K.R., Yip, P.: Discrete Cosine Transform: Algorithms, Advantages. Applications. Academic press, New York (2014)
17. Breiman, L.: Random forests. Mach. Learn. **45**(1), 5–32 (2001)
18. Baraniuk, R.G.: Compressive sensing [lecture notes]. IEEE Signal Process. Mag. **24**, 118–121 (2007)
19. Reitan, R.M.: Validity of the Trail Making Test as an indicator of organic brain damage. Perceptual Motor Skills **8**(3), 271–276 (1958)
20. Tan, C., Sun, F., Kong, T., Zhang, W., Yang, C., Liu, C.: A survey on deep transfer learning. In: International Conference on Artificial Neural Networks, pp. 270–279. Springer, Cham (2018)

21. Deng, J., Dong, W., Socher, R., Li, L.J., Li, K., Fei-Fei, L.: Imagenet: a large-scale hierarchical image database. In: 2009 IEEE Conference on Computer Vision and Pattern Recognition, pp. 248–255. IEEE, June 2009

22. Chollet, F.: Keras. Keras documentation: Keras Applications. Keras.io (2020). https://keras.io/api/applications/. Accessed 30 Oct 2020

23. Zoph, B., Vasudevan, V., Shlens, J., Le, Q.V.: Learning transferable architectures for scalable image recognition. In: Proceedings of the IEEE Conference on Computer Vision and Pattern Recognition, pp. 8697–8710 (2018)

24. He, K., Zhang, X., Ren, S., Sun, J.: Deep residual learning for image recognition. In: Proceedings of the IEEE Conference on Computer Vision and Pattern Recognition, pp. 770–778 (2016)

25. Szegedy, C., Vanhoucke, V., Ioffe, S., Shlens, J., Wojna, Z.: Rethinking the inception architecture for computer vision. In: Proceedings of the IEEE Conference on Computer Vision and Pattern Recognition, pp. 2818–2826 (2016)

26. Szegedy, C., Ioffe, S., Vanhoucke, V., Alemi, A.: Inception-v4, inception-resnet and the impact of residual connections on learning. arXiv preprint arXiv:1602.07261 (2016)

27. Liwicki, M., Graves, A., Fernàndez, S., Bunke, H., Schmidhuber, J.: In Proceedings of the 9th International Conference on Document Analysis and Recognition, ICDAR 2007 (2007)

28. Bahdanau, D., Cho, K., Bengio, Y.: Neural machine translation by jointly learning to align and translate. arXiv preprint arXiv:1409.0473 (2014)

29. Pereira, C.R., Weber, S.A., Hook, C., Rosa, G.H., Papa, J.P.: Deep learning-aided Parkinson's disease diagnosis from handwritten dynamics. In: 2016 29th SIBGRAPI Conference on Graphics, Patterns and Images (SIBGRAPI), pp. 340–346. IEEE, October 2016

30. Isenkul, M., Sakar, B., Kursun, O.: Improved spiral test using digitized graphics tablet for monitoring Parkinson's disease. In: Proceedings of the International Conference on e-Health and Telemedicine, pp. 171–175, May 2014

31. Impedovo, D., et al.: Writing generation model for health care neuromuscular system investigation. In: International Meeting on Computational Intelligence Methods for Bioinformatics and Biostatistics, pp. 137–148. Springer, Cham, June 2013

32. Likforman-Sulem, L., Esposito, A., Faundez-Zanuy, M., Clémençon, S., Cordasco, G.: EMOTHAW: a novel database for emotional state recognition from handwriting and drawing. IEEE Trans. Hum. Mach. Syst. 47(2), 273–284 (2017)

33. Diaz, M., Ferrer, M.A., Impedovo, D., Pirlo, G., Vessio, G.: Dynamically enhanced static handwriting representation for Parkinson's disease detection. Pattern Recogn. Lett. 128, 204–210 (2019)

34. Impedovo, D., Pirlo, G., Vessio, G., Angelillo, M.T.: A handwriting-based protocol for assessing neurodegenerative dementia. Cognit. Comput. 11(4), 576–586 (2019)

35. Puigcerver, J.: Are multidimensional recurrent layers really necessary for handwritten text recognition? In: 2017 14th IAPR International Conference on Document Analysis and Recognition (ICDAR), vol. 1, pp. 67–72. IEEE, November 2017

36. Doetsch, P., Zeyer, A., Ney, H.: Bidirectional decoder networks for attention-based end-to-end offline handwriting recognition. In: 2016 15th International Conference on Frontiers in Handwriting Recognition (ICFHR), pp. 361–366. IEEE, October 2016

37. Dentamaro, V., Impedovo, D., Pirlo, G.: Gait analysis for early neurodegenerative diseases classification through the Kinematic Theory of Rapid Human Movements. IEEE Access (2020)

38. Dentamaro, V., Impedovo, D., Pirlo, G.: Sit-to-stand test for neurodegenerative diseases video classification. In: International Conference on Pattern Recognition and Artificial Intelligence, pp. 596–609. Springer, Cham, October 2020

# Robust End-to-End Offline Chinese Handwriting Text Page Spotter with Text Kernel

Zhihao Wang[1], Yanwei Yu[1,2(✉)], Yibo Wang[1], Haixu Long[1],
and Fazheng Wang[1]

[1] School of Software Engineering, University of Science and Technology of China,
Hefei, China
{cheshire,wrainbow,hxlong,wfzc}@mail.ustc.edu.cn
[2] Suzhou Institute for Advanced Research, University of Science and Technology
of China, Suzhou, China
ywyu@ustc.edu.cn

**Abstract.** Offline Chinese handwriting text recognition is a long-standing research topic in the field of pattern recognition. In previous studies, text detection and recognition are separated, which leads to the fact that text recognition is highly dependent on the detection results. In this paper, we propose a robust end-to-end Chinese text page spotter framework. It unifies text detection and text recognition with text kernel that integrates global text feature information to optimize the recognition from multiple scales, which reduces the dependence of detection and improves the robustness of the system. Our method achieves state-of-the-art results on the CASIA-HWDB2.0-2.2 dataset and ICDAR-2013 competition dataset. Without any language model, the correct rates are 99.12% and 94.27% for line-level recognition, and 99.03% and 94.20% for page-level recognition, respectively. Code will be available at GitHub.

**Keywords:** Offline Chinese handwriting text page spotter ·
End-to-end · Robust · Text kernel · Multiple scales

## 1 Introduction

Offline Chinese handwriting text recognition remains a challenging problem. The main difficulties come from three aspects: the variety of characters, the diverse writing styles, and the problem of character-touching. In recent years, methods based on deep learning have greatly improved recognition performance. There are two types of handwriting text recognition methods, page-level recognition methods and line-level recognition methods. For offline Chinese handwriting text recognition, most studies are based on line-level text recognition.

For line-level recognition, it is mainly classified into two research directions: over-segmentation methods and segmentation-free methods. Over-segmentation methods first over-segment the input text line image into a sequence of primitive

© Springer Nature Switzerland AG 2021
E. H. Barney Smith and U. Pal (Eds.): ICDAR 2021 Workshops, LNCS 12917, pp. 21–35, 2021.
https://doi.org/10.1007/978-3-030-86159-9_2

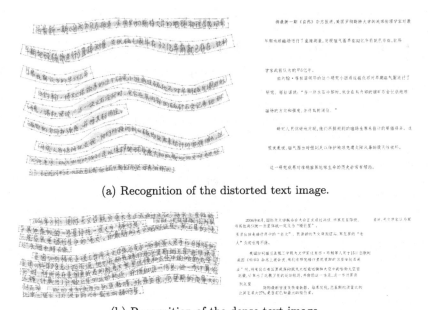

(a) Recognition of the distorted text image.

(b) Recognition of the dense text image.

**Fig. 1.** Results of End-to-end detection and recognition. The pink areas are the segmentation results of kernel areas and the text boxes are generated by the center points generation algorithm. (Color figure online)

segments, then combine the segments to generate candidate character patterns, forming a segmentation candidate lattice, and classify each candidate pattern to assign several candidate character classes to generate a character candidate lattice. Wang Q F et al. [1] first proposed the over-segmentation method from the Bayesian decision view and convert the classifier outputs to posterior probabilities via confidence transformation. Song W et al. [2] proposed a deep network using heterogeneous CNN to obtain hierarchical supervision information from the segmentation candidate lattice.

However, the over-segmentation method has its inherent limitations. If the text lines are not correctly segmented, it brings great difficulties to subsequent recognition. The segmentation-free method based on deep learning is proposed to solve this problem. Messina et al. [3] proposed multidimensional long-short term memory recurrent neural networks (MDLSTM-RNN) using Connectionist Temporal Classifier [4] (CTC) as loss function for end-to-end text line recognition. Xie et al. [5] proposed a CNN-ResLSTM model with a data preprocessing and augmentation pipeline to rectify the text pictures to optimize recognition. Xiao et al. [6] proposed a deep network with Pixel-Level Rectification to integrate pixel-level rectification into CNN and RNN-based recognizers.

For page-level recognition, it can be classified into two-stage recognition methods or end-to-end recognition methods. The two-stage recognition methods apply two models for text detection and recognition respectively. End-to-end methods gradually compress the picture into several lines or a whole line of feature maps for recognition. Bluche et al. [10] proposed a modification of MDLSTMRNNs to recognize English handwritten paragraphs. Yousef et al. [11] proposed the OrigamiNet which constantly compresses the width of the feature map to 1 dimension for recognition. The two-stage methods generally detect the text line first and then cut out the text line for recognition. Li X et al. [7] proposed segmentation-based PSENet with progressive scale expansion to detect arbitrary-shaped text lines. Liu Y et al. [8] proposed ABCNet based on third-order Bezier curve for curved text line detection. Liao M et al. [9] proposed Mask TextSpotter v3 with a Segmentation Proposal Network (SPN) and hard RoI masking for robust scene text spotting.

It can be noted that the methods of page-level recognition without detecting text lines lose the information of the text location. If the text layout is complicated, it is difficult to correctly recognize text images with these methods. The method of Liao M et al. [9] applying hard RoI masking into the RoI features instead of transforming text lines may process very big feature maps and loses information when resizing feature maps. Regardless of the line-level or the two-stage recognition methods, the location information of the text line is obtained first, and the text line is segmented and then recognized, which actually separates the detection and recognition.

We think the detection and recognition should not be separated. The detection can only provide the local information of text lines, which makes it difficult to utilize the global text image information during recognition. Whether the detection box is larger or smaller than the ground truth box, it causes difficulties for subsequent recognition. This is because the alignment of text line images is in the original text page image, and the robustness is not enough for the recognition. And the detection of text lines is so important that if the text line cannot be detected well, the text line image cannot be accurately recognized. We believe that the key to text recognition lies in accurately recognizing text and we just need to know the approximate location of the text, instead of precise detection.

In this paper, we propose a robust end-to-end Chinese text page recognition framework with text kernel segmentation and multi-scale information integration to unify text detection and recognition. Our main contributions are as follows:

1) We propose a novel end-to-end text page recognition framework, which utilizes global information to optimize detection and recognition.
2) We propose a method to align text lines with the text kernel, which is based on center-lines to extract text line strips from feature maps.
3) We propose a text line recognition model with multi-scale information integration, which uses TCN and Self-attention instead of RNNs.
4) We have done a series of experiments to verify the effectiveness of our model and compare it with other state-of-the-art methods. We achieve state-of-the-art performance on both the CASIA-HWDB dataset and the ICDAR-2013 dataset. The page-level recognition performance of our method is even better than the line-level recognition methods.

## 2    Method

The framework of our method is depicted in Fig. 2. This framework consists of three modules for text detection and recognition. The segmentation module is used to generate the segmentation map of the kernel area of the text line and the feature map of the text page. The connection module is introduced to extract the text line feature map according to the segmentation map. The recognition module is used for text line recognition which is based on DenseNet [15] with TCN and Self-attention. The segmentation and recognition results are shown in Fig. 1.

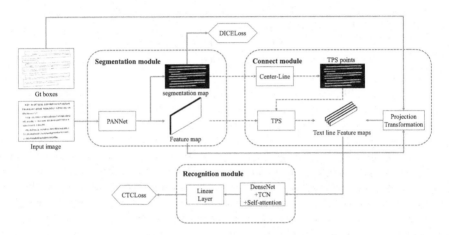

**Fig. 2.** Overview of our text page recognition framework. The dashed arrow denotes that the text line feature maps are transformed by the center lines with the kernel areas when predicting.

### 2.1    Segmentation Module

The segmentation module processes the input image to generate a feature map and a segmentation map that are one-quarter the size of the original image. In this part, We mainly utilize the network structure of PANNet [12] for its good performance on text segmentation of arbitrary shapes. We use ResNet34 [13] as its backbone and change the strides of the 4 feature maps generated by backbone to 4, 4, 8, 8 with respect to the input image. We extract larger feature maps because we need fine-grained feature information for text recognition. We retain the Feature Pyramid Enhancement Module (FPEM) and Feature Fusion Module (FFM) of PANnet for extracting and fusing feature information of different scales and the number of repetitions of the FPEM is 4. The size of the text line kernel area we set is 0.6 of the original size, which is enough to distinguish different text line regions.

## 2.2 Connection Module

The connection module is used to extract the text line feature map according to the segmentation map. We believe that the feature map contains high-dimensional information than the original image, and the feature map can gather the text information in the kernel area, which makes the extraction of the text line feature map more robust. We transform the feature map randomly with the kernel area as the center, such as perspective transformation, so that the text feature information is concentrated in the kernel area. We scale all the text line feature maps to a height of 32-pixel for subsequent recognition.

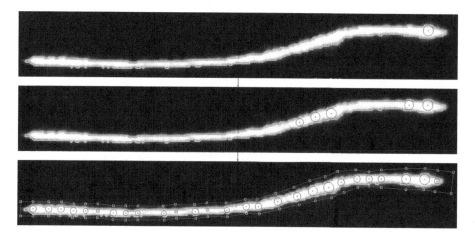

**Fig. 3.** The center points are generated by continuously finding the center of the current largest inscribed circle.

We assume that the text line is a strip, and its center-line can be considered as passing through the center of each character. Generally speaking, the length of a text line is greater than its height, and each text line haves one center-line. We use the inscribed circle of the text line strip to find the text center-line. We assume that the center of the largest inscribed circle of the text line strip falls on the text center-line.

During training, we use perspective transformation to align the text line according to the ground truth text box. And the aligned text line feature maps are randomly affine transformed in the direction of the interior for data enhancement which allows the model to learn to concentrate the text feature information in the kernel area.

In the evaluation, we align text lines based on the segmentation map. But in this way we can only get the contour of the text line kernel region, we also need to get the trajectory of the text line. Here we propose the Algorithm 1 to generate the center-line based on the contour. We need to calculate the shortest distance between each inside point and the contour boundary, so we can get the maximum inscribed circle radius of each point in the contour. *Distance* represents the calculation of the Euclidean distance between two points. *Area* and *Perimeter* represent the calculation of the area and perimeter of the contour.

---

**Data**: *contour*
**Result**: $Points_{center}$
Calculate the shortest distance between each inside point and the contour boundary as *distances*;
$Points_{center} = []$;
$min_R = Area(contour)/Perimeter(contour)$;
**while** $Max(distances) > min_R$ **do**
    $distance_{max} = Max(distances)$;
    Delete $distance_{max}$ from *distances*;
    Add the *point* with its $distance_{max}$ to $Points_{center}$ ;
    **for** $point_i \in contour$ **do**
        **if** $Distance(point, point_i) < 4 * min_R$ **then**
         |  distance of $point_i = 0$
        **else**
         |  pass
        **end**
    **end**
**end**

**Algorithm 1:** Center points generation

---

This algorithm continuously finds the largest inscribed circle in the contour and adds its center to the point set of the contour center-line and we also add the points that extend at both ends of the center-line to the center point set. Figure 3 shows the process of generating the center point.

We assume that the abscissa of the start point is smaller than the endpoint. And We can reorder them by the distance between the center points, as shown in Algorithm 2. We use thin-plate-spline (TPS) [14] interpolation to align the images, which can match and transform the corresponding points and minimize the bending energy generated by the deformation.

According to the points and radius of the center-line, we can generate coordinate points for TPS transformation, as shown in Fig. 4. Through TPS transformation, we can transform irregular text line strips into rectangles.

## 2.3   Recognition Module

This module is roughly equivalent to traditional text line recognition, except that we replaced the text line image with the text line feature map. Most methods use RNNs to construct semantic relations for time-series feature maps. We think that RNNs have three main disadvantages: 1) it is prone to gradient disappearance; 2) the calculation speed is slow; 3) the effect of processing long text is not good.

**Data:** $Points_c$ of length $l$, $min_R$

**Result:** $Points\_reorder_c$

$Points\_reorder_c = []$;

**if** *Len(Points$_c$)==1* **then**

   | $Points\_new_c =[Points_c]$ ;

**else**

   | $Points\_new_c = Points_c[:2]$;

   | **for** $i \leftarrow 2$ **to** $l$ **do**

      | $left_d = Distance(point_i, Points\_new_c[0])$;

      | $right_d = Distance(point_i, Points\_new_c[-1])$;

      | $left\_right_d = Distance(Points\_new_c[0], Points\_new_c[-1])$;

      | **if** $right_d > left\_right_d$ *and* $right_d > left_d$ **then**

         | Insert $point_i$ to $Points\_new_c$ in position 0;

      | **else**

         | **if** $left_d > left\_right_d$ *and* $right_d < left_d$ **then**

            | Insert $point_i$ to $Points\_new_c$ in position $-1$;

         | **else**

            | $mid = \arg\min_{j \in [1,l]}(Distance(point_i, Points_c[j]) + Distance(point_i, Points_c[j-1]))$;

            | Insert $point_i$ to $Points\_new_c$ in position $mid$;

         | **end**

      | **end**

   | **end**

**end**

**Algorithm 2:** Center points reordering

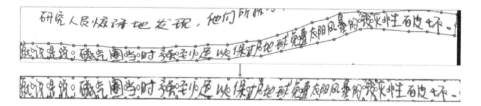

**Fig. 4.** TPS transformation on the text picture. Actually, our method performs transformation operations in the feature map.

We designed a multi-scale feature extraction network for text line recognition without using RNNs. And compared with CNN, TCN and Self-attention have stronger long-distance information integration capabilities. The structure of the recognition module is shown in Fig. 5.

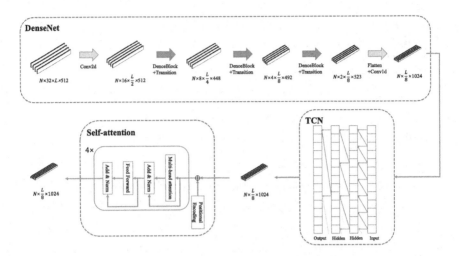

**Fig. 5.** Network architecture of recognition module.

For the task of text line recognition, we think there are three levels of recognition. The first is to extract features of each character based on each character's image information. The second is to perform auxiliary feature extraction based on several characters surrounding every single character. The third is to optimize features based on global text information.

For the first level, we use DenseNet as the backbone to perform further feature extraction on the feature map, gradually compress the height of the feature map to 1, and convert it into a time-sequential feature map. The multiplicative factor for the number of bottleneck layers is 4. There are 32 filters added to each Dense Layer. We added the CBAM [16] layer before each Dense Block to exploit both spatial and channel-wise attention. And the other detailed configuration of DenseNet is presented in Table 1.

We use the Temporal Convolutional Network (TCN) [17] for the second level of feature extraction. The dilated convolution is used on TCNs to obtain a larger receptive field than CNN. Compared with RNNs, it performs parallel calculations, and the calculation speed is faster, and the gradient is more stable. A 4-layer TCN is introduced in our model and the maximum dilation is 8 which makes the model has a receptive field of size 32.

For the third level, we adopt the structure of Self-attention for global information extraction. We only use the Encoder layer of Transformer [18], and the multi-head self-attention mechanism is the core to build global feature connections. The self-attention mechanism is mainly used in the field of natural

**Table 1.** The structure of DenseNet.

| Moudle | Output size | Config |
|---|---|---|
| Convolution 1 | $16 \times \frac{W}{2}$ | $3 \times 3\ conv,\ stride\ 2, 512 \rightarrow 512$ |
| Dense block 1 | $16 \times \frac{W}{2}$ | $\begin{bmatrix} 1 \times 1\ conv \\ 3 \times 3\ conv \end{bmatrix} \times 4,\ 512 \rightarrow 640$ |
| Transition layer 1 | $8 \times \frac{W}{4}$ | $3 \times 3\ conv,\ stride\ 2 \times 2,\ 640 \rightarrow 448$ |
| Dense block 2 | $8 \times \frac{W}{4}$ | $\begin{bmatrix} 1 \times 1\ conv \\ 3 \times 3\ conv \end{bmatrix} \times 8,\ 448 \rightarrow 704$ |
| Transition layer 2 | $4 \times \frac{W}{8}$ | $3 \times 3\ conv,\ stride\ 2 \times 2,\ 704 \rightarrow 492$ |
| Dense block 3 | $4 \times \frac{W}{8}$ | $\begin{bmatrix} 1 \times 1\ conv \\ 3 \times 3\ conv \end{bmatrix} \times 8,\ 492 \rightarrow 748$ |
| Transition layer 3 | $2 \times \frac{W}{8}$ | $3 \times 3\ conv,\ stride\ 2 \times 1,\ 748 \rightarrow 523$ |
| Flatten | $\frac{W}{8}$ | $523 \rightarrow 1046$ |
| Convolution 2 | $\frac{W}{8}$ | $1\ conv,\ 1046 \rightarrow 1024$ |

language processing, but the text obtained by text recognition also has semantic features. We can use Self-attention to construct semantic association to assist recognition. A 4-layer Self-attention encoder is applied in our model, and the hidden layer dimension is 1024 with 16 parallel attention heads.

## 2.4 Loss Function

The loss function is as follows:

$$L = L_{text} + \alpha L_{kernel} \qquad (1)$$

$L_{text}$ is the character recognition loss, calculated with CTC, and $L_{kernel}$ is the loss of the kernel, and $\alpha$ is to balance the importance of the two, we set $\alpha$ to 0.1.

Considering the imbalance between the text area and the non-text area, we use the dice coefficient as the method to evaluate the segmentation results.

$$L_{kernel} = 1 - \frac{2 \sum_i P_{text}(i) G_{text}(i)}{\sum_i P_{text}(i)^2 + \sum_i G_{text}(i)^2} \qquad (2)$$

$P_{text}(i)$ and $G_{text}(i)$ represent the value of the $i$th pixel in the segmentation result and the ground truth of the text regions respectively. The ground truth of the text regions is a binary image, in which the text pixel is 1 and the non-text pixel is 0.

## 3   Experiment

### 3.1   Dataset

We use CASIA-HWDB [19] as the main dataset, which has been divided into the train set and test set. It includes offline handwriting isolated character pictures and offline unconstrained handwritten text page pictures. CASIA-HWDB1.0-1.2 contains 3,721,874 isolated character pictures of 7,356 classes. CASIA-HWDB2.0-2.2 contains 5,091 text page pictures, containing 1,349,414 characters of 2,703 classes. CASIA-HWDB2.0-2.2 is divided into 4,076 training samples and 1,015 test samples. ICDAR-2013 Chinese handwriting recognition competition Task 4 [20] is a dataset containing 300 test samples, containing 92,654 characters of 1,421 classes. The isolated character pictures of CASIA-HWDB1.0-1.2 are used to synthesize 20,000 text page pictures, containing 9,008,396 characters. The corpus for synthesizing is a dataset containing 2.5 million news articles.

We regenerate the text line boxes by using the smallest enclosing rectangle method according to the outline of the text line, as shown in Fig. 6. We use the same method as PANNet to generate the text kernels that adopt the Vatti clipping algorithm [21] to shrink the text regions by clipping d pixels. The offset pixels d can be determined as $d = A(1 - r^2)/L$, where $A$ and $L$ are the area and perimeter of the polygon that represents the text region, and $r$ is the shrink ratio, which we set to 0.6.

We convert some confusing Chinese symbols into English symbols, such as quotation marks and commas.

**Fig. 6.** Regenerate the text boxes with the smallest enclosing rectangle.

### 3.2   Experimental Settings

We use PyTorch to implement our system. We train the whole system with the Adam optimizer and the batch size is set to 4. We resize the pictures with a length not greater than 1200 or a width not greater than 1600. No language model is used to optimize the recognition results. We conducted two main experiments, and the rest of the experimental settings are as follows:

1) Train set: train set of CASIA-HWDB2.0-2.2;
   Test set: test set of CASIA-HWDB2.0-2.2;
   Learning rate: initialized to $1 \times 10^{-4}$ and multiplied by 0.9 every 2 epochs;
   Train epoch: 50.
2) Train set: train set of CASIA-HWDB2.0-2.2 and 20,000 synthesized text page pictures;
   Test set: test set of CASIA-HWDB2.0-2.2 and test set of ICDAR-2013;
   Learning rate: initialized to $5 \times 10^{-5}$ and multiplied by 0.9 every 2 epochs;
   Train epoch: 50;
   The model weight of Experiment 1 is used for initialization, except for the final fully connected layer.

**Data**: $kernel\_boxers$ with length $m$, $gt\_boxes$ with length $l$
**Result**: $kernel\_boxers\_group$
$kernel\_boxers\_group$ is a list with length $l$;
**for** $i \leftarrow 0$ **to** $m$ **do**
$\quad | \quad index = \arg\max_{j \in [0, l-1]}(IOU(kernel\_boxers[i], gt\_boxes[j]));$
$\quad | \quad$ add $kernel\_boxers[i]$ to $kernel\_boxers\_group[index]$
**end**

**Algorithm 3:** Kernel boxes grouping

### 3.3 Experimental Results

**With Ground Truth Box**

The ground truth boxes are used for segmentation so that our model is equivalent to the text line recognition model. But compared with the previous text line recognition model, our model can utilize the global information of the text page for recognition, which has stronger anti-interference and robustness. The experimental results are shown in Table 2. Notably, our method is superior to the previous results by a large absolute margin of 1.22% CR and 2.14% CR on CASIA-HWDB2.0-2.2 and ICDAR-2013 dataset without the language model. The recognition performance of our model in CASIA-HWDB2.0-2.2 dataset is even better than the method using the language model.

Even when the text boxes do not include the whole text line, our model can recognize the text line correctly, as shown in Fig. 7. We also conduct an experiment with the size of the gt boxes randomly changed that the outline points of gt boxes are randomly moved by $-0.2$ to $0.2$ of the width in the vertical direction and moved by $-1$ to $1$ of the width in the horizontal direction, and the CR only dropped by about 0.5%, as shown in Table 2, which shows that our model is very robust. Because we perform segmentation and transformation of text lines in the feature map of the text page, we expand or shrink the transformation box so that the text information is concentrated in the core area of the

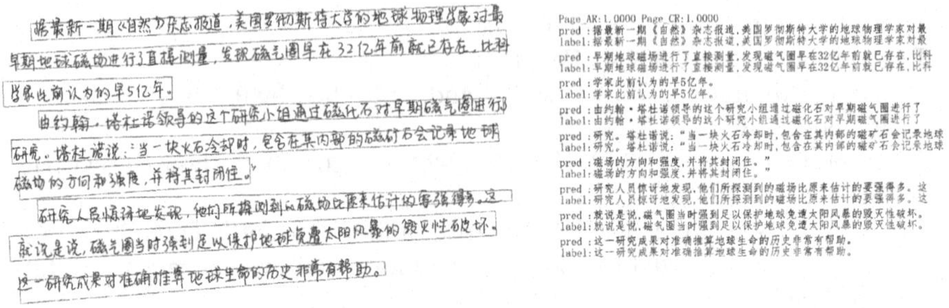

**Fig. 7.** Recognize text picture when the text box is not right.

text. Low-dimensional information is more sensitive, if such data enhancement is performed directly on the original picture, some text will become incomplete and cannot be correctly recognized

**Table 2.** Comparison with the start-of-the-art methods.

| Method | Without LM | | | | With LM | | | |
|---|---|---|---|---|---|---|---|---|
| | HWDB2 | | ICDAR-2013 | | HWDB2 | | ICDAR-2013 | |
| | CR (%) | AR (%) | CR (%) | AR (%) | CR (%) | AR (%) | CR (%) | AR (%) |
| Wu [23] | – | – | – | – | 95.88 | 95.88 | 96.20 | 96.20 |
| Peng [22] | – | – | 89.61 | 90.52 | – | – | 94.88 | 94.88 |
| Song [2] | 93.24 | 92.04 | 90.67 | 88.79 | 96.28 | 95.21 | 95.53 | 94.02 |
| Xiao [6] | 97.90 | 97.31 | – | – | – | – | – | – |
| Xie [5] | 95.37 | 95.37 | 92.13 | 91.55 | 97.28 | 96.97 | 96.99 | 96.72 |
| Ours, with gt box[a] | **99.12** | **98.84** | – | – | – | – | – | – |
| Ours, with center-line[a] | 99.03 | 98.64 | – | – | – | – | – | – |
| Ours, with changed gt box[c] | 98.58 | 98.22 | – | – | – | – | – | – |
| Ours, line level[d] | 98.38 | 98.22 | – | – | – | – | – | – |
| Ours, with gt box[b] | 98.55 | 98.21 | **94.27** | **93.88** | – | – | – | – |
| Ours, with center-line[b] | 98.38 | 97.81 | 94.20 | 93.67 | – | – | – | – |

[a] Experiment 1. [b] Experiment 2.
[c] Experiment 1 with the size of the gt boxes randomly changed.
[d] Line level recognition with modified recognition module.

## With Center-Line

It can be noted that the CR calculating with center-line segmentation only drops by about 0.1%, which fully proves the effectiveness of center-line segmentation.

Since one line of text may be divided into multiple text lines during segmentation, it is necessary to correspond the text box obtained by the segmentation with the ground truth box. We use the method of calculating the intersection over union (IOU) for the match, as shown in Algorithm 3. Before performing this algorithm, it is necessary to sort all the divided boxes according to the abscissa, so that multiple text lines are added to the corresponding group in a left-to-right manner. Each group corresponds to one text line label. If there are extra boxes, that is, all IOUs are calculated as 0, they can be added to any group, which does not affect the calculation of CR and AR.

After all the segmentation boxes are divided into groups, the detection result is considered correct when the length of the recognition result of the group is greater than or equal to 90% of the label length. Table 3 shows the detection performance of our model.

**Effectiveness of the Segmentation Module**
We can slightly change the structure of the recognition module to make it a text line recognition model. The main modification is to change the 3-layer Dense Block to the 5-layer Dense Block, from $[4, 8, 8]$ to $[1, 4, 8, 8, 8]$. The input data is changed to a 64-pixel text line image with a segmented height and batch size is set to 8, and the other settings are similar to experiment 1. The recognition results show that the segmentation module can make good use of the global text information to optimize the text feature extraction, thereby improving the recognition performance.

**Table 3.** Text detection results on CASIA-HWDB2.0-2.2 and ICDAR-2013 dataset.

|  | Precision (%) | Recall (%) | F-measure (%) |
|---|---|---|---|
| HWDB2 | 99.97 | 99.79 | 99.88 |
| ICDAR-2013 | 99.91 | 99.74 | 99.83 |

**Effectiveness of TCN and Self-attention**
We conduct a series of ablation experiments on CASIA-HWDB2.0-2.2 dataset without data augmentation to verify the effectiveness of TCN and Self-attention, with results shown in Table 4. On the test set, 1.08% and 1.66% CR improvements are obtained by the TCN module and the Self-attention module, respectively. And CR is increased by 1.96% when the TCN module and Self-attention module are applied together.

**Table 4.** Ablation experiment results on CASIA-HWDB2.0-2.2.

| Method | With gt box | | With center-line | |
|---|---|---|---|---|
| | CR (%) | AR (%) | CR (%) | AR (%) |
| Baseline | 97.18 | 96.81 | 97.09 | 96.63 |
| +Self-attention | 98.74 | 98.42 | 98.66 | 98.28 |
| +TCN | 98.26 | 97.97 | 98.03 | 97.62 |
| +TCN and self-attention | 99.12 | 98.84 | 99.03 | 98.64 |

## 4    Conclusion

In this paper, we propose a novel robust end-to-end Chinese text page spotter framework. It utilizes global information to concentrate the text features in the kernel areas, which allows the model only need to roughly detect the text area for recognition. TPS transformation is used to align the text lines with center points. TCN and Self-attention are introduced into the recognition module for multi-scale text information extraction. It can perform end-to-end text detection and recognition, or optimize recognition when ground truth text boxes are provided. This architecture can be easily modified that the segmentation module and recognition module can be replaced by better models. Experimental results on CASIA-HWDB2.0-2.2 and ICDAR-2013 datasets show that our method achieves state-of-the-art recognition performance. Future work will be to investigate the performance of our method on the English dataset.

## References

1. Wang, Q., Yin, F., Liu, C.: Handwritten Chinese text recognition by integrating multiple contexts. IEEE Trans. Pattern Anal. Mach. Intell. **34**, 1469–1481 (2012)
2. Wang, S., Chen, L., Xu, L., Fan, W., Sun, J., Naoi, S.: Deep knowledge training and heterogeneous CNN for handwritten Chinese text recognition. In: 2016 15th International Conference on Frontiers in Handwriting Recognition (ICFHR), pp. 84–89 (2016)
3. Messina, R.O., Louradour, J.: Segmentation-free handwritten Chinese text recognition with LSTM-RNN. In: 2015 13th International Conference on Document Analysis and Recognition (ICDAR), pp. 171–175 (2015)
4. Graves, A., Fernández, S., Gomez, F., Schmidhuber, J.: Connectionist temporal classification: labelling unsegmented sequence data with recurrent neural networks. In: Proceedings of the 23rd International Conference on Machine Learning (2006)
5. Xie, C., Lai, S., Liao, Q., Jin, L.: High Performance Offline Handwritten Chinese Text Recognition with a New Data Preprocessing and Augmentation Pipeline. DAS (2020)
6. Xiao, S., Peng, L., Yan, R., Wang, S.: Deep network with pixel-level rectification and robust training for handwriting recognition. In: 2019 International Conference on Document Analysis and Recognition (ICDAR), pp. 9–16 (2019)
7. Li, X., Wang, W., Hou, W., Liu, R., Lu, T., Yang, J.: Shape robust text detection with progressive scale expansion network. In: 2019 IEEE/CVF Conference on Computer Vision and Pattern Recognition (CVPR), pp. 9328–9337 (2019)
8. Liu, Y., Chen, H., Shen, C., He, T., Jin, L., Wang, L.: ABCNet: real-time scene text spotting with adaptive Bezier-curve network. In: 2020 IEEE/CVF Conference on Computer Vision and Pattern Recognition (CVPR), pp. 9806–9815 (2020)
9. Liao, M., Pang, G., Huang, J., Hassner, T., Bai, X.: Mask TextSpotter v3: Segmentation Proposal Network for Robust Scene Text Spotting. Arxiv, arXiv:abs/2007.09482 (2020)
10. Bluche, T.: Joint line segmentation and transcription for end-to-end handwritten paragraph recognition. In: NIPS (2016)

11. Yousef, M., Bishop, T.E.: OrigamiNet: weakly-supervised, segmentation-free, one-step, full page text recognition by learning to unfold. In: 2020 IEEE/CVF Conference on Computer Vision and Pattern Recognition (CVPR), pp. 14698–14707 (2020)
12. Wang, W., et al.: Efficient and accurate arbitrary-shaped text detection with pixel aggregation network. In: 2019 IEEE/CVF International Conference on Computer Vision (ICCV), pp. 8439–8448 (2019)
13. He, K., Zhang, X., Ren, S., Sun, J.: Deep residual learning for image recognition. In: 2016 IEEE Conference on Computer Vision and Pattern Recognition (CVPR), pp. 770–778 (2016)
14. Bookstein, F.: Principal warps: thin-plate splines and the decomposition of deformations. IEEE Trans. Pattern Anal. Mach. Intell. **11**, 567–585 (1989)
15. Huang, G., Liu, Z., Weinberger, K.Q.: Densely connected convolutional networks. In: 2017 IEEE Conference on Computer Vision and Pattern Recognition (CVPR), pp. 2261–2269 (2017)
16. Woo, S., Park, J., Lee, J.-Y., Kweon, I.S.: CBAM: convolutional block attention module. In: Ferrari, V., Hebert, M., Sminchisescu, C., Weiss, Y. (eds.) ECCV 2018. LNCS, vol. 11211, pp. 3–19. Springer, Cham (2018). https://doi.org/10.1007/978-3-030-01234-2_1
17. Lea, C.S., Flynn, M.D., Vidal, R., Reiter, A., Hager, G.: Temporal convolutional networks for action segmentation and detection. In: 2017 IEEE Conference on Computer Vision and Pattern Recognition (CVPR), pp. 1003–1012 (2017)
18. Vaswani, A., et al.: Attention is All you Need. ArXiv, arXiv:abs/1706.03762 (2017)
19. Liu, C., Yin, F., Wang, D., Wang, Q.: CASIA online and offline Chinese handwriting databases. In: 2011 International Conference on Document Analysis and Recognition, pp. 37–41 (2011)
20. Yin, F., Wang, Q., Zhang, X., Liu, C.: ICDAR 2013 Chinese handwriting recognition competition. In: 2013 12th International Conference on Document Analysis and Recognition, pp. 1464–1470 (2013)
21. Vatti, B.R.: A generic solution to polygon clipping. Commun. ACM **35**, 56–63 (1992)
22. Peng, D., Jin, L., Wu, Y., Wang, Z., Cai, M.: A fast and accurate fully convolutional network for end-to-end handwritten Chinese text segmentation and recognition. In: 2019 International Conference on Document Analysis and Recognition (ICDAR), pp. 25–30 (2019)
23. Wu, Y., Yin, F., Liu, C.: Improving handwritten Chinese text recognition using neural network language models and convolutional neural network shape models. Pattern Recognit. **65**, 251–264 (2017)

# Data Augmentation vs. PyraD-DCNN: A Fast, Light, and Shift Invariant FCNN for Text Recognition

Ahmad-Montaser Awal[✉], Timothée Neitthoffer, and Nabil Ghanmi

Research Department - ARIADNEXT, Cesson-Sévigné, France
{montaser.awal,timothee.neithoffer,nabil.ghanmi}@ariadnext.com

**Abstract.** Modern OCRs rely on the use of recurrent deep neural networks for their text recognition task. Despite its known capacity to achieve high accuracies, it is shown that these networks suffer from several drawbacks. Mainly, they are highly dependent on the training data-set and particularly non-resistant to shift and position variability. A data augmentation policy is proposed to remedy this problem. This policy allows generating realistic variations from the original data. A novel fast and efficient fully convolutional neural network (FCNN) with invariance properties is also proposed. Its structure is mainly based on a multi-resolution pyramid of dilated convolutions. It can be seen as an end-to-end 2D signal to sequence analyzer that does not need recurrent layers. In this work, extensive experiments have been held to study the stability of recurrent networks. It is shown that data augmentation significantly improves network stability. Experiments have also confirmed the advantage of the PyraD-DCNN based system in terms of performance but also in terms of stability and resilience to position and shift variability. A private data-set composed of more than 600k images has been used in this work.

**Keywords:** Text recognition · FCNN · Dilated convolutions · Deep learning · OCR · Data augmentation

## 1 Introduction

Optical character recognition (OCR) systems have become more and more necessary in our days. Although they first appeared in the 80s [16], text recognition is still a challenging problem and new approaches are continuously emerging. This is motivated by the explosion of the use of digital images producing different formats of text, varying from high quality scanned documents up to Text in the Wild. In this work, we are interested in text recognition in the context of Know Your Customer (KYC) systems where service providers need to extract users' information from their Identity Documents (ID). It is worth noting that the text localization and the layout analysis are out of the scope of this paper. The text fields are already localized thanks to a document knowledge base. Despite this

E. H. Barney Smith and U. Pal (Eds.): ICDAR 2021 Workshops, LNCS 12917, pp. 36–50, 2021.
https://doi.org/10.1007/978-3-030-86159-9_3

pre-localization, text fields recognition is still challenging as they are incorporated with a complex textured background, as shown in the Fig. 1. In addition, in constraint-free capture conditions, the quality of text fields is very variable which makes this task even harder.

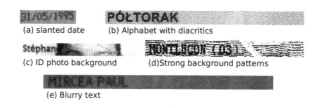

(a) slanted date     (b) Alphabet with diacritics

(c) ID photo background     (d)Strong background patterns

(e) Blurry text

**Fig. 1.** Examples of text fields from various IDs in challenging conditions.

Nowadays, hybrid neural networks provide state-of-the-art performance on the task of text recognition. They are basically constructed by stacking two types of network allowing them to simultaneously handle two types of information: spatial and temporal as shown in Fig. 2(a). The first stack acts as an optical feature extractor, based on convolutional layers (CNN). The second one is a temporal (sequential) modeling part based on a recurrent neural network (RNN, ex: BLSTM) that produces a sequence of character probabilities interpreted by a Connectionist Temporal Classification heuristic (CTC) [10]. However, this type of network presents a major drawback when it comes to storage and computational costs. In fact, CNN-BLSTM-like text recognition systems are too costly as a recurrent neuron has the same cost as 4 dense ones. Furthermore, the recurrence implies non-parallelizable computing over the text-reading axis. In addition, traditional convolution layers combined with the BLSTM layers make such system dependent on the training data set and sensitive to variations present in the real text field images.

We propose in this context the following contributions:

- an experimental study allowing to measure the sensitivity of a given text recognition system,
- a data-augmentation method to overcome sensitivity issues and increase the system's invariance,
- a novel text recognition system based on dilated convolutions,
- an empirical demonstration of the advantage of using FCNN to overcome data dependency (increase the network invariance).

It is then shown how this design provides a very good alternative to the costly CNN-BLSTM-like systems while reducing the system sensitivity.

The rest of this paper is organized as follows. First, related works and a brief stat-of-the-art are presented in Sect. 2. The proposed data augmentation policy is then explained in Sect. 3 together with our novel FCNN architecture for text recognition. Results and experiments on a real data-set are finally discussed in Sect. 4 before concluding the paper and giving some perspectives.

## 2   Related Work

Early techniques of OCR systems were based on the use of hand-crafted features followed by traditional classifiers (SVM, HMM, ...) applied to segmented words or even characters. Segmentation-free approaches have become more popular allowing significant improvements in OCR accuracy. Such systems employed a simultaneous segmentation/recognition approach [7]. More recently, deep neural networks have emerged as a way to decode text once localized, often through convolutional layers, or using a sliding window to simultaneously localize and decode the text. Recurrent neural networks (RNN) have been massively used in this context as explained in the following section.

### 2.1   Bidirectional Short-Long Term Memory Neural Networks (BLSTM) for OCR

Basically, our benchmark system is based on a CNN-BLSTM neural network as described in [13]. The main idea is to learn the relationship between successive inputs by using the output of neurons as part of their own inputs. This is usually done by considering time as a discrete dimension and feeding input piece by piece. Thus, a RNN would be able to use the temporal structure of the data as an additional piece of information.

In practice, it allows the modelling of the sequential nature of the text with the writing direction presenting the temporal dimension of the input signal. The input field image is cut into overlapping windows and then fed to a CNN that acts as a feature extractor. The features extracted at this stage are purely image dependent (ex. lines, curves, ...) and no temporal information is yet used. A BLSTM restores the temporal dimension of the extracted feature maps. This construction allows to read the features from both left to right and right to left. Finally, a softmax layer is used to predict probabilities for each possible character for each window. A beam-search algorithm [10] is then employed to convert obtained probabilities to a text string. An overview of the used structure is shown in Fig. 2(b) with the same CNN part (OFE) as in Fig. 4. However, besides their high cost (in terms of time and storage), BSLTMs suffer from high sensitivity to text position in the field images as it will be shown in Sect. 4. And even small input shifts can significantly change the OCR result. To overcome this problem, a natural idea is to incorporate data variability in the training process.

### 2.2   Data Augmentation

Data augmentation techniques are mostly used in image classification tasks. The direct intuitive solution is to artificially add samples with various changes (translation, rotation, etc.) to the training set. This can be done by applying specific transformations, such as adding blocks to simulate occlusions [6], or by augmenting the training data based on a parameterized distribution of applicable transformations [1]. In the latter, an adapted loss function is proposed to learn invariance from augmented data. The individual outputs obtained from

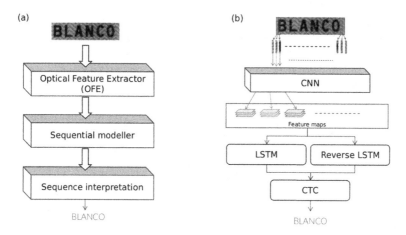

**Fig. 2.** (a) Global structure of NN architecture for OCR, (b) CNN-BLSTM structure for OCR

the augmented samples of the same label are then averaged to converge together to the correct label. While most of data augmentation techniques are performed at the network entrance, a data augmentation policy inside the network itself is proposed in [20]. To this end, the intermediate feature maps are transformed to simulate data augmentation. Applying the transformation on the lower layers feature maps (at the beginning of the network) was proven to be more efficient than applying it to higher layers (towards the end of the network).

As we can notice, there are several augmentation operations, each of which makes the trained model less sensitive to one or more variations. In order to properly apply these techniques and avoid potential negative effects, several strategies are developed. *KeepAugment* [8] tries to find the important image components in order to avoid degrading them when generating additional examples for data augmentation. *Fast(er) AutoAugment* [12] searches for an effective augmentation policy from a combination of augmentation techniques and the corresponding parameters to achieve augmentation on the data-set. In [21], the authors propose three algorithms (*WeMix*, *AugDrop*, and *MixLoss*) to overcome the data bias introduced by the augmentation as augmented data distribution can be different from the original one.

Despite their capacity to improve trained systems sensibility, most of the aforementioned solutions increase the data quantity and though make the training process harder (more difficult convergence) and longer.

## 2.3   Fully Convolutional Neural Networks in Modern OCR Systems

As mentioned earlier, recurrent layers are the costly part of OCR systems preventing their use in embedded environments, such as smartphones for example. Using straight CNN models could be thought as the simplest way to overcome recurrence layers by transforming the problem to a multi-class classification one.

While this solution might be efficient to handle short strings, it does not scale to real contexts as it might be re-trained for each new text string to consider [9]. A more generic approach is proposed in [17,19] based on a sliding shared output layer. This solution, however, requires to build a secondary system to estimate text-lines length.

Text recognition could also be seen as a combination of 2D signal processing (as image segmentation) and sequence processing (as language modeling). Both fields have seen recently common advancements based on the use of dilated convolution neural networks (DCNN). Such networks have been first introduced to expand observation windows within the convolution without impacting the input high-resolution features [22]. In fact, stacking dilated convolutional layers has allowed to simultaneously analyze multi-scale features. Networks with parallel branches of stacked dilated convolutions have been widely used in various applications, such as counting objects in real scenes [4], real-time brain tumor segmentation [3], or even text detection [18]. On the other hand, DCNNs have also been used in sequence processing systems. In this context, convolutions are applied to the sequence axis such as for modelling DNA sequence [11]. Dilated convolutions have shown a very good ability to handle very large sequences and have also been successfully applied for text classification as in [15]. Multi-resolution approaches have also been inspired by those works. It has been shown that a progressive merging of layers is necessary to avoid unwanted artifacts from the dilations sub-sampling [5]. Finally, a similar design to our proposed architecture has been presented in [2] but using Gated Residual Dilated convolutions instead of densely connected ones. On the other hand, sharing weights among stacked convolutional layers allows to reproduce a recurrence mechanism. This type of mechanism is implemented in Recurrent Convolutional Neural Networks (RCNN) using Recurrent Convolution Layers (RCL) [2].

## 3    Propositions

In this section, our data augmentation approach is first detailed. Independently from the neural network architecture, it is intended to increase the text recognition robustness to real life variation of the handled data. In a second time, our FCNN neural network architecture, namely PyraD-DCNN, is presented. In addition to be faster and lighter than traditional CNN-BLSTM networks, it also significantly increases the OCR robustness to variations in regards to the training data-set.

### 3.1    Data Augmentation: A Probabilistic Approach

Generally, training data-sets contain samples issued from representative examples of the real-life problem. In our context, training samples are extracted from identity documents by pre-segmenting field images through the document analysis process. This implies that the extracted field images have fixed margins around the centered text as defined in the layout model of the document.

However, as the segmentation step is not perfect, this process results with some variability that was not seen during the training stage. In addition, it is strongly dependent on the analysis process and any change results in margins variability. The above two situations lead to unstable recognition systems with unexpected behaviour when shift variability is introduced (i.e. margins different from those seen during training).

In fact, data augmentation can be seen as the intuitive solution to improve the system's robustness to unexpected variations. Thus, we propose an augmentation policy to add field images with different margin configurations applied during the segmentation step. An augmentation policy that randomly sample margins to each side of the original images is able to incorporate *natural* variations, allowing the system to learn to deal with such cases.

The augmentation also needs to preserve the original text content to avoid relabeling the newly generated images. To that aim, it must not incorporate text from other text lines nor cut characters from the current one. In order to respect this constraint, the *random* added margins must fall in the interval between a fixed threshold and the distance to the closest neighbour text line.

It is also important to keep the local context of the image. Indeed, in this study, we are working on highly textured backgrounds that are difficult to reproduce. As a result, it is not relevant to use padding techniques that modulate the dimensions by adding white pixels (for example) as they disrupt the background. To keep this important background, each text line is segmented using the largest possible margin as defined beforehand and the coordinates of the text position in this image are also kept. Then, several images are produced by applying the augmentation policy on the original enlarged image and coordinates (Fig. 3).

**Enlarged field zone**

(originally detected text in black rectangle)

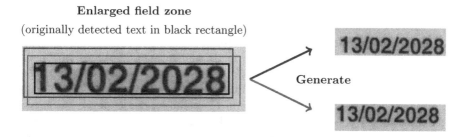

**Fig. 3.** Representation of the augmentation policy (the exact sizes are not respected for the sake of comprehension).

Now that the general concept of the augmentation is defined, there are multiple distributions to sample the margin from. In the real data, the segmentation is in general close to the optimal one, shifting from only some pixels following a normal law. It is then interesting to use such distribution to generate more realistic augmented data in regards to the real data. This raw distribution only focuses on small optimal variations. Additional variations are thus necessary as we want our system to be invariant to extremum. This is done by combining the normal distribution with the uniform distribution using a Bernouilli trial selector as shown in Eq. 1.

$$Margin = Bern(p) * N(\mu, \sigma) + (1 - Bern(p)) * U(0, max\_margin) \quad (1)$$

Here, $Margin \in [0, max\_margin]$ is the sampled size of the margin. $max\_margin$ is the maximum size of the margin as discussed previously. $Bern(p)$ is the Bernoulli trial of probability $p.N(\mu, \sigma)$ is the normal distribution of parameters $\mu$ the designed optimal margin of the system and $\sigma$ the variance in the real data distribution. $U(0, max\_margin)$ is the uniform distribution. Such distribution can also be tuned using $p$ to adapt it to the desired distribution.

Nevertheless, data augmentation increases dramatically the quantity of the training data (up to ×5 to have significant improvements). This growth lengthens training time and makes network convergence more difficult. In addition, data augmentation could be seen as a workaround to the main problem, neural network sensitivity. The PyraD-DCNN is proposed as a more efficient solution to this problem as we will see in next section.

### 3.2 PyraD-DCNN: A Fast, Light, and Shift Invariant FCNN for Text Recognition

Differently from designs discussed in Sect. 2.3, the proposed model relies on a Pyramid of Densely connected Dilated Convolutional layers [14]. The main goal of this design is to reproduce the LSTM layers ability to handle long/short-term spatial dependencies by using lighter alternatives with a lower storage footprint. Indeed, introducing dilated convolutions allows to dramatically reduce networks size. This is possible thanks to the dilated kernels that allow processing of very large windows with relatively shallow stacks of convolutional layers. In addition, they provide long-term dependencies perfectly replicating LSTM layers behavior leading to accuracy improvements.

Moreover, the pyramid-like design allows to combine downscale and upscale dilated layers. This is particularly helpful to include different levels of feature maps. Thus we use a stack of dilated convolutions in increasing order of dilation (1-2-4), then in decreasing order (4-2-1) to operate a smooth fusion of the highly dilated features produced in the first half. It is expected that this design will help producing accurate and robust features by making the best use of the DCNN and thus naturally learn to be invariant to text position variability.

Furthermore, the depth of the network has required the introduction of skip connections that have been shown more efficient than residual connections. The main idea is to forward the concatenation of the input and the output of a given layer. In such a way, the following layers are receiving features with different levels of abstraction or scales of the visual input field.

As the different levels of dilation induce different context views and that they are all important to the decision, the pyramid needs be densely connected to be able to compose the different upstream steps. This latter configuration helps to manage both short and long term dependencies, and so, completely reproducing the LSTM behavior.

The drawback of using skip connections is the increasing number of channels when stacking skipping layers. The direct effect of this is the production of a growing number of features that may cause the network to require an unaffordable number of parameters. Pointwise $1 \times 1$ convolutional layers are used as bottlenecks on the input features by reducing their number of channels to *conv_depth*, Fig. 5(b). A $1 \times 3$ dilated channel-wise convolutional is then used to produce the spatial dimension, *row*, followed by a ReLU activation and a dropout layer. The bottleneck is also introduced in the middle and at the end of the pyramid to limit any possible channel expansion by compressing the features down to a fixed number of channels, the *feature_depth*. Overall, the proposed architecture is composed of:

- **An optical feature extractor (OFE)** producing the feature maps of the input image that will be then analyzed by the sequence modeller. The architecture of the implemented OFE is detailed the Fig. 4.
- **A sequence modeller network** transforming features into character probabilities. Three dilation units (Fig. 5(a)) are stacked twice to build a symmetrical multi-resolution pyramid, Fig. 5(c).
- **A connectionist temporal classifier** aligning characters into text strings. It is used to interpret the output per-frame predictions and compute a loss that bridges the spatially aligned prediction to the text string.

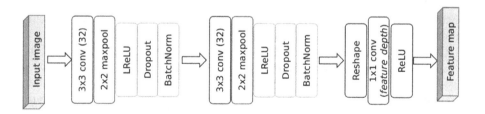

**Fig. 4.** Optical Feature Extractor architecture

## 4   Experiments

### 4.1   Data-Sets

We focus in this paper on the recognition of text fields in identity documents. Fields can be alphabetic (names, birth places, ...), numerical (ex. personal numbers), or alphanumerical (ex. addresses). Each training data-set is split as follows: 60% for training, 20% for validation and the last 20% for evaluation within the training cycle. The following data-sets are considered in our experiments.

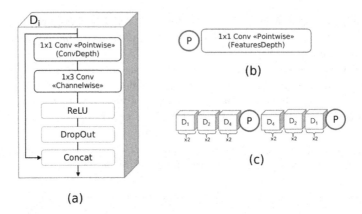

**Fig. 5.** a) Dilated Convolution unit, b) Pointwise bottleneck unit, c) proposed sequence modeller architecture

**Italian Driving License (IDL).** This data-set is composed of 2790 training documents (22320 field images) and 252 documents (2016 field images) for the test. Eight fields concerning the ID holder information are considered (named in Fig. 6). Each field, except for the dates and the document number, can be composed of multiple words. Two sets are extracted from the training documents. *IDL-Train-Origin* is composed of field images extracted with a fixed 5 px margin; and *IDL-Train-Aug* which is generated via the augmentation policy applied on the Original training set (with $max\_margin = 10$ px).

Similarly, two sets are used for the test stage. *IDL-Test-Origin* is composed of field images extracted with a fixed 5 px margin; and *IDL-Test-Var* constituted of field images with the margins drawn from a normal distribution centered on 5 with a standard deviation of 2.

This distribution is chosen to simulate the "real-world" data variability as measuring the real distribution would require intensive human annotation of a huge amount of real data. Even though the standard deviation would be smaller in reality, it allows the test set to sample extremum margins and be more representative of the possible variations.

**EU-ID.** Composed from seven different European ID documents, EU-ID counts a total of 115 characters including digits, accented letters, and punctuation representing some 15 different field types. Totaling 600 000 samples, this data-set offers high variability in field formats, fonts, and backgrounds.

## 4.2 Experimental Setup

**Metrics.** Both perfect field match rate (*accuracy*) and character error rate (*CER*) are considered during training cycles. At each training epoch the network is tested against the validation set, the best network in terms of *accuracy* is then evaluated on the test set to get the final scores.

**Hyper-parameters.** All trainings were held on a GeForce GTX 1080 Ti. The inference (prediction) is run on the following setup: a CPU-only i7-7820HQ CPU @ 2.90 GHz/32 Go RAM, which can be representative of a industrial deployment platform.

The hyper-parameters of the optical feature extractor (4) have been set experimentally. The input images are gray-scale, resized to 32 pixels in height, and padded to 800 pixels in width. The features are normed to 64 channels to feed the second part of the OFE. The *alpha* parameter of the LReLU is set to 0.3. The output features have 144 frames with depth related to the data-set characters dictionary. The batch size is set to 16 and the training is stopped at the 50th epoch. In addition, the *features_depth* and *channels_depth* have been set to 128 and 64 respectively. Dropout in the Optical Feature Extractor is set to 0.05 and in the Sequence Modeler to 0.2. We use the SGD algorithm to regress the loss. Learning rate evolves between a minimum value of 0.0001 and a maximum of 0.01 using a periodic triangular profile.

**Augmentation Setup.** The use of a data augmentation policy induces an increase of the storage space as well as the time needed to pass through the data-set. In order to allow for faster experiments, we decided on generating 5 variants per text field and did not explore the evolution of the performance with respect to this number.

Some fields are under-represented and thus difficult to learn (ex. document number, birth date). In order to alleviate this problem, a larger amount of variants for those fields is generated (here 10 variants).

We sample the margins of the variants via a custom distribution that draw a margin either from a normal distribution or a uniform distribution as presented in Eq. 1. This choice is made given a parameter $p$. In this work, we performed several experiments to chose $p$ and selected the one giving the best results: $p = 0.5$.

### 4.3   Sensitivity of CNN-BLSTM Systems and the Use of Data Augmentation

Recurrent based systems view images as time series and model probabilities of the next step from what they have seen in the previous steps. Thus, they are intrinsically sensitive to unseen variations. Here, we demonstrate this sensitivity to margin size and text position as well as its impact on the OCR system through two experiments. We also show that the use of an adapted data augmentation method can alleviate the impact of such variations.

**Fixed Margin.** In this experiment, the *IDL-Test-Origin* set is replicated by using a fixed margin $\Delta x$. In each copy, the zone of each extracted field is enlarged by the same margin $\Delta x$ (regarding its original size as defined in the corresponding model). This experiment intends to show and compare the accuracy evolution on the different sizes when the system is trained with *IDL-Train-Origin* (using a fixed margin $x_t = 5$ px) and with the augmented set *IDL-Train-Aug*.

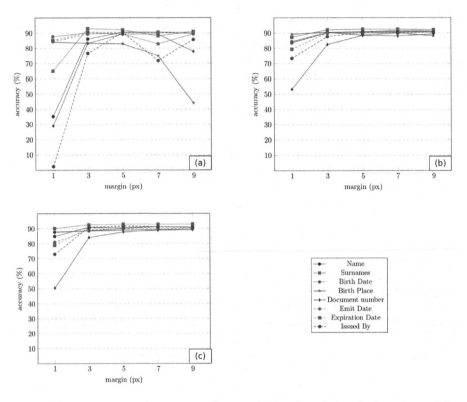

**Fig. 6.** Field accuracy with respect to the margin size (5 px being the margin used for *IDL-Train-Original*) for (a) CNN-BLSTM system trained on *IDL-Train-Original*, (b) CNN-BLSTM system trained on *IDL-Train-Aug*, (c) PyraD-DCNN system trained on *IDL-Train-Origin*.

We can observe from Fig. 6(a) that the CNN-BLSTM, trained with *IDL-Train-Origin*, has an erratic behaviour on the different fields when shifting from the "original" 5 px margin. The date fields, however, maintain a good accuracy thanks to their specific format. It can also be seen that the accuracy drops faster for the smaller margins than it does for the bigger margins.

On the other hand, training the CNN-BLSTM system using the *IDL-Train-Aug* set produces a more stable behavior as seen in Fig. 6(b). The accuracy of the different fields are kept at the same level on the different margin sizes. However, the system still suffers from some loss when facing really low margin size (1 px).

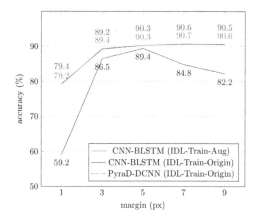

**Fig. 7.** Comparison of the mean accuracy with respect to the margin size of the different systems: CNN-BLSTM trained on *IDL-Train-Origin*, CNN-BLSTM trained on *IDL-Train-Aug* and PyraD-DCNN trained on *IDL-Train-Origin*.

The Fig. 7 compares the mean accuracy of each system with respect to the margin size (mean accuracy over the fields). It confirms that the use of data augmentation techniques has an important impact on the LSTM-based system invariance. Indeed, training the system with the augmented data-set improves accuracy for every margin size compared to the system trained with the Original data-set.

**Real-World Test Set.** In this experiment, the impact of a real-world like situation is studied. To that end, the two CNN-BLSTM systems (trained as explained in the previous section) are evaluated on the *IDL-Test-Var* set.

One can observe from Table 1 that in the simulated real-world situation, the augmentation allows a significant improvement of the system's stability. If we also compare the mean loss of accuracy between the ideal case (5 px margin for every field) and the real-world case (Variable test set), the system trained on the Original set looses 5.4 points while the system trained on the Augmented set looses 1.45 points. These results demonstrate the sensitivity of CNN-BLSTM based systems to variability in the position of the text in the image as well as the size of the margins, making them highly dependent to segmentation. They also show that the use of a suitable augmentation policy can correct such sensitivity.

## 4.4   The Use of PyraD-DCNN for Text Recognition

The proposed FCNN is a new type of architecture that holds the same long/short-term properties than LSTM systems. However, as it gets rid of the slow recurrent part of the network, it is both faster and lighter. In addition, as it uses dilated convolutional layers in its design and aggregate global and local context, it is expected to be less sensitive to position shift and size.

**Table 1.** Accuracy of the CNN-BLSTM vs. PyraD-DCNN when tested on the *IDL-Test-Var*

| Text recognition system | CNN-BLSTM | | PyraD-DCNN |
|---|---|---|---|
| Fields | Training set | | |
| | DL-Train-Origin | IDL-Train-Aug | DL-Train-Origin |
| Names | 86.5 | 89.7 | **90.1** |
| Surnames | 89.7 | **92.5** | 92.1 |
| Birth date | 87.3 | **90.1** | 89.7 |
| Birth place | 79.4 | 87.7 | **89.7** |
| Document number | 82.1 | **84.5** | 82.5 |
| Emit date | 88.5 | **89.3** | 87.7 |
| Expiration date | 88.9 | 88.1 | **89.7** |
| Issued by | 69.4 | 88.5 | **88.5** |

**Performance Comparison with CNN-BLSTM Systems.** We can observe from Table 2 that the training time is greatly reduced (on GPU) when using PyradD-DCNN. In addition the proposed structure achieves a very competitive scores in terms of accuracy and CER while reducing the inference time on CPU (up to three times faster).

**Table 2.** Proposed FCNN structure vs. LSTM based system evaluated on EU-ID. *The LSTM network training has been stopped after maximum 50 h of training

| Model | Accuracy (%) | CER | Inference time (in ms on CPU) | Number of weights (× 1000) | Training time on GPU |
|---|---|---|---|---|---|
| *FCNN* | **94.65** | **0.81** | *13* | *454* | *13 h 20* |
| *LSTM* | *94.20* | *0.84* | *44* | *360* | *50 h** |

**Shift Invariance of PyraD-DCNN System.** In order to evaluate the sensitivity of the PyraD-DCNN system, we compare it to the CNN-BLSTM system trained on the Augmented database on the different setups presented previously.

*Fixed Margin.* From Fig. 6(c), it can be seen that the PyraD-DCNN has a similar behaviour to the CNN-BLSTM trained on *IDL-Train-Aug* with a very stable accuracy across the various margin sizes and some loss on the smaller margins. Moreover, from Fig. 7(c), the curves for the CNN-BLSTM system trained on *IDL-Train-Aug* and the PyraD-DCNN system are overlapping, showing that they bear the same invariance to position shifts and margin sizes.

*Real-World Test Set.* From the previous experiment, it has been shown that the PyraD-DCNN system is resilient to margin size modification and naturally has a behavior similar to the one obtained through augmentation. This experiment aims to verify the generalization of this property to a real case scenario. To that end, the accuracy of the PyraD-DCNN is evaluated on the *IDL-Test-Var* set.

From Table 1, it can be seen that both systems are equivalent. Moreover, they both are stable in the scenario of shift position as their performances in this setting only drop by a little when compared to the ideal case of 5 px margin (1.45 points for the CNN-BLSTM system trained on *IDL-Train-Aug* and 1.5 points for the PyraD-DCNN on average).

Then, from the previous experiments, it can be observed that the PyraD-DCNN possesses a degree of shift invariance equal to a CNN-BLSTM system trained using data augmentation for that purpose.

## Conclusion and Perspectives

The high sensitivity of CNN-BSLTM OCR systems to position shift has been shown in this paper. Such systems are highly dependent on the training dataset. The use of the proposed data augmentation policy allows to overcome this drawback. Furthermore, a new FCNN architecture has been proposed as a more generic solution by tackling the problem itself and avoid duplicating training data. Using dilated convolutions can replace traditional CNN-BLSTM while increasing the system's performance in terms of time and storage. In addition, the multi-resolution build-up allows to naturally increase the OCR invariance to position shifts and size variability. The aforementioned observations have been confirmed via experiments held on an industrial real data-set. More generally, the proposed architecture can be applied to a wide range of fields using 2D or even 3D kernels such as for end-to-end scene text detection and recognition, document localization, sound processing, or video processing. Future work will be focused on testing the data augmentation policy in the context of the FCNN architecture. In addition, more experiments are to be held on public data-sets to confirm the generic aspects of the proposed approaches.

## References

1. Benton, G., Finzi, M., Izmailov, P., Wilson, A.G.: Learning invariances in neural networks. arXiv preprint arXiv:2010.11882 (2020)
2. Chang, S.-Y., et al.: Temporal modeling using dilated convolution and gating for voice-activity-detection. In: IEEE International Conference on Acoustics, Speech and Signal Processing, pp. 5549–5553. IEEE (2018)
3. Chen, C., Liu, X., Ding, M., Zheng, J., Li, J.: 3d dilated multi-fiber network for real-time brain tumor segmentation in MRI
4. Deb, D., Ventura, J.: An aggregated multicolumn dilated convolution network for perspective-free counting. CoRR
5. Devillard, F., Heit, B.: Multi-scale filters implemented by cellular automaton for retinal layers modelling. Int. J. Parallel Emergent Distrib. Syst. **35**(6), 1–24 (2018)

6. Devries, T., Taylor, G.W.: Improved regularization of convolutional neural networks with cutout. CoRR, abs/1708.04552 (2017)
7. Elagouni, K., Garcia, C., Mamalet, F., Sébillot, P.: Combining multi-scale character recognition and linguistic knowledge for natural scene text OCR
8. Gong, C., Wang, D., Li, M., Chandra, V., Liu, Q.: KeepAugment: a simple information-preserving data augmentation approach. arXiv preprint arXiv:2011.11778 (2020)
9. Goodfellow, I.J., Bulatov, Y., Ibarz, J., Arnoud, S., Shet, V.: Multi-digit number recognition from street view imagery using deep convolutional neural networks (2013)
10. Graves, A., Fernández, S., Gomez, F.J., Schmidhuber, J.: Connectionist temporal classification: labelling unsegmented sequence data with recurrent neural networks. In: International Conference on Machine Learning (2006)
11. Gupta, A., Rush, A.M.: Dilated convolutions for modeling long-distance genomic dependencies (2017)
12. Hataya, R., Zdenek, J., Yoshizoe, K., Nakayama, H.: Faster autoaugment: learning augmentation strategies using backpropagation (2019)
13. He, P., Huang, W., Qiao, Y., Loy, C., Tang, X.: Reading scene text in deep convolutional sequences. In: Proceedings of the AAAI Conference on Artificial Intelligence, vol. 30 (2016)
14. Jouanne, J., Dauchy, Q., Awal, A.M.: PyraD-DCNN: a fully convolutional neural network to replace BLSTM in offline text recognition systems. In: International Workshop on Computational Aspects of Deep Learning (2021)
15. Lin, J., Su, Q., Yang, P., Ma, S., Sun, X.: Semantic-unit-based dilated convolution for multi-label text classification (2018)
16. Mori, S., Suen, C.Y., Yamamoto, K.: Historical review of OCR research and development. Proc. IEEE **80**(7), 1029–1058 (1992)
17. Ptucha, R., Such, F.P., Pillai, S., Brockler, F., Singh, V., Hutkowski, P.: Intelligent character recognition using FCNN. Pattern Recogn. **88**, 604–613 (2019)
18. Renton, G., Soullard, Y., Chatelain, C., Adam, S., Kermorvant, C., Paquet, T.: Fully convolutional network with dilated convolutions for handwritten text line segmentation (2018)
19. Such, F.P., Peri, D., Brockler, F., Paul, H., Ptucha, R.: Fully convolutional networks for handwriting recognition (2018)
20. Sypetkowski, M., Jasiulewicz, J., Wojna, Z.: Augmentation inside the network (2020)
21. Xu, Y., Noy, A., Lin, M., Qian, Q., Li, H., Jin, R.: WEMIX: how to better utilize data augmentation (2020)
22. Yu, F., Koltun, V.: Multi-scale context aggregation by dilated convolutions (2015)

# A Handwritten Text Detection Model Based on Cascade Feature Fusion Network Improved by FCOS

Ruiqi Feng[1], Fujia Zhao[1], Shanxiong Chen[1(✉)], Shixue Zhang[2], and Dingwang Wang[1]

[1] Southwest University, Chongqing, China
csxpml@163.com
[2] Guizhou University of Engineering Science, Bijie, Guizhou, China

**Abstract.** In this paper, we propose a method for detecting handwritten ancient texts. The challenges in detecting this type of data are: the complexity of the layout of handwritten ancient texts, the varying text sizes, mixed arrangement of pictures and texts, the high number of hand-drawn patterns and the high background noise. Unlike general scene text detection tasks (ICDAR, TotalText, etc.), the texts in the images of ancient books are more densely distributed. For the features of the dataset, we propose a detection model based on cascade feature fusion called DFCOS, which aims to improve the fusion of localization information in lower layers. Specifically, bottom-up paths are created to use more localization signals from low-levels, and we incorporate skip connections to better extract information in the backbone, and then improve our model by parallel cascading. We verified the effectiveness of our DFCOS on HWAD (Handwritten Ancient-Books Dataset), a dataset containing four languages - Yi, Chinese, Tibetan and Tangut - provided by the Institute of Yi of Guizhou University of Engineering Science and National Digital Library of China, and its precision, recall and F-measure outperformed most of the existing text detection models.

**Keywords:** Scene text detection · Handwritten text detection · FCOS · Ancient books

## 1 Introduction

Scene text detection, which refers to locate the position of text regions in an image, has attracted a lot of attention in the field of computer vision in recent years as the first step in scene text reading. With the rapid development of deep learning and convolutional neural networks, researchers have proposed an increasing number of excellent frameworks for scene text detection. Most of these CNN-based methods are built on top of successful generic object detection frameworks or semantic segmentation technologies. However, at this stage, scene text detection remains challenging even with the support of deep neural

© Springer Nature Switzerland AG 2021
E. H. Barney Smith and U. Pal (Eds.): ICDAR 2021 Workshops, LNCS 12917, pp. 51–65, 2021.
https://doi.org/10.1007/978-3-030-86159-9_4

networks due to various factors such as diverse text fonts and styles, variable text sizes and complex image backgrounds. However, at this stage, scene text detection remains challenging even with the support of deep neural networks, due to various distracting factors such as diverse text fonts and styles, variable text sizes and complex image backgrounds.

As an important carrier of Chinese culture, handwritten ancient books are of great historical value. The digitization of handwritten ancient text images is vital for the preservation of cultural heritage. However, the existing text detection methods do not perform well on handwritten ancient text documents for the following reasons: 1) The layout structure of handwritten ancient texts is complex, which means it is common to find mixed texts and illustrations, arbitrary text sizes, various text arrangements, a large number of hand-painted patterns and noisy background; 2) The text is densely distributed and the overall text target is smaller than the scene text; 3) The backgrounds of different kinds of handwritten ancient texts and writing styles differ. Therefore, we consider proposing a new text detection method for the characteristics of ancient texts.

Anchor-free object detection methods have emerged in recent years. Compared with traditional anchor-based methods, this detection method is more suitable for our research, because the ancient text objects have greater variability in aspect ratio and orientation compared to the objects in object detection tasks, which can be very demanding for the design of anchors and can increase a huge amount of workload, causing the decrease in the overall efficiency of the detection task. Another defect of anchor-based methods is that it does not handle multi-directional and curved texts well. The paper Fully Convolutional One-Stage Object Detection (FCOS) [1] proposed a one-stage object detection algorithm based on FCN [2] and FPN [3], which is an excellent representative of the anchor-free method. Its biggest advantage is that the accuracy of detection is maintained while eliminating anchors. However, the FPN structure used in the FCOS model has an obvious drawback, i.e. it has only top-down paths with horizontal information fusion (add operation), which makes the network obtain mainly information from the high-levels and under-utilize the information from lower layers. This defect makes it less effective on ancient books with small and dense text objects. To overcome these problems, we propose a novel feature fusion network, which we design to integrate into FCOS's framework to improve the detection of textual objects in our ancient books.

The main contributions of this paper are as follows: 1) We constructed a Handwritten Ancient-Books Dataset (HWAD) containing 8k images in four languages [4], which laid the data foundation for the subsequent research on digitization of handwritten ancient books; 2) For the case of smaller and denser objects of ancient books, we are inspired by FPN and propose a bottom-up information fusion path for its tendency to miss low-level localization information, and at the same time, we add an additional skip connection path between the backbone and the bottom-up outputs; 3) In order to more fully reuse the obtained feature information, we form a feature cascade fusion network by overlaying multiple bottom-up and skip-connected structures in a parallel cascade manner; 4) Since the fixed sampling points in standard convolutional layers will restrict the receptive field to a fixed position, which is not conducive to the detection of ancient texts, we introduce a deformable convolution [5] for adaptive sampling; 5) After an in-depth study of the advantages and disadvantages of the FCOS, we introduce Gaussian weighted Soft-NMS [6] instead of NMS in the post-processing part, and replace the loss function in regression branch with a more suitable CIoU [7]. Experiments have proved that our DFCOS performs well on HWAD, surpassing most existing text detection methods.

## 2    Related Work

### 2.1    Scene Text Detection

In recent years, scene text detection methods can be roughly divided into two categories: regression-based methods and segmentation-based methods.

Regression-based methods are often improved from commonly used object detection frameworks, such as Faster R-CNN [8] and SSD [9]. TextBoxes [10] modified the scale of the anchors and the convolution kernels for text detection on the basis of SSD. EAST [11] used FCN to directly predict the score map, rotation angle and b-box for each pixel. LOMO [12] proposed an iterative optimization module and introduced an instance-level shape expression module to solve the problem of detecting scene texts of arbitrary shapes.

Regression-based methods often achieve certain results, but are highly dependent on anchor sizes which are manually set, and many regression-based text detectors are designed for specific text detection scenarios, resulting in low robustness.

According to segmentation-based methods, each pixel in the original image is segmented into text or non-text to determine the approximate text region. It has become the mainstream method for detecting multi-directional and arbitrary-shaped texts. PixelLink [13] predicted the connections between pixels and localized text regions by separating links belonging to different text instances. PSENet [14] adopted progressive scale expansion network using ground-truths to generate a series of masks of different sizes, ultimately improving the detection of irregular text.

Although PSENet works well, the post-processing process is quite time-consuming, leading to slow inference, which is also a common problem of segmentation-based detectors. To address this phenomenon, Liao et al. [15] proposed DBNet, which incorporates the binarization process into training and removes it during inference, resulting in a speedup in model inference.

### 2.2    Text Detection of Ancient Books

Over the past few years, a number of studies have been conducted on the detection of texts in ancient Chinese and minority languages. Su et al. [16] first binarized the Mongolian ancient texts by OTSU, then used the vertical projection information to locate the text columns, and finally got the individual Mongolian characters by analyzing connected components. However, the datasets involved in the study are neatly arranged, with the images are less polluted and noisy. Yang et al. [17] proposed a single-character detection framework for Chinese Buddhist scriptures using recognition results to guide detection. Shi et al. [18] detected and segmented oracle bones written on bone fragments by a connected component-based approach. Han et al. [19] presented a binarization algorithm based on the combination of Lab color space channels as well as local processing, starting from the different colors of Tibetan ancient documents.

The detection of ancient texts in China started relatively late, and most of the current studies are completed under the condition of standardized printed characters or well-arranged layout with less noise, which does not fit the characteristics of most extant ancient texts. As a result, the models obtained do not have good generalization. Our images are of different styles of handwritten texts with noise due to age and poor preservation, etc. A study on this dataset would be more meaningful.

## 3   Methodology

In Fig. 1 shows the architecture of our proposed network. It is based on FCOS, but to better avoid the problem of gradient disappearance and gradient explosion caused by too many layers, Dense Network (DesNet) [20] was chosen to build our backbone, i.e. each layer is connected to all previous layers by concatenate. This structure of DesNet is also capable of multiplexing the low-level information while ensuring the complete transmission of them. We propose a new feature fusion network that is independent of the feature extraction process, at the same time, we improve the feature extraction method, the loss function of the regression part and the post-processing algorithm, while keeping the other parts of the FCOS framework. Below, we will separately introduce the specific improvement methods of each part.

### 3.1   Improved Multi-scale Feature Fusion Network

FPN, compared with using a single feature map for prediction, makes use of the features of neural networks at each stage and can handle the multi-scale variation in object detection (i.e. the fusion of low-resolution feature maps with strong semantic information and high-resolution feature maps with weak semantic information but rich spatial information) with a small increase in computational effort. This is why FCOS chose FPN as their feature fusion network. However, as the targets in our research are smaller and more densely arranged than those in object detection and scene text detection, we need to focus more on the low-level spatial localization signals, while the top-down fusion of FPN is rich in information but many detailed features will be lost after layer-by-layer pooling, which makes the fused feature maps focus more on the abstract information from higher layers and less on low-level spatial information, which is not exactly suitable for our images.

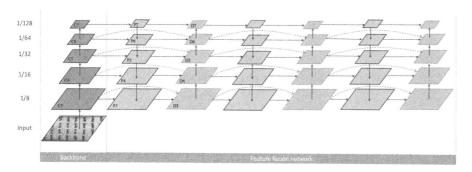

**Fig. 1.** Architecture of our model, we omit the structure of the subsequent forecast part because we adopted the same structure as FCOS. {C3, C4, C5, C6, C7} represent the features extracted using CNN, {P3, P4, P5, P6, P7} represent all levels of Feature maps generated by FPN, among which P6 and P7 are directly upsampled by P5; {D3, D4, D5, D6, D7} represent the feature maps generated by the bottom-up and skip connection structure.

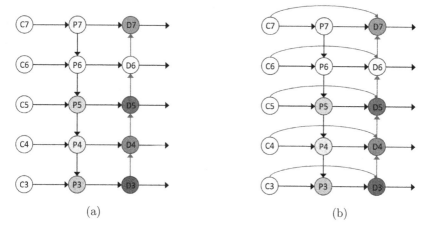

**Fig. 2.** (a) Bottom-up feature fusion method; (b) Feature fusion structure with skip connection.

**Bottom-Up Paths.** The top-down structure pays more attention to the abstract information of higher layers, while the bottom-up network contains more low-level localization information. Thus, we consider combining the two to improve the performance of the model. As shown in Fig. 2a, bottom-up feature fusion paths are created to incorporate more localization signals from lower layers. Similar to FPN, the method of feature combination still uses add operation with less computation, that is, to keeping the image dimension unchanged and adding elements correspondingly. Similar to FPN, the size of feature map of each layer among the bottom-up paths corresponds to the same layer of the previous level. The following shows how to calculate the bottom-up fusion:

$$\begin{cases} D_3 = Conv\,(P_3) \\ D_4 = Conv\,(P_4 + Resize\,(D_3)) \\ \cdots \\ D_7 = Conv\,(P_7 + Resize\,(D_6)) \end{cases} \tag{1}$$

where *Resize* is an upsampling operation, which is used to make the size of Feature maps consistent; *Conv* represents a convolution operation for feature processing.

**Skip Connection Paths.** We connect the inputs and outputs in a skip connection manner in both the top-down and bottom-up paths, as shown in Fig. 2b, which has the advantage of fusing more features without introducing additional parameters. Just as the original intention of ResNet [21], skip connection can solve the problem of gradient disappearance and gradient explosion, while helping with backpropagation and speeding up the training process. By transferring the feature maps of convolution layer to the bottom-up stage, we can get more details from images and improve the final detection accuracy. Since the information in the backbone is the most important during the entire feature fusion process, we hope to combine the information in the backbone with the top-down features. The concrete calculation process of combining bottom-up and skip connection structure is shown in (2):

$$\begin{cases} D_3 = Conv\,(P_3 + C_3) \\ D_4 = Conv\,(P_4 + Resize\,(D_3) + C_4) \\ \cdots \\ D_7 = Conv\,(P_7 + Resize\,(D_6) + C_7) \end{cases} \tag{2}$$

**Cascade Feature Fusion Structure.** With the addition of the bottom-up and skip connection structures, there is a visible improvement in the performance of the model, but to more fully reuse the feature information, we introduce the idea of cascade (i.e. multiple identical structures are connected in series, and the output of the previous part is used as the input of the next stage). Compared with the single-layer feature fusion network, it can make full use of the information in backbone. However, it is worth mentioning that with the deepening of cascade layers, it will inevitably lead to more computation and a higher model complexity. How to balance the model performance and complexity is also an issue of great concern (see the experiments in the following parts of this paper).

## 3.2 Deformable Convolution

There are many types of handwritten ancient books with complex and variable text sizes, resulting in different aspect ratios of the texts. Since the geometrical structure of standard convolution (CNNs) is fixed (that is, the convolutional unit samples a fixed position of the input feature map), considering that different kinds and sizes of texts require different sizes of perceptual fields, modulated deformable convolutions are applied in all the convolutional layers to extract features, which enables the convolutional layers to adapt to the input images, thus enhancing the overall network's ability to model geometric transformations.

### 3.3   Other Improvements

GIoU [22] was used as the loss function for the regression part in FCOS. Through our research, we found that GIoU tends to make a larger intersection area between the bounding boxes and ground-truth boxes by increasing the size of the anchor boxes during the regression process, which leads to a slow decrease in the average loss and more iterations for training. In this paper, we introduce CIoU as an alternative, which not only converges faster, but also takes into account the aspect ratio of the texts. We retain the loss functions for other parts of FCOS, so that the total loss function can be expressed as (3):

$$L\left(\{P_{x,y}\}, \{t_{x,y}\}\right) = \frac{1}{N_{pos}} \sum_{x,y} L_{cls}\left(P_{x,y}, c^*_{x,y}\right)$$
$$+ \frac{\lambda}{N_{pos}} \sum_{x,y} 1\left\{c_{x,y} > 0\right\} L_{re.g.}\left(t_{x,y}, t^*_{x,y}\right) \qquad (3)$$
$$+ BCE(Centerness)$$

$$L_{reg} = Centerness * L_{CIoU} \qquad (4)$$

Equation (3) comes from FCOS, and the part we change (i.e. $L_{cls}$, represents the regression loss) is shown in (4).

Due to the "one-size-fits-all approach" of NMS algorithm according to scores, the texts tend to be overcut (a single character is divided into multiple characters) and undercut (the bounding box fails to cover the entire character) during prediction process, thus Soft-NMS with Gaussian weighting is chosen to optimize the post-processing module.

## 4   Dataset

The dataset contains the scanned pictures of ancient books provided by Research Institute of Yi Nationality Studies of Guizhou University of Engineering Science and National Digital Library of China, including the four languages of Yi, Chinese, Tibetan and Tangut. After manual organization and screening, 600 Yi scriptures, 500 Chinese scriptures, 300 Tangut scriptures and 200 Tibetan scriptures were obtained. These images were divided into simple layout (as shown in Fig. 3a) and complex layout (as shown in Fig. 3b) according to whether the layout structure was neat. Among them, 1,250 (78%) of the antiquities were in simple layout and 350 (22%) were in complex layout. Referring to ICDAR2015 [23], we annotated them in the following way: starting from the top left corner, annotate four points clockwise to obtain the result.

(a)                                    (b)

**Fig. 3.** (a) Images in simple layout; (b) Images in complex layout.

As our images are not quantitatively sufficient, data augmentation is introduced in this paper. In addition to regular methods of panning, rotation, flipping, scaling, colour dithering and adding noise, we propose a stitching-based data augmentation referring to CutMix [24], which is explained in detail below:

For samples with simple layout, we directly divided them into four equal parts by centre point through a script; for those with complex layout, we manually cropped them, keeping only the geometrically shaped text layouts, hand-drawn graphics and irregular parts of them. The cut images were brought together and four of them were taken at a time, each with conventional data enhancement before being stitched together (the gaps in the background are filled with grey), as shown in Fig. 4.

The 8k images obtained after data augmentation are our entire dataset. There are 4000 simple pages and 4000 complex pages, which are named HWAD-s and HWAD-c respectively for convenience of writing. We randomly selected 70% of these images as the training set and the rest as the test set.

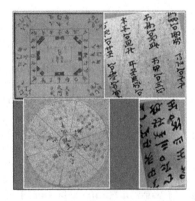

**Fig. 4.** Image stitching method.

# 5 Experiments

## 5.1 Implementation Details

The proposed model is implemented in PyTorch [25] framework using the open source toolkit for object detection, Mmdetection [26]. We conduct experiments on a workstation with 3.6 GHz CPU, RTX 2080 Super 8G GPU and Ubuntu 64-bit OS. SGD is chosen as the optimizer with momentum set to 0.9 and weight decay of 0.0001. We also use a warm up strategy, making the learning rate increase from 0 to 0.0025, so that the model can be stabilized quickly, and to approach this point by reducing the learning rate through cosine annealing as the Loss approaches a global minimum.

## 5.2 Ablation Study

**Bottom-Up and Skip Connection Fusion.** In this part, we applied the bottom-up and skip connection paths to the feature fusion module of FCOS. Considering that the effect will not be obvious if separate experiments are conducted for these two structures, we combined them for testing. In Table 1, we can see that our proposed structure significantly improves the performance for FCOS with ResNet-50, ResNet-101 and DesNet. For the ResNet-50 backbone, performance gain in terms of F-measure of 2.9% and 11.3% is achieved on HWAD-s and HWAD-c respectively. For the ResNet-101 backbone, it brings 1.9% (on HWAD-s) and 3.4% (on HWAD-c) improvements. For the DesNet backbone, the module results in 3.3% (on HWAD-s) and 6.9% (on HWAD-c) improvements.

**Table 1.** Effect of bottom-up and skip connection fusion structure on HWAD-s and HWAD-c

| Method | HWAD-s | | | HWAD-c | | |
|---|---|---|---|---|---|---|
| | P | R | F1 | P | R | F1 |
| DesNet | 0.801 | 0.785 | 0.793 | 0.640 | 0.595 | 0.617 |
| DesNet+skip | 0.824 | **0.828** | **0.826** | **0.679** | **0.694** | **0.686** |
| ResNet50 | 0.698 | 0.675 | 0.687 | 0.432 | 0.365 | 0.395 |
| ResNet50+skip | 0.716 | 0.716 | 0.716 | 0.498 | 0.519 | 0.508 |
| ResNet101 | 0.814 | 0.773 | 0.793 | 0.513 | 0.432 | 0.469 |
| ResNet101+skip | **0.824** | 0.800 | 0.812 | 0.547 | 0.465 | 0.503 |

**Cascade Feature Fusion Network.** Based on the experiment above, we further verify the effectiveness of cascade feature fusion network, and explore whether there is saturation in cascade operations. Because the experiment will become very time-consuming as the cascade network deepens, we only conducted the experiment on HWAD-s. As shown in Fig. 5, with the increase of cascade layers, the values of the three evaluation metrics P, R and F1 rise first, reach a peak, and then start to decline instead. Take the DesNet for example, when the number of cascade layers reaches 3, the value of P reaches its maximum. Even though the recall increases when the cascade layers are 4, the F-measure is almost unchanged. The improvement in recall is important, but every additional layer of cascade increases the complexity of the model, and the weak increase in one metric alone is of little significance. When the cascade layers reach 5, the performance even starts to degrade. According to the experimental results, we decided to set the cascade layers as 3 to balance the performance and complexity of the model.

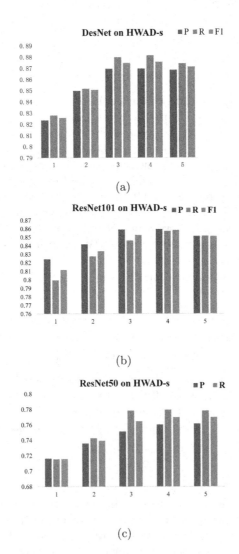

**Fig. 5.** Experimental results of cascade feature fusion on different backbones: (a) DesNet; (b) ResNet-101; (c) ResNet-50.

**Deformable Convolution.** We tested the effects of deformable convolution network (DCN) on the DesNet and ResNet. As shown in Table 2, DCN brings a relatively limited improvement on HWAD-s, but on HWAD-c, it generates significant improvements of 9.9% (ResNet-50), 3.1% (ResNet-101) and 6.6% (DesNet) respectively.

**Table 2.** Performance grows with DCN

| Backbone | DCN | HWAD-s | | | HWAD-c | | |
|---|---|---|---|---|---|---|---|
| | | P | R | F1 | P | R | F1 |
| DesNet | × | 0.801 | 0.785 | 0.793 | 0.640 | 0.595 | 0.617 |
| DesNet | ✓ | 0.818 | **0.802** | 0.810 | **0.679** | **0.687** | **0.683** |
| ResNet50 | × | 0.698 | 0.675 | 0.687 | 0.432 | 0.365 | 0.395 |
| ResNet50 | ✓ | 0.723 | 0.743 | 0.733 | 0.512 | 0.477 | 0.494 |
| ResNet101 | × | 0.814 | 0.773 | 0.793 | 0.513 | 0.432 | 0.469 |
| ResNet101 | ✓ | **0.828** | 0.801 | **0.814** | 0.525 | 0.459 | 0.490 |

**Comprehensive Experiment.** Combining the modules from the previous three experiments makes up our final feature fusion network. We verified the performance of DFCOS and FCOS incorporating this feature fusion network on HWAD-s and HWAD-c. As we can see in Table 3, DFCOS outperforms FCOS on both datasets, and has a substantial lead on the HWAD-c dataset (ResNet-50 18.9%, ResNet-101 18.8%).

**Table 3.** Comparison of DFCOS and FCOS on HWAD-s and HWAD-c

| Method | HWAD-s | | | HWAD-c | | |
|---|---|---|---|---|---|---|
| | P | R | F1 | P | R | F1 |
| DFCOS (DesNet) | **0.870** | **0.880** | **0.875** | **0.736** | **0.762** | **0.749** |
| FCOS (ResNet50) | 0.751 | 0.778 | 0.765 | 0.553 | 0.568 | 0.560 |
| FCOS (ResNet101) | 0.859 | 0.846 | 0.853 | 0.593 | 0.532 | 0.561 |

**Loss Function.** As shown in Fig. 6, we compared the performance of GIoU, DIoU [7] and CIoU during training. The horizontal axis represents the number of iterations, 99,500 iterations were performed for the three experiments, and the vertical axis is the loss value. It can be seen that the initial value of CIoU is the lowest, the convergence speed is the fastest, and the convergence value is the lowest. At the same time, we respectively compared the total loss value of the model when using these three regression loss functions. As shown in the figure, it can be seen that CIoU is still the best in convergence speed and convergence value.

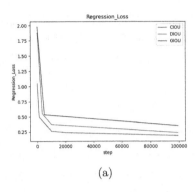 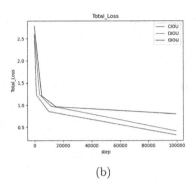

(a)                                              (b)

**Fig. 6.** (a) Comparison of three regression loss functions; (b) Comparison of the total loss with three loss functions respectively.

**Post-processing Module.** Under the same experimental environment, the Soft-NMS algorithm has a higher recall than the NMS algorithm, whether using linear or Gaussian weighting function, and the improvement is more obvious when using Gaussian weighting function: when the threshold is 0.5, the improvement is 0.4% and 1.1% on HWAD-e and HWAD-d, respectively.

### 5.3    Comparisons with Previous Methods

We compare the model we proposed with previous methods for scene text detection on the two datasets, and the experimental results are shown in Table 4 and Table 5. Specifically, our method achieves better performance than the other models on both HWAD-s and HWAD-c in terms of P, R and F1 (and leads more on HWAD-c). Among the methods for comparison, DBNet is the fastest to inference because the DB module is removed when predicting. Additionally, some test examples are visualized in Fig. 7.

**Table 4.** Comparison with prior arts on HWAD-s

| Method | P | R | F1 | FPS |
|---|---|---|---|---|
| CTPN [27] | 0.593 | 0.587 | 0.590 | 7.1 |
| SegLink [28] | 0.603 | 0.594 | 0.598 | 8.4 |
| RRD [29] | 0.633 | 0.658 | 0.645 | 4.5 |
| PixelLink [13] | 0.588 | 0.573 | 0.590 | 13.6 |
| EAST [11] | 0.625 | 0.675 | 0.649 | 8.7 |
| PSENet [14] | 0.744 | 0.724 | 0.734 | 3.7 |
| CRAFT [30] | 0.858 | 0.842 | 0.850 | 7.4 |
| DBNet [15] | 0.880 | 0.883 | 0.881 | **15.7** |
| DFCOS | **0.926** | **0.918** | **0.922** | 12.2 |

**Table 5.** Comparison with prior arts on HWAD-c

| Method | P | R | F1 | FPS |
|--------|-----|-----|-----|-----|
| CTPN | 0.403 | 0.365 | 0.383 | 6.5 |
| SegLink | 0.463 | 0.322 | 0.380 | 6.3 |
| RRD | 0.433 | 0.338 | 0.380 | 3.2 |
| PixelLink | 0.388 | 0.350 | 0.368 | 9.6 |
| EAST | 0.429 | 0.401 | 0.415 | 5.7 |
| PSENet | 0.524 | 0.492 | 0.507 | 1.8 |
| CRAFT | 0.645 | 0.602 | 0.623 | 6.3 |
| DBNet | 0.722 | 0.694 | 0.707 | **11.6** |
| DFCOS | **0.872** | **0.883** | **0.877** | 8.4 |

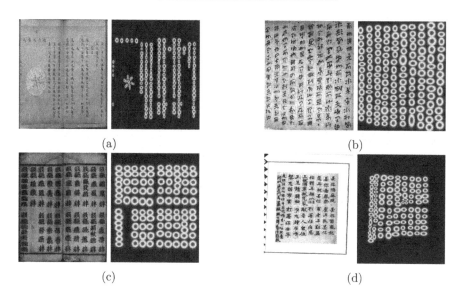

(a)  (b)

(c)  (d)

**Fig. 7.** Detecting results: (a) Chinese; (b) Yi; (c) Tangut; (d) Tibetan.

## 6    Conclusion and Future Work

In this paper, we propose an effective framework for detecting handwritten ancient texts called DFCOS. According to the characteristics of the text data of ancient books, we propose a new feature fusion network referring to FCOS. We have experimentally demonstrated that our method works well on ancient books. In the future, we hope to enlarge the dataset to improve the generalization performance and optimize the network structure to improve the inference and training speed.

# References

1. Tian, Z., Shen, C., Chen, H., He, T.: FCOS: fully convolutional one-stage object detection. In: ICCV (2019)
2. Long, J., Shelhamer, E., Darrell, T.: Fully convolutional networks for semantic segmentation. In: CVPR (2015)
3. Lin, T.-Y., Dollár, P., Girshick, R., He, K., Hariharan, B., Belongie, S.: Feature Pyramid Networks for Object Detection, arXiv preprint. arXiv: 1612.03144 (2017)
4. Handwritten Ancient-Books Dataset: HWAD. Unpublished Data
5. Dai, J., et al.: Deformable convolutional networks. In: ICCV (2017)
6. Bodla, N., Singh, B., Chellappa, R., Davis, L.: Improving object detection with one line of code. In: ICCV (2017)
7. Zheng, Z., Wang, P., Liu, W., Li, J., Ye, R., Ren, D.: Distance-IoU loss: faster and better learning for bounding box regression. In: AAAI (2020)
8. Ren, S., He, K., Girshick, R., Sun, J.: Faster R-CNN: towards real-time object detection with region proposal networks. In: NIPS (2015)
9. Liu, W., et al.: SSD: single shot multibox detector. In: Leibe, B., Matas, J., Sebe, N., Welling, M. (eds.) ECCV 2016. LNCS, vol. 9905, pp. 21–37. Springer, Cham (2016). https://doi.org/10.1007/978-3-319-46448-0_2
10. Liao, M., Shi, B., Bai, X., Wang, X., Liu, W.: Textboxes: a fast text detector with a single deep neural network. In: AAAI (2017)
11. Zhou, X., et al.: East: an efficient and accurate scene text detector. In: CVPR (2017)
12. Zhang, C., et al.: Look more than once: an accurate detector for text of arbitrary shapes. In: CVPR (2019)
13. Deng, D., Liu, H., Li, X., Cai, D.: PixelLink: detecting scene text via instance segmentation. In: AAAI, pp. 6773–6780 (2018)
14. Wang, W., et al.: Shape robust text detection with progressive scale expansion network. In: CVPR (2019)
15. Liao, M., Wan, Z., Yao, C., Chen, K., Bai, X.: Real-time scene text detection with differentiable binarization. In: AAAI (2020)
16. Su, X., Gao, G.: A knowledge-based recognition system for historical Mongolian documents. Int. J. Document Anal. Recogn. Neural Netw. **124**, 117–129 (2020)
17. Shi, X., Huang, Y., Liu, Y.: Text on oracle rubbing segmentation method based on connected domain. In: Proceedings of IEEE Advanced Information Management, Communicates Electronic and Automation Control Conference, pp. 414–418. IEEE Computer Society Press, Anyang (2016)
18. Hailin, Y., Lianwen, J., Weiguo, H., et al.: Dense and tight detection of Chinese characters in historical documents: datasets and a recognition guided detector. IEEE Access **6**, 30174–30183 (2018)
19. Han, Y.H., Wang, W.L., Wang, Y.Q.: Research on automatic block binarization method of stained Tibetan historical document image based on Lab color space. In: International Forum on Management, Education and Information Technology Application, pp. 327–338 (2018)
20. Huang, G., Liu, Z., Maaten, L., Weinberger, Q.: Densely connected convolutional networks. In: CVPR (2017)
21. He, K., Zhang, X., Ren, S., Sun, J.: Deep residual learning for image recognition. In: ICCV (2016)
22. Rezatofighi, H., Tsoi, N., Gwak, J.Y., Sadeghian, A., Reid, I., Savarese, S.: Generalized intersection over union: a metric and a loss for bounding box regression. In: CVPR (2019)

23. Karatzas, D., et al.: ICDAR 2015 competition on robust reading. In: ICDAR 2015 (2015)
24. Yun, S., Han, D., Oh, S.J., Chun, S., Choe, J., Yoo, Y.: CutMix: regularization strategy to train strong classifiers with localizable features. In: ICCV (2019)
25. Paszke, A., et al.: Automatic differentiation in PyTorch (2017)
26. Chen, K., et al.: MMDetection: Open MMLab Detection Toolbox and Benchmark, arXiv preprint. arXiv: 1906.07155 (2019)
27. Tian, Z., Huang, W., He, T., He, P., Qiao, Y.: Detecting text in natural image with connectionist text proposal network. In: Leibe, B., Matas, J., Sebe, N., Welling, M. (eds.) ECCV 2016. LNCS, vol. 9912, pp. 56–72. Springer, Cham (2016). https://doi.org/10.1007/978-3-319-46484-8_4
28. Shi, B., Bai, X., Belongie, S.: Detecting oriented text in natural images by linking segments. In: Proceedings of CVPR, pp. 3482–3490 (2017)
29. Liao, M., Zhu, Z., Shi, B., Xia, G.-S., Bai, X.: Rotation-sensitive regression for oriented scene text detection. In: CVPR (2018)
30. Baek, Y., Lee, B., Han, D., Yun, S., Lee, H.: Character region awareness for text detection. In: CVPR (2019)

# Table Structure Recognition Using CoDec Encoder-Decoder

Bhanupriya Pegu, Maneet Singh$^{(\boxtimes)}$, Aakash Agarwal, Aniruddha Mitra, and Karamjit Singh

AI Garage, Mastercard, Gurgaon, India
{bhanupriya.pegu,maneet.singh,aakash.agarwal,aniruddha.mitra,
karamjit.singh}@mastercard.com

**Abstract.** Automated document analysis and parsing has been the focus of research since a long time. An important component of document parsing revolves around understanding tabular regions with respect to their structure identification, followed by precise information extraction. While substantial effort has gone into table detection and information extraction from documents, table structure recognition remains to be a long-standing task demanding dedicated attention. The identification of the table structure enables extraction of structured information from tabular regions which can then be utilized for further applications. To this effect, this research proposes a novel table structure recognition pipeline consisting of row identification and column identification modules. The column identification module utilizes a novel Column Detector Encoder-Decoder model (termed as *CoDec* Encoder Decoder) which is trained via a novel loss function for predicting the column mask for a given input image. Experiments have been performed to analyze the different components of the proposed pipeline, thus supporting their inclusion for enhanced performance. The proposed pipeline has been evaluated on the challenging ICDAR 2013 table structure recognition dataset, where it demonstrates state-of-the-art performance.

**Keywords:** Table structure recognition · Encoder-Decoder · Document analysis

## 1 Introduction

The volume of digital information getting generated is growing at an astonishing rate, where text documents correspond to a major portion of it. Parsing such documents and extracting the required information is a challenging task since many such documents contain tables with varying layouts and colour schemes. For example, Fig. 1 show sample tabular regions of various layouts in different document types, such as invoices, research papers, and reports. To enable automated processing of these documents, accurate tabular parsing methodology is required. Significant efforts have been made in the past to extract this tabular information from documents using automated processes [6,7,12,14,20,23].

© Springer Nature Switzerland AG 2021
E. H. Barney Smith and U. Pal (Eds.): ICDAR 2021 Workshops, LNCS 12917, pp. 66–80, 2021.
https://doi.org/10.1007/978-3-030-86159-9_5

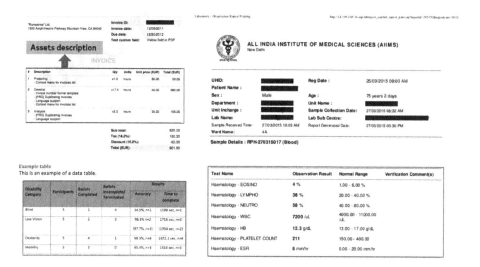

**Fig. 1.** Table structure recognition has applications involving automated extraction of tabular content for further analysis. For example, extraction of related fields from invoices, research publications, or reports.

The problem of successful table parsing can be decomposed into two sub-problems [22]: (i) table detection and (ii) structure recognition. The first sub-problem of table detection can be solved by detecting the pixels representing the tabular region in a document. Several methods have been proposed in the past to solve this problem [6,20,23] which have shown high detection results on publicly available datasets. Once a tabular region is successfully detected, the next sub-problem is to identify the structure of a table by understanding its layout and detecting the cell region in it [7]. Detection of cell regions can further be broken down into row and column identification which can ultimately be combined to discover the corresponding cells in a table [20]. The problem of structure recognition is extremely challenging due to significant intra-class variability, e.g., tables can have different layouts, several colour schemes, the erratic use of ruling lines for tables, structure delineation, or simply due of diverse table contents [2]. While recent techniques such as the CascadeTabNet [18] have shown almost near perfect results for table detection, the task of table structure identification still requires dedicated attention. To this end, this paper focuses on the table structure recognition sub-problem.

In this paper, an end-to-end pipeline is proposed for table structure recognition containing two components: (i) column identification module, and (ii) row identification module. The column identification module utilizes a novel Column Detector Encoder-Decoder model (termed as *CoDec* Encoder-Decoder) which is trained via a novel loss function containing *Structure loss* and *Symmetric loss*. The intuition of the proposed method is to develop a small and compact deep learning architecture which can be used to train models with limited training data and have split second inference time to enable real-time applications. In particular, the contributions of this research are as follows:

- This research proposes an end-to-end pipeline for table structure recognition using a small and compact deep learning architecture. The relatively lower trainable parameters enables model training with limited data and split second inference time for applicability in real-world scenarios.
- The proposed pipeline utilizes a novel column identification module, termed as the *CoDec Encoder-Decoder* model. The CoDec model is trained with a novel loss function consisting of a Structure and Symmetric loss for faster and accurate learning.
- The performance of the proposed pipeline has been evaluated on the challenging ICDAR 2013 dataset [7]. The proposed pipeline demonstrates improvement from the state-of-the-art networks even without explicitly training or fine-tuning on the ICDAR 2013 dataset, thus supporting the generalizability of the proposed technique. Further, analysis has been performed on the proposed pipeline via an ablation study which supports the inclusion of different components.

The rest of the paper is organized as follows: Sect. 2 outlines an overview of the related work on tabular structure recognition. Section 3 presents the detailed description of the proposed pipeline. Section 4 elaborates upon the details of the experiments and datasets. Section 5 presents the results and analysis of the proposed pipeline, and Sect. 6 presents the concluding remarks.

## 2   Related Work

The concept of recognising table structure has evolved gradually from pre-Machine-Learning (ML) era, when it used to be completely heuristic based, to the recent age of deep learning. Comprehensive summary of the algorithmic evolution can be traced in the surveys available describing and summarizing the state-of-the-art in the field [2,3,11,24,29]. One of the earliest successful developments in table understanding could be found in T-RECS by Kieninger and Dengel [12], where they built a framework to group words into columns on basis of its horizontal overlaps, followed by dividing those word-groups into cells with respect to the column's margin structure. In the same period many handcrafted features based algorithms were introduced [5,9,28], which were task specific and demonstrated heavy utilization of the domain knowledge. Another early data driven approach by Wang et al. [27] proposes a seven step formulation based on probability optimization. Considering the high intra-class variability, Shigarov et al. [21] proposed a table decomposing algorithm with sets of domain-specific rules, where they also rely on PDF metadata like font and character bounding boxes as well as ad-hoc heuristics.

The proposed table structure recognition pipeline is conceptually connected with recent Deep Learning (DL) based developments on this subject. The remainder of this Section thus focuses on recent works which set benchmarks utilizing deep-learning techniques. Though research related to table detection in PDF documents can be traced back to the technique published by Hao et al. [8], research on table structure recognition still remains limited owing to the challenging and complex nature of the problem. Schreiber et al. [20] tackle the problem of scarce labelled data, which hinders high parameterized DL training, by leveraging Domain Adaptation and Transfer Learning. The authors used Fully-Convolution Networks (FCN) based general object detection models to adapt to the domain of documents using Transfer Learning. However, their performance metric restricts itself to identifying rows/columns instead of using the cell-level information. Siddiqui et al. [22] propose to constrain the problem space for obtaining improved performance. Qasim et al. [19] modelled the problem of table recognition with Graph Neural Networks where Convolution Neural Network (CNN) and Optical Character Recognition (OCR) engine are employed to extract the feature maps and word positions, respectively. The representation of the features are learned through an interaction network. The representations are then concatenated and fed into a dense neural network to classify the vertex pairs. Recently, TableNet [16] architecture was proposed which is a multi-task network built on a VGG based encoder followed by task specific decoders, to model the inter-dependency between the twin tasks of table detection and table structure identification. Further, recently, CascadeTabNet [18] was proposed, where the tasks of table detection and structure recognition are accomplished by a single CNN model utilizing cascade mask region-based CNN and a high-resolution network.

In literature, the closest technique to the current manuscript is the TableNet architecture [16]. The TableNet model utilizes a large-scale pre-trained VGG network as the back-bone architecture, and performs semantic segmentation on the given input image for generating a column mask, along with the utilization of domain knowledge for generating the row co-ordinates. In comparison, the proposed table structure recognition pipeline utilizes a novel light-weight CoDec Encoder-Decoder model which is trained using a novel loss function for column detection. Further, as opposed to semantic segmentation which involves generating a mask for each class, the proposed CoDec Encoder-Decoder constructs a single image containing the column mask, resulting in further reduction in the number of trainable parameters. Detailed description of the proposed pipeline is provided in the next Section.

## 3    Table Structure Recognition

In order to perform table structure recognition, this research follows a top-down approach. A two-step approach is followed: (i) identification of columns, followed by (ii) identification of rows. Figure 2 presents a broad over-view of the proposed table structure recognition pipeline. A given input table image is processed to identify the column details via the proposed Column Detector Encoder-Decoder model (termed as the *CoDec* Encoder-Decoder), followed by the identification of different rows in the table using the domain knowledge and different image

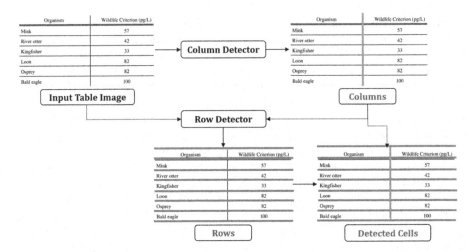

**Fig. 2.** Proposed pipeline for table structure recognition. The input tabular image is provided to the row detection and column detection modules, which return the row and column co-ordinates. The information from the two modules is then fused together to generate the cell co-ordinates.

processing rules. The row and column information is then combined to generate the cell co-ordinates. Detailed explanation of each component is provided in the following subsections.

### 3.1 Column Identification via Proposed CoDec Encoder-Decoder

Column identification involves identifying the columns in a given tabular image. As shown in Fig. 3, the task suffers from several challenges such as varying tabular formats, presence/absence of columns lines, differing space between different columns, etc. Existing techniques in literature have either utilized hand-crafted techniques or focused on specific table designs only. In order to develop a more generalized solution, this research proposes a novel deep learning based CoDec Encoder-Decoder formulation for identifying columns in the given input table.

Figure 4 presents a diagrammatic overview of the proposed CoDec Encoder-Decoder model. Given an input image, the model outputs a mask with the column identifiers. The loss function of the proposed CoDec Encoder-Decoder model utilizes a (i) Structure loss and a (ii) Symmetric loss for identifying the columns from the given tabular image. For $n$ training samples, the loss function of the proposed CoDec Encoder-Decoder model is given as follows:

$$\mathcal{L}_{CoDec} = \frac{1}{n} \sum_{i=1}^{n} \left( \underbrace{\|f(g(x^i)) - x^i_{mask}\|_2^2}_{Structure\ Loss} + \underbrace{\lambda \|f(g(x^i)) - \mathcal{P}(f(g(x^i)))\|_2^2}_{Symmetric\ Loss} \right) \quad (1)$$

where, $x^i$ and $x^i_{mask}$ refer to the $i^{th}$ training image and the corresponding column mask. $g(.)$ and $f(.)$ refer to the Encoder and Decoder modules, respectively, while

(a)

| Instance | Size | Algorithm | $r_{best}$ | $\bar{r}$ | $r_\sigma$ | $t$ (s) |
|---|---|---|---|---|---|---|
| 1 | 5 | SA | 12.776 | 13.693 | 0.62 | 0.82 |
| | | CA-PSLS | 10.942 | 11.704 | 0.49 | 4.31 |
| | | PSO | 11.046 | 11.716 | 0.49 | 4.89 |
| 2 | 6 | SA | 16.004 | 17.377 | 0.59 | 1.66 |
| | | CA-PSLS | 14.686 | 15.590 | 0.67 | 7.76 |
| | | PSO | 14.320 | 15.349 | 0.56 | 8.28 |
| 3 | 9 | SA | 20.849 | 22.328 | 0.87 | 6.84 |
| | | CA-PSLS | 18.157 | 19.797 | 1.03 | 21.07 |
| | | PSO | 18.579 | 19.205 | 0.49 | 22.84 |
| 4 | 20 | SA | 29.969 | 31.680 | 0.98 | 92.01 |
| | | CA-PSLS | 27.927 | 33.129 | 5.48 | 125.52 |
| | | PSO | 32.596 | 34.426 | 2.23 | 138.00 |

(b)

| Language | Names | Avg. Len. |
|---|---|---|
| German | 3153 | 15.1 |
| English | 1660 | 13.6 |
| Serbocroatian | 1474 | 14.3 |
| Italian | 1151 | 16.2 |
| French | 1141 | 15.8 |
| Polish | 1057 | 16.0 |
| Spanish | 1031 | 14.0 |
| Danish | 817 | 15.7 |
| Dutch | 809 | 15.1 |
| Swedish | 746 | 15.7 |
| Czechoslovak | 653 | 13.6 |
| Norwegian | 622 | 16.2 |
| Portuguese | 600 | 11.1 |
| Total | 14914 | 14.8 |

(c)

| File | Words | Dice | Jaccard | Overlap | WN1 |
|---|---|---|---|---|---|
| Bra14 | 931 | 0.699 | 0.701 | 0.711 | 0.742 |
| Bra02 | 959 | 0.637 | 0.685 | 0.697 | 0.753 |
| Brb20 | 930 | 0.672 | 0.674 | 0.693 | 0.731 |
| Bra15 | 1071 | 0.653 | 0.651 | 0.684 | 0.732 |
| Bra13 | 924 | 0.667 | 0.673 | 0.682 | 0.735 |
| Bra01 | 1033 | 0.650 | 0.648 | 0.674 | 0.714 |
| Brb13 | 947 | 0.649 | 0.650 | 0.674 | 0.722 |
| Bra12 | 1163 | 0.626 | 0.622 | 0.649 | 0.717 |
| Brn11 | 1043 | 0.634 | 0.639 | 0.648 | 0.708 |
| Brc01 | 1100 | 0.625 | 0.627 | 0.638 | 0.688 |

(d)

| | Census-Income | | | | DBGEN | | |
|---|---|---|---|---|---|---|---|
| | cardinalités | sans tri | d1 ... d10 | d10 ... d1 | cardinalités | sans tri | d1 ... d10 | d10 ... d1 |
| d1 | 7 | 42 427 | 32 | 42 309 | 2 | $0.75 \times 10^6$ | 24 | $0.75 \times 10^6$ |
| d2 | 8 | 36 980 | 200 | 36 521 | 3 | $1.11 \times 10^6$ | 38 | $1.11 \times 10^6$ |
| d3 | 10 | 34 257 | 1 215 | 28 975 | 7 | $2.58 \times 10^6$ | 150 | $2.78 \times 10^6$ |
| d4 | 47 | $0.13 \times 10^6$ | 12 118 | $0.13 \times 10^6$ | 9 | $0.37 \times 10^6$ | 100 6 | $3.37 \times 10^6$ |
| d5 | 51 | 35 203 | 17 789 | 28 803 | 11 | $4.11 \times 10^6$ | 10 824 | $4.11 \times 10^6$ |
| d6 | 91 | $0.27 \times 10^6$ | 75 065 | $0.25 \times 10^6$ | 50 | $13.60 \times 10^6$ | $0.44 \times 10^6$ | $1.42 \times 10^6$ |
| d7 | 113 | 12 199 | 9 217 | 12 178 | 2 526 | $23.69 \times 10^6$ | $22.41 \times 10^6$ | $23.69 \times 10^6$ |
| d8 | 132 | 20 028 | 14 062 | 19 917 | 20 000 | $24.00 \times 10^6$ | $24.00 \times 10^6$ | $22.12 \times 10^6$ |
| d9 | 1 240 | 29 223 | 24 313 | 28 673 | 400 000 | $24.84 \times 10^6$ | $24.84 \times 10^6$ | $19.14 \times 10^6$ |
| d10 | 99 800 | $0.50 \times 10^6$ | $0.48 \times 10^6$ | $0.30 \times 10^6$ | 984 298 | $27.36 \times 10^6$ | $27.31 \times 10^6$ | $0.88 \times 10^6$ |
| total | - | $1.11 \times 10^6$ | $0.64 \times 10^6$ | $0.87 \times 10^6$ | - | $0.122 \times 10^9$ | $0.009 \times 10^9$ | $0.079 \times 10^9$ |

(e)

| location | Dimensions | | | Measures | |
|---|---|---|---|---|---|
| | time | salesman | product | cost | profit |
| Montreal | March | John | shoe | 100$ | 10$ |
| Montreal | December | Smith | shoe | 150$ | 30$ |
| Quebec | December | Smith | dress | 175$ | 45$ |
| Ontario | April | Kate | dress | 90$ | 10$ |
| Paris | March | John | shoe | 100$ | 20$ |
| Paris | March | Marc | table | 120$ | 10$ |
| Paris | June | Martin | shoe | 120$ | 5$ |
| Lyon | April | Claude | dress | 90$ | 10$ |
| New York | October | Joe | chair | 100$ | 10$ |
| New York | May | Joe | chair | 90$ | 10$ |
| Detroit | April | Jim | dress | 90$ | 10$ |

**Fig. 3.** Sample tabular regions having different formats, varying column identifiers (lines/no lines), and varying spacing. The presence of different formats makes the problem of structure identification a challenging task.

$\lambda$ is the weight for controlling the contribution of the Symmetric Loss. $\mathcal{P}(.)$ refers to the flip operator such that the input image is mirrored across the x-axis.

As mentioned above, the proposed CoDec Encoder-Decoder model is trained via a combination of the (i) Structure loss and the (ii) Symmetric loss. As shown in Fig. 4, the *Structure loss* minimizes the distance between the decoded sample and the column mask for the input tabular image. That is, unlike traditional Encoder-Decoder models, the input is not reconstructed at the output, instead the Decoder is trained to generate the column mask for the input by recognizing the columns in the given input. Thus, the Encoder learns a latent representation for the given image, which is then up-sampled by the Decoder for generating the corresponding column mask. It is our belief that the different convolutional filters are able to encode the variations observed in the varying tabular formats, making it possible to recognize the column structure from the given image. The second component of the CoDec Encoder-Decoder model corresponds to the *Symmetric loss* which attempts to benefit from the symmetric structure observed in tabular regions. As observed from Fig. 3, the column information is mostly symmetric across the x-axis, and thus the Symmetric loss attempts to ensure that the decoded column mask also maintains the same property by being symmetric across the x-axis. Therefore, as part of the CoDec loss function, the distance between the generated column mask and the flipped column mask is also minimized, which ensures stricter boundaries across the length of the column. In the CoDec loss function, $\mathcal{P}(.)$ (from Eq. 1) refers to the flip operator, which flips the provided image across the x-axis. Mathematically, for an input image $x$, the $\mathcal{P}(.)$ operator is represented as:

$$\mathcal{P}(x) = (x^R)^R \tag{2}$$

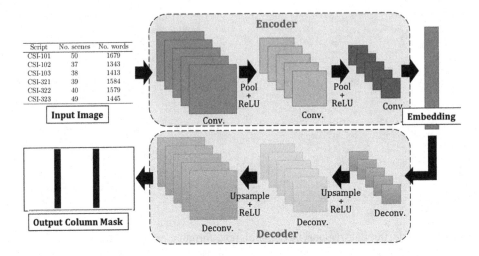

**Fig. 4.** Diagrammatic representation of the CoDec Encoder-Decoder model for extracting the column information from a given tabular image. The model is trained via a combination of Structure loss and Symmetric loss for identifying the columns.

where, $x^R$ refers to rotating the image by $90°$. The Symmetric Loss is thus developed using the available domain knowledge for tabular regions for extracting an accurate column mask.

During training, the CoDec Encoder-Decoder is optimized using the Structure loss and the Symmetric loss (Eq. 1). At inference, the trained model is used to output the column mask for the given tabular image. The generated column mask is then post-processed via binary thresholding to identify the columns. Given the extracted column information, the input image is processed via the row identification module for row recognition. Details regarding the row identification module are provided in the next subsection.

## 3.2 Row Identification Module

In literature, row identification of tabular regions has mostly been performed via the use of domain knowledge and business rules. Similarly, in this research, a combination of different rules is used for the identification of rows in tabular regions. As observed in Fig. 3, a new row is often signified by the presence of an entry in the first column of the table. In order to utilize this domain knowledge, once the column mask has been extracted, the input image and the mask are provided to the row extraction module for the identification of rows. Further, the co-ordinates of each word are also extracted using the Tesseract OCR [26] which are also utilized for identifying the row details in the given tabular image. The following process is followed by the row identification module:

1. Image processing based line detection is applied on the input image. The given image is converted into grayscale, followed by Canny edge detection [1].

Hough transform [10] based line detection is then applied on the processed image for detecting the horizontal lines (lines with a large gap between their $y_1$ and $y_2$ co-ordinates are eliminated as vertical lines). Post-processing is performed wherein detected lines with a gap of less than a chosen threshold of pixels along the $y$-axis are removed. This is done to eliminate duplicate row lines and double boundaries.

2. Since all tabular regions do not contain row boundaries (lines), parallely, the row co-ordinates are also estimated using the $y$-coordinates obtained with the extracted words. Initially, each $y$-coordinate beyond a chosen threshold is identified as a new row line. The column information is then utilized for modeling multi-line cells in un-bordered tables. If a given row contains text in very few columns (less than 80% of the total columns), it is deemed as the continuation of the previous row, and the information of that $y$-coordinate is updated.

3. As a final step, the rows identified by the above two techniques are fused together to generate the final row coordinates.

### 3.3    Table Structure Recognition and Data Extraction

The row and column coordinates obtained by the two modules are fused together to generate the overall structure of the table. Parallely, as mentioned above, data extraction is performed from the tabular region using the Tesseract OCR [26]. The OCR returns the content in the given image along with the coordinates of each word containing the $x$ co-ordinate, $y$ co-ordinate, and width of the word. These coordinates are then used to divide the content into the corresponding cells created using the row and column coordinates obtained via the proposed table structure recognition pipeline.

Once the content has been split into the different cells of the table, the information is then post-processed for comparison with the ground-truth. Similar to the existing techniques [16], 1-D tuples are generated for each cell containing the content of the neighbouring cells (upper, lower, immediate left, and immediate right cells). These tuples are then compared with the ground-truth information provided with the datasets. The datasets available for table structure recognition often contain an XML file for each table containing the details regarding the structure and the content (coordinates of each cell along with the content). The XML files are thus used for generating the ground-truth 1-D tuples for each cell, following which matching is performed with the tuples generated using the proposed table structure recognition pipeline.

## 4    Datasets and Protocols

Two datasets have been used for experiments: (i) Marmot dataset [4] and the (ii) ICDAR 2013 dataset [7]. Details regarding each are as follows:

– **Marmot Dataset**[1] [4]: The Marmot dataset contains over 1000 PDF documents in English and Chinese languages with tabular regions. The ground-truth annotations of the table structure (row and column co-ordinates) have also been provided as an XML file for each document. As part of this research, the Marmot dataset has been used for training the novel CoDec Encoder-Decoder model. Specifically, the 509 English documents are pre-processed for the extraction of the tabular region (input of the Encoder) along with the creation of the column mask (output of the Decoder) based on the ground-truth provided with the dataset. Owing to the limited training data, data augmentation has been performed on the tabular regions, specifically, mirroring along the $y-axis$ and incorporation of minor Gaussian noise.

– **ICDAR 2013 Dataset**[2] [7]: The ICDAR 2013 dataset is one of the most popular and commonly used dataset for table structure recognition. In this research as well, the ICDAR 2013 dataset has been used for evaluating the proposed pipeline for table structure recognition. The dataset contains a total of 67 PDF documents with tabular regions. The ICDAR dataset contains both vertical and horizontal tables - over 30% of the total tables are vertical in nature. The dataset provides XML files for each document containing ground-truth annotations with respect to the table position and its structure. Information such as the cell co-ordinates and the content has also been provided. Consistent with existing techniques and in order to compare with recent state-of-the-art algorithms [16,20], the standard protocol is followed on this dataset, wherein 34 images are used for evaluating the model. Existing techniques often utilize the remaining samples for fine-tuning the pre-trained architecture, however, we do not utilize the ICDAR 2013 dataset for training or fine-tuning.

### 4.1   Implementation Details

As elaborated in the previous Section, column identification has been performed using the proposed CoDec Encoder-Decoder model which consists of an Encoder module and a Decoder module. The Encoder is composed of four convolutional layers with $3 \times 3$ kernels and filter sizes of $[32, 16, 8, 4]$. We use *ReLU* [15] as the activation function after each convolution layer. Max-pooling is also applied post each convolution layer for reducing the dimension of the feature. The Decoder model is the mirror of the encoder architecture with four transposed convolutional layers having $3 \times 3$ kernels and filter sizes of $[4, 8, 16, 32]$. After each layer, *ReLU* activation function has been used. An image of dimension $224 \times 224$ is provided as input to the CoDec Encoder-Decoder model. The model is implemented in PyTorch [17], and the Adam optimizer [13] has been used to train the model with an initial learning rate of 0.01. The weight for the Symmetric loss ($\lambda$ in Eq. 1) is set to 0.01. In the row identification module, a minimum gap of 20

---

[1] https://www.icst.pku.edu.cn/cpdp/sjzy/index.htm.
[2] http://www.tamirhassan.com/html/competition.html.

**Table 1.** Table structure recognition performance on the ICDAR 2013 dataset, along with comparison with recent state-of-the-art algorithms. Owing to the same protocol, results have directly been taken from the published manuscript [16].

| Algorithm | Recall | Precision | F1-Score |
|---|---|---|---|
| TableNet + Semantic features (fine-tuned) [16] | <u>0.9001</u> | 0.9307 | <u>0.9151</u> |
| TableNet + Semantic features [16] | 0.8994 | 0.9255 | 0.9122 |
| TableNet [16] | 0.8987 | 0.9215 | 0.9098 |
| DeepDeSRT (fine-tuned) [20] | 0.8736 | **0.9593** | 0.9144 |
| **Proposed** | **0.9289** | <u>0.9337</u> | **0.9304** |

pixels is maintained between each detected row in order to eliminate duplicate line boundaries and incorrect row co-ordinates. Default parameters have been used for the Canny detector which remain consistent across all grayscale input images. For the Hough transform, the 'minLineLength' is a function of the size of the image, and the 'maxLineGap' is set to 10. Other parameters are kept as default. The parameters of the row-identification module are configured once and then used consistently across all the images. That is, once trained/configured, the entire pipeline is automated in nature without any manual intervention. Given a tabular image, the pipeline outputs the table structure using the (i) column detector followed by the (ii) row detector, resulting in an end-to-end framework. During the generation of the 1-D tuples, the extracted words (obtained via the ground-truth XML file or the Tesseract OCR), are converted to lower case after removing any trailing or preceding white spaces. All special characters are replaced with a '_', followed by matching between the words.

## 5    Results

Table 1 presents the results obtained on the ICDAR 2013 dataset for table structure recognition. The top result is presented in bold, while the second best result is underlined. As per the existing research, performance has been reported in terms of the recall, precision, and F-1 score:

$$Recall = \frac{TP}{TP + FP}; \ Precision = \frac{TP}{TP + FN}; \tag{3}$$

$$F1 \ Score = \frac{2 \times Precision \times Recall}{Precision + Recall} \tag{4}$$

It can be observed that the proposed technique achieves a recall of 0.9289, thus demonstrating an increase of over 2% as compared to the state-of-the-art technique (0.9001 obtained by TableNet [16]). The proposed technique also obtains the second best performance for precision by reporting 0.9337, while DeepDeSRT [20] obtains 0.9593. It is important to note that the best results of DeepDeSRT and the TableNet architecture are obtained after fine-tuning on the ICDAR 2013

**Table 2.** Ablation study on the proposed table structure recognition pipeline. Analysis has been performed by modifications of the row detection module (removal of the image processing (IP) based row detection and co-ordinate based row detection) and the column detection module (removal of the Symmetric loss).

| Algorithm | Recall | Precision | F1-Score |
|---|---|---|---|
| Proposed - IP based row detection | 0.7832 | 0.7984 | 0.7882 |
| Proposed - Coordinate based row detection | 0.8545 | 0.8910 | 0.8636 |
| Proposed - $\mathcal{L}_{Symmetric}$ | 0.8335 | 0.8918 | 0.8556 |
| **Proposed** | **0.9289** | **0.9337** | **0.9304** |

dataset, while the proposed technique is only trained on the Marmot dataset and does not utilize the ICDAR 2013 dataset during training or fine-tuning. The improved performance obtained without explicit training/fine-tuning on the ICDAR 2013 dataset supports the generalizable behavior of the proposed pipeline. Finally, an overall F1 score of 0.9304 is obtained via the proposed technique demonstrating an improvement from the state-of-the-art TableNet (0.9151). The improved results obtained on the standard benchmark dataset demonstrates the efficacy of the proposed algorithm. Further, Fig. 5 presents sample images from the ICDAR 2013 dataset, along with the raw column mask generated by the proposed CoDec Encoder-Decoder model. The model is able to identify the column demarcations in the absence of column lines (Fig. 5(a)), as well as in the presence of line demarcations (Fig. 5(b–c)). The high performance obtained on the ICDAR 2013 dataset, without explicitly training or fine-tuning on it further promotes the utility of the column detection model on unseen datasets not used during the training of the model.

**Ablation Study and Effect of $\lambda$:** In order to analyze the proposed pipeline for table structure recognition, an ablation study has been performed on the same. Table 2 presents the results obtained by removing different components from the structure recognition pipeline. Experiments have been performed by removing: (i) image processing based row detection from the row detection module, (ii) co-ordinate based row detection from the row detection module, and (iii) symmetric loss from the column detection module. As demonstrated from Table 2, removal of any component from the proposed pipeline results in a reduction in the precision, recall, and F1-score performance. Specifically, maximum drop in performance is observed by removing the image processing based row detection component (almost 15% drop in F1 score). Drop in performance is also observed upon removing the Symmetric Loss from the proposed CoDec Encoder-Decoder model for column detection (Eq. 1). The drop in performance further reinstates the benefit on incorporating the Symmetric loss across the $x$-axis. Further, experiments were also performed to analyze the impact of $\lambda$ (Eq. 1) which controls the contribution of the Symmetric Loss in the CoDec model. With a much smaller value ($\lambda = 0.001$), the F-1 score reduces to 84.78%, while a larger $\lambda$ ($\lambda = 0.1$)

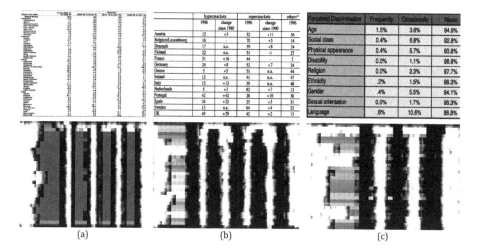

**Fig. 5.** Sample images from the ICDAR 2013 dataset and the corresponding column mask generated by the proposed CoDec Encoder-Decoder. The model is able to process images with/without column lines and is able to identify column details well.

results in a F-1 score of 86.16%. The reduction in performance upon the removal of different components demonstrates the benefit of each component in the final table structure recognition pipeline, along with the choice of appropriate hyper-parameters.

**Comparison on Number of Trainable Parameters:** The proposed table structure recognition pipeline utilizes a light-weight Encoder Decoder architecture for column extraction. The proposed column detection model contains only 9,269 trainable parameters, whereas existing state-of-the-art column detection models such as the TableNet [16] and the DeepDeSRT [20] contain at least 1.38M (VGG-19 and VGG-16 architectures [25], respectively). The light-weight nature of the proposed framework results in lesser number of trainable parameters and also reduced size of the model. This enables the proposed pipeline to be trained with lesser number of images, and also makes it deployable in real world scenarios with less resource requirement. The lesser number of parameters prevents the model from over-fitting on the training dataset (Marmot dataset), thus resulting in a generalized behavior on a new dataset as well (ICDAR 2013 dataset). Further, the entire pipeline takes less than 1 s. for inference on an input image.

## 6   Conclusion

The requirement of automated detection and identification of tables from document images has been increasing over the last two decades. Information extraction from tabular regions is useful for automated content generation and summarization. Extraction of relevant content from tabular regions also requires table structure recognition, which corresponds to identifying the exact structure of the table along with the cell information. Despite the wide-spread applicability, table structure recognition has received limited attention and continues to be a challenging task due to the complexity and diversity in the structure and style of different tables. To this effect, this paper presents an end to end pipeline for table structure recognition containing two components: (i) column identification module, and (ii) row identification module. The column identification module utilizes a novel Column Detector Encoder-Decoder model (termed as *CoDec* Encoder-Decoder) which is trained via a novel loss function containing Structure loss and Symmetric loss. The detection of the columns is followed by the identification of different rows in the table using domain information and different image processing rules. The performance of the proposed pipeline is evaluated on the ICDAR 2013 dataset, where it demonstrates improvement from the state-of-the-art networks even without explicitly training or fine tuning on the ICDAR 2013 dataset, thereby suggesting a generalizable behavior of the proposed pipeline. Further, as part of this research, ablation study has also been performed by removing different components from the structure recognition pipeline, and results of each experiment have been discussed in this paper. Another key contribution of this research revolves around the limited number of trainable parameters of the proposed CoDec Encoder-Decoder model as compared to existing techniques. There are only 9,269 trainable parameters (in comparison to over 1.39M of existing techniques) in the proposed column detection model which makes the proposed framework trainable with limited number of images, and also makes it deployable with less resource requirement in real world scenarios. As part of future work, the proposed technique can be improved to better model multi-column variations for table structure recognition.

## References

1. Canny, J.: A computational approach to edge detection. IEEE Trans. Pattern Anal. Mach. Intell. **6**, 679–698 (1986)
2. Coüasnon, B., Lemaitre, A.: Recognition of tables and forms. In: Handbook of Document Image Processing and Recognition (2014). https://doi.org/10.1007/978-0-85729-859-1
3. Embley, D.W., Hurst, M., Lopresti, D., Nagy, G.: Table-processing paradigms: a research survey. Int. J. Doc. Anal. Recogn. **8**(2–3), 66–86 (2006)
4. Fang, J., Tao, X., Tang, Z., Qiu, R., Liu, Y.: Dataset, ground-truth and performance metrics for table detection evaluation. In: International Workshop on Document Analysis Systems, pp. 445–449 (2012)
5. Gatos, B., Danatsas, D., Pratikakis, I., Perantonis, S.J.: Automatic table detection in document images. In: Singh, S., Singh, M., Apte, C., Perner, P. (eds.) ICAPR 2005. LNCS, vol. 3686, pp. 609–618. Springer, Heidelberg (2005). https://doi.org/10.1007/11551188_67
6. Gilani, A., Qasim, S.R., Malik, I., Shafait, F.: Table detection using deep learning. In: International Conference on Document Analysis and Recognition, pp. 771–776 (2017)
7. Göbel, M., Hassan, T., Oro, E., Orsi, G.: ICDAR 2013 table competition. In: International Conference on Document Analysis and Recognition, pp. 1449–1453 (2013)

8. Hao, L., Gao, L., Yi, X., Tang, Z.: A table detection method for pdf documents based on convolutional neural networks. In: IAPR Workshop on Document Analysis Systems, pp. 287–292 (2016)
9. Hu, J., Kashi, R.S., Lopresti, D.P., Wilfong, G.: Medium-independent table detection. In: Document Recognition and Retrieval VII, vol. 3967, pp. 291–302 (1999)
10. Illingworth, J., Kittler, J.: A survey of the Hough transform. Comput. Vis. Graph. Image Process. 44(1), 87–116 (1988)
11. Khusro, S., Latif, A., Ullah, I.: On methods and tools of table detection, extraction and annotation in pdf documents. J. Inf. Sci. 41(1), 41–57 (2015)
12. Kieninger, T., Dengel, A.: The T-Recs table recognition and analysis system. In: International Workshop on Document Analysis Systems, pp. 255–270 (1998)
13. Kingma, D.P., Ba, J.: Adam: a method for stochastic optimization. arXiv preprint arXiv:1412.6980 (2014)
14. Li, M., Cui, L., Huang, S., Wei, F., Zhou, M., Li, Z.: Tablebank: table benchmark for image-based table detection and recognition. In: Language Resources and Evaluation Conference, pp. 1918–1925 (2020)
15. Nair, V., Hinton, G.E.: Rectified linear units improve restricted Boltzmann machines. In: International Conference on International Conference on Machine Learning, pp. 807–814 (2010)
16. Paliwal, S.S., Vishwanath, D., Rahul, R., Sharma, M., Vig, L.: TableNet: deep learning model for end-to-end table detection and tabular data extraction from scanned document images. In: International Conference on Document Analysis and Recognition, pp. 128–133 (2019)
17. Paszke, A., et al.: Automatic differentiation in PyTorch (2017)
18. Prasad, D., Gadpal, A., Kapadni, K., Visave, M., Sultanpure, K.: CascadeTabNet: an approach for end to end table detection and structure recognition from image-based documents. In: IEEE/CVF Conference on Computer Vision and Pattern Recognition Workshops, pp. 572–573 (2020)
19. Qasim, S.R., Mahmood, H., Shafait, F.: Rethinking table recognition using graph neural networks. In: International Conference on Document Analysis and Recognition, pp. 142–147 (2019)
20. Schreiber, S., Agne, S., Wolf, I., Dengel, A., Ahmed, S.: DeepDesrt: deep learning for detection and structure recognition of tables in document images. In: International Conference on Document Analysis and Recognition, vol. 1, pp. 1162–1167 (2017)
21. Shigarov, A., Mikhailov, A., Altaev, A.: Configurable table structure recognition in untagged pdf documents. In: ACM Symposium on Document Engineering, pp. 119–122 (2016)
22. Siddiqui, S.A., Khan, P.I., Dengel, A., Ahmed, S.: Rethinking semantic segmentation for table structure recognition in documents. In: International Conference on Document Analysis and Recognition, pp. 1397–1402 (2019)
23. Siddiqui, S.A., Malik, M.I., Agne, S., Dengel, A., Ahmed, S.: DeCNT: deep deformable CNN for table detection, vol. 6, pp. 74 151–74 161 (2018)
24. e Silva, A.C., Jorge, A.M., Torgo, L.: Design of an end-to-end method to extract information from tables. Int. J. Doc. Anal. Recogn. (IJDAR) 8(2–3), 144–171 (2006)
25. Simonyan, K., Zisserman, A.: Very deep convolutional networks for large-scale image recognition. arXiv preprint arXiv:1409.1556 (2014)
26. Smith, R.: An overview of the Tesseract OCR engine. In: International Conference on Document Analysis and Recognition, vol. 2, pp. 629–633 (2007)

27. Wang, Y., Phillips, I.T., Haralick, R.M.: Table structure understanding and its performance evaluation. Pattern Recogn. **37**(7), 1479–1497 (2004)
28. Wang, Y., Phillips, I., Haralick, R.: Automatic table ground truth generation and a background-analysis-based table structure extraction method. In: International Conference on Document Analysis and Recognition, pp. 528–532 (2001)
29. Zanibbi, R., Blostein, D., Cordy, J.R.: A survey of table recognition. Doc. Anal. Recogn. **7**(1), 1–16 (2004)

# Advertisement Extraction Using Deep Learning

Boraq Madi$^{(\boxtimes)}$, Reem Alaasam$^{(\boxtimes)}$, Ahmad Droby$^{(\boxtimes)}$, and Jihad El-Sana$^{(\boxtimes)}$

Ben-Gurion University of the Negev, Beersheba, Israel
{borak,rym,drobya}@post.bgu.ac.il, el-sana@cs.bgu.ac.il

**Abstract.** This paper presents a novel deep learning model for extracting advertisements in images, PTPNet, and multiple loss functions that capture the extracted object's shape. The PTPNet model extracts features using Convolutional Neural Network (CNN), feeds them to a regression model to predict polygon vertices, which are passed to a rendering model to generate a mask corresponding to the predicted polygon. The loss function takes into account the predicted vertices and the generated mask. In addition, this paper presents a new dataset, AD dataset, that includes annotated advertisement images, which could be used for training and testing deep learning models. In our current implementation, we focus on quadrilateral advertisements. We conducted an extensive experimental study to evaluate the performance of common deep learning models in extracting advertisement from images and compare their performance with our proposed model. We show that our model manages to extract advertisements at high accuracy and outperforms other deep learning models.

**Keywords:** Ads extraction · Loss function · Segmentation model · Regression model

## 1 Introduction

Advertisements play a significant role in various aspects of our daily life. Commercial organizations advertise their products and services for the general public, and governments utilize the advertisement framework to deliver messages and announcements for educating the public on a wide range of issues. As a result, they occupy none trivial portions of our buildings, streets, and media (printed and digital). Detecting advertisements in a view, an image, or a document has numerous applications. For example, local municipalities are interested in detecting advertisements on the streets for taxing issues, and advertisements agencies want to measure the exposure of their posted advertisements in various media.

Advertisement extraction can be viewed as an image segmentation problem, which is the task of assigning the pixels of each object, in the image, the same label. The segmentation output is usually represented as a pixel map, a graph, or a list of polygons that form the boundaries of these segments. Many recent

© Springer Nature Switzerland AG 2021
E. H. Barney Smith and U. Pal (Eds.): ICDAR 2021 Workshops, LNCS 12917, pp. 81–97, 2021.
https://doi.org/10.1007/978-3-030-86159-9_6

research [1,2,7] focus on predicting polygons instead of explicit pixel-wise labeling, due to their simple representation. In this work, we adopt this scheme and focus on extracting the boundary of objects in a polygon representation.

Object extraction research has made impressive progress over the recent year [14]. However, they are far from obtaining pixel-level accuracy. In this work, we limit our work to convex quadrilateral (Tetragon) objects and obtain high pixel-level accuracy. These objects are very common in our daily views and include advertisements, billboards, frames, paintings, screens, etc. We show that by narrowing the search domain, we can obtain more accurate results than the state-of-the-art general segmentation methods.

This paper explores extracting advertisements from images. It introduces a new dataset (AD dataset), presents a novel deep learning model for advertisement extraction, and develops several geometry-based loss functions for the presented model. Our experimental study shows that our model outperforms other general segmentation methods.

We introduce a new dataset for advertisement extraction, AD dataset. It includes around six thousand images containing various forms of advertisements. The images were collected from the internet and manually annotated by determining the boundary of each advertisement. The label of each advertisement includes a boundary polygon and a mask. In addition, we present and evaluate a novel deep learning model called PTPNet, which is trained using a loss function that considers the polygon and mask labels in the AD dataset. The PTPNet is composed of a regression model and a mask generator model. The regression model extracts CNN features, which are fed to fully connected regression layers to output the vertices of a polygon. These vertices are passed to the generator model to produce a mask, which is used by the loss function. The current PTPNet model outputs the vertices and the mask of a quadrilateral.

The vertex-based loss function is the vertex-wise $L_1$ distance between the predicted and the ground-truth polygon. The drawback of these loss functions is the assumption that the object's vertices are independent variables. However, these vertices are highly correlated. Considering the area of the polygon reduces the dependency and strengthens the loss function. To overcome this limitation, we introduce a novel loss function that takes into account the polygon and its mask representations. The mask loss function is the Dice distance between the ground-truth mask and the mask generated from the predicted polygon. The final PTPNet loss function combines these two loss functions. According to our experimental study, the loss function contributes to higher accuracy and accelerates the model convergence; these findings align with this work [23].

To summarize, the main contributions of this paper are:

- A new dataset of quadrilateral objects, AD dataset, which includes advertisements images labeled by boundary contours and binary mask.
- Novel loss functions for regression models that take into account the geometric shape and the distance between the vertices of quadrilateral objects.
- A novel regression model, PTPNet, that utilizes boundary contours and binary masks of elements in the AD dataset to learn predicting the boundary polygon of an advertisement.

The rest of the paper is organized as follows: we review the related work in Sect. 2, then describe the new dataset in Sect. 3. In Sect. 4 we present our method and in Sect. 5 we describe the different loss functions. Section 6 presents our experiments and results. Finally, we draw concluding remarks in Sect. 7.

## 2   Related Work

Extracting the polygon representation of an object can be done by first segmenting the object in the image and extracting its contour, which is often simplified to form a compact polygon. Classic object contour extraction methods are based on superpixel grouping [15], Grabcut [20], and saliency detection [4,22].

One of the popular methods for polygon extraction using deep learning is Polygon-RNN [3] and its improved version [1]. This method provides semi-automatic object annotation using polygons. The RNN's decoder considers only three preceding vertices when predicting a vertex at each step which may produce polygons with self-intersections and overlaps. PolyCNN [8] extracts rectangular building roofs from images. However, this method does not handle perspective transformations and its accuracy does not reach pixel level. PolyMapper [16] presents a more advanced solution by using CNNs and RNNs with convolutional long-short term memory modules. It provides good results for aerial images of residential buildings.

In recent years, many deep learning methods and architectures have been proposed for semantic segmentation and they lead to outstanding progress in semantic segmentation. The most two relative groups of methods for our work are Regional proposal and Upsampling/Deconvolution models.

Regional proposal models detect regions according to similarity metrics, determine whether an object is present in the region or not, and applying segmentation methods for positive regions. He et al. [9] proposed a Mask Regional Convolutional Neural Network (Mask-RCNN). It extends Faster R-CNN [19] abilities for object detection and segmentation.

The deconvolution models focus on extracting high-level features via layer-to-layer propagation and obtain segmentation by upsampling and deconvolution. The reconstruction techniques for obtaining a segmentation map include refinement methods that fuse low and high-level features of convolutional layers. Long et al. [18] proposed the first Fully Convolutional Network (FCN), which constructs the segmentation by adding skip architecture that combines the semantic and appearance for precise segmentation results.

Many other segmentation models [14] are used to provide benchmark results for new datasets, Such as Xception [5], Resenet [10], MobileNet [11] and DenseNet [12]. These models are regression models which output polygon, while Mask-RCNN [9] provides an object's mask. In this paper, we apply these models to test various loss functions and compare their performance with our proposed model.

## 3    Dataset

This section presents a new dataset of advertisement images called AD dataset[1]. It includes about six thousand RGB-images where each image contains at least one quadrilateral advertisement. The AD dataset includes a wide range of advertisements types and sizes, starting from billboards on the highway to advertisements in malls. Figure 1 shows example images from the dataset.

(a)                              (b)

**Fig. 1.** Example of images in AD dataset.

### 3.1    Collecting and Labeling Images

The images of the AD dataset were collected automatically from the internet and labeled manually. We collected images that include advertisements using a python script called 'Google images download'[2] which downloads images from google using given keywords. The resolution of the images varies from $256 \times 144$ to $2048 \times 1080$. The advertisements within these images have diverse perspective views.

We label the images using Labelbox[3], the annotators went over each of the downloaded images and traced the boundary of each advertisement in the image. We define the boundary of an advertisement by its vertices in a counter-clockwise (CCW) order, as shown in Fig. 2b. The labeling task was conducted by several undergraduate and graduate students in our lab. In addition, we generate a binary mask (See Fig. 2c) using the annotated polygon for each image, where the pixels of the advertisement are marked by 1 and the background by 0. Hence, we have two types of labels for each image.

---

[1] https://github.com/BorakMadi/Ads-Extraction-using-Deep-learning.

[2] https://github.com/hardikvasa/google-images-download.

[3] https://labelbox.com/.

**Fig. 2.** Element example of AD dataset: (a) the original image, (b) the vector of vertexes label, and (c) the binary mask label

## 4   The PTPNet Model

Yu *et al.* [23] introduced Intersection-Over-Union (IOU) loss function to regress four vertices of a box. This loss function considers the four vertices as a whole unit, i.e. the bounding box. Their trained model manages to obtain higher accuracy than those using $L_2$ (Euclidean) distance loss function. Their IOU loss function assumes axis aligned bounding boxes, which are calculated directly from the predicted and ground-truth boxes. This calculation scheme of the IOU does not suit our task, as the perspective view of the quadrilateral is not expected to be axis aligned. In addition, the scalar difference between the size of the predicted and the ground-truth polygons is an inappropriate approximation for the IOU loss.

To overcome these limitations, we constructed a novel learning model called *PTPNet*, which enables applying advanced geometrical loss functions to optimize the prediction of the vertices of a polygon. The PTPNet outputs a polygon and its mask representation and manages to combine classical regression and advanced geometrical loss functions, such as IOU.

### 4.1   PTPNet Architecture

PTPNet network includes two sub-networks: Regressor and Renderer. The Regressor predicts the four vertices of a quadrilateral. It utilizes the Xception model as its backbone. The classification component of the Xception model is removed and replaced by a regression component that outputs a vector of eight scalars that represent the four vertices. The regression component includes Global Average Pooling layers followed by 4-Fully-Connected layers with different sizes, as shown in Fig. 9.

The Renderer (rendering model) generates a binary mask corresponding to the Regressor's predicted polygon. It is trained separately from the regression model using the quadrilaterals' contours from the AD dataset (see Render Datasets). The trained model is concatenated to the regression model and its weights are frozen during the training of the regression model (Fig. 3).

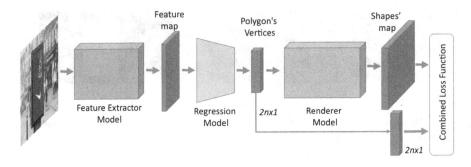

**Fig. 3.** The architecture of PTPNet

## 4.2   Rendering Models

We experimented with various rendering decoder networks, which are based on fully connected and deconvolution networks. The results obtained using the fully connected decoders were inadequate, while the deconvolutional decoders produce promising results (see Sect. 6). Therefore, we developed a novel rendering model based on the Progressive Growing of GANs [13], which is a form of deconvolutional decoders. We removed the discriminator network and modified the generator network (the deconvolutional decoder) to act as a supervised based learning decoder that accepts a polygon representation and outputs its corresponding mask. We shall refer to this decoder as *GenPTP*.

To train and test the rendering models we build two datasets, *synthesized-quads* and *ads-quads*. The synthesized-quads dataset was constructed by randomly sampling convex quadrilaterals, i.e. sampling four vertices. In typical images, the convex quadrilaterals are the perspective transformation of rectangular advertisements. Figure 4(b) shows an example of such quadrilaterals. The ads-quads dataset was generated by considering only the contours of the advertisements in the AD-Dataset (the manually annotated quadrilaterals), as shown in Fig. 4(a). The labels of these contours are the binary masks of the corresponding advertisement, which are part of the AD-Dataset. We evaluated the performance of the Rendering models using the two datasets, i.e. synthesized-quads and ads-quads. We trained two different instances of each rendering model, one for each dataset and compare their performances. For faithful comparison, we evaluate the performance of these instances using test sets from the ads-quads dataset.

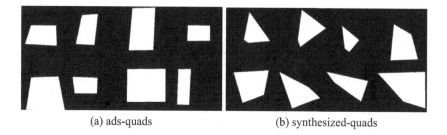

(a) ads-quads                    (b) synthesized-quads

**Fig. 4.** Sample from ads-quads and synthesized-quads.

## 5   Loss Functions

In this section, we present several loss functions for optimizing our models. These loss functions are categorized into vertex-based regression functions, areas-based loss functions, and hybrid methods.

### 5.1   Vertex-Based Loss Functions

These loss functions compute the sum of the distances between the corresponding vertices of predicted and ground-truth polygons. To determine the corresponding among the vertices, we compute all the possible circular shifts of vertices of the predicted polygon and choose the shift with minimum distance with respect to the ground truth vertices. The distance between vertices is calculated using $L_1$ and $L_2$ metric. In this approach, the model learns to regress to polygon vertices independent of their order, similar to [8]. Equation 1 and Eq. 2 formulate the loss functions with respect to $L_1$ and $L_2$, receptively, where $n$ is the number of vertices, $V_{gt}$ and $V_{pred}$ represent vertices of the ground truth and predicted polygons, and $R$ is the circular shift function applied to the $V_p$ with step $r$. Since we limit our domain to quadrilaterals, $n$ is equal to four.

$$MinR_{L_2} = \min_{\forall r \in [0,3]} \frac{1}{n} \sum_{i=1}^{n} \|V_g - R(V_p, 2*r)\|_2 \tag{1}$$

$$MinR_{L_1} = \min_{\forall r \in [0,3]} \frac{1}{n} \sum_{i=1}^{n} \|V_g - R(V_p, 2*r)\|_1 \tag{2}$$

### 5.2   Area Loss Functions

Area based loss functions, such as the difference between the area of prediction and ground-truth polygons, are insufficient. These functions motivate the model to learn to regress to four vertices that have the same area as the ground-truth, but without considering the location of these vertices, which is incorrect. We propose an alternative approach that considers the locations of the vertices and treats them as representatives of geometric shapes rather than independent vertices, similar to IOU in [23].

Theoretically, calculating the similarity between two quadrilateral can be performed using advanced geometrical loss functions such as IOU. However, calculating such similarity based only on polygon representation is complicated. Thus, we approximate the difference between the intersection and the union between the two polygons using two trapezoids. Among possible shifts of the predicted vertices, we choose the permutation with minimum distance with respect to the ground truth according to $L_1$ criterion (see Eq. 3). Let us denote this permutation, by $r_{min}$. We calculate the two trapezoid areas by utilizing four vertices, two from the $r_{min}$ permutation and two from the ground truth. The final loss value is the sum of these two trapezoid areas (see Eq. 4 and Fig. 5c).

$$r_{min} = \min_{\forall r \in [0, \frac{n}{2}-1]} \frac{1}{n} \sum_{i=1}^{n} \|V_g - R(V_p, 2*r)\|_1 \tag{3}$$

$$Trapezoids_{loss} = TrapezoidArea(p1, p2, g1, g2) +$$
$$TrapezoidArea(p3, p4, g3, g4) \tag{4}$$

### 5.3 Hybrid Loss Functions

The vertex-based and area-based approaches suffer from vertex-independent optimization and overlook the importance of boundary vertices, respectively. Since the vertices of a quadrilateral are correlated, the independent optimization can not provide the best results. The distance between vertices does not describe the difference in the induced areas faithfully. For example, the ground-truth (See Fig. 6a) and predicted (See Fig. 6b) share the same three vertices and differ in one (See Fig. 6c). As seen, the loss is the green region (See Fig. 6d),

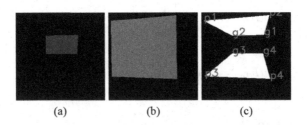

(a)          (b)          (c)

**Fig. 5.** The trapezoids between the two polygons

the loss based only on the vertices is small and gives insufficient feedback. The area-based loss functions provide uniform weights for all quadrilateral points and neglect the importance of vertices and edges.

To overcome these limitations, we combine independent vertex regression and geometrical loss functions to take into account the appropriate weight of the vertices and the correlation between them. Practically, we define the loss function as a sum of the quadrilateral area and vertex-distance loss functions.

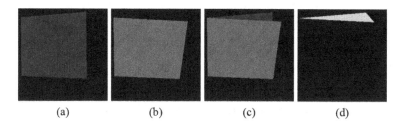

(a)              (b)              (c)              (d)

**Fig. 6.** Intersection and loss regions

*Vertex and area difference* based loss functions combine the distance between the corresponding vertices using $L_1$ or $L_2$ and the difference between the size of the quadrilaterals, i.e. subtracting the two scalars. The size difference is not an adequate optimization parameter in our case (see the discussion earlier). Nevertheless, it improves performance when combined with vertex distance loss components, i.e. this combination guides the loss to choose the nearest points with the same area. Practically, we combine Eq. 1 or Eq. 2 with area difference. The distance between ground-truth and predicted vertices is performed using Eq. 1 or Eq. 2. We calculate the area based on Shoelace or Gauss's area formula. The area difference is computed by subtracting the areas of polygons spanned by the ground truth $V_{gt}$ and the predicted $V_{psh}$ vertices. This combination is formally expressed by Eq. 5 and Eq. 6

$$Area\_MinR_{L_2} = MinR_{L2} + AreaSub(V_{psh}, V_{gt}) \tag{5}$$

$$Area\_MinR_{L_1} = MinR_{L_1} + AreaSub(V_{psh}, V_{gt}) \tag{6}$$

### 5.4   Loss Function for PTPNet

Above we have shown how to integrate vertex-based loss function with simple geometrical loss, such as areas. The geometrical element in previous hybrid-loss functions is limited to scalar representation. This limitation prohibits using sophisticated loss functions that accurately describe the geometrical relation between the two quadrilateral shapes. The PTPNet model overcomes this by integrating a rendering component, which generates a binary mask that resembles the quadrilateral corresponding to the predicted vertices.

The PTPNet loss function, Eq. 7, combines vertex-regression, Eq. 2, with Sørensen–Dice coefficient (Dice) loss function. The Dice is applied to the generated and ground-truth mask.

$$PTPNet\_loss = MinR_{L_1} + (1 - Dice) \tag{7}$$

## 6    Experiments

In this section, we overview data preparation, models construction, and evaluation results.

### 6.1    Dataset Preparation

In Sect. 3 we discussed collecting and annotating the dataset. Next, we overview preparing the train and test dataset, the augmentation process, and the sanity-check dataset.

The train and test sets are selected from the AD-dataset (see Sect. 3). In this study, we focus on images that include one quadrilateral advertisement and subdivide them to 70% and 30% for training and testing, respectively. The train and test images are resized to $256 \times 256$. The annotations are modified according to the resized images. Recall that the annotation of each advertisement includes its boundary contours and a binary mask. The contour (polygon) is represented by a normalized vector.

We augment the training dataset by applying the basic geometric transformation, i.e. rotation, scale, and sheer. We apply random rotations between $0°$ to $90°$, scale between 0.8 to 1.2, and shear from $0°$ to $20°$. The same transformation is applied to the image and its labels, i.e. contour and mask.

We build a sanity-check dataset for the initial evaluation of our regression methods. This dataset is generated from the contours of advertisements within the images of the AD-dataset, i.e. a contour (quadrilateral polygon) is embedded within a black image creating a binary image with a quadrilateral polygon (see Fig. 7) (Fig. 8).

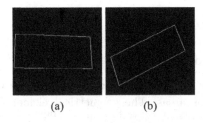

(a)                    (b)

**Fig. 7.** Samples from sanity-check dataset.

| (a) | (b) | (c) | (d) |

**Fig. 8.** Examples of augmented images

## 6.2 Study Models

Toward our evaluation study, we experimented with various types of deep learning architectures, which we discuss next.

We explored applying Mask-RCNN [9] to detect and localize advertisements within images and outputting a binary mask. The MaskRCNN was pre-trained on COCO-dataset [17]. We define the performance of the pre-trained MaskRCNN as the baseline for our experiments.

Our regression networks accept a color image that contains quadrilateral advertisement and predict the coordinates of its four vertices. These neural networks consist of a features extractor and a regression model. The model regresses to four vertices (vector) using the latent vector of the feature extractor, similar to PolyCNN [8], but with a different truncation style and model architecture.

We choose a set of *study models* that includes 13 network architecture, which are variations[4] of Xception [5], Resenet [10], MobileNet [11] and DenseNet [12]. We refer to these 13 models as the *study models*, which are pre-trained on ImageNet [6] and truncated to act as feature extractors. To build a regression model, we append to each feature extractor a component composed of Global Average Pooling layers followed by 4-Fully-Connected layers with different sizes as shown in Fig. 9.

We trained the study models on $10k$ synthesized images using the Mean Square Error (MSE) loss function. As shown in Table 1 the top five models are the modified (replacing the fully connected component with a regression model)

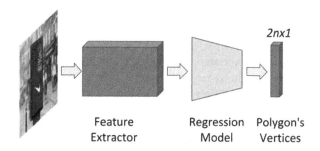

**Fig. 9.** The modified regression models network

---

[4] The full list appears in Table 1.

versions of Xception [5], ResNet50 [10], ResNet101 [10], DensNet121 [12] and MobileNetV2 [21]. Therefore, we adopted these models to study the influence of the various loss functions, i.e. we explore the performance of these models using the loss functions we discussed in Sect. 5.

**Table 1.** The accuracy of the regression models on the sanity-check dataset

| Regression model | Loss function | Accuracy | Regression model | Loss function | Accuracy |
|---|---|---|---|---|---|
| Xception | $MSE$ | **0.961** | InceptionV3 | $MSE$ | 0.937 |
| InceptionResNetV2 | $MSE$ | 0.927 | VGG16 | $MSE$ | 0.947 |
| ResNet50 | $MSE$ | **0.965** | ResNet50V2 | $MSE$ | 0.949 |
| ResNet101 | $MSE$ | **0.964** | ResNet101V2 | $MSE$ | 0.951 |
| ResNet152V2 | $MSE$ | 0.951 | MobileNet | $MSE$ | 0.936 |
| MobileNetV2 | $MSE$ | **0.964** | DenseNet121 | $MSE$ | **0.955** |
| DenseNet201 | $MSE$ | 0.943 | | | |

In this experiment we compare the performance of the modified regression models using Eq. 1 and Eq. 2 as mentioned in Sect. 5. The training of the models utilizes the polygon label only.

**Table 2.** The regression models results using $MinR_{L_2}$ and $MinR_{L_1}$ loss functions on AD dataset.

| Regression model | Loss function | Accuracy | Regression model | Loss function | Accuracy |
|---|---|---|---|---|---|
| Xception | $MinR_{L_1}$ | 0.843 | Xception | $MinR_{L_2}$ | 0.822 |
| ResNet50 | $MinR_{L_1}$ | 0.814 | ResNet50 | $MinR_{L_2}$ | 0.775 |
| ResNet101 | $MinR_{L_1}$ | 0.837 | ResNet101 | $MinR_{L_2}$ | 0.751 |
| MobileNetV2 | $MinR_{L_1}$ | 0.839 | MobileNetV2 | $MinR_{L_2}$ | 0.808 |
| DenseNet121 | $MinR_{L_1}$ | 0.850 | DenseNet121 | $MinR_{L_2}$ | 0.78 |

The results are shown in Table 2. As seen, Eq. 2 gives better results than Eq. 1 for the five models. It outperforms Eq. 1 by at least 2.5% for all the models. However, these two loss functions consider the four vertices independently, i.e. do not take into account the correlation among the vertices. To overcome this limitation, we integrate area difference in the loss functions.

The area difference, $d_{area}$, measures the size difference between the area of the ground truth and predicated quadrilaterals. We add $d_{area}$ to the $MinR_{L_2}$ and $MinR_{L_1}$ and get hybrid loss functions that consider the vertices independently and takes into account the area of the polygon they define. The loss functions we use are $Area\_MinR_{L_2}$ and $Area\_MinR_{L_1}$.

**Table 3.** The performance of regression models using $Area\_MinR_{L_2}$ and $Area\_MinR_{L_1}$

| Regression model | Loss function | Accuracy | Regression model | Loss function | Accuracy |
|---|---|---|---|---|---|
| Xception | $Area\_MinR_{L_1}$ | 0.863 | Xception | $Area\_MinR_{L_2}$ | 0.8303 |
| ResNet50 | $Area\_MinR_{L_1}$ | 0.825 | ResNet50 | $Area\_MinR_{L_2}$ | 0.816 |
| ResNet101 | $Area\_MinR_{L_1}$ | 0.840 | ResNet101 | $Area\_MinR_{L_2}$ | 0.818 |
| MobileNetV2 | $Area\_MinR_{L_1}$ | 0.8366 | MobileNetV2 | $Area\_MinR_{L_2}$ | 0.801 |
| DenseNet121 | $Area\_MinR_{L_1}$ | 0.8578 | DenseNet121 | $Area\_MinR_{L_2}$ | 0.8177 |

As shown in Table 3 adding the area difference to $MinR_{L_2}$ and $MinR_{L_1}$ improves the performance of all the models by 2% on average, expect MobileNetV2. The accuracy of MobileNetV2 deteriorates by less than 0.05%. We believe this is due to the lack of training data, as our dataset is not big enough.

$Area\_MinR_{L_2}$ and $Area\_MinR_{L_1}$ loss functions consider the vertices and the area separately. They capture the correlation between the vertices as a geometric shape; i.e. the predicted vertices aim at producing a shape, which area is equal to that of the ground truth.

The $Trapezoids_{loss}$ loss function aims at capturing the geometric correlation between the predicted vertices and the shape they form without separating the two, as we explained in Sect. 5. As shown in Table 4 the trapezoid loss function gives better results than $MinR_{L_2}$ for all the study models. In addition, it gives better results than $MinR_{L_1}$ for both Xception and MobileNetV2 models. This indicates its usability for extracting objects in a similar way to $MinR_{L_2}$ and $MinR_{L_1}$ with a focus on the object as a whole.

The $Trapezoids_{loss}$ loss function considers the shape as one entity, vertices and area together, thus restricting the independent movement of the vertices. To evaluate the role of $MinR_{L_1}$ in the independent movement of the vertices and the overall performance, we added the $MinR_{L_1}$ to the $Trapezoids_{loss}$ loss function. We refer to this loss function, as $Trapezoids\_MinR_{L_1}$.

As shown in Table 5, using $Trapezoids\_MinR_{L_1}$ outperform $Trapezoids_{loss}$ for all the study models, expect ResNet101. The accuracy of the remaining study

**Table 4.** The performance of the regression models using $Trapezoids_{loss}$

| Regression model | Loss function | Accuracy |
|---|---|---|
| Xception | $Trapezoids_{loss}$ | 0.8671 |
| ResNet50 | $Trapezoids_{loss}$ | 0.7949 |
| ResNet101 | $Trapezoids_{loss}$ | 0.7923 |
| MobileNetV2 | $Trapezoids_{loss}$ | 0.8461 |
| DenseNet121 | $Trapezoids_{loss}$ | 0.8068 |

**Table 5.** The performance of the regression models using $Trapezoids\_MinR_{L_1}$

| Regression model | Loss function | Accuracy |
|---|---|---|
| Xception | $Trapezoids\_MinR_{L_1}$ | 0.8674 |
| ResNet50 | $Trapezoids\_MinR_{L_1}$ | 0.8349 |
| ResNet101 | $Trapezoids\_MinR_{L_1}$ | 0.7852 |
| MobileNetV2 | $Trapezoids\_MinR_{L_1}$ | 0.840 |
| DenseNet121 | $Trapezoids\_MinR_{L_1}$ | 0.820 |

models increased by an average of 2%, which indicates that combing the vertices with the geometric shape loss function improves performance.

### 6.3   PTPNet

We refer to the rendering model composed of Fully-Connected (FC) layers as n-FCGenNet, where $n$ refers to the number of FC layers in the network. In this experiment we compare the performance of n-FCGenNet and GenPTP (see Sect. 4). We experiment with two n-FCGenNet instance models: 3-FCGenNet and 6-FCGenNet. The two models input a vertex vector and output a pixel vector, which represents a $256 \times 256$ mask. We choose the Dice coefficient loss function to train the rendering models, in which accuracy is measured using IOU and DICE.

Table 6 summarized the comparison of the three rendering models. As seen, the GenPTP outperforms the 3-FCGenNet and 6-FCGenNet models. The GenPTP trained using ads-quads outperforms the same model trained on the synthesized-quads dataset by %12 and %7 using IOU and DICE metrics, respectively. We train GenPTP separately and combine it with the Regressor.

**Table 6.** The performance of different rendering models

| Render model | Dataset | IOU | DICE |
|---|---|---|---|
| GenPTP | ads-quads | 91.18 | 95.33 |
| GenPTP | synthesized-quads | 79.64 | 88.21 |
| 3-FCGenNet | ads-quads | 59.73 | 73.3 |
| 3-FCGenNet | synthesized-quads | 49.23 | 63.38 |
| 6-FCGenNet | ads-quads | 75.59 | 85.95 |
| 6-FCGenNet | synthesized-quads | 65.4 | 78.15 |

**Table 7.** Comparing the accuracy of PTPNet and MaskRCNN

| Model | Loss function | Accuracy |
|-------|---------------|----------|
| PTPNet | $PTPNet\_loss$ | 0.851 |
| MaskRCNN | $MaskRCNNLossfunction$ | 0.810 |

We decided to compare the performances of PTPNet with MaskRCNN. As shown in Table 7, PTPNet outperforms MaskRCNN by 4%. The performance of PTPNet is similar to the top regression model, which is the Xception architecture with $Trapezoids\_MinR_{L_1}$ loss function. However, PTPNet is easier to generalize for any polygon since its loss function does not assume a prior geometric shape. In addition, having the mask and the polygon in the training phase enables handling more complex tasks.

# 7 Conclusion

In this paper, we presented various deep learning models with different loss functions for advertisement extraction. We introduce AD dataset, which is a dataset of quadrilateral advertisements. Modified versions of several regression models are explored and their performances are studied using various loss functions. We use $L_2$ and $L_1$ loss functions and add the area difference to improve performance. We introduce the trapezoid loss function that considers the vertices as representatives of a shape instead of focusing on the predicted vertices independently and show that adding $L_1$ loss to trapezoid loss gives the best results in most of the modified regression models. In addition, we introduce the PTPNet model with its own loss function that combines the results of a rendering model and a regression model. We conduct an extensive experimental study to evaluate the performance of common deep learning models in extracting advertisements from images and compare their performance with our proposed model. We show that our proposed model manages to extract advertisements at high accuracy and outperforms common deep learning models. The scope of future work includes extending our approach to handle general polygons.

**Acknowledgment.** This research was supported in part by Frankel Center for Computer Science at Ben-Gurion University of the Negev. One of the authors, Reem Alaasam, is a fellow of the Ariane de Rothschild Women Doctoral Program, and would like to thank them for their support.

# References

1. Acuna, D., Ling, H., Kar, A., Fidler, S.: Efficient interactive annotation of segmentation datasets with polygon-RNN++. In: Proceedings of the IEEE Conference on Computer Vision and Pattern Recognition, pp. 859–868 (2018)

2. Bauchet, J.P., Lafarge, F.: KIPPI: kinetic polygonal partitioning of images. In: Proceedings of the IEEE Conference on Computer Vision and Pattern Recognition, pp. 3146–3154 (2018)
3. Castrejon, L., Kundu, K., Urtasun, R., Fidler, S.: Annotating object instances with a polygon-RNN. In: Proceedings of the IEEE Conference on Computer Vision and Pattern Recognition, pp. 5230–5238 (2017)
4. Cheng, M.M., Mitra, N.J., Huang, X., Torr, P.H., Hu, S.M.: Global contrast based salient region detection. IEEE Trans. Pattern Anal. Mach. Intell. **37**(3), 569–582 (2014)
5. Chollet, F.: Xception: deep learning with depthwise separable convolutions. In: Proceedings of the IEEE Conference on Computer Vision and Pattern Recognition, pp. 1251–1258 (2017)
6. Deng, J., Dong, W., Socher, R., Li, L.J., Li, K., Fei-Fei, L.: ImageNet: a large-scale hierarchical image database. In: 2009 IEEE Conference on Computer Vision and Pattern Recognition, pp. 248–255. IEEE (2009)
7. Duan, L., Lafarge, F.: Image partitioning into convex polygons. In: Proceedings of the IEEE Conference on Computer Vision and Pattern Recognition, pp. 3119–3127 (2015)
8. Girard, N., Tarabalka, Y.: End-to-end learning of polygons for remote sensing image classification. In: IGARSS 2018-2018 IEEE International Geoscience and Remote Sensing Symposium, pp. 2083–2086. IEEE (2018)
9. He, K., Gkioxari, G., Dollár, P., Girshick, R.: Mask R-CNN. In: Proceedings of the IEEE International Conference on Computer Vision, pp. 2961–2969 (2017)
10. He, K., Zhang, X., Ren, S., Sun, J.: Deep residual learning for image recognition. In: Proceedings of the IEEE Conference on Computer Vision and Pattern Recognition, pp. 770–778 (2016)
11. Howard, A.G., et al.: MobileNets: efficient convolutional neural networks for mobile vision applications. arXiv preprint arXiv:1704.04861 (2017)
12. Huang, G., Liu, Z., Van Der Maaten, L., Weinberger, K.Q.: Densely connected convolutional networks. In: Proceedings of the IEEE Conference on Computer Vision and Pattern Recognition, pp. 4700–4708 (2017)
13. Karras, T., Aila, T., Laine, S., Lehtinen, J.: Progressive growing of GANs for improved quality, stability, and variation. arXiv preprint arXiv:1710.10196 (2017)
14. Lateef, F., Ruichek, Y.: Survey on semantic segmentation using deep learning techniques. Neurocomputing **338**, 321–348 (2019)
15. Levinshtein, A., Sminchisescu, C., Dickinson, S.: Optimal contour closure by superpixel grouping. In: Daniilidis, K., Maragos, P., Paragios, N. (eds.) ECCV 2010. LNCS, vol. 6312, pp. 480–493. Springer, Heidelberg (2010). https://doi.org/10.1007/978-3-642-15552-9_35
16. Li, Z., Wegner, J.D., Lucchi, A.: Topological map extraction from overhead images. In: Proceedings of the IEEE International Conference on Computer Vision, pp. 1715–1724 (2019)
17. Lin, T.-Y., et al.: Microsoft COCO: common objects in context. In: Fleet, D., Pajdla, T., Schiele, B., Tuytelaars, T. (eds.) ECCV 2014. LNCS, vol. 8693, pp. 740–755. Springer, Cham (2014). https://doi.org/10.1007/978-3-319-10602-1_48
18. Long, J., Shelhamer, E., Darrell, T.: Fully convolutional networks for semantic segmentation. CoRR abs/1411.4038 (2014). http://arxiv.org/abs/1411.4038
19. Ren, S., He, K., Girshick, R.B., Sun, J.: Faster R-CNN: towards real-time object detection with region proposal networks. CoRR abs/1506.01497 (2015). http://arxiv.org/abs/1506.01497

20. Rother, C., Kolmogorov, V., Blake, A.: "GrabCut" interactive foreground extraction using iterated graph cuts. ACM Trans. Graph. (TOG) **23**(3), 309–314 (2004)
21. Sandler, M., Howard, A., Zhu, M., Zhmoginov, A., Chen, L.C.: MobileNetV2: inverted residuals and linear bottlenecks. In: Proceedings of the IEEE Conference on Computer Vision and Pattern Recognition, pp. 4510–4520 (2018)
22. Wang, L., Wang, L., Lu, H., Zhang, P., Ruan, X.: Saliency detection with recurrent fully convolutional networks. In: Leibe, B., Matas, J., Sebe, N., Welling, M. (eds.) ECCV 2016. LNCS, vol. 9908, pp. 825–841. Springer, Cham (2016). https://doi.org/10.1007/978-3-319-46493-0_50
23. Yu, J., Jiang, Y., Wang, Z., Cao, Z., Huang, T.: UnitBox: an advanced object detection network. In: Proceedings of the 24th ACM International Conference on Multimedia, pp. 516–520 (2016)

# Detection and Localisation
# of Struck-Out-Strokes
# in Handwritten Manuscripts

Arnab Poddar[1]([✉]), Akash Chakraborty[2], Jayanta Mukhopadhyay[3],
and Prabir Kumar Biswas[2]

[1] Advanced Technology Development Centre, Indian Institute of Technology
Kharagpur, Kharagpur, India
arnabpoddar@iitkgp.ac.in
[2] Department of Electronics and Electrical Communication Engineering,
Indian Institute of Technology Kharagpur, Kharagpur, India
pkb@ece.iitkgp.ac.in
[3] Department of Computer Science and Engineering, Indian Institute of Technology
Kharagpur, Kharagpur, India
jay@cse.iitkgp.ac.in

**Abstract.** The presence of struck-out texts in handwritten manuscripts adversely affects the performance of state-of-the-art automatic handwritten document processing systems. The information of struck-out words (STW) are often important for real-time applications like handwritten character recognition, writer identification, digital transcription, forensic applications, historical document analysis etc. Hence, the detection of STW and localisation of struck-out strokes (SS) are crucial tasks. In this paper, we introduce a system for simultaneous detection of STWs and localisation of the SS using a single network architecture based on Generative Adversarial Network (GAN). The system requires no prior information about the type of SS stroke and it is also able to robustly handle variant of strokes like straight, slanted, cris-cross, multiple-lines, underlines and partial STW as well. However, we also present a methodology to generate STW with high variability of SS for network learning. We have evaluated the proposed pipeline on publicly available IAM dataset and also on struck-out words collected from real-world writers with high variability factors like age, gender, stroke-width, stroke-type etc. The evaluation metrics show robustness and applicability in real-world scenario.

**Keywords:** Handwritten document image · Struck-out words · Handwritten manuscripts · Generative Adversarial Networks (GAN)

The work is partially supported by the project entitled as "Information Access from Document Images of Indian Languages" sponsored by IMPRINT, MHRD, Govt. of INDIA.

# 1   Introduction

Automatic handwritten text recognition is a prime topic of research in digital document image analysis [20,23,24]. A prime limitation in most published reports on handwriting recognition is that they consider the documents containing no writing error or struck-out texts. However, a free-form handwritten manuscript often contains struck-out words. Some typical examples of struck-out words (STW) are shown in Fig. 1. Such word produces nonsense output in optical character recognition (OCR) or writer verification framework [12]. To tackle the problem, we need an automatic module to identify STW and analyse further if required.

The meta-information of STWs are often important for real-time applications like writer identification, digital transcriptions, handwritten character recognition, etc. The recent digital transcriptions of famous writers like G. Washington, J. Austen, H. Balzac, G. Flaubert etc. use annotations of the STWs [1,3,8,10]. In forensic applications also, a quick and automatic detection of struck-out texts and examining their patterns may provide important clues like behavioral and psychological pattern of the suspect [13] and mentally challenged patients [14]. Automatic detection and localisation of STWs and strokes may be helpful in such cases as well.

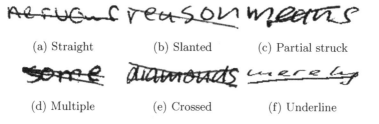

(a) Straight          (b) Slanted          (c) Partial struck

(d) Multiple          (e) Crossed          (f) Underline

**Fig. 1.** Some examples of striked-through words with various kinds of strokes.

A few previous approaches dealt with the detection of struck-out texts. Tuganbaev and Deriaguine [26] registered a US patent for their crossed-out character recognizer using a feature-based classifier. The work in [19] presented a HMM (Hidden Markov Model) based crossed-out word recognition. A graph-based solution for detection of STW and localisation of SS from handwritten manuscripts was reported in [11]. More recently, a modified version of [11] was presented in [15], where morphological and graph based features were extracted from STW which was followed by an SVM classifier. In [12], a CNN-SVM based approach is proposed to detect STW in context of writer identification.

The approach presented in literature use separate modules for detection of SS and localisation of SS which allows error propagation from detection module to localisation modules [11,12,15]. The works are also dependent on prior information about the type of the SS and prescribe rule-based solutions for each type of SS like straight, slanted, crossed, etc. which attracts manual intervention [11,12,15]. In this paper, we present an earliest attempt to tackle both the problems i.e., detection of STWs and localising the SS simultaneously without prior knowledge of the type of stroke. We use a single network architecture based on Generative Adversarial Network (GAN) [17,22] for localising the struck-out region. Further the system extracts concatenated features from the input image and the image localising the struck-out region. Finally, an SVM based classifier is used to classify between clean and struck-out word-images. The system uses no additional computation for localisation after the detection of SS. The system robustly handles different variant of strokes like straight, slanted, cris-cross, multiple-lines etc. and is also able to detect partially STWs. The system takes the handwritten word image $I_{HW}$ and generates a mask image $I_{SS}$ localising the potential SS. The $I_{SS}$ is further used for detecting whether $I_{HW}$ is a STW or a clean word. A simple block diagram of the proposed system is shown in Fig. 2. Since we require training samples of STWs to train deep learning architectures, we also present a technique to generate struck out words of various kind. We use the proposed pipeline on publicly available IAM dataset [21] and achieved encouraging results. The system is also tested on struck-out words from real world scenario which indicates the robustness and applicability of the system in challenging variability conditions.

The contribution of the paper are listed as follows

1. A single network architecture solution for simultaneous detection of struck-out words and localisation of strike-through stroke in handwritten manuscripts.
2. The proposed system requires no prior information about the type of strokes and is also able to handle partially struck-out handwritten words.
3. We introduce methods to generate STWs and corresponding ground truths from clean handwritten words and hand-drawn stroke templates.

The rest of the paper is organized as follows. The details of the proposed method are described in Sect. 2. The experimental set-up and datasets are described in Sect. 3. Then the experimental results are presented in Sect. 4. Finally, Sect. 5 summarizes the pros and cons of this approach and suggests the scope of future work.

## 2    Proposed Methodology

We propose the localisation of struck-out regions of the input word image and the detection of strike-through or clean word-image in two steps. Initially, the input image is passed through a localisation network, where the struck-out regions (if any) of the input word-image is localised. Further few simple features are calculated from input image and the corresponding struck-out region localised image. Further an SVM classifier is used to discriminate between clean words and words with struck-out strokes. Figure 2 shows an overall workflow of the proposed system.

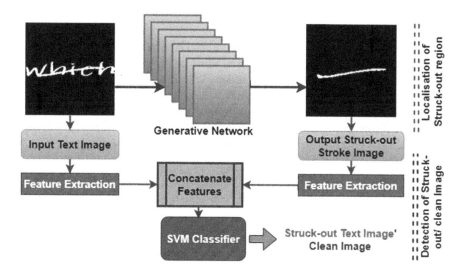

**Fig. 2.** Block diagram for the workflow of proposed system

We conceptualise the problem of localising the struck-out-stroke in handwritten word as an image to image translation problem, where we generate the SS image from a given image of a handwritten word. We directly feed the segmented word image to the generator network irrespective of the presence of SS (Fig. 2). The network is supposed to generate a mask image $I_{SS}$ localising SS. Our generator model learns a mapping function from handwritten word images to their corresponding SS. The proposed network is expected to generate a uniform black image ideally for clean handwritten word input. The loss function used in the network simultaneously uses data loss, measured from mean square error and the structural loss (SSIM) of $I_{SS}$ and expected ground-truth outcome $I_{GT}$. The detector of the adversarial model is implemented such that it differentiates between the real struck-out-strokes and its fake counterparts.

The generated $I_{SS}$ is used for detection of STW. For this, the foreground pixels of the mask are considered which also appear as foreground in input image. We then consider the largest axis parallel bounding box of the connected components in the mask and the features of the contour is considered for detection of struck-out and clean words. The extracted features are finally fed into an SVM for the final decision of the clean and struck-out word (Fig. 2).

## 2.1 GAN Preliminary

In our proposed system, we use conditional GANs to learn a mapping from observed condition image $x$ and random noise vector $z$, to $y$, $G : x, z \longrightarrow y$ [17,22]. The generator network $G$ is trained to deliver outputs that are not differentiable from "real" images by its adversary discriminator, $D$.

## 2.2  Objective

In this work, we design the network to generate the SS image $I_{SS}$ from a handwritten word image $I_{HW}$. First, we crop the word image from IAM dataset. Then, we resize binarised handwritten word and pad zeros to fit into size $128 \times 128$ as an input to the generator network. The discriminator network $D$ acts as adversary to detect fake samples. Both the ground truth of real SS images ($I_{GT}$) and the generated ones ($I_{SS}$) are fed into discriminator $D$. During training, the discriminator $D$ enforces the generator to produce realistic images SS. The objective function of the GAN network, conditioned on the input handwritten word images, can be expressed as:

$$\mathcal{L}_{cGAN}(G, D) = \mathbb{E}_{I_{HW} \sim p_{train}(I_{HW}), I_{GT} \sim p_{train}(I_{GT})}[\log D(I_{HW}, I_{GT})]$$
$$+ \mathbb{E}_{I_{HW} \sim p_{train}(I_{HW}), z}[\log(1 - D(I_{HW}, G(I_{HW}, z)))] \quad (1)$$

where $G$ tries to minimize this objective against an adversarial $D$ that tries to maximize it, i.e.

$$G* = \arg\min_G \max_D \mathcal{L}_{cGAN}(G, D) \quad (2)$$

We mix the GAN objective with a more traditional $\ell_1$ loss, which encourages less blurring:

$$\mathcal{L}_{\ell 1}(G) = \mathbb{E}_{I_{HW} \sim p_{train}(I_{HW}), I_{GT} \sim p_{train}(I_{GT})}[\| I_{GT} - G(I_{HW}) \|_1] \quad (3)$$

We incorporate noise as dropout, applied on several layers of our generator in both training and evaluation phase. Here, we are interested to generate the struck-out strokes having structural similarity with ground-truth SS. Contrary to the $\ell 1$ loss, the structural similarity (SSIM) index provides a measure of the similarity by comparing two images based on luminance, contrast and structural similarity information. In our case, the handwritten word images are supposed to contain the SS inside the word. Hence structural similarity may play a vital role. Hence we also introduce the SSIM loss in the objective of the network.

For two images $I^{HW}$ and $I^{GT}$, the luminescence similarity $l(I^{HW}, I^{GT})$, contrast similarity $c(I^{HW}, I^{GT})$ and structural similarity $s(I^{HW}, I^{GT})$ are calculated and the overall structural loss for the generator can be defined as

$$\mathcal{L}_{SSIM}(G) = 1 - \mathbb{E}_{I_{HW} p_{train}(I_{HW}), I_{GT} p_{train}(I_{GT}), z}[l(I^{HW}, I^{GT})$$
$$\times c(I^{HW}, I^{GT}) \times s(I^{HW}, I^{GT})] \quad (4)$$

Our final objective is

$$G* = \arg\min_G \max_D \mathcal{L}_{cGAN}(G, D) + \lambda \mathcal{L}_{\ell 1}(G) + \alpha \mathcal{L}_{SSIM}(G) \quad (5)$$

where $\lambda$ and $\alpha$ are ratio control parameter for data-loss and structure-loss respectively.

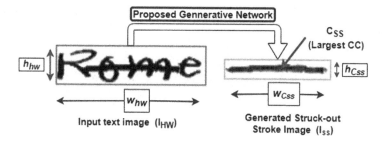

**Fig. 3.** An Example of localisation of struck-out stroke using proposed generative architecture. It shows various parameters for extraction of features from struck-out stroke image and input image for detection of struck-out and clean words.

## 2.3   Generator Architecture

The work adopts network architectures from Johnson et al. [18]. Here, 9 residual blocks are used in the generator network to train $128 \times 128$ size images. Let, c7s1-k denote a $7 \times 7$ Convolution-InstanceNorm-ReLU layer with k filters and stride 1. The parameter dk denotes a $3 \times 3$ Convolution-InstanceNorm-ReLU layer with k filters and stride 2. Reflection padding was used to reduce artifacts. Whereas Rk denotes a residual block that contains two $3 \times 3$ convolutional layers with the same number of filters on both layers. uk denotes a $3 \times 3$ fractional-strided-Convolution-InstanceNorm-ReLU layer with k filters and stride $\frac{1}{2}$. The network with 9 residual blocks consists of:

   c7s1-64, d128, d256, R256, R256, R256, R256, R256, R256, R256, R256, R256, u128, u64, c7s1-3

## 2.4   Discriminator Architecture

For discriminator networks, $70 \times 70$ PatchGAN is used [16]. Let Ck denote a $4 \times 4$ Convolution-InstanceNorm-LeakyReLU layer with $k$ filters and stride 2. After the last layer, a convolution layer is applied to produce a 1-dimensional output. The proposed system does not use InstanceNorm for the first c64 layer. It uses leaky ReLUs with a slope of 0.2. The discriminator architecture is: C64, C128, C256, C512.

## 2.5   Detection of Struck-Out Word

We use the generated mask images ($I_{SS}$) and input image $I_{HW}$ to train an SVM for detection of STWs or clean words. We compute features from both $I_{HW}$ and ($I_{SS}$) an d concatenate them to train an SVM classifier for detection of STws and clean words. We consider the foreground pixels which are also present in $I_{HW}$. The residual pixels in $I_{SS}$ are discarded as noise and a new image $I'_{SS}$ is generated. We compute simple feature-vector from $I'_{SS}$ to train the SVM for detection of STW. Furthermore, the largest connected component

$C_{I'_{SS}}$ is selected from $I'_{SS}$. The pixel-width $w_{ss}$ from leftmost foreground pixel to rightmost foreground pixel of $I'_{SS}$ is calculated. The horizontal and vertical pixel-span $w_{c_{ss}}$ and $h_{c_{ss}}$ of contour $C_{I'_{SS}}$ is calculated to extract features. Furthermore, from image $I_{HW}$ horizontal and vertical pixel-span of foreground pixels $w_{hw}$ and $h_{hw}$ are also calculated. An illustrative diagram is shown in Fig. 3 describing the procedure to extract features from $I'_{SS}$ images for detection of struck-out words. We consider the following features from $C_{I'_{SS}}$, $I_{SS}$ and $I_{HW}$ to formulate SVM feature vector for detection:

1. The axis parallel bounding box and contour area of $C_{I'_{SS}}$
2. The ratio $w_{c_{ss}}/w_{ss}$ and $w_{c_{ss}}/w_{hw}$
3. The values of $h_{c_{ss}}$, $h_{hw}$, $h_{ss}$
4. set all values to 0 if $C_{I'_{SS}}$ is $NULL$.

Thus we train and test the SVM for detection of struck-out images from the features mentioned in above list.

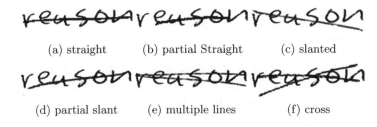

(a) straight        (b) partial Straight        (c) slanted

(d) partial slant        (e) multiple lines        (f) cross

**Fig. 4.** Various types of STWs generated from single clean word.

## 3    Database and Experimental Set-Up

The previous approaches mostly used classical image processing and machine learning techniques and used private databases with limited examples [11,12,15]. However, deep learning based methods require large number of training samples with high variability of writer, writing style, word-length, word-content, stroke-width etc. However there is no publicly available database to deal with STW in handwritten manuscripts. Here, we use original clean words from IAM database [21], and simultaneously collect separate stroke images various types like straight, slanted, cross, etc. A total of 2400 hand-drawn strokes are collected to generate the STWs for training and separate 1230 hand-drawn strokes are collected for testing. We have used 81412 handwritten word-images for training and 22489 word-images for testing the network for the task of SS localisation.

## 3.1  Generation of Struck-Out Words

The STWs are generated using the words from IAM dataset and collected hand-drawn strokes. Both the inputs, i.e., the word-images and stroke-images are taken from separate pool for training and testing. We describe the struck-out word generation procedure in Algorithm 1.

---

**Algorithm 1:** Generation of struck-out words

---

**Result**: Struck-out word

1. Compute height ($h$) and width ($w$) of input word;
2. Select a pool $P$ of hand-drawn strokes having width within the range $w \pm r\%$;
3. Select randomly a stroke ($s$) from the pool $P$;
4. Give a rotation to $s$ of degree $\pm\theta$;
5. Perturb the vertical center of $s$ randomly within the range $\pm hg\%$ of the word-height $h$.;
6. Superimpose the stroke $s$ on the word.;

---

In word generation, we introduce randomised rotation ($\pm\theta$), shift of vertical centre of stroke ($\pm hg\%$), and select random strokes from $P$ in Algorithm 1 to ensure different variability like slant of strokes, vertical and horizontal position, type variability etc. We have varied these parameters $r\%, \pm\theta, \pm hg\%$ to generate various types of strokes like straight, slanted, cris-cross, underline etc. The strokes are used in a way to generate both full and partial STWs. The generation technique allows to generate various types of STWs using single clean word as shown in Fig. 4.

## 4  Experimental Results and Discussion

The models are trained in mixed setting with equal proportion of straight, slanted, partial straight, partial slanted, Multiple strike and crossed strokes with a total of 81412 STWs. The single trained model is used for testing with various types of strokes separately and also with mixed type accumulating all types of STWs. The control parameters $\lambda, \alpha$ used in Eq. 5 are given the values 50 for both and the learning rate is used 0.0002 for training.

### 4.1  Localisation of Struck Out Stroke

Localisation result of STWs using our proposed system is shown in Fig. 8, 9, 10, 11 and 12. Figure 13 shows the effectiveness of our proposed method also on underline strokes. We consider the generated $I'_{SS}$ and use foreground pixels count in SS for performance evaluation. We compute precision (P), recall (R) and F-measure (FM) for each $I_{SS}$ from input $I_{HW}$, and finally take the average (harmonic mean) of them. The performance metrics are measured in pixel-to-pixel setting. For a strike-through component true positive (TP), false positive (FP) and false negative (FN) are measured as follows:

**Table 1.** Performance comparison of objective functions: $\ell 1, SSIM$ $and$ $\ell 1 + SSIM$ on STW with straight strokes only

| Word length | Precision (%) | | | Recall (%) | | | F1-Score | | |
|---|---|---|---|---|---|---|---|---|---|
| | $\ell 1$ | SSIM | $\ell 1$ + SSIM | $\ell 1$ | SSIM | $\ell 1$ + SSIM | $\ell 1$ | SSIM | $\ell 1$ + SSIM |
| 1 | 73.11 | 74.97 | 83.47 | 75.76 | 77.59 | 75.39 | 74.41 | 76.26 | **79.22** |
| 2 | 83.58 | 80.10 | 87.51 | 86.12 | 84.56 | 85.25 | 84.83 | 82.27 | **86.37** |
| 3 | 88.07 | 84.71 | 91.64 | 83.96 | 79.21 | 82.31 | 85.97 | 81.87 | **86.73** |
| 4 | 91.00 | 87.80 | 93.40 | 85.09 | 79.07 | 82.11 | 87.94 | 83.21 | **87.39** |
| 5 | 92.41 | 88.98 | 94.26 | 87.20 | 80.46 | 83.86 | 89.73 | 84.50 | **88.76** |
| $\geq 6$ | 90.81 | 86.88 | 93.68 | 89.94 | 84.24 | 88.51 | 90.37 | 85.54 | **91.02** |
| All | 89.91 | 86.38 | 92.86 | 87.19 | 81.78 | 85.17 | 88.53 | 84.02 | **88.85** |

**Table 2.** Results of localisation of struck-out region by the proposed system on IAM dataset. The network localising the struck-out region is trained with mixed type of struck-out words such as straight, partial straight, slanted, partial slanted, crossed, and multiple in equal proportion. Separate performance evaluation on different types of struck-out words are reported below.

| Stroke type | Precision (%) | Recall (%) | F1 score |
|---|---|---|---|
| Straight | 92.86 | 85.17 | 88.85 |
| Partial straight | 89.95 | 89.18 | 89.52 |
| Slanted | 96.54 | 84.60 | 90.17 |
| Partial slanted | 92.01 | 87.65 | 89.78 |
| Crossed | 95.51 | 82.72 | 88.65 |
| Multiple | 95.33 | 81.21 | 87.70 |
| Mixed | 93.97 | 83.92 | 88.66 |

- **TP**: number of black (object) pixels in SS (correctly classified),
- **FP**: number of black pixels those are incorrectly labeled as SS (unexpected),
- **FN**: number of black pixels of SS those are not labeled (missing result).

Table 1 presents the comparison of performance of localisation of SS with data loss $\mathcal{L}_{\ell 1}(G)$, structural loss $\mathcal{L}_{SSIM}(G)$ and fusion of both. However, Table 1 also depicts the performance measures in varying character length (1, 2, 3, 4, 5, greater than 5), which show more robust and reliable performance to spot the stroke for larger character length. Fusing SSIM loss information with $\ell 1$ loss improves performance of localisation of SS. The overall performance of the network is presented for various types of STWs in Table 2. We obtain consistent performance for various types of struck-out like straight, slanted, multiple strokes, crossed strokes, mixed etc. The system is also found to perform consistently well on partial STWs.

## 4.2   Detection of Strike-Out Textual Component

For detection of STWs we compute features from input image and $I_{HW}$ generated $I'_{SS}$ and concatenate them for classification. We compute the features

**Table 3.** Results of detection performance of struck-out and clean words on IAM dataset.

| Stroke type | Precision (%) | Recall (%) | F1 score |
|---|---|---|---|
| Straight | 98.60 | 99.37 | 98.98 |
| Partial straight | 96.40 | 98.03 | 97.21 |
| Slanted | 86.20 | 98.31 | 91.86 |
| Partial slanted | 95.56 | 97.85 | 96.74 |
| Crossed | 97.12 | 98.02 | 97.57 |
| Multiple | 95.81 | 97.24 | 96.52 |
| Mixed | 95.78 | 97.11 | 96.44 |

**Table 4.** Performance comparison of detection of struck-out words between proposed system and Lenet-SVM network [12] on IAM dataset

| System type | Precision (%) | Recall (%) | F1 score |
|---|---|---|---|
| Lenet-SVM [12] | 88.60 | 76.98 | 82.65 |
| Proposed | 95.78 | 97.11 | **96.44** |

for training and testing of SVM as presented in Sect. 2.5. The region localised images, i.e., $I_{SS}$ are used to evaluate the detection performance. We subdivide the test set of localisation task, i.e., 22489 images into 70:30 ratio for training and testing the SVM for detection of struck-out and clean words. The performance metrics for detection performance for various strokes are presented in Table 3. We obtained very high values of accuracy for detection upto 98.93% for straight strokes and 97.31% for mixed strokes. We consistently obtained significant performance for other variant of strokes as depicted in Table 3 which indicates robustness of the system in different types of STWs (Figs. 5, 6 and 7).

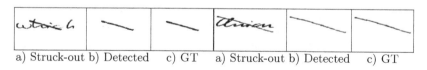

a) Struck-out b) Detected    c) GT    a) Struck-out b) Detected    c) GT

**Fig. 5.** Generated output images with partially-slanted (above left) and fully slanted (above right) strokes.

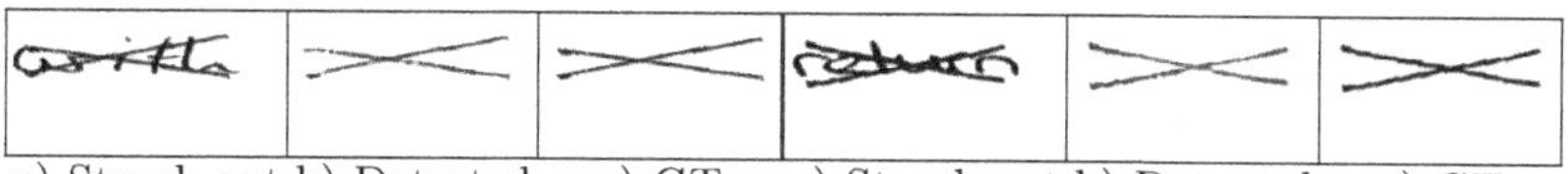

a) Struck-out b) Detected    c) GT    a) Struck-out b) Detected    c) GT

**Fig. 6.** Generated output images cross struck-out-strokes

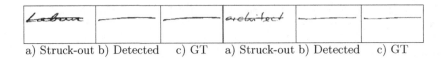

a) Struck-out  b) Detected      c) GT      a) Struck-out  b) Detected      c) GT

**Fig. 7.** Generated output images straight struck-out strokes

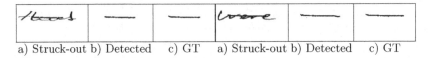

a) Struck-out  b) Detected      c) GT      a) Struck-out  b) Detected      c) GT

**Fig. 8.** Generated output images partially-straight struck-out strokes

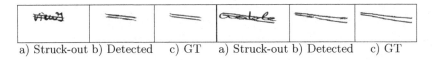

a) Struck-out  b) Detected      c) GT      a) Struck-out  b) Detected      c) GT

**Fig. 9.** Generated output images multiple struck-out strokes

### 4.3    Performance Comparison

In our proposed framework, we have encountered two tasks i.e., localisation of struck-out region of a word image and detection of struck-out and clean words. We have uniquely proposed a struck-out region localisation network and reported the performance metrics pixel-wise. However, the work in [12] and [15] reported struck-out word detection performance in a privately prepared data-set. The work in [12] uses a Lenet-SVM based deep-network architecture for detection performance. Here we compare the detection performance of the proposed system with that of [12] in publicly available IAM dataset [21]. The struck-out words are generated as described in Sect. 3.1. We have used word images from IAM data-set in 70:30 ratio for training and testing respectively. The struck-out words and clean words are both present in 50:50 ratio for evaluation. We have compared the performance of proposed system with that of the Lenet-SVM framework as in [12] with mixed type of struck-out words. The mixed type of struck-out words include struck-out stroke types like straight, partial straight, slanted, partial slanted, crossed, multiple as described in previous section. The Table 4 shows that the proposed system performs significantly better than the state-of-the-art system in terms of all precision, recall and F-1Score. As the performance is measured on widely used and publicly available IAM database [21], the system shows significant performance in presence of challenges like writer variability, age variability, cursiveness, etc.

### 4.4    Performance on Real Word Images

The results in Table 3 depict the performance of the proposed system on IAM database. Here we intent to show the performance of the system on the image

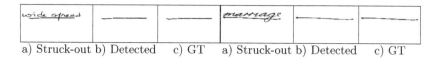

a) Struck-out  b) Detected     c) GT     a) Struck-out  b) Detected     c) GT

**Fig. 10.** Generated output images underline strokes

**Table 5.** Results for detection performance of struck-out and clean words on real unconstrained handwritten data-set.

| Training data | Test data | Precision (%) | Recall (%) | F1 score |
|---|---|---|---|---|
| IAM | Real struck-out | 87.70 | 99.39 | 93.18 |

collected from real world to evaluate the robustness of the proposed system. The IAM database inherently contains significant variability in terms of writing style, age of the writer, gender, texture of the page, ink of writing, stroke-width etc. The proposed struck-out region localisation network is trained on struck-out words from IAM data-set. However it would be informative to evaluate the performance of the proposed system on other data, collected from real-world writers. We have collected English handwriting from 45 individuals. The writers were of both the genders in the age group of 19–56 years with various regional and spoken-language background. Each writer provided 1 full page of handwriting sample. The content of handwriting is selected independently by the writers and written in running hand. The writers were requested to strike-through some of the words in their running handwriting style. The writers were given an A4 sized 75 GSM $(g/m^2)$ white paper and instructed to use any pen of their choice with black or blue colored ink. Thus we collected 443 unconstrained various types of struck-out words. We also collected 1661 clean words to measure the detection performance. The Table 5 presents the performance metrics on the collected real data.

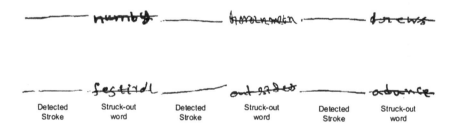

| Detected Stroke | Struck-out word | Detected Stroke | Struck-out word | Detected Stroke | Struck-out word |

**Fig. 11.** Performance on real struck-out word images

The proposed struck-out region localisation network is trained on mixed simulated struck-out images as mentioned in previous section. The test results are obtained from collected real manuscripts of the writers. Figure 11 depicts few examples on real handwritten struck-out word images. The collected test dataset

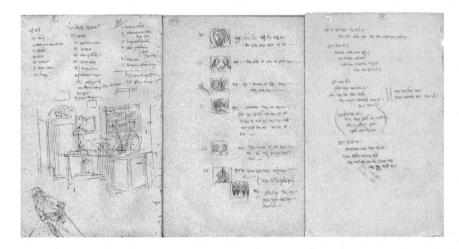

**Fig. 12.** Unconstrained offline document images from Satyajit Ray's movie-scripts of Goopi Gyne Bagha Byne and Apu Triology.

contains various types of struck-out words in running hand. We have obtained a high precision and f1-score on real world scenario. The proposed system seem to work robustly on the real world scenario.

### 4.5    Performance on Satyajit Ray's Manuscript

Here we present the result-images of struck-out strokes on Film-maker/writer Satyajit Ray's [2,7] manuscripts of his movie-scripts. We collected the scanned images of Satyajit Ray's movie-manuscripts with consent from National Digital Library of India [6,9]. The struck out words are collected from the script of the movies like 'Goopi Gayen Bagha Bayen' [4] and 'Apu Triology' [5]. Few pages of Satyajit Ray's manuscript from aforementioned movie-scripts are shown in the Fig. 12. The struck-out word-images are collected from annotated pages using MultiDIAS Annotation Tool [25].

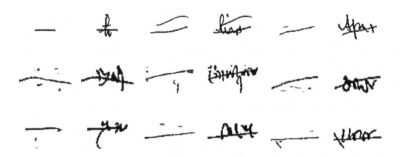

**Fig. 13.** Result images from Satyajit Ray's movie scripts of Goopi Gyne Bagha Byyne and Apu Triology on Both English (top row) and Bengali (middle and bottom row) language.

Satyajit Ray's manuscript images are written in nearly 1960–1980. The manuscript contains highly cursive words. The texture of the page and pen ink is quite different from that of images from IAM data-set. Our system is trained on IAM data-set with mixed types of struck-out words. Here we have shown the localisation of struck-out regions by the proposed localisation network. However, we have presented few result images on various environment to evaluate the robustness of the system. The resultant struck-out strokes are obtained from manuscript of Satyajit Ray as shown in Fig. 13. The performance is displayed on both cursive English and Bengali script. This shows that our system is useful in various scripts.

## 5    Conclusion

We present a single network architecture solution for simultaneous detection of STW and localisation of SS in handwritten manuscripts. It is the earliest attempt where the proposed system requires no prior information about the type of strokes and is also able to handle partially struck-out handwritten words. The experimentation is done on wide variability of SS types like straight, slanted, multiple strikes, underlines, crossed etc. We observed very high performance metrics for both detection of STWs and localisation of SS. The robustness and applicability of the proposed system is evaluated with real-time free form handwritten manuscripts. In future, the work can be extended with improvements in architectures and with other challenging and complex strike through texts. Further the localisation of struck-out regions can be used to reconstruct the clean word from struck-out word.

## References

1. Bibliothèque de Rouen (Rouen Library), Rouen Cedex-76043, France. http://www. bovary.fr. Accessed 19 July 2008
2. Ray, S.: https://en.wikipedia.org/wiki/Satyajit_Ray. Accessed 22 Feb 2021
3. The Morgan Library & Museum, New York, USA-10016. http://www.themorgan. org. Accessed 19 July 2008
4. A brief about the movie Goopi Gyne Bagha Byne. https://en.wikipedia.org/wiki/ Goopy_Gyne_Bagha_Byne. Accessed 22 Feb 2021
5. A brief about the movie series Apu Triology. https://en.wikipedia.org/wiki/The_ Apu_Trilogy. Accessed 22 Feb 2021
6. A brief description about 'National Digital Library of India' on Wikipedia. https:// en.wikipedia.org/wiki/National_Digital_Library_of_India. Accessed 22 Feb 2021
7. A brief description of 'Satyajit Ray' by Satyajit Ray Organisation. https:// satyajitray.org/. Accessed 22 Feb 2021
8. George Washington Papers, The Library of Congress, USA. http://memory.loc. gov/ammem/gwhtml/gwhome.html. Accessed 19 July 2008
9. National Digital Library of India. https://www.ndl.gov.in/. Accessed 22 Feb 2021
10. Queensland State Archive, Australia-4113. http://www.archivessearch.qld.gov.au. Accessed 19 July 2008

11. Adak, C., Chaudhuri, B.B.: An approach of strike-through text identification from handwritten documents. In: 2014 14th International Conference on Frontiers in Handwriting Recognition, pp. 643–648. IEEE (2014)
12. Adak, C., Chaudhuri, B.B., Blumenstein, M.: Impact of struck-out text on writer identification. In: 2017 International Joint Conference on Neural Networks (IJCNN), pp. 1465–1471. IEEE (2017)
13. Brink, A., Schomaker, L., Bulacu, M.: Towards explainable writer verification and identification using vantage writers. In: Ninth International Conference on Document Analysis and Recognition (ICDAR 2007), vol. 2, pp. 824–828. IEEE (2007)
14. Caligiuri, M.P., Mohammed, L.A.: The Neuroscience of Handwriting: Applications for Forensic Document Examination. CRC Press, Boca Raton (2012)
15. Chaudhuri, B.B., Adak, C.: An approach for detecting and cleaning of struck-out handwritten text. Pattern Recogn. **61**, 282–294 (2017)
16. Gatys, L.A., Ecker, A.S., Bethge, M.: Image style transfer using convolutional neural networks. In: Proceedings of the IEEE Conference on Computer Vision and Pattern Recognition, pp. 2414–2423 (2016)
17. Isola, P., Zhu, J.Y., Zhou, T., Efros, A.A.: Image-to-image translation with conditional adversarial networks. In: Proceedings of the IEEE Conference on Computer Vision and Pattern Recognition, pp. 1125–1134 (2017)
18. Johnson, J., Alahi, A., Fei-Fei, L.: Perceptual losses for real-time style transfer and super-resolution. In: Leibe, B., Matas, J., Sebe, N., Welling, M. (eds.) ECCV 2016. LNCS, vol. 9906, pp. 694–711. Springer, Cham (2016). https://doi.org/10.1007/978-3-319-46475-6_43
19. Likforman-Sulem, L., Vinciarelli, A.: Hmm-based offline recognition of handwritten words crossed out with different kinds of strokes (2008)
20. Liu, C.L., Yin, F., Wang, D.H., Wang, Q.F.: Online and offline handwritten Chinese character recognition: benchmarking on new databases. Pattern Recogn. **46**(1), 155–162 (2013)
21. Marti, U.V., Bunke, H.: The IAM-database: an English sentence database for offline handwriting recognition. Int. J. Doc. Anal. Recogn. **5**(1), 39–46 (2002)
22. Mirza, M., Osindero, S.: Conditional generative adversarial nets. arXiv preprint arXiv:1411.1784 (2014)
23. Pal, U., Jayadevan, R., Sharma, N.: Handwriting recognition in Indian regional scripts: a survey of offline techniques. ACM Trans. Asian Lang. Inf. Process. (TALIP) **11**(1), 1 (2012)
24. Plamondon, R., Srihari, S.N.: Online and off-line handwriting recognition: a comprehensive survey. IEEE Trans. Pattern Anal. Mach. Intell. **22**(1), 63–84 (2000)
25. Poddar, A., Mukherjee, R., Mukhopadhyay, J., Biswas, P.K.: MultiDIAS: a hierarchical multi-layered document image annotation system. In: Sundaram, S., Harit, G. (eds.) DAR 2018. CCIS, vol. 1020, pp. 3–14. Springer, Singapore (2019). https://doi.org/10.1007/978-981-13-9361-7_1
26. Tuganbaev, D., Deriaguine, D.: Method of stricken-out character recognition in handwritten text, 25 June 2013. uS Patent 8,472,719

# Temporal Classification Constraint for Improving Handwritten Mathematical Expression Recognition

Cuong Tuan Nguyen$^{(\boxtimes)}$ ⓘ, Hung Tuan Nguyen ⓘ, Kei Morizumi,
and Masaki Nakagawa ⓘ

Tokyo University of Agriculture and Technology, Tokyo 184-8588, Japan
fx4102@go.tuat.ac.jp, s171962w@st.go.tuat.ac.jp,
nakagawa@cc.tuat.ac.jp

**Abstract.** We present a temporal classification constraint as an auxiliary learning method for improving the recognition of Handwritten Mathematical Expression (HME). Connectionist temporal classification (CTC) is used to learn the temporal alignment of the input feature sequence and corresponding symbol label sequence. The CTC alignment is trained with the encoder-decoder alignment through a combination of CTC loss and encoder-decoder loss to improve the feature learning of the encoder in the encoder-decoder model. We show the effectiveness of the approach in improving symbol classification and expression recognition on the CROHME datasets.

**Keywords:** Temporal classification · Mathematical expression · Encoder-decoder model

## 1 Introduction

Recently, there are many applications of Information and Communications Technology in mathematics education, such as the Intelligent Tutoring System (ITS) [1], which provides rich benefits for students from learning at their own pace, tailor feedback, step-by-step hints, and curriculum. As handwriting is a natural way for learning and practising math, Handwritten Mathematical Expression (HME) recognition plays an important role as a user interface that bridges information exchanges between students and ITS.

Recognizing mathematical expression is a challenging problem composed of two main tasks: symbol detection/recognition and structure recognition through analyzing spatial relations. Recently, the encoder-decoder-based end-to-end recognition approach [2–4], which benefits from the global context, is getting a high advantage comparing with the multi-stage approach [5, 6], which is shown in the latest recognition competitions [7, 8]. The approach use encoder for detecting symbol recognition features and an attention-based decoder to learn the alignment constraint between input and latex transcription. The attention-based approach, however, does not apply any constraint on sequential order, which may reduce the performance of the encoder-decoder model.

© Springer Nature Switzerland AG 2021
E. H. Barney Smith and U. Pal (Eds.): ICDAR 2021 Workshops, LNCS 12917, pp. 113–125, 2021.
https://doi.org/10.1007/978-3-030-86159-9_8

Extensive research has been focused on improving the encoder-decoder model for HME recognition. To improve the encoder, a large model of Convolutional Neural Network (CNN) such as densely connected CNN (DenseNet) is applied to learn high-level features [9], Adversarial Networks are applied to learn robust features [10]. To improve the decoder, coverage-based attention is applied to avoid over-decoding or under-decoding [3]. Moreover, multi-scaled attention is used to deal with small symbol recognition [9]. Another approach is to apply the auxiliary learning method to improve the encoder of the model. Truong et al. used weakly supervised learning as auxiliary learning, which improves symbol detection and recognition [11]. Weakly supervised learning, however, suffers from the problem of learning multiple instances of a symbol class in a mathematical expression.

In literature, there are many works to use temporal classification constraints for learning alignment in the sequential classification task. Connectionist temporal classification learns alignment without segmentation and is widely applied for handwriting recognition and speech recognition [12, 13].

In this work, we use temporal classification constraints as auxiliary learning to improve the encoder-decoder model. The encoder benefits from the alignment of both the temporal classification constraint and the attention decoder constraint. We show the effect of our approach to improve encoder classification performance and improve the mathematical expression recognition rate.

The rest of the paper is as follows: Sect. 2 introduces the related works, Sect. 3 describes our methods, Sect. 4 shows the experiment results and discussion, Sect. 5 draws our conclusion.

## 2    Related Works

The encoder-decoder model is a neural network model that has achieved high accuracy in machine translation [14]. It consists of an encoder component that extracts features and a decoder component that outputs a sequence of words or characters according to the task. In detail, the encoder component transforms the input character sequence into a feature vector, and the decoder component takes this vector as input to generate the sequential result. On the other hand, this encoder-decoder model is also used in other tasks. The encoder-decoder model is extended to extract features from the image using a CNN-based encoder. This extension model has been widely employed in offline HME recognition [2, 3, 11, 15].

Watch, Attend and Parse (WAP) [3] is an encoder-decoder model for mathematical image recognition. In addition, coverage-based attention [16] is used together with the decoder. The conventional attention mechanism focuses on the location of a single symbol at different decoding steps. The attentive region should be distinct so that the outputs of different decoding steps are not duplicated. Thus, coverage-based attention is proposed to accumulate the attention maps from the previous decoding steps, eliminating the attention map's duplications. The recognition rate is 46.55%, which is about 9% points higher than the conventional method without using the coverage attention mechanism. An extension of WAP applied the DenseNet encoder and two levels of multi-scaled attention (MSA) [9], which further improve the performance of the system.

Paired Adversarial Learning (PAL) [10] is a recognition model for HMEs. There are two main methods for improvement, one is hostile learning, and the other is an improvement of feature extraction of the encoder part by Multi-Dimensional Long Short-term Memory (MDLSTM). Due to the high variants of HME, this method focuses on extracting invariant features even in different writing styles. A pair of input images is prepared with a handwritten mathematical expression image and a similar mathematical expression image rendered by the LaTeX generator. Each image in this pair is processed through the same network to extract the discriminative features. A discriminator is used to identify whether the extracted features are from a handwritten image or a rendered image. It assists the feature extractor in focusing on the invariant features which do not depend on writers.

Moreover, there are corresponding symbols in HME, such as parentheses which are essential to judge the context between symbols. In PAL, an MDLSTM network implemented by applying LSTM in four directions (up, down, left, and right) is incorporated into the encoder. Thus, the encoder of PAL may access the context of a 2D image and extract local features that match the handwritten mathematical formulas. For improving the decoder, a convolution sequence-to-sequence with attention (Conv-Attention) combined with Position Embedding is employed to generate the corresponding LaTeX sequences from the encoded features [6]. The recognition rate of PAL is improved by about 2.5% percentage points compared to the conventional method.

Truong et al. used Weakly Supervised Learning (WSL) as an auxiliary learning method for the encoder-decoder of WAP. The method showed its effectiveness in improving the high-level feature extraction of the encoder. WSL learns to predict whether each symbol existing in an input image or not. The features learned by WSL benefit from detecting symbols in HME images. However, since WSL only learns the existence of symbols, it may hard to learn when many symbols of the same class existed in the HME image. WSL also suffers from discriminating co-occurrence symbols such as a pair of brackets in the HME image.

## 3 Methodology

### 3.1 Encoder-Decoder Architecture in WAP

The encoder-decoder model in WAP is based on the image-to-sequence model, where the model learns to transcribe an input image to a label sequence of Latex symbols. The WAP architecture consists of a CNN encoder to encode the 2-dimensional (2D) input image into a 2D feature image, a GRU decoder to transcribe the serialized 2D feature image into the label sequence.

Let X the input image, the feature image F by CNN encoder is obtained as follows:

$$F = CNN(X) \tag{1}$$

Let $f_t$ the serialized feature of F as timestep $t$, the decoder output $y_t$ is calculated as follows:

$$h_t = GRU\left(h_{t-1}, \left[f_t, c_t\right]\right) \tag{2}$$

$$y_t = FC(h_t) \tag{3}$$

where $c_t$ is the coverage-based attention context vector at each decoding timestep $t$, $FC$ denotes the fully connected layer to transform the GRU hidden output $h_t$ into label output $y_t$.

### 3.2 Temporal Classification Constraint for WAP

Apply the auxiliary learning method helps the encoder learn better features along with the main learning method [11]. Auxiliary learning uses a constraint as a loss function to learn the features. Truong et al. use a constraint on the existence of symbols in the HME. Due to its limitations, we propose a new constraint of temporal classification, which could overcome the problem by WSL constraint.

The temporal classification constraint method considers the output features of the encoder as a feature sequence of temporal order. The sequential feature could be obtained by tracing the two-dimensional features row-by-row or column-by-column. The sequential features are then classified by a temporal classifier. In this sequence, the temporal classified label order is constrained by their positions in HME input images. To learn the temporal classification constraint, we use Connectionist Temporal Classifier (CTC) loss [12]. Briefly, the CTC loss provides another way to learn symbol classification from the label sequence. It is combined with sequence decoder loss to form a multi-task learning method that improves the symbol classification and expression rates.

Our model with CTC auxiliary learning is shown in Fig. 1. The model consists of three modules: CNN encoder, GRU decoder, and a temporal classification block. CNN encoder and GRU decoder architecture are inspired in WAP and its extension using DenseNet [9]. We named the model DenseWAP. The temporal classification block is a sequence of two fully connected layers for symbol classification. In the example, the temporal classification constraint is applied to the sequential features obtained by tracing the two-dimensional feature column-to-column. Then, the label sequence is constrained by the sequence of 'a', '-' (fraction), ' +', 'c'.

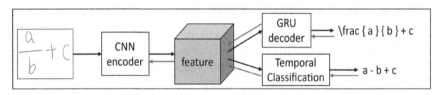

**Fig. 1.** Temporal classification auxiliary learning for WAP.

### 3.3 Learning of the Model

Let $F_{seq}$, the serialized output feature of the CNN encoder, obtained by serializing the two-dimensional feature $F$ row-by-row or column-by-column.

Let $C$ is a set of characters (mathematical symbols), $T$ is the length of $F_{seq}$, the temporal classification block output a temporal classification probability for each timestep $t \in T$ in $F_{seq}$ as follows:

$$y_k^t = p(k, t|F_{seq}), \forall k \in C \cup blank \tag{4}$$

where $y_k^t$ is the output $y_t$ for a class $k$ and 'blank' denotes no label.

The probability of a label sequence $l$ is calculated by the total probabilities of all the paths that produce the label sequence $l$:

$$p(l|x) = \sum\nolimits_{\pi_{1:T} \in B^{-1}(l)} p(\pi_{1:T}|F_{seq}) \tag{5}$$

where $p(\pi_{1:T}|F_{seq}) = \prod_{t=1}^{T} p(k_t, t|F_{seq})$ is the probability over the path $\pi_{1:T}$ of the temporal classification output $y_k^t$.

For a pair of an input feature sequence $F_{seq}$ and an output sequence $l$ from the training dataset, the network is trained by minimizing the CTC loss obtained by Eq. (6).

$$loss_{CTC} = -log(p(l|F_{seq})) \tag{6}$$

We add the temporal classification constraint to the WAP model as an auxiliary loss function. The model trains two tasks: Latex transcription from an image as the default WAP model and label transcription from that image by temporal classification constraint. Therefore, the model is learned by a combination of two-loss functions: the encoder-decoder loss for learning to generate Latex sequence and the CTC loss for learning to generate the temporal label sequence.

The encoder-decoder loss is the cross-entropy loss for generating the Latex sequence. Let $l_s \in L$ the Latex label sequence of length $S$, two-dimensional input feature $F$, then the cross-entropy loss is shown in (7):

$$loss_{decode} = \sum\nolimits_{s=1}^{S} log(p(l_s|F)) \tag{7}$$

Then, the combined loss of the CTC WAP system $Loss_{all}$ is shown in (8):

$$loss_{combine} = (1 - \alpha) \cdot loss_{decode} + \alpha \cdot loss_{CTC} \tag{8}$$

where $\alpha \in [0, 1]$ is a parameter for balancing between $loss_{decode}$ and $loss_{CTC}$. If $\alpha$ is high, the model is mainly learned from $loss_{CTC}$, on the contrary, the model is mainly learned from $loss_{decode}$.

### 3.4   CTC Auxiliary Learning

We apply the temporal classification constraint by the location of the symbols in the mathematical expression. Considering the output of the encoder highlights a location for each symbol, we obtain the sequential order constraint of symbols by tracing through the 2D feature maps vertically or horizontally. In this work, we process the constraint on temporal location vertically, as shown in Fig. 2.

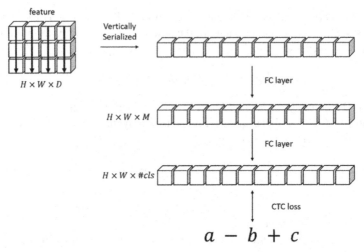

**Fig. 2.** Temporal classification auxiliary model for WAP.

From the 2D encoded features, we obtain the 1D encoded features by serializing them vertically. The features are fed to the temporal classification by a stack of two fully connected layers and learn from label sequence by CTC loss.

The label sequence is also obtained by tracing the symbols vertically with symbol localization ground truth. From label ground truth with the location of symbols in an HME image, we first create a grid on an input image as the same size with the feature by the encoder, i.e., 1/16 scaled size of the input image in the current setting of the DenseNet encoder (see Sect. 4.3). The center of each symbol is assigned to a single cell of the grid, as shown in Fig. 3. If two or more centers are in the same cell, we assign them to the neighbor cells where they are not occupied. In the current setting of the output feature scaled by the DenseNet encoder, there are 3.22% of symbols need to be re-assigned in the training datasets. We could reduce the number by learning features at a higher resolution.

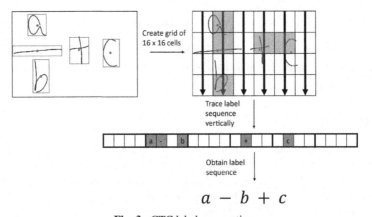

**Fig. 3.** CTC label preparation.

# 4   Evaluation

We validated our method using the official datasets of Competition on Recognition of Online Handwritten Mathematical Expressions (CROHME) [17, 18]. We also analyze how temporal classification improves end-to-end HME recognition.

## 4.1   Datasets

We conducted the experiments on two datasets of CROHME 2014 and CROHME 2016 [17, 18] for mathematical expression recognition. The number of symbol's classes is 101. There are 8,836 samples for training. Note that the Latex sequences for training are normalized by removing redundant braces symbols.

## 4.2   Setting and Metrics for Evaluation

We evaluate our method according to the two following metrics: Expression Recognition Rate (ExpRate) and Word Error Rate (WER), which are commonly used in other research in HME recognition. ExpRate is evaluated as in (9).

$$\text{ExpRate} = \frac{N_{truepredicted}^{Exp}}{N^{Exp}} \tag{9}$$

where:

- $N_{truepredicted}^{Exp}$: number of successful predicted HMEs.
- $N^{Exp}$: number of HMEs in the dataset.

Next, WER is the edit distance between the decoded Latex sequence and the ground truth Latex sequence.

In the experiments, we tuned the parameter $\alpha$ of the combined loss in Eq. (8) from 0 to 1 with a step of 0.25 and obtained $\alpha = 0.5$ based on the best validation result.

## 4.3   Network Configuration

The configurations of the model are shown in Table 1. We used DenseNet architecture in the encoder since it was successful employed for HME recognition as DenseWAP [9]. In general, DenseNet consists of three Dense blocks with a Transition layer between two Dense blocks. Each Dense block has 16 pairs of convolution layers using kernel sizes of $1 \times 1$ and $3 \times 3$, with a growth rate of 24. Note that the convolutional layers in Dense and Transition blocks are equipped with batch normalization and Rectified Linear Unit activation layers. The Dense blocks might extract visual features, while the Transition block seems to generalize those extracted features so that the extracted features would be invariant.

The encoder block is followed by a classification block which is designed for temporal classification. Our classification block consists of two Fully Connected (FC) layers of 256 nodes and 101 nodes. The FC layer with 256 nodes is used for the GRU decoder, while the FC layer with 101 nodes is used with CTC loss to predict the character classes.

**Table 1.** Network architecture.

| Module | Type | Configurations | Output shape |
|---|---|---|---|
| Encoder block | Convolution | $7 \times 7$ conv, 48, s $2 \times 2$ | (h/2), (w/2), 48 |
| | Max pooling | $2 \times 2$ max pooling, s $2 \times 2$ | (h/4), (w/4), 48 |
| | Dense block | $\begin{bmatrix} 1 \times 1,\ \text{conv} \\ 3 \times 3,\ \text{conv} \end{bmatrix} \times 16$, G 24, s $1 \times 1$ | (h/4), (w/4), 432 |
| | Transition layer | $1 \times 1$ conv, 216, s $1 \times 1$ | (h/4), (w/4), 216 |
| | | $2 \times 2$ avg pooling, s $2 \times 2$ | (h/8), (w/8), 216 |
| | Dense block | $\begin{bmatrix} 1 \times 1,\ \text{conv} \\ 3 \times 3,\ \text{conv} \end{bmatrix} \times 16$, G 24, s $1 \times 1$ | (h/8), (w/8), 600 |
| | Transition layer | $1 \times 1$ conv, 300, s $1 \times 1$ | (h/8), (w/8), 300 |
| | | $2 \times 2$ avg pooling, s $2 \times 2$ | (h/16), (w/16), 300 |
| | Dense block | $\begin{bmatrix} 1 \times 1,\ \text{conv} \\ 3 \times 3,\ \text{conv} \end{bmatrix} \times 16$, G 24, s $1 \times 1$ | (h/16), (w/16), 684 |
| Classification block | FC layer | 256 units | (h/16) × (w/16), 256 |
| | FC layer | 101 units, softmax | (h/16) × (w/16), 101 |

s: stride, G: growth rate, (h, w): input shape.

### 4.4  Experiments

This section presents the results of the proposed model with the previous models on both CROHME 2014 and 2016 datasets. First, we would compare the proposed model with the WAP and DenseWAP models to demonstrate the effect of CTC on the DenseWAP model. Secondly, we compare with the previous models, including the state-of-the-art models in HME recognition. Thirdly, we conduct experiments to analyze the effect of hyperparameter, $\alpha$ and present the way to optimize the hyperparameter. Finally, we visualize the attention maps during the decoding stage of the proposed model to support our conclusions on the effectiveness of CTC for improving recognition accuracy.

**Effect of CTC on Recognition Rate.** We Compare the Results of the DenseWAP Model with and Without Applying CTC in Table 2.

In terms of ExpRate, the model with CTC achieves more than 3 percentage points on CROHME 2014 and more than 2.5 percentage points on CROHME 2016 compared to the model without CTC. For WER, applying CTC also helps DenseWAP reduce WER by 2.3 and 1.3 percentage points on CROHME 2014 and 2016 datasets. ExpRate has been improved further than the WER, suggesting that some expressions with a single error have been correctly recognized.

These results show the effectiveness of CTC loss on the DenseWAP model, which improves both the symbol classification (reduced WER) and ExpRate. Although the integrated CTC loss is simple, it seems helpful to improve the state-of-the-art model's accuracy, such as DenseWAP.

**Table 2.** Expression rate (ExpRate) and word error rate (WER) (%).

| Models | CROHME 2014 | | CROHME 2016 | |
|---|---|---|---|---|
| | ExpRate | WER | ExpRate | WER |
| WAP [3] | 46.55 | 17.73 | 44.55 | - |
| DenseWAP | 47.97 | 14.50 | 47.17 | 14.46 |
| DenseWAP + CTC | **50.96** | **12.20** | **49.96** | **13.10** |

**Compare with State-Of-The-Arts.** We Compare Our Method with Other Methods on Both CROHME 2014 and CROHME 2016 Datasets in Table 3.

As compared to other extensions of WAP models, our model is closed to the Dense-WAP + MSA* [9], where an ensemble of five models with different initialized states was used. The model performs worse than DenseWAP + WSL [11], which has a gap of 2 points lower on CROHME 2014 and CROHME 2016. Here, CTC loss may make the model overfit to the sequential order of symbols and propagate that error to the convolution features.

**Table 3.** Expression rate (ExpRate) on CROHME testing sets (%).

| | Systems | CROHME 2014 | CROHME 2016 |
|---|---|---|---|
| CFG-based parser, graph-based model | València / Wiris [17, 18] | 37.22 | 49.61 |
| | Tokyo [17, 18] | 25.66 | 43.94 |
| | Sao Paolo [17, 18] | 15.01 | 33.39 |
| | Nantes [17, 18] | 26.06 | 13.34 |
| Data augmentation | SCUT [15] | 56.59 | 54.58 |
| Ensemble of models | WAP* [3] | 44.40 | 44.55 |
| | DenseWAP + MSA* [9] | 52.80 | 50.10 |
| Single model, no data augmentation | PAL_v2 [10] | 48.78 | 49.35 |
| | DenseWAP + WSL [11] | 53.65 | 51.96 |
| | DenseWAP + CTC | **50.96** | **49.96** |

" *" denotes an ensemble of five differently initialized recognition models.

Compared to other deep neural network-based models, our proposed model outperforms the single PAL_v2 [10] model but lower than SCUT [15] since the SCUT model benefits from large training data by scaled data augmentation.

**Effect of Hyperparameter.** We Evaluated the DenseWAP + CTC Model According to ExpRate, WER of the Latex Recognition Result, and WER for the CTC Output Label Sequence with Different Settings of $\alpha$ Hyperparameter for the Combined Loss Function in Eq. (5). Figure 4 Shows the Results. Increasing $\alpha$ Makes the Model More Focusing on the CTC Loss (or the Constraint of Temporal Order) While Decreasing $\alpha$ Reduces the Constraint. Compared to the Baseline Model Without CTC ($\alpha = 0$), the Proposed Model Improves the ExpRate and Reduces WER. The Best ExpRate is Achieved with $\alpha$ of 0.5. Although the $\alpha$ Hyperparameter Has a High Impact on ExpRate and WER of the Latex Result, It Does not Affect WER by CTC.

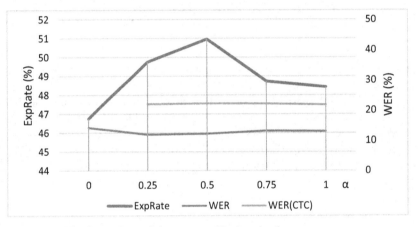

**Fig. 4.** ExpRate of the system with changing hyperparameter.

**Effect of CTC to Feature.** The Model with CTC Produces More Precise Attention Visualization Than the Model Without CTC, as Shown in Fig. 5. The Attentions by the Model with CTC on the Symbols 'cos', '2', and 'theta' (Fig. 5a) Show Their High Strength as Compared to the Attention by the Model Without CTC (Fig. 5b). The Model with CTC also Suppresses the Attention on the Non-symbol Output Such as '{', Which Encourages the Decoder to Learn to Output the Non-symbol from Sequential Context Instead of Visual Features. The Effect Comes from the Better Symbol Detection Features Learned in the Encoder, Which Helps the Attention Decoder Attend the Symbol with High Probability and Suppress the Attention on Non-symbol Output.

Figure 6 shows an example of prediction using the proposed model so that the symbols are recognized through the temporal features of a 2D image. The model correctly recognizes all the symbols and match them with the temporal order constraint. However, some predictions are away from the location of the corresponding symbols. For example, the ' =' symbol and the second 'x' symbol from the left.

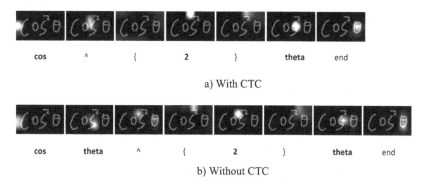

a) With CTC

b) Without CTC

**Fig. 5.** Attention visualization of the system.

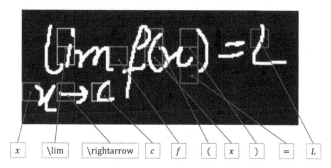

**Fig. 6.** CTC prediction of a test sample.

## 5    Conclusion

We proposed to apply temporal classification constraint as auxiliary learning for the encoder-decoder model. The temporal classification constraint has been employed by integrating the CTC loss. By employing a stack of two fully connected layers, the temporal classification constraint seems to help to generalize the extracted features by the encoder since these features are optimized for multiple objectives. In experiments, we applied the proposed constraint on DenseWAP, a variant of the state-of-the-art model, WAP, for Handwritten Mathematical Expression recognition [3]. Compared with a pure DenseWAP model, the model with our proposed constraint achieved a higher expression recognition rate (ExpRate) by more than 3 and 2.5 percentage points in CROHME 2014 and CROHME 2016, respectively. In general, the recognition model with auxiliary learning improved the ExpRate. Although it is simple to integrate auxiliary learning such as temporal classification constraint into a deep neural network, there is a good effect on improving the recognition accuracy. Thus, the auxiliary learning might open future researches to improve even the state-of-the-art models for higher accuracy.

**Acknowledgement.** This work is being partially supported by the Grant-in-Aid for Scientific Research (A) 19H01117 and that for Early-Career Scientists 21K17761.

# References

1. Anthony, L., Yang, J., Koedinger, K.R.: Toward next-generation, intelligent tutors: adding natural handwriting input. IEEE Multimed. **15**, 64–68 (2008)
2. Deng, Y., Kanervisto, A., Ling, J., Rush, A.M.: Image-to-Markup generation with coarse-to-fine attention. In: Proceedings of the 34th International Conference on Machine Learning, pp. 980–989 (2017)
3. Zhang, J., et al.: Watch, attend and parse: an end-to-end neural network based approach to handwritten mathematical expression recognition. Pattern Recognit. **71**, 196–206 (2017)
4. Zhang, J., Du, J., Dai, L.: Track, Attend, and Parse (TAP): an end-to-end framework for online handwritten mathematical expression recognition. IEEE Trans. Multimed. **21**, 221–233 (2019)
5. Le, A.D., Nakagawa, M.: A system for recognizing online handwritten mathematical expressions by using improved structural analysis. International Journal on Document Analysis and Recognition (IJDAR) **19**(4), 305–319 (2016). https://doi.org/10.1007/s10032-016-0272-4
6. Álvaro, F., Sánchez, J.A., Benedí, J.M.: Recognition of on-line handwritten mathematical expressions using 2D stochastic context-free grammars and hidden Markov models. Pattern Recognit. Lett. **35**, 58–67 (2014)
7. Mahdavi, M., Zanibbi, R., Mouchère, H.: ICDAR 2019 CROHME + TFD : competition on recognition of handwritten mathematical expressions and typeset formula detection. In: Proceedings of the 15th International Conference on Document Analysis and Recognition, pp. 1533–1538 (2019)
8. Wang, D.H., et al.: ICFHR 2020 competition on offline recognition and spotting of handwritten mathematical expressions-OffRaSHME. In: Proceedings of the 17th International Conference on Frontiers in Handwriting Recognition, pp. 211–215 (2020)
9. Zhang, J., Du, J., Dai, L.: Multi-scale attention with dense encoder for handwritten mathematical expression recognition. In: Proceedings of the 24th International Conference on Pattern Recognition, pp. 2245–2250 (2018)
10. Wu, J.W., Yin, F., Zhang, Y.M., Zhang, X.Y., Liu, C.L.: Handwritten mathematical expression recognition via paired adversarial learning. Int. J. Comput. Vis. **128**, 2386–2401 (2020)
11. Truong, T.N., Nguyen, C.T., Phan, K.M., Nakagawa, M.: Improvement of end-to-end offline handwritten mathematical expression recognition by weakly supervised learning. In: Proceedings of the 17th International Conference on Frontiers in Handwriting Recognition, pp. 181–186 (2020)
12. Graves, A., Liwicki, M., Fernández, S., Bertolami, R., Bunke, H., Schmidhuber, J.: A novel connectionist system for unconstrained handwriting recognition. IEEE Trans. Pattern Anal. Mach. Intell. **31**, 855–868 (2009)
13. Graves, A., Mohamed, A., Hinton, G.: Speech recognition with deep recurrent neural networks. In: Proceedings of the 38th IEEE International Conference on Acoustics, Speech and Signal Processing, pp. 6645–6649 (2013)
14. Luong, T., Pham, H., Manning, C.D.: Effective approaches to attention-based neural machine translation. In: Proceedings of the 12th Conference on Empirical Methods in Natural Language Processing, pp. 1412–1421 (2015)
15. Li, Z., Jin, L., Lai, S., Zhu, Y.: Improving attention-based handwritten mathematical expression recognition with scale augmentation and drop attention. In: Proceedings of the 17th International Conference on Frontiers in Handwriting Recognition, pp. 175–180 (2020)
16. Tu, Z., Lu, Z., Yang, L., Liu, X., Li, H.: Modeling coverage for neural machine translation. In: Proceedings of the 54th Annual Meeting of the Association for Computational Linguistics, pp. 76–85 (2016)

17. Mouchere, H., Viard-Gaudin, C., Zanibbi, R., Garain, U.: ICFHR 2014 competition on recognition of on-line handwritten mathematical expressions (CROHME 2014). In: Proceedings of the 14th International Conference on Frontiers in Handwriting Recognition, pp. 791–796 (2014)
18. Mouchère, H., et al.: ICFHR 2016 CROHME : competition on recognition of online handwritten mathematical expressions. In: Proceedings of the 15th International Conference on Frontiers in Handwriting Recognition, pp. 607–612 (2016)

# Using Robust Regression to Find Font Usage Trends

Kaigen Tsuji[(✉)], Seiichi Uchida[ID], and Brian Kenji Iwana[ID]

Kyushu University, Fukuoka, Japan
`kaigen.tsuji@human.ait.kyushu-u.ac.jp`, {uchida,iwana}`@ait.kyushu-u.ac.jp`

**Abstract.** Fonts have had trends throughout their history, not only in when they were invented but also in their usage and popularity. In this paper, we attempt to specifically find the trends in font usage using robust regression on a large collection of text images. We utilize movie posters as the source of fonts for this task because movie posters can represent time periods by using their release date. In addition, movie posters are documents that are carefully designed and represent a wide range of fonts. To understand the relationship between the fonts of movie posters and time, we use a regression Convolutional Neural Network (CNN) to estimate the release year of a movie using an isolated title text image. Due to the difficulty of the task, we propose to use of a hybrid training regimen that uses a combination of Mean Squared Error (MSE) and Tukey's biweight loss. Furthermore, we perform a thorough analysis on the trends of fonts through time.

**Keywords:** Font analysis · Year estimation · Movie poster · Regression neural network · Tukey's biweight loss

## 1 Introduction

After Gutenberg invented the printing press with metal types in the fifteenth century, a huge variety of fonts have been created. For example, MyFonts[1] provides more than 130,000 different fonts and categorizes them into six types (Sans-Serif, Slab-Serif, Serif, Display, Handwriting, Script) for making it easy to search their huge collection. Some of them are very famous and versatile (e.g., Helvetica, Futura, Times New Roman) and some are very minor and rarely used.

An important fact about fonts is that they have trends in their history. It is well-known that sans-serif fonts were originally called "grotesque" when serif fonts were the majority (around the early twentieth century). However, sans-serif fonts are not grotesque for us anymore. Moreover, many sans-serif font styles (neo-grotesque, geometric, humanist) have been developed after grotesque and had big trends in each era. It is also well-known that psychedelic fonts were often used in the 1970s and recent high-definition displays allow us to use very thin fonts, such as Helvetica Neue Ultralight.

---

[1] https://www.myfonts.com/.

ⓒ Springer Nature Switzerland AG 2021
E. H. Barney Smith and U. Pal (Eds.): ICDAR 2021 Workshops, LNCS 12917, pp. 126–141, 2021.
https://doi.org/10.1007/978-3-030-86159-9_9

In this paper, we attempt to find the trends in font usage using a robust regression technique and a large collection of font images. More specifically, we first train a regression function $y = f(x)$, where $x$ is the visual feature of a font image and $y$ is the year that the image was created. Then, if the trained regression function can estimate the year of unseen font images with reasonable accuracy, we can think that the function catches the expected trend—in other words, there were historical trends in the font usage.

The values of this attempt are twofold. First, it will give objective and reliable evidence about the font usage trends, which are very important for typographic designers to understand past history. It should be noted that not only the trends (i.e., main-stream fonts) but also the exceptional usages of fonts are also meaningful. Second, it will help to realize an application to support typography designs. If we can evaluate the "age" of the font image by using the regression function, it helps the typographers on their selection of appropriate fonts for their artworks.

It is easy to imagine that this regression task is very difficult. For example, Helvetica, which was created in the 1950s, is still used today. In addition, many early typographic designers tried to use futuristic fonts, and current designers often use "retro" fonts. In short, font usage is full of exceptions (i.e., outliers) and trends become invisible and weak by those exceptions. In addition, non-experts use the fonts without careful consideration and disturb the trends. Moreover, the ground truth (i.e., the year that a font image was created) is often uncertain or not available, especially for old images.

We tackle this difficult regression task with two ideas. First, we use font images collected from movie posters in this paper, as shown in Fig. 1. Since we know when the movie was created and published, we also know the accurate year that the image was made. In addition, movie posters are created by professional designers, who are fully careful of the impressions given by fonts and trends. Moreover, the recent movie poster dataset, Internet Movie Database (IMDb), allows us to use 14,504 images at maximum. This large collection is helpful to catch the large trends while weakening the effect of the exceptional usages.

Second, we developed a new robust regression method based on deep neural networks with a loss function, Tukey's biweight loss. Deep neural networks realize heavily nonlinear regression functions and thus are appropriate to our difficult task. Tukey's biweight loss has been utilized in robust statistical estimation and, roughly speaking, discounts the large deviations from the average, whereas a typical loss function, such as L2, gives a larger loss for a larger deviation. By this loss function, we can suppress the effect of exceptional samples.

One more point that we must consider is what font image features should be used in our year estimation task. Of course, deep neural networks will have a representation learning function, which automatically determines appropriate features. However, it might learn some uninteresting features such as catching the printing quality of old movie posters, like resolution or color fading. Even though the regression might become accurate by these printing quality features, it does not catch the trends of font usages.

poster:

title image:

release year:    1942          1975              2014

**Fig. 1.** Example movie posters and their title text and release year.

We, therefore, employ a second feature representation using the font shape features. As for shape features, we use two features; edge images and Scale-Invariant Feature Transform (SIFT) [17] features. The former is straightforward; by extracting edges from a font image, it is possible to remove colors and textures that suggest rough printing. The latter is a classical local descriptor and expected to capture the corner and curve shapes of font outlines. By following a standard Bag-of-Visual Words (BoVW) approach, we can have a fixed-dimensional feature representation even though each font image will give a different number of SIFT descriptors. The main contributions of this paper are summarized as follows:

- To the authors' best knowledge, this is the first attempt to reveal font usage trends based on regression functions trained on a large number of font images that have the created year as ground truth.
- The learned regression function shows a reasonable year estimation accuracy and compares different features as the base of regression.
- From a technical aspect, this paper proposes a novel robust regression method based on deep neural networks with Tukey's biweight loss, which helps to realize a regression function robust to exceptional samples (i.e., outliers).

## 2    Related Work

### 2.1    Deep Regression

Deep regression is the use of neural networks for regression tasks and it is a wide field of study with many different architectures and applications [14]. Some example applications of deep regression include housing price prediction from house images [22], television show popularity prediction based on text [12], image orientation and rotation estimation [5,18], estimation of wave velocity through rocks [9], stock prices [16,20,24], and age prediction [21,32].

In these methods, usually, the network is trained using Mean Squared Error (MSE) or sometimes Mean Absolute Error (MAE). However, using MSE can lead to a vulnerability to outliers [2]. Thus, Belagiannis et al. [2] proposed using Tukey's biweight loss function in order to perform deep regression that is robust to outliers. Huber loss [7] and Adaptive Huber loss [3] are other losses that used to be less sensitive to outliers. DeepQuantreg [8] is one example of a deep regression model that uses Huber loss. However, Lathuiliere et al. [14] found MSE loss to be better than Huber loss across all four of their tested datasets.

### 2.2   Movie Poster Analysis

Studying and analyzing design choices of movie posters is of interest to researchers in art and design. In particular, research on the relationship between movie poster design and the fonts and text used has been performed. For example, Wang [30] describes the fonts of movie posters and identified relationships between the tonality, use, and design of the fonts. They explain that the fonts on a movie poster contain a large amount of information about the film by communicating artistic expression and symbolic functions to the viewer. As part of a larger study on movie taglines, Mahlknecht [19] looked at the relationship between the tagline and poster image. In relationship to the design of the tagline text, Mahlkecht found that the size of the tagline on the poster image carries meaning. Also, many studies have been done that look at the cultural aspects of the design of text in movie posters across the world. Kim et al. [10] compared the differences between title text (as well as characters, backgrounds, and color contrast) between localized Korean movies and Western movies. In addition, there have been studies on title design of Indian movies [25, 26] and calligraphy of Chinese movies [23, 27].

The difference between these works and ours is that we take a quantitative approach using machine learning. In this regard, there have only been attempts that classify the genre [4, 6, 13, 28, 29, 31] and predict box office revenues [33] from movie posters. Compared to these, we analyze the movie poster fonts from their relationship with time.

## 3   Movie Posters and Title Text

The elements that make up a movie poster can be roughly divided into two categories: visual information that represents the theme of the movie, and textual information such as the title, actors, production companies, and reviews. The visual information is mainly placed in the center of the poster and conveys the characters in the film and the atmosphere of the film to the viewer. The textual information is often placed at the top or bottom of the poster to avoid overlapping with the visual information. It conveys detailed information to the viewer who is interested in the movie after seeing the visual information. The upper part of Fig. 1 shows examples of movie posters. Looking at the textual information, the movie title tends to be larger than other textual information

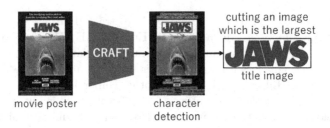

movie poster                    character
                                detection

**Fig. 2.** The process of extracting the title image. The text regions on the movie poster are detected using CRAFT [1] and then the largest area region is cropped to be the title image.

(e.g., the names of the director and actors). It is also more elaborately designed than other textual information.

In this paper, a title image is defined as a cutout of the part of a movie poster where the title is placed. The title image is a rectangle, and the background information within the rectangle is retained. The lower part of Fig. 1 shows examples of title images. In some posters, the title is placed vertically or curved which might cause vertical title images or images with significant portions of the background, respectively.

To extract the title image, we use a popular scene text detection method called Character Region Awareness for Text Detection (CRAFT) [1]. Figure 2 shows the workflow used to create a title image using CRAFT. The poster is input to CRAFT and text regions in the poster are detected then cropped. As mentioned above, the movie titles tend to be larger than other textual information. Thus, the title image is created by cropping the word proposal with the largest area.

## 4    Year Prediction by Robust Regression

In this paper, we propose three methods for estimating the year of movie release from title images: Image-based Estimation, Shape-based Estimation, and Feature-based Estimation. In Image-based Estimation, a regression CNN is constructed to estimate the year of movie release given a title image of a movie poster. In the case of Shape-based Estimation, edge detection is first applied to the title image in order to remove background and color information. Then a regression CNN is used on the title outline image. In Feature-based Estimation, an MLP is constructed to estimate the year of movie release based on SIFT features [17].

### 4.1    Image-Based Estimation

In Image-based Estimation, a regression CNN is constructed to estimate the year of movie release when a title image of a movie poster is input. Figure 3 shows the

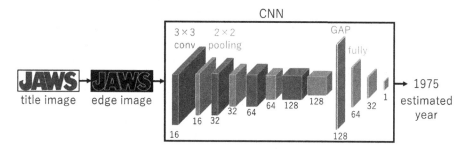

**Fig. 3.** The regression CNN used for image-based estimation and shape-based estimation.

architecture of the proposed Image-based Estimation. The Image-based Estimation regression CNN is a standard CNN except with a single output node representing the year.

In a standard CNN, a fixed image size is required because of the fixed number of parameters in the fully connected layers. In order to input title images of any size to the proposed regression CNN, a Global Average Pooling (GAP) layer [15] is incorporated before the fully connected layer. The GAP layer outputs the average value of each feature map. Since the number of features output by the GAP layer depends only on the number of feature maps output by the feature extraction layer, title images can be input to the CNN even if they have different sizes.

### 4.2   Shape-Based Estimation

For Shape-based Estimation, we use edge detection in a pre-processing step before using the same structure as Image-based Estimation (Fig. 3). The reason for using edge images as inputs is to remove the color information from the title image. For the purpose of this paper, it is not desirable that features other than font affect the estimation of the movie release year. When raw pixels are used, it is possible for color to have more effect on the regression CNN than the font itself. By using edge images without color information as input, we attempt to estimate the movie release year based on font features alone. It also removes the background image and noise.

### 4.3   Feature-Based Estimation

In Feature-based Estimation, an MLP is constructed to estimate the movie release year when the local features of the title image are input. Figure 4 shows the architecture of Feature-based Estimation. SIFT features are extracted from the title image. Next, the SIFT features are embedded into a Bag-of-Visual-Words (BoVW) representation, where the bags are found through $k$-means clustering.

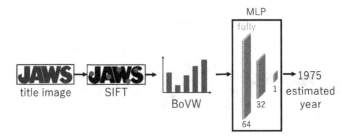

**Fig. 4.** The procedure for performing feature-based estimation.

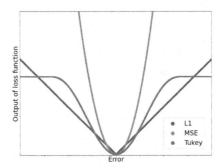

**Fig. 5.** A comparison of different loss functions.

We use SIFT as a method to extract local features from title images. SIFT, which can detect rotation- and scale-invariant feature points, is effective for feature extraction of character designs with various inclinations and sizes. Each detected feature point is represented by a 128-dimensional vector, but since the number of feature points varies depending on the image, it cannot be directly inputted to the MLP.

By converting the local features to a BoVW representation, we can make the input data have the same number of dimensions. The method of converting local features to BoVW representation is as follows. First, the local features of all images are divided into $k$ number of clusters using the $k$-means method. Next, the feature points in each title image are grouped into the nearest cluster and a histogram for each image is created. The $k$-dimensional vector created by this process is the BoVW representation.

The possibility of removing color information from the title image is another reason for using the BoVW representation as input. As mentioned above, there is a correlation between the font color and the movie release year, but the color information is not included in the features extracted by SIFT. By using the BoVW representation converted from these features, we attempt to estimate the movie release year based on font features alone.

## 4.4   Robust Regression

Since there are many variations in designs, some title images have feature values that are far from the distribution in the feature space. For example, some of the designs are bizarre, such as using objects instead of simple lines to compose the text. Such data can be regarded as outliers because it is difficult to estimate the movie release year and the residuals are large.

In some cases, the designer intentionally chooses a design that matches the setting year of the movie. For example, the movie release year is in the 2000s, but the poster image uses a retro font design reminiscent of the 1950s to match the setting year of the movie. Such data can also be regarded as an outlier because it is difficult to estimate the movie release year and the residuals are large.

To reduce the impact of outliers on learning, we use Tukey's biweight loss, which is a loss function that is robust to outliers [2]. Figure 5 compares a plot of Tukey's biweight loss with other common loss functions. The horizontal axis is the residuals of the Ground Truth and Prediction, and the vertical axis is the output value of the loss function. For MSE loss and L1 loss, the output value always increases as the residuals increase, so outliers have a large impact on learning. On the other hand, for Tukey's biweight loss, the outliers do not affect the learning as much because the output value remains constant when the residual exceeds the threshold value.

## 4.5   Automatic Loss Switch

In the early stage of training, the output values of the neural network are unstable. At this time, if Tukey's biweight loss is used as the loss function, the residuals become large and many data are treated as outliers. Therefore, the weights are not updated properly and the learning process does not progress.

We propose the use of MSE loss in the early stages of training and change to Tukey's biweight loss after the output of the neural network has stabilized. In order to determine when the output of the neural network becomes stable, it is possible to check the learning curve of training using only MSE loss in advance. However, there are some disadvantages, such as the need to train once and the need to determine heuristically.

Therefore, we propose a method to determine the stability of the output of a neural network and automatically switch the loss function. Specifically, if the validation loss does not improve for a certain number of epochs, the output is judged to have stabilized. However, there is a certain delay between the time when the output is judged to be stable and the time when the loss function is switched. This prevents switching the loss function when the weights of the neural network fall into a local solution.

# 5    Experimental Results

## 5.1    Dataset

The movie poster images and release year were obtained from the Internet Movie Database (IMDb) dataset available on Kaggle[2]. For the purpose of language unification, we focused on movies released in the U.S. and obtained data for 14,504 movies. Of these, we selected movies released during the 85-year period from 1932 to 2016. We randomly selected 56 movies from each year to make the final dataset. The dataset was randomly divided into 20 training images, 8 validation images, and 28 test images. The total dataset consisted of 1,700 training images, 680 validation images, and 2,380 test images.

## 5.2    Architecture and Settings

In Image-based Estimation and Shape-based Estimation, Regression CNNs were constructed with four convolutional and pooling layers, one Global Average Pooling (GAP) layer, and three fully-connected layers. The filter size of all convolutional layers was set to $3 \times 3$, stride 1. The number of filters was 16, 32, 64, and 128 respectively, and Rectified Linear Units (ReLU) were used as the activation functions. Max pooling was used for all pooling layers, and the filter size was $2 \times 2$ with a stride of 2. The GAP layer was added after the last pooling layer and it computes the average value for each channel, so the number of channels is 128. The number of nodes in the fully-connected layers is 64, 32, and 1, and ReLU is used as the activation function. Dropout was used after the first two fully-connected layers with a dropout rate of 0.5.

In Feature-based Estimation, the maximum number of local features to be extracted from the title images using SIFT was 500. The number of clusters $k$ was set to 128 and clustering was performed using the $k$-means method using only the local features extracted from the training images. Then, all title images were converted to BoVW representation. Regression MLPs were constructed with three fully-connected layers. The number of nodes in all the fully-connected layers is 64, 32, and 1, respectively, and ReLU is used as the activation function. Dropout was used after the first two fully-connected layers with a dropout rate of 0.5.

The learning procedure is as follows. The number of epochs was set to 3,000. The batch size was set to 128. For the proposed loss function, *MSE+Tukey (Prop)*, we used MSE loss in the early stage of learning and Tukey's biweight loss after learning convergence. The convergence of learning was judged when the minimum validation loss was not updated for 50 consecutive epochs and automatically switch after 450 epochs if convergence was not reached. We also conducted comparison experiments using only MSE loss, Tukey biweight loss, and a comparative L1 loss. Adam [11] was used as the optimization method, and the learning coefficient was set to $10^{-4}$. The range of the target variable in the regression, the movie release year was normalized so that the first year starts at 0. In other words, we changed the movie release years from 1932–2016 to 0–85, respectively.

---

[2] https://www.kaggle.com/neha1703/movie-genre-from-its-poster.

**Table 1.** Test results comparison

| Method | Loss function | MAE ↓ | $R^2$ score ↑ | Corr. ↑ |
|---|---|---|---|---|
| Image-based estimation (CNN w/RGB images) | L1 | 16.90 | 0.2472 | 0.5112 |
| | MSE | 17.14 | **0.2580** | 0.5093 |
| | Tukey | 16.80 | 0.1784 | 0.5079 |
| | MSE+Tukey (Prop) | **16.61** | 0.2002 | **0.5264** |
| Shape-based estimation (CNN w/Edge images) | L1 | 18.24 | 0.0667 | 0.3793 |
| | MSE | 18.50 | 0.1760 | 0.4209 |
| | Tukey | 18.48 | 0.0389 | 0.3642 |
| | MSE+Tukey (Prop) | 18.35 | 0.0845 | 0.3995 |
| Feature-based estimation (MLP w/SIFT+BoVW) | L1 | 18.78 | 0.0679 | 0.3491 |
| | MSE | 18.83 | 0.1364 | 0.3712 |
| | Tukey | 18.75 | 0.0434 | 0.3397 |
| | MSE+Tukey (Prop) | 18.71 | 0.0653 | 0.3577 |
| Linear regression (SIFT+BoVW) | | 19.18 | 0.0768 | 0.3287 |
| Constant prediction | | 21.25 | 0.0000 | 0.0000 |

## 5.3   Results

To evaluate the proposed method, we use three comparison losses for each estimation type. The losses used are the *L1* loss, *MSE* loss, *Tukey*'s biweight loss, and the proposed *MSE+Tukey* with the automatic loss switching. We also include two more comparisons, *Linear Regression* and Constant Prediction. Linear Regression uses the SIFT features represented by BoVW and Constant Prediction is a baseline where the average year of the training dataset is used for all predictions. We evaluated the losses using three metrics, *MAE*, $R^2$ *score*, and the correlation coefficient (*Corr.*).

Table 1 shows the evaluation indices of each method. The performance of Image-based Estimation is better than that of the other methods, indicating that there is a correlation between the color information of the title image and the movie release year. In addition, the performance of Shape-based Estimation is better than that of Feature-based Estimation, indicating that there are other factors in the title image that are related to the year estimation besides local features. Linear Regression using the SIFT features in a BoVW representation generally performed worse than the neural network based methods.

When comparing the loss functions, the proposed MSE+Tukey generally performed the best using the MAE metric and MSE loss generally performed the best on the other metrics. As mentioned before, this is because MSE loss puts a heavy emphasis on the outliers while Tukey's biweight loss and L1 loss do not. Also, in every case, MSE+Tukey performed better than Tukey's biweight loss alone. Thus, the benefit of the proposed method is that the loss is a balance between the robustness and emphasis on the outliers.

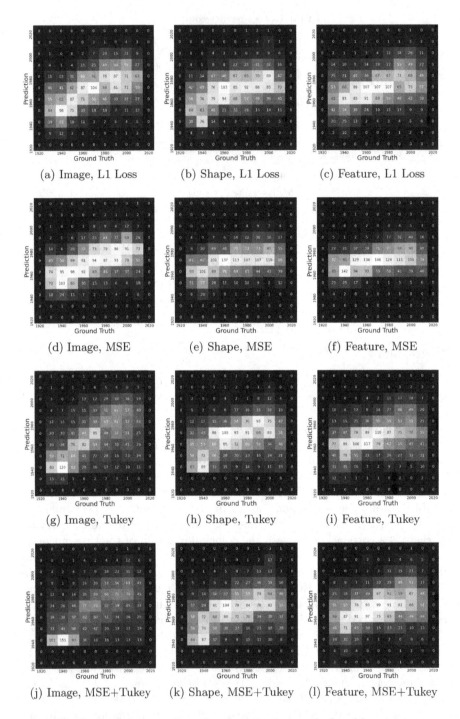

**Fig. 6.** Confusion matrices of each estimation method and loss type.

**Fig. 7.** Image-based estimation with MSE+Tukey loss: top four smallest residual title images of each decade. In each case, the year was successfully estimated to the year.

Figure 6 shows confusion matrices of each estimation method and loss type. In all the methods, the prediction is skewed to the center when MSE or L1 loss is used. When Tukey's biweight loss and the proposed combination of MSE and Tukey's biweight loss is used, the output is spread out much more. While the evaluation metrics might show that there is an overall worse year estimation, having a wider distribution of predictions is desirable so that analysis can be performed. As mentioned before, Tukey+MSE gets the best characteristics of both MSE and Tukey's biweight loss.

### 5.4   Relationship Between Color and Year

The results from Table 1 demonstrated that the Image-based Estimation resulted in the highest correlation between the predicted results and the year. One reason for this is because the color information in the RGB images conveys a lot of information about the year. As shown in Fig. 7, older movie posters tended to have warmer colors, such as red or yellow fonts. But, the trend changed around the 1970s to the 1980s, and achromatic and cold colors were used more frequently afterward. In addition, dark backgrounds became prominent in the 1980s to the 2000s. Figure 8 shows example title images with large residuals. The "Brooklyn" and "U.N.C.L.E." titles, in particular, are warm colors and accordingly were estimated to be older.

### 5.5   Relationship Between Shape and Year

Figure 9 shows the four title images with the smallest residuals for each year in the Shape-based Estimation with MSE+Tukey's biweight loss. In the 1950s and earlier, thicker lines were used more frequently in characters. When examining the presence or absence of serifs, most titles for every decade are sans-serif typefaces, with exception to the 1990s and one from the 1980s. This trend can also be seen in other methods, as shown in Figs. 7 and 10. This captures the

(a) Image-based Estimation

(b) Shape-based Estimation

(c) Feature-based Estimation

**Fig. 8.** Example title images that have large residuals. Each method uses the proposed MSE+Tukey loss.

**Fig. 9.** Shape-based estimation with MSE+Tukey loss: top four smallest residual title images of each decade. In each case, the year was successfully estimated to the year.

fact that serif fonts were frequently used for movie posters in the 1990s. Another observation that can be found is that for the 1960s and before, the fonts of the movie titles are often skewed, curved, different sizes, or italic. This trend is lost after the 1980s, where the fonts are generally straight and uniform.

## 5.6 Relationship Between Local Features and Year

Even though the results of Feature-based Estimation performed the worst, there are still clear trends in the SIFT features that can be observed. As shown in Fig. 10, thicker lines were mostly used before the 1970s, but the trend changed and thinner lines were mostly used after the 1990s. In the 1960s and 1970s, there were also a lot of fonts with decorations. As mentioned in Sect. 5.5, in the 1990s, most the title images are serif typefaces. However, unlike Shape-based Estimation, there are still title images with serif typefaces in the 2000s and later when using Feature-based Estimation.

**Fig. 10.** Feature-based Estimation with MSE+Tukey loss: top four smallest residual title images of each decade. In each case, the year was successfully estimated to the year.

## 6 Conclusion

In this paper, we attempt to find the trends in font usage using robust regression on a large collection of text images. We use the font images collected from movie posters which are used because they are created by professional designers and portray a wide range of fonts. We found that it is very difficult to find the trends in font usage due to being full of exceptions. We tackle this difficult regression task by a robust regression method based on deep neural networks with a combination of MSE and Tukey's biweight loss.

We experimented using three methods: Image-based Estimation, Shape-based Estimation, and Feature-based Estimation. Using these different estimation methods, we compared L1 loss, MSE loss, Tukey's biweight loss, and the proposed MSE+Tukey with the automatic loss switching. As a result, It becomes clear that the benefit of the proposed method is that the loss is a balance between the robustness and emphasis on the outliers. In addition, It is found that the font usage, such as the thickness, slant of characters, and the presence or absence of serifs, has changed over time. We hope that the findings in this paper will contribute to a better understanding of the historical changes in font design.

## References

1. Baek, Y., Lee, B., Han, D., Yun, S., Lee, H.: Character region awareness for text detection. In: CVPR (2019)
2. Belagiannis, V., Rupprecht, C., Carneiro, G., Navab, N.: Robust optimization for deep regression. In: ICCV (2015)
3. Cavazza, J., Murino, V.: Active regression with adaptive Huber loss. arXiv preprint arXiv:1606.01568 (2016)
4. Chu, W.T., Guo, H.J.: Movie genre classification based on poster images with deep neural networks. In: MUSA2 (2017)

5. Fischer, P., Dosovitskiy, A., Brox, T.: Image orientation estimation with convolutional networks. In: Gall, J., Gehler, P., Leibe, B. (eds.) GCPR 2015. LNCS, vol. 9358, pp. 368–378. Springer, Cham (2015). https://doi.org/10.1007/978-3-319-24947-6_30

6. Gozuacik, N., Sakar, C.O.: Turkish movie genre classification from poster images using convolutional neural networks. In: ICEEE (2019)

7. Huber, P.J.: Robust estimation of a location parameter. The Ann. Math. Stat. **35**(1), 73–101 (1964)

8. Jia, Y., Jeong, J.H.: Deep learning for quantile regression: deepquantreg. arXiv preprint arXiv:2007.07056 (2020)

9. Karimpouli, S., Tahmasebi, P.: Image-based velocity estimation of rock using convolutional neural networks. Neural Netw. **111**, 89–97 (2019)

10. Kim, J., Kim, J., Suk, H.J.: Layout preference for Korean movie posters: contextual background or character dominance. In: IASDR (2019)

11. Kingma, D.P., Ba, J.: Adam: a method for stochastic optimization. arXiv preprint arXiv:1412.6980 (2014)

12. Krishnamoorthy, N., Ramya, K.S., Pavithra, K., Naveenkumar, D.: TV shows popularity and performance prediction using CNN algorithm. J. Adv. Res. Dyn. Control Syst. **12**(SP7), 1541–1550 (2020)

13. Kundalia, K., Patel, Y., Shah, M.: Multi-label movie genre detection from a movie poster using knowledge transfer learning. Augment. Hum. Res. **5**(1), 1–9 (2020)

14. Lathuiliere, S., Mesejo, P., Alameda-Pineda, X., Horaud, R.: A comprehensive analysis of deep regression. IEEE Trans. Pattern Anal. Mach. Intell. **42**(9), 2065–2081 (2020)

15. Lin, M., Chen, Q., Yan, S.: Network in network. In: ICLR (2014)

16. Lin, M., Chen, C.: Short-term prediction of stock market price based on GA optimization LSTM neurons. In: ICDLT (2018)

17. Lowe, D.: Object recognition from local scale-invariant features. In: ICCV (1999)

18. Mahendran, S., Ali, H., Vidal, R.: 3d pose regression using convolutional neural networks. In: CVPR Workshops (2017)

19. Mahlknecht, J.: Three words to tell a story: the movie poster tagline. Word Image **31**(4), 414–424 (2015)

20. Mehtab, S., Sen, J.: Stock price prediction using CNN and LSTM-based deep learning models. In: DASA (2020)

21. Niu, Z., Zhou, M., Wang, L., Gao, X., Hua, G.: Ordinal regression with multiple output CNN for age estimation. In: CVPR (2016)

22. Piao, Y., Chen, A., Shang, Z.: Housing price prediction based on CNN. In: ICIST (2019)

23. Qiang, Z.: Calligraphy font is analysed in the level of new appearance in the movie poster design. In: Art and Design, p. 11 (2016)

24. Selvin, S., Vinayakumar, R., Gopalakrishnan, E.A., Menon, V.K., Soman, K.P.: Stock price prediction using LSTM, RNN and CNN-sliding window model. In: ICACCI (2017)

25. Shahid, M.: Title design in bollywood movie posters: design features, trends, and the role of technology. The Int. J. Vis. Design **14**(4), 15–33 (2021)

26. Shahid, M., Bokil, P., Kumar, D.U.: Title design in bollywood film posters: a semiotic analysis. In: ICoRD (2014)

27. Shi, Z.: Preliminary exploration of calligraphy elements in Chinese movie poster design. In: IELSS (2019)

28. Sirattanajakarin, S., Thusaranon, P.: Movie genre in multi-label classification using semantic extraction from only movie poster. In: ICCCM (2019)

29. Supriya, K., Bharathi, P.: Movie classification based on genre using map reduce. Int. J. Appl. Res. Inform. Tech. Comput. **11**(1), 27 (2020)
30. Wang, L.: The art of font design in movie posters. In: ICASSEE (2019)
31. Wi, J.A., Jang, S., Kim, Y.: Poster-based multiple movie genre classification using inter-channel features. IEEE Access **8**, 66615–66624 (2020)
32. Zhang, C., Liu, S., Xu, X., Zhu, C.: C3AE: exploring the limits of compact model for age estimation. In: CVPR (2019)
33. Zhou, Y., Zhang, L., Yi, Z.: Predicting movie box-office revenues using deep neural networks. Neural Comput. Appl. **31**(6), 1855–1865 (2017). https://doi.org/10.1007/s00521-017-3162-x

# Binarization Strategy Using Multiple Convolutional Autoencoder Network for Old Sundanese Manuscript Images

Erick Paulus[1,3]([✉]) [ID], Jean-Christophe Burie[2] [ID], and Fons J. Verbeek[1] [ID]

[1] Leiden Institute of Advanced Computer Science, Leiden University,
Leiden, The Netherlands
{e.paulus,f.j.verbeek}@liacs.leidenuniv.nl
[2] Laboratoire Informatique Image Interaction (L3i) University of La Rochelle,
Avenue Michel Crépeau, 17042 La Rochelle Cedex 1, France
jcburie@univ-lr.fr
[3] Computer Science Department, Universitas Padjadjaran, Bandung, Indonesia
erick.paulus@unpad.ac.id

**Abstract.** The binarization step for old documents is still a challenging task even though many hand-engineered and deep learning algorithms have been offered. In this research work, we address foreground and background segmentation using a convolutional autoencoder network with 3 supporting components. The assessment of several hyper-parameters including the window size, the number of convolution layers, the kernel size, the number of filters as well as the number of encoder-decoder layers on the network is conducted. In addition, the skip connections approach is considered in the decoding procedure. Moreover, we evaluated the summation and concatenation function before the up-sampling process to reuse the previous low-level feature maps and to enrich the decoded representation. Based on several experiments, we determined that kernel size, the number of filters, and the number of encoder-decoder blocks have a little impact in term of binarization performance. While the window size and multiple convolutional layers are more impactful than other hyper-parameters. However, they require more storage and may increase computation costs. Moreover, a careful embedding of batch normalization and dropout layers also provides a contribution to handle overfitting in the deep learning model. Overall, the multiple convolutional autoencoder network with skip connection successfully enhances the binarization accuracy on old Sundanese palm leaf manuscripts compared to preceding state of the art methods.

**Keywords:** Binarization · Autoencoder · Palm leaf manuscript · Deep learning

## 1 Introduction

In document image analysis (DIA), the binarization task is responsible for determining which pixels belong to the character and which to the background. As a

© Springer Nature Switzerland AG 2021
E. H. Barney Smith and U. Pal (Eds.): ICDAR 2021 Workshops, LNCS 12917, pp. 142–157, 2021.
https://doi.org/10.1007/978-3-030-86159-9_10

result, binarization plays a key role on the performance of the OCR. Nonetheless, there are numerous difficulties in binarizing old document images, including fading text, age color change, crack, and bleed-through noises [1]. Furthermore, the binarization problems are exacerbated by a variety of tools and acquisition methods. Owing to inadequate lighting and camera blitz, the image quality may suffer from blur, low contrast, and uneven illumination [2].

Many works to solve these problems have been proposed in the literature. Initially, several researchers advocated hand-engineered solutions that relied on image characteristics. The global thresholding procedure is one of the most basic binarization methods. The Otsu algorithm [3] is the most widely used thresholding technique and is commonly used as a benchmarking tool. When dealing with degraded images, a single threshold seldom produces acceptable binarization results. Since 1986, researchers have been pursuing local adaptive thresholding, for example, Niblack's [4] and Bernsen's [5]. It has an advantage in finding the adaptable threshold per patch or window. Some authors utilize histograms and statistics to gain a better threshold by calculating the minimum, maximum, mean, and standard deviation of intensity for each window. Another advantage of these handcrafted methods is that they are less expensive to compute because they only use the image characteristics and do not need a labelled image to be trained. They also successfully yield a good binarized image over uniform and low contrast problems. Nonetheless, local adaptive thresholding algorithms are not enough to tackle images suffering from uneven illumination. For example, the F-measure of hand-engineered binarization approaches on old Sundanese manuscripts, only reached a score of 46.79 [6].

Another challenge of traditional adaptive thresholding is that there are several parameters to be adjusted to achieve better performance. As a result, it is regarded as a disadvantage. Several methods for automatically adjusting those parameters have been proposed and evaluated. Howe improved his previous algorithm by incorporating automatic parameter tuning with pairwise Canny-based approach [7]. Later, Mesquita optimized Howe's method using a racing algorithm [8]. Kaur et al. improved Sauvola's thresholding [9] by applying stroke width transform (SWT) to automatically calculate window size.

The advent of deep learning algorithms allowing machine work with or without any supervision has opened a new gate to address more complex problems and tune automatically parameter. In 2015, Pastor et al. first introduced the CNN implementation on binarization task by classifying every pixel into background and foreground based on a sliding window [10]. In the meantime, a transformation of CNN into a Fully Convolutional Networks (FCN) topology was announced for the semantic segmentation task where every pixel is determined as one of K classes [11,12]. Later, Tensmeyer et al. defined binarization likewise a semantic segmentation with K = 2 and gained a better performance compared to CNN [13]. Recently, a combination of an hourglass-shaped deep neural network (DNN) and convolution network was proposed to handle binarization challenges. It is named as Convolutional Auto Encoder (CAE) network that has the main advantage to convolve image-to-image as well as to learn compressed and distributed encoding (representation) from a training set. Furthermore, in [14,15], a composite of residual block and CAE outperformed several prior DNN especially on LRDE DBD [16], DIBCO [17], and palm leaf manuscript [18] datasets.

In this paper, we firstly aimed to assess the contribution of five parameters in CAE and three supporting components including the window size, the number of encoder-decoder layers, the number of convolution layers, the number of filters, the kernel size, the skip connector, the dropout size, and the existence of the batch normalization to achieve the better binarization performance. Second, we aimed to assess the effectiveness of the CAE network to deal with binarization challenges on old Sundanese palm leaf manuscripts. The main contribution of this work is to propose a strategy to determine the best parameters to configure CAE. We present a set of comprehensive experiments analyzing in details the influence of both hyper-parameters of CAE and supporting components. So that, the efficacy of each hyper-parameter can be exploited to build an appropriate CAE network for processing the images of our Old Sundanese palm leaf manuscripts. As far we know, to the proposed CAE network in this paper is the first architecture with several convolution layers for each encoder-decoder block in a hour-glass model.

This paper is organized as follow. The second section describes the related works in terms of CAE network, skip connection, dropout, and batch normalization. Experiment procedures and evaluation metrics are detailed in the third section. Then, the fourth section presents the results of the experiments. Finally, the conclusions and future research topics are discussed in the last section.

## 2    Related Work

### 2.1    Convolutional Autoencoder Network

Autoencoder is a type of unsupervised neural network which can learn latent representation from input data [19]. In this case, it does not need a labeled training set. Autoencoder is also well-known as a dimensionality reduction algorithm, feature detector, generative model, or unsupervised pretraining of deep learning model. Several implementations of this neural network in computer vision area are image denoising [20], augmentation data [20], anomaly detection [21,22], image restoration [23] and invertible grayscale [24]. In another case, the execution of autoencoder in a two-level segmentation task or binarization task still needs labeled image or ground truth binary image. In other words, the weights of this deep neural network architecture are gained from supervised learning.

In general, an autoencoder is made up of three parts: an encoder part, an internal hidden layer (the code part), and a decoder part. The encoder is responsible for compressing the input and producing the code or latent representation, whereas the decoder is responsible for rebuilding or unzipping the code part provided by the hidden layer. Furthermore, autoencoders use fully-connected feedforward neural networks for training, following the same procedure as back-propagation.

Several types of autoencoders have been proposed successively, including simple autoencoders, convolutional autoencoders (CAE), and variational autoencoders. The similarity of which is that the dimension of output and input are always same. Nonetheless, the different types of autoencoders can be distinguished by the manner in which the latent representation (the code part) is constructed with the input data, then used to reconstruct the output data. The simple autoencoder employs a dense network, which means that each output is linearly connected to each input. The convolution autoencoder can use convolution operators to generate a low-level representation feature map, which can then be used to analyze the images. So, the spatial information of the image is preserved, which can be useful during the training step. Variational autoencoders propose a learning process based on a probability distribution function by imposing additional constraints on the encoded representations using Bayesian inference [25]. The purpose of a variational autoencoder is to serve as a generative model with the ability to generate new data instances. Nonetheless, the principle is much easier to understand than those of Generative Adversarial Networks.

In this study, we focus the use of the convolutional autoencoder on the binarization task. We decompose the CAE for experimental purposes into convolution blocks, deconvolution blocks, encoder blocks, and decoder blocks. As shown in Fig. 1, the decoder block is generally mirrored to the encoder block. Each encoder block employs a convolution block with a stride of 2, in order to compress the input by dividing the dimension of the output by two. During the model training process, each decoder block uses transposed convolution with the same stride as the encoder to up sample the input and learn how to fill in missing parts. In addition, some researchers demonstrated the use of a pooling layer for down sampling in the encoder and an up-sampling method in the decoder, as explained by [22], allows to solve the problem of anomaly image detection. In the case of image restoration using the Residual Encoder-Decoder Network (RED-Net), a pooling layer is not recommended for down sampling because deconvolution in the decoder does not work well [23]. However, the use of strided convolution and pooling layers is still debatable. Because each brings its own set of benefits and drawbacks. [26] proposes another study on strided convolution layer and pooling layer.

Aside from stride, several parameters in the convolution and deconvolution blocks must be configured: the number of filters, kernel size, padding type, and activation function. RED-Net advised to configure layers with a kernel size of 3 × 3 and 64 filters [23]. Meanwhile, Selectional Auto-Encoder (SAE) recommended that the convolution block and deconvolution be fixed with kernel sizes of 5 × 5 and 64 filters [15]. In addition, both blocks have a rectified linear activation ReLU and padding "same".

In the context of the encoder-decoder block, we must consider the number of encoder and decoder blocks, as well as the number of convolution/deconvolution within them. In a typical CAE, one convolution is used for each encoder and decoder block. Indeed, in convolution/deconvolution, these parameters are closely related to the input size and stride size. For instance, if there are five encoder blocks and the size of the squared input is 256, the size of the encoded representation after the fifth encoding with stride 2 in each convolution layer is $8 \times 8$. If the same configuration is used, but the number of encoders increases to eight, the size of the encoded representation in the final encoding is $1 \times 1$. The information in the latent representation is now meaningless. Fixing the larger patch size, on the other hand, will increase the computation cost. Furthermore, the patch should be smaller in size than the image. As a result, we need to manage them carefully.

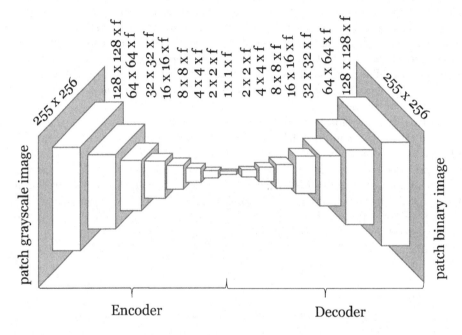

**Fig. 1.** The standard architecture of convolutional autoencoder.

## 2.2   Skip Connections Method

The skip connection is used for two reasons [23]. First, the feature banks delivered by skip connections contain significantly more information, which will be useful for obtaining an enriched representation in the next layer. Second, the skip connection in the deeper network allows training easier.

Concatenation and summation methods are the two most common skip connection variants. ResNet [27], for example, uses element-wise summation as a standard skip-connection architecture. It enables the features to be refined as it moves through the network's various layers. The number of features is preserved by using additional skip-connections in which the feature dimensions are the same before and after the skip-connection. Residual Encoder Decoder Network

(RED-Net) [23] and Selectional Auto Encoder (SAE) [15] are two other deep learning architectures that use the summation method.

Concatenated skip connections, on the other hand, increase the dimension of the representations. Nonetheless, it allows to reuse representations from previous layers. Concatenation's main advantage is that it provides more information, which is useful in deep supervision and may result in better gradient propagation. DenseNet [28] and Inception [29] are two samples of deep learning architecture which utilize the concatenation approach.

### 2.3  Dropout Regularization

Aside from incurring higher computation costs, training in a deeper network with large hyperparameters may result in an overfitting problem. Dropout is a regularization technique that is commonly used to overcome this problem. The main idea is to drop units at random during training to avoid excessive coadaptation [30]. The probability of dropping a unit is a float value in range 0 and 1. The drop value of 0 indicates that all units are preserved. However, the higher the dropout, the smaller the unit. Dropout can be placed in either the input or hidden layers. Dropout close to 0 is ideal for the input layer, while dropout between 0.2 and 0.5 is ideal for the hidden layer.

### 2.4  The Batch Normalization

To address the training speed issue, Ioffe and Szegedy developed a batch normalization (BN) mechanism that reduced internal covariate shift. BN has several advantages, including the use of higher learning rates, a low intention on initialization, and, in some cases, a reduction in the dropout layer [31]. They demonstrated some promising results with the corresponding to BN and Dropout experiments into Inception model. Many current deep learning architectures, including [32] and [33], use BN because of its ability to reduce training time while avoiding overfitting. An empirical study of BN over several deep neural networks was also presented in [34].

Other arguments with respect to BN and Dropout mechanisms are found in the literature. Garbin et al. presented a comprehensive comparison study between them on multilayer perceptron networks and convolutional neural networks (ConvNet) [35]. As practical guidelines for ConvNet, BN can improve accuracy with a short training period and should be the first consideration when optimizing a CNN. Dropout regularization, on the other hand, should be used with caution because it can reduce accuracy.

## 3  Dataset and Methodology

### 3.1  Datasets

In our experimental study, we used old Sundanese palm leaf manuscripts written by different scribers. The data set consists in 61 images split in a training set of 31 images and a testing set of 30 images. This dataset is the one used for the ICFHR 2018 competition [36]. It allows to compare our results to the methods presented in the competition. In addition, each image has its own ground truth image, which was manually created using Pixlabeler software by a group of philologists and students from the Sundanese Department at Universitas Padjadjaran in Indonesia.

### 3.2    Experiment Procedures

We investigate the impact of five CAE architecture parameters and three other components on binarization performance. We conduct the experiments on old Sundanese palm leaf manuscripts. Note that these experiments do not use augmented data. As a starting point, we followed the SAE configuration recommendation [15]. Instead of processing the image as a whole, image data is trained locally using a patch-wise approach. The weights of network are initialized based on Xavier weight initialization to manage the variance of activations and back-propagated gradients, resulting in faster convergence and higher accuracy. They are then optimized using stochastic gradient descent, a batch size of 10, and an adaptive learning rate of 0.001 [37]. The training step iteration is set to 200 epochs. However, if there is no decrease in training loss after 20 epochs, the early stopping strategy is executed. Several configurations are available, including a square window with a side size of 256, 64 filters and a kernel size of five for each convolution layer, five encoder blocks and five decoder blocks, a skip connection for each decoder block, BN, and a dropout of 0.

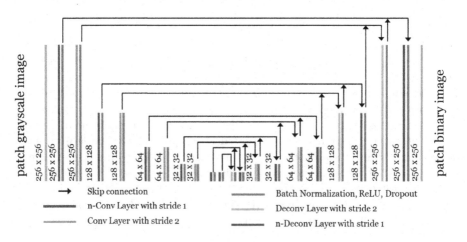

**Fig. 2.** The architecture of CAE network with its supporting components.

The first experiment scenario focus on CAE hyperparameters. We set five window sizes, which is successively to 64, 128, 256, 320, and 384. Then, we take the best two window sizes and change the number of filters to 16, 32, 64, 96, and 128 while keeping the kernel size at $3 \times 3$ and $7 \times 7$. Several modifications to the number of encoder-decoder blocks is simulated based on the best two previous experiments. Afterward, we evaluated the number of strided convolution/deconvolution inside encoder-decoder block. The rectified linear activation function (ReLU) and padding "same" are applied in all convolution and deconvolution in this scenario. The position of each CAE layer and its supporting components is depicted in Fig. 2.

Secondly, a review of supporting components is carried out. We begin by varying the dropout value between 0.1 and 0.5 with and without BN. The existence of skip-connections for both summation and concatenation is then evaluated using the best two results of the previous configurations.

### 3.3  Evaluation Method

During the training process, F-measure (Fm) is used to evaluate the CAE architecture, because this measure is a popular metric for imbalanced classification [38]. In the context of binarization, this means that the number of background pixels is much greater than the number of foreground pixels.

## 4  Result and Discussion

All experiments are written in Python programming language and supported by packages such as Scikit-Learn, Numpy, Keras, and Tensorflow. In terms of hardware, we use a GeForce 2x RTX 2070 (8 GB/GPU) Graphics Processing Unit (GPU) and 48 GB RAM. All discussions are based on the binarized images predicted by the best-weighted model on the testing dataset.

### 4.1  Discussion on Five Hyper Parameters Experiments

In the first experiment we assess the influence of window size of input image and the numbers of filters in CAE model which apply three supporting components. As shown in Fig. 3, we evaluated five different squared window sizes ranging from $64 \times 64$ to $384 \times 384$ in four different numbers of filters (f16, f32, f64, and f96). The box size of 384 had the lowest accuracy. Because, the boxes are too wide or long in comparison to the original image which may also result in poor performance. While boxes with sides of 64, 128, 256, and 320 produce comparable results. It suggests that those four window sizes can be used to binarize images, particularly for old Sundanese palm leaf manuscripts. However, we can see that a window size of 320 provides the most stable performance, with F-measure greater than 68 for all number of filters.

Furthermore, the highest measurement score on this experiment is obtained for a window size of 320 and 96 filters, with a F-measure of 68.44. Another important consideration is that using a larger window requires more storage and has an impact on computation costs. Afterward, the kernel size is evaluated using the previous best configuration. As shown in Fig. 4, three kernel sizes (3, 5, and 7)

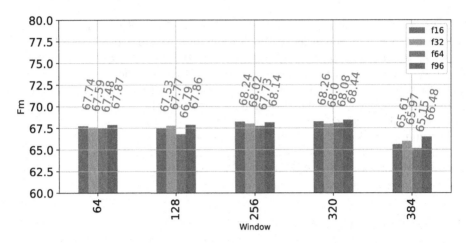

**Fig. 3.** The binarization performance of autoencoder network based on the number of filters across window size.

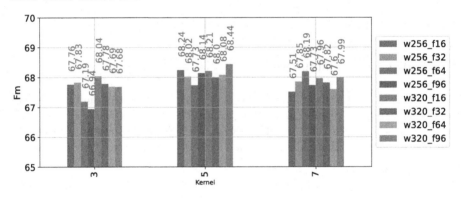

**Fig. 4.** The binarization performance of autoencoder network based on the number of filters across kernel.

are evaluated in eight CAE models by combining two window sizes (256 and 320) and four number of filters (16, 32, 64, and 96). The impact of kernel size varies depending on the size of the window and the number of filters. The best performance is reached with a window size of 320, 96 filters, and a kernel size of 5, which is the same as in the previous experiment.

The number of encoder-decoder blocks are evaluated in the following experiment. We reevaluated the best model from the previous experiment and a baseline configuration by varying the number of encoder and decoder blocks. As shown in Fig. 5, the performance rate fluctuates as the encoder-decoder block is added. Nonetheless, the strided convolution layer within the encoder and decoder blocks must be considered. In a configuration with a squared window size of 320 and strided convolution of 2, for example, we cannot use more than 6 encoder blocks (blue line). Because the dimension of the encoded representation from the last encoder and the dimension of the decoded representation from the first decoder are not the same. This indicates that we must set this parameter with care.

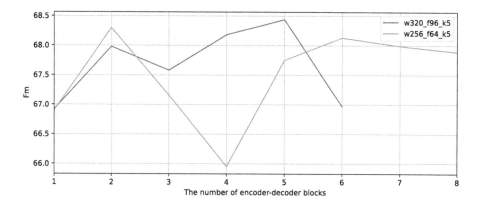

**Fig. 5.** The binarization performance of autoencoder network based on the number of encoder and decoder blocks. (Color figure online)

Table 1 shows how the presence of multiple convolution layers in an autoencoder network affects computation cost and performance rate. With each additional layer, the computation requires more storage to handle larger trainable and non-trainable parameters. We did not consider storage costs in this study, instead focusing on the impact of multiple convolution layers in binarization performance. We only evaluate multiple convolution layer (MCL) in CAE model with window size of 256, 64 filters, and a kernel size of 5 which is the default configuration proposed in [15]. The use of MCL may improve accuracy, but it requires more computation time and storage. According to the results of this experiment, 15 MCL in the encoder block and 15 MCL in the decoder block performed better than other settings.

**Table 1.** The binarization performance of autoencoder network (W256_f64_k5) based on the number of multiple convolution layers.

| MCL in encoder | MCL in decoder | Trainable params | Non-trainable params | Fm |
|---|---|---|---|---|
| 5 | 5 | 926,721 | 1,280 | 67.73 |
| 10 | 10 | 1,952,641 | 2,560 | 68.57 |
| 15 | 15 | 2,978,561 | 3,840 | 69.58 |
| 20 | 20 | 4,004,481 | 5,120 | 69.36 |
| 25 | 25 | 5,030,401 | 6,400 | 69.45 |

### 4.2   Experiments on Batch Normalization, Dropout, Skip-Connection in CAE Network

To assess the impact of the three supporting components, we set several static hyper-parameters as follows: 5 encoder/decoder blocks, each with one convolution layer and a kernel size of 5. Figure 6 and Fig. 7 depict the types of accuracy performance measures performed by CAE networks with various supporting components, covering a dropout rate range of 0 to 0.5 in the binarization task of old Sundanese palm leaf manuscripts. We can see that the accuracy of the CAE networks increases in general. The configuration with added skip connection and no batch normalization provides the best performance overall if many filters are occupied.

First, a discussion of dropout regularization is presented. Based on 4 CAE models, we discovered that the existence of dropout layer relatively increases the accuracy rate. Without dropout layer (dropout = 0.0), the F-measure reaches most of the time the lowest level for a given CAE model. On contrary, the dropout layer in range 0.2 and 0.5 generally may obtain the better accuracy. In order to obtain a global optimum performance, dropout rate could be different for each model. This conclusion corresponds to the suggestion made by the authors who proposed the dropout concept [30]. It means that dropout rate needs to be set carefully.

In terms of skip-connection, CAE models with no residual blocks cannot outperform CAE models with either added or concatenated skip-connection. The highest accuracy score for a CAE model with a residual block is based on a window size of $320 \times 320$, 96 filters, a kernel size of 5, and a dropout rate of

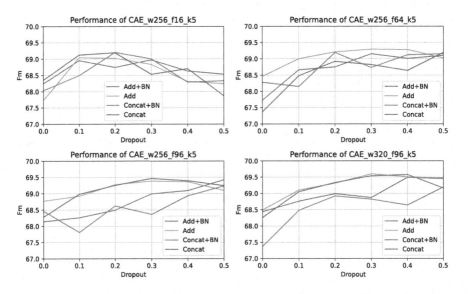

**Fig. 6.** An accuracy performance comparison of different dropout rates, skip connection functions and batch normalization on CAE network.

0.3. In many cases, such as the number filters of 96, the summation function (blue and orange lines) outperforms the concatenation function (green line and red line). We recommend using the residual block with the summation function. Another important point is that transferring meaningful information from the previous layer via the skip method may improve accuracy. Figure 6 depicts the performance of a residual CAE network with an accuracy rate higher than 67. In contrast, as shown in Fig. 7, a CAE model with a non-residual block can accurately segment below this rate.

As shown in Fig. 7, the BN has a significant impact on the CAE network with no skip-connection. On all dropout rates, BN within non-residual CAE (blue line) consistently outperforms the configuration without BN (orange line). Additionally, the model of residual CAE with BN and 50% dropout after each convolution block performs slightly better than the model without BN, as depicted in Fig. 6. In contrast, the residual CAE with a dropout rate of 0.4 or less in some configuration produces less performance than the model without BN. Meanwhile, the model without residual block and batch normalization fails to segment foreground and background because the F-measure is most of the time lower than 55%.

Figure 8 illustrates the qualitative results of several binarization methods applied on testing dataset of old Sundanese palm leaf manuscripts. Several experiments on supporting components in CAE architecture revealed that the highest accuracy rate (F-measure) is 69.60 for a $320 \times 320$ window size, 96 filters, $5 \times 5$ kernel, with added skip-connection, no batch normalization, and dropout rate of 0.3. Another intriguing aspect is the possibility of improving accuracy by using multiple convolutional layers. The accuracy improvement of CAE_add_BN_w256_f64_k5

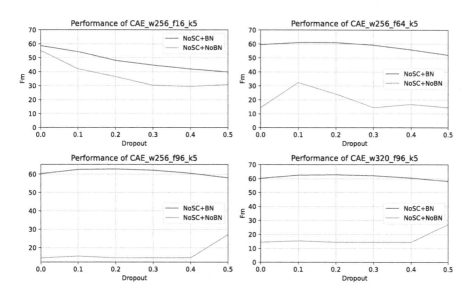

**Fig. 7.** A comparison of the dropout and batch normalization (BN) influence on CAE network without skip-connection (SC). (Color figure online)

model can be seen in the configuration of CAE_add_BN_30MCL_w256_f64_k5. Overall, the evaluation metric of the CAE model also validates the promising results and outperforms the winner of ICFHR 2018 [36]. Moreover, from the image visualization of a sample image, the binarized images produced by CAE models are cleaner than prior state of the art approaches in term of separate foreground from random distributed noise, as shown in Fig. 9.

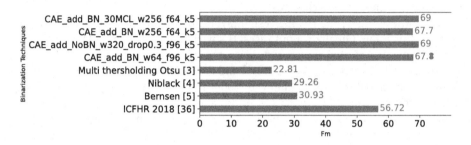

**Fig. 8.** Evaluation metric on several binarized images.

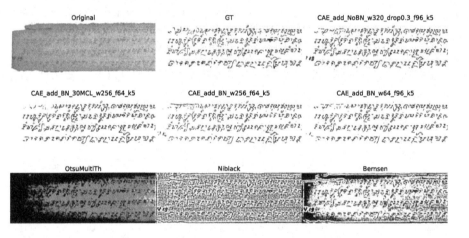

**Fig. 9.** A sample of original palm leaf manuscript image, ground truth image (GT) and the binarization results.

## 5   Conclusion

Several experiments on the CAE network's hyper-parameters and three supporting components were carried out for the binarization task on old Sundanese palm leaf manuscripts. Tuning the parameters can improve the performance of the binarization. Nevertheless, some parameters such as kernel size, number of filters and the number of encoder-decoder blocks, don't play a significant role on accuracy measurement. However, in this network, window size and the multiple convolution layer have a significant impact. Furthermore, skip-connection of both summation and concatenation functions that supply the encoded representation from the previous layer can increase accuracy. To handle overfitting in deep network, dropout layer needs to be set up in range 0.2 and 0.5. Moreover, the batch normalization also provides a big impact if there is no residual block in CAE network. Therefore, to achieve good performance, we need to choose carefully the hyper-parameters and the supporting components. These experimental results will help us for future works on document image analysis, specifically for evaluating the CAE network with larger datasets of palm leaf manuscripts.

**Acknowledgments.** We would like to thank Riki Nawawi and Undang Darsa as Sundanese philologists for their fruitful discussions regarding old Sundanese manuscripts. This work is supported by the Directorate General of Higher Education, Ministry of Education, Culture, Research and Technology, the Republic of Indonesia through the funding of the BPPLN-DIKTI Indonesian Scholarship Program.

# References

1. Sulaiman, A., Omar, K., Nasrudin, M.F.: Degraded historical document binarization: a review on issues, challenges, techniques, and future directions. J. Imaging **5**(4), 48 (2019)
2. Tensmeyer, C., Martinez, T.: Historical document image binarization: a review. SN Comput. Sci. **1**(3), 1–26 (2020). https://doi.org/10.1007/s42979-020-00176-1
3. Otsu, N.: A threshold selection method from gray-level histograms. IEEE Trans. Syst. Man Cybern. **9**(1), 62–66 (1979)
4. Niblack, W.: An Introduction to Digital Image Processing. Prentice-Hall, Englewood Cliffs (1986)
5. Bernsen, J.: Dynamic thresholding of gray level image. In: ICPR 1986: Proceedings of International Conference on Pattern Recognition, Berlin, pp. 1251–1255 (1986)
6. Kesiman, M.W.A., et al.: Benchmarking of document image analysis tasks for palm leaf manuscripts from Southeast Asia. J. Imaging **4**(2), 43 (2018)
7. Howe, N.R.: Document binarization with automatic parameter tuning. Int. J. Doc. Anal. Recognit. **16**(3), 247–258 (2013)
8. Mesquita, R.G., Silva, R.M.A., Mello, C.A.B., Miranda, P.B.C.: Parameter tuning for document image binarization using a racing algorithm. Expert Syst. Appl. **42**(5), 2593–2603 (2015)
9. Sauvola, J., Seppanen, T., Haapakoski, S., Pietikainen, M.: Adaptive document binarization. In: Proceedings of the Fourth ICDAR, pp. 147–152 (1997)
10. Pastor-Pellicer, J., España-Boquera, S., Zamora-Martínez, F., Afzal, M.Z., Castro-Bleda, M.J.: Insights on the use of convolutional neural networks for document image binarization. In: Rojas, I., Joya, G., Catala, A. (eds.) IWANN 2015. LNCS, vol. 9095, pp. 115–126. Springer, Cham (2015). https://doi.org/10.1007/978-3-319-19222-2_10
11. Long, J., Shelhamer, E., Darrell, T.: Fully convolutional networks for semantic segmentation. In: 2015 IEEE Conference on Computer Vision and Pattern Recognition (CVPR), pp. 3431–3440 (2015)
12. Shelhamer, E., Long, J., Darrell, T.: Fully convolutional networks for semantic segmentation. IEEE Trans. Pattern Anal. Mach. Intell. **39**(4), 640–651 (2017)

13. Tensmeyer, C., Martinez, T.: Document image binarization with fully convolutional neural networks. In: 2017 14th IAPR ICDAR, pp. 99–104 (2017)
14. Peng, X., Cao, H., Natarajan, P.: Using convolutional encoder-decoder for document image binarization. In: 2017 14th IAPR ICDAR, pp. 708–713 (2017)
15. Calvo-Zaragoza, J., Gallego, A.J.: A selectional auto-encoder approach for document image binarization. Pattern Recognit. **86**, 37–47 (2019)
16. Lazzara, G., Levillain, R., Geraud, T., Jacquelet, Y., Marquegnies, J., Crepin-Leblond, A.: The SCRIBO module of the Olena platform: a free software framework for document image analysis. In: 2011 ICDAR, pp. 252–258 (2011)
17. Pratikakis, I., Zagoris, K., Barlas, G., Gatos, B.: ICFHR 2016 handwritten document image binarization contest (H-DIBCO 2016). In: Proceedings of the International Conference on Frontiers in Handwriting Recognition (ICFHR), pp. 619–623 (2016)
18. Burie, J.C., et al.: ICFHR2016 competition on the analysis of handwritten text in images of balinese palm leaf manuscripts. In: 2016 15th ICFHR, pp. 596–601 (2016)
19. Géron, A.: Hands-on Machine Learning with Scikit-Learn, Keras, and TensorFlow?: Concepts, Tools, and Techniques to Build Intelligent Systems. O'Reilly, Sebastopol (2019)
20. Feng, X., Jonathan, W., Q.M., Yang, Y., Cao, L.: An autuencoder-based data augmentation strategy for generalization improvement of DCNNs. Neurocomputing **402**, 283–297 (2020)
21. Ribeiro, M., Lazzaretti, A.E., Lopes, H.S.: A study of deep convolutional auto-encoders for anomaly detection in videos. Pattern Recognit. Lett. **105**, 13–22 (2018)
22. Heger, J., Desai, G., Zein El Abdine, M.: Anomaly detection in formed sheet metals using convolutional autoencoders. In: Procedia CIRP, pp. 1281–1285. Elsevier B.V. (2020)
23. Mao, X.-J., Shen, C., Yang, Y.-B.: Image restoration using very deep convolutional encoder-decoder networks with symmetric skip connections. In: Proceedings of the 30th International Conference on Neural Information Processing Systems, pp. 2810–2818. Curran Associates Inc., Red Hook (2016)
24. Xia, M., Liu, X., Wong, T.T.: Invertible grayscale. ACM Trans. Graph. **37**, 1–10 (2018)
25. Kingma, D.P., Welling, M.: Auto-encoding variational bayes. In: 2nd International Conference on Learning Representations, ICLR 2014, Canada, pp. 1–14 (2014)
26. Springenberg, J.T., Dosovitskiy, A., Brox, T., Riedmiller, M.: Striving for simplicity: the all convolutional net. In: ICLR-2015 (2015)
27. He, K., Zhang, X., Ren, S., Sun, J.: Deep residual learning for image recognition. In: 2016 IEEE Conference on Computer Vision and Pattern Recognition (CVPR), pp. 770–778 (2016)
28. Huang, G., Liu, Z., Van Der Maaten, L., Weinberger, K.Q.: Densely connected convolutional networks. In: 2017 IEEE Conference on CVPR, pp. 2261–2269 (2017)
29. Szegedy, C., Vanhoucke, V., Ioffe, S., Shlens, J., Wojna, Z.: Rethinking the inception architecture for computer vision. In: 2016 IEEE Conference on CVPR, pp. 2818–2826 (2016)
30. Srivastava, N., Hinton, G., Krizhevsky, A., Sutskever, I., Salakhutdinov, R.: Dropout: a simple way to prevent neural networks from overfitting. J. Mach. Learn. Res. **15**, 1929–1958 (2014)

31. Ioffe, S., Szegedy, C.: Batch normalization: accelerating deep network training by reducing internal covariate shift. In: The 32nd International Conference on Machine Learning, Lille, France (2015)
32. Szegedy, C., Ioffe, S., Vanhoucke, V., Alemi, A.A.: Inception-v4, inception-ResNet and the impact of residual connections on learning. In: Proceedings of the Thirty-First AAAI Conference on Artificial Intelligence, pp. 4278–4284. AAAI Press (2017)
33. Sun, S., Pang, J., Shi, J., Yi, S., Ouyang, W.: FishNet: a versatile backbone for image, region, and pixel level prediction. In: 32nd Conference on Neural Information Processing Systems, pp. 754–764 (2018)
34. Thakkar, V., Tewary, S., Chakraborty, C.: Batch normalization in convolutional neural networks - a comparative study with CIFAR-10 data. In: 2018 Fifth International Conference on Emerging Applications of Information Technology, pp. 1–5 (2018)
35. Garbin, C., Zhu, X., Marques, O.: Dropout vs. batch normalization: an empirical study of their impact to deep learning. Multimed. Tools Appl. **79**, 12777–12815 (2020)
36. Kesiman, M.W.A., et al.: ICFHR 2018 competition on document image analysis tasks for southeast Asian palm leaf manuscripts. In: Proceedings of ICFHR, pp. 483–488. IEEE (2018)
37. Kingma, D.P., Ba, J.: Adam: a method for stochastic optimization. In: International Conference on Learning Representations (ICLR) (2015)
38. Pastor-Pellicer, J., Zamora-Martínez, F., España-Boquera, S., Castro-Bleda, M.J.: F-measure as the error function to train neural networks. In: Rojas, I., Joya, G., Gabestany, J. (eds.) IWANN 2013. LNCS, vol. 7902, pp. 376–384. Springer, Heidelberg (2013). https://doi.org/10.1007/978-3-642-38679-4_37

# A Connected Component-Based Deep Learning Model for Multi-type Struck-Out Component Classification

Palaiahnakote Shivakumara[1(✉)], Tanmay Jain[2], Nitish Surana[2], Umapada Pal[2], Tong Lu[3], Michael Blumenstein[4], and Sukalpa Chanda[5]

[1] Faculty of Computer Science and Information Technology, University of Malaya, Kuala Lumpur, Malaysia
shiva@um.edu.myss
[2] Computer Vision and Pattern Recognition Unit, Indian Statistical Institute, Kolkata, India
umapada@isical.ac.in
[3] National Key Lab for Novel Software Technology, Nanjing University, Nanjing, China
lutong@nju.edu.cn
[4] University of Technology Sydney, Sydney, Australia
michael.blumenstein@uts.edu.au
[5] Department of Information Technology, Østfold University College, Halden, Norway
sukalpa@ieee.org

**Abstract.** Due to the presence of struck-out handwritten words in document images, the performance of different methods degrades for several important applications, such as handwriting recognition, writer, gender, fraudulent document identification, document age estimation, writer age estimation, normal/abnormal behavior of person analysis, and descriptive answer evaluation. This work proposes a new method which combines connected component analysis for text component detection and deep learning for classification of struck-out and non-struck-out words. For text component detection, the proposed method finds the stroke width to detect edges of texts in images, and then performs smoothing operations to remove noise. Furthermore, morphological operations are performed on smoothed images to label connected components as text by fixing bounding boxes. Inspired by the great success of deep learning models, we explore DenseNet for classifying struck-out and non-struck-out handwritten components by considering text components as input. Experimental results on our dataset demonstrate the proposed method outperforms the existing methods in terms of classification rate.

**Keywords:** Handwriting recognition · Writer identification · Connected component analysis · Deep learning · Struck-out words

## 1 Introduction

It is common to see writing errors such as crossing out words when a writer finds mistakes during writing answers or filling forms, writing letters, etc. [1]. As a result, handwritten documents are vulnerable to different struck-out words and noise during handwriting. It

E. H. Barney Smith and U. Pal (Eds.): ICDAR 2021 Workshops, LNCS 12917, pp. 158–173, 2021.
https://doi.org/10.1007/978-3-030-86159-9_11

can be noted that appearance of struck-out words varies with respect to writer to writer. Therefore, the presence of such struck-out handwritten words causes problems for several important applications, such as handwriting recognition [2], automatic answer script evaluation [3], gender identification [4], writer identification [2], document age estimation [5], writer age estimation [6], normal, abnormal behavior of writer identification, fraudulent document identification and forged text detection [7–9]. For all the above-mentioned applications, most of these methods require 'clean' handwritten documents without writing errors and crossed out words to achieve better results.

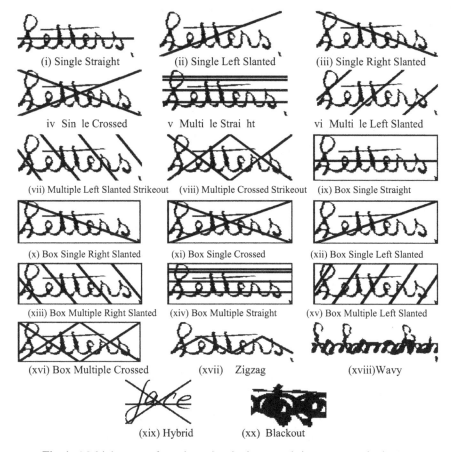

(i) Single Straight
(ii) Single Left Slanted
(iii) Single Right Slanted
iv  Sin le Crossed
v  Multi le Strai ht
vi  Multi le Left Slanted
(vii) Multiple Left Slanted Strikeout
(viii) Multiple Crossed Strikeout
(ix) Box Single Straight
(x) Box Single Right Slanted
(xi) Box Single Crossed
(xii) Box Single Left Slanted
(xiii) Box Multiple Right Slanted
(xiv) Box Multiple Straight
(xv) Box Multiple Left Slanted
(xvi) Box Multiple Crossed
(xvii)  Zigzag
(xviii)Wavy
(xix) Hybrid
(xx) Blackout

**Fig. 1.** Multiple types of struck-out handwritten words in answers script images

However, in reality, obtaining clean documents without errors for all the situations is challenging. Thus, the performance of methods for the above-mentioned applications degrades. This has motivated us to address the challenge of struck-out and non-struck-out handwritten word classification in this work. Since crossing out handwritten text depends on a person's mindset, context, and mistake, one cannot expect a particular type of struck-out words. This makes the problem more complex and challenging. The possible type of struck-out words are listed in Table 1, and the sample images are shown in Fig. 1, where it is observed that classification of struck-out words from non-struck out words is not easy because of multiple types of struck-out words.

Therefore, this work focuses on developing a new method for classifying struck-out and non-struck-out words such that the performance of handwriting analysis improves. The main contributions of the proposed work are as follows. (i) Use of stroke width information for detecting text components is new for classification. (ii) Exploring the DenseNet model for classifying struck-out and non-struck-out words is novel in contrast to the existing methods, which generally use handcrafted features such as intersection points and properties of cross-out lines. (iii) The way the proposed work combines connected component analysis and deep learning for successful classification is the key contribution compared to existing work.

**Table 1.** Multi-type struck-out handwritten words

| (i)     | Single Straight Strikeout         | (xi)    | Box Single Slanted Right Strikeout   |
|---------|-----------------------------------|---------|--------------------------------------|
| (ii)    | Single Slanted Left Strikeout     | (xii)   | Box Single Crossed Strikeout         |
| (iii)   | Single Slanted Right Strikeout    | (xiii)  | Box Multiple Straight Strikeout      |
| (iv)    | Single Crossed Strikeout          | (xiv)   | Box Multiple Slanted Left Strikeout  |
| (v)     | Multiple Straight Strikeout       | (xv)    | Box Multiple Slanted Right Strikeout |
| (vi)    | Multiple Slanted Left Strikeout   | (xvi)   | Box Multiple Crossed Strikeout       |
| (vii)   | Multiple Slanted Right Strikeout  | (xvii)  | Zigzag Strikeout                     |
| (viii)  | Multiple Crossed Strikeout        | (xviii) | Wavy Strikeout                       |
| (ix)    | Box Single Straight Strikeout     | (xix)   | Hybrid Strikeout                     |
| (x)     | Box Single Slanted Left Strikeout | (xx)    | Blackout Strikeout                   |

# 2   Related Work

Since struck-out and non-struck-out handwritten word classification is at the infancy stage, we hardly found any methods for classification. Brink et al. [2] proposed a method for automatic removal of crossed-out handwritten text and studied the effect on writer verification and identification. The method uses connected component analysis, which involves branching, size features of text and a classifier for distinguishing crossed-out words in document images. However, the method does not focus on multiple types of struck-out words as considered in our work. Adak et al. [10] used a connected component analysis approach, which constructs a graph using intersection points to study degrees of straightness to identify struck-out words in handwritten document images. However, the method is not robust to noise and degraded documents because the performance depends on finding the correct intersection points. Chaudhuri et al. [1] focused on cleaning of struck-out handwritten words in document images based on the combination of pattern classification and a graph-based approach. The method uses feature-based two-class classification for detecting struck-out words in the images. Then a graph-based approach is used for identifying pixels that represent struck-out lines over text. Since the method involves several stages and proposes many heuristics, the method may not perform well for multiple struck-out words. The success of the method depends on the success of many stages, which is feasible for different applications and situations. Adak et al. [11] proposed a hybrid method, which combines an SVM classifier and convolutional neural network for classifying struck-out handwritten words. In addition, the method studies the impact of struck-out words on writer identification. However, it is not clear whether the method works for different types of struck-out words or focuses on a particular type of struck-out words.

In summary, it is noted from the above review that the existing methods primarily use handcrafted features and the combination of deep learning and handcrafted features for struck-out word classification. The main purpose of the methods is to improve recognition of handwriting by removing struck-out words from a document image. However, Nisa et al. [12] proposed a deep learning-based method for recognizing handwritten text in the presence of struck-out text. It is achieved by exploring convolutional recurrent neural networks. The method works well for improving recognition performance but not for other handwriting-based applications, such as answer script evaluation, gender identification, and fraudulent document identification. In the same way, Qi et al. [13] proposed a deep learning-based method for weakly-supervised ink artifact removal in document images. The method works based on the fact that the text region could have random ink smudges or spurious strokes. For the purpose of removing such ink, the method explores deep learning called DeepErase. However, the scope of the method is confined to simple struck-out handwritten words but not complex struck-out words such as those shown in Fig. 1. For online documents, Bhattacharya et al. [14] proposed a method for struck-out detection based on different density-based features. The method focusses on two class classification, namely, relevant and unwanted. How-ever, the method is applicable for online cross out words detection but not for offline.

Overall, although methods have been developed for struck-out handwriting word detection and classification, none of them considered multiple struck-out word types for classification. Most of the methods are confined to particular types of struck-out words. In addition, the methods work well for images affected by horizontal strike-out lines. This is not true in reality because strike-out lines depend on a person's mindset. Hence, we propose a novel method for classifying struck-out handwritten words, irrespective of struck-out line types in this work.

## 3 Proposed Model

In the case of handwritten documents, usually characters of each word are connected and hence we may not find spaces between characters. But one can expect clear spaces between words. In addition, if a word is struck-out, strike-out lines may connect characters in each word. This observation motivated us to propose a connected component analysis approach for word components as text components. The proposed method finds stroke width and then performs smoothing vertically and horizontally to remove noise. Furthermore, we perform morphological operations over smoothed images to fix bounding boxes for each component. This step detects text components irrespective of struck-out or non-struck-out words. Inspired by the deep learning models that have great success in solving complex classification problems [11], we explore DenseNet [15, 16] for classification of struck-out and non-struck-out words. This step classifies struck-out and non-struck-out handwriting components irrespective of strike-out lines and types. This is the advantage of the proposed model.

### 3.1 Text Component Detection

For a given input handwritten document image, the proposed method finds Stroke Width (SW) by counting black pixels across the horizontal axis in images, which is the mode of the run-value of black pixels. This gives a clue about the thickness of stroke width. While finding SW, there is a chance of introducing noise pixels. Therefore, to remove noise pixels, the proposed method performs smoothing operations, as follows.

**Vertical Smoothing:** If any run value of black pixels present in the image when traversing vertically is less than the stroke width, those pixels are considered as noise pixels and hence are discarded. Otherwise, pixels are considered as text ones (Black-Flag) as defined in Eq. (1).

$$Black_{Flag(x)} = \begin{cases} 1, Run\_Black\_Pixels(x) \leq SW \\ 0, Run_{Black_{Pixels(x)}} > SW \end{cases} \quad (1)$$

**Horizontal Smoothing:** If any run of white pixels present in the image when traversing horizontally is less than five times the stroke width, the pixels are considered as text ones(White-Flag). Otherwise, the pixels are considered as noise pixels as defined in Eq. (2). The value of five is determined empirically by testing on samples chosen randomly.

$$White_{Flag(x)} = \begin{cases} 1, Run_{White_{Pixels}} \leq 5 \times SW \\ 0, Run_{White_{Pixels}} > 5 \times SW \end{cases} \quad (2)$$

**Morphological Operations:** The proposed model performs morphological opening operation using a kernel of (SW, SW × 6) to merge all the characters of each word as a text component. The value of six is estimated based on our experiments. The effectiveness of the above steps is illustrated in Fig. 2, where for the non-struck-out and different struck-out words shown in Fig. 2(a), the results of smoothing and morphological operations are shown in Fig. 2(b) and Fig. 2(c), respectively. It is noted from Fig. 2 that the steps work well for both struck-out and non-struck-out words, and they merge the characters of words as a single component.

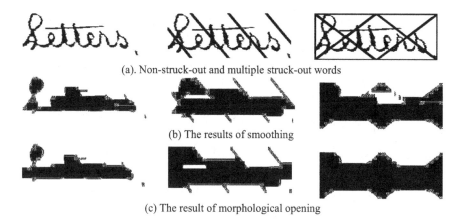

(a). Non-struck-out and multiple struck-out words

(b) The results of smoothing

(c) The result of morphological opening

**Fig. 2.** Text component detection

## 3.2 Deep Learning Model for Struck-Out Component Classification

The step presented in the previous section outputs text components regardless of different struck-out and non-struck-out components. As mentioned in the Introduction Section, one cannot expect a particular type of struck-out words, and classification of struck-out components from non-struck-out components is challenging. In order to balance the number of struck-out and non-struck-out components for training the deep learning model, we create a synthetic Struck-out dataset using the IAM Handwriting [17] and Washington Manuscript Dataset [18]. The images are resized to the standard size of 224 × 224. In total, our dataset contains nearly 100,000 samples, including struck-out and non-struck-out components for training the proposed deep learning model. The data is split into a ratio of 80:10:10, i.e., 80% of the data is used for training the model, 10% for validation, and the remaining 10% is used for testing the model to perform the experiments. In this work, we explore the classic DenseNet121 architecture with a few modifications as shown in Fig. 3, where we can see layers of our model. Here, we replace the last fully-connected layers present in the classic DenseNet architecture with a fully-connected layer of 1024 units. This fully connected layer is followed by a single output unit using the sigmoid activation function that gives the probability of the component being struck-out or not. We use the weights pre-trained on the ImageNet dataset, Stochastic Gradient Descent (SGD) as the optimizer with a learning rate of 0.001, and binary cross-entropy as the loss. The details of proposed DenseNet architecture are as follows.

The dense block shown above consists of a stack of densely-connected convolutional layers. It can extract more efficient features than plain and residual convolutional networks. It consists of a batch normalization layer, a ReLU layer and a convolutional layer. Dense Block#1 contains the BN-ReLU-Conv(1 x 1)-BN-ReLU-Conv(3 x 3) architecture repeated 6 times, whereas Dense Block contains the same architecture 12 times, Dense Block #3 24 times and Dense Block#4 16 times, respectively. The feature maps pass through a dense block which keeps its size the same. But it is important to reduce the feature map size in a convolutional network, so we apply a transition block between

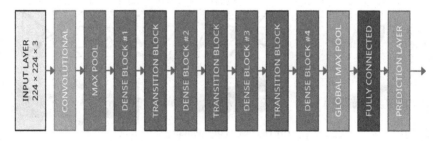

**Fig. 3.** The proposed deep learning DenseNet architecture for struck-out word classification

two dense blocks to decrease the size. All the transition layers shown in the architecture above consist of a batch normalization layer, a 1x1 convolutional layer followed by an average pooling layer with kernel size 2 × 2.

The input image given to the architecture is first passed through a convolution layer containing a 7 × 7 kernel followed by a 3 × 3 max pooling layer. After this, the image is passed through multiple dense blocks and transition blocks before passing it through a 7 × 7 global max pooling layer. Next, the image is passed through a fully-connected layer containing 1024 units with a ReLU activation function and finally sent to the prediction layer containing one unit with a sigmoid activation function that outputs the probability, where the given connected component is either struck-out or not.

## 4   Experimental Results

There is no benchmark dataset for experimentation. In this work, we create our own dataset by collecting it from answer scripts written by different students at different levels.

### 4.1   Dataset Creation and Evaluation

We randomly chose 50 answer sheets for experimentation in this work. For creating the ground truth, we manually fixed a bounding box around each word to label components as either struck-out or non-struck out components for all the 50 answer sheets. It is expected that the number of struck-out components is lower than the number of non-struck-out components. To balance the sizes of struck-out and non-struck-out components, we use a synthetic dataset for generating struck-out components using the IAM Handwriting Dataset [17] and Washington Manuscript dataset [18]. The IAM Handwriting Dataset provides forms of handwritten English text that are widely used by text recognition applications as the training dataset. It contains over 1,500 pages of scanned text written by over 650 writers. The Washington Manuscript Dataset provides data from the George Washington Papers present at the Library of Congress. It contains over 4000 word-level segmented text images.

The main challenge during the creation of a synthetic dataset for struck-out components was replicating how students perform strike-out lines on the actual answer sheet. Many factors need to be considered while creating this dataset, such as the stroke width used by students, different patterns for strike-outs that are followed, placement of strike-outs, etc. For example, in the case of a single straight strike-out, the student does not always strike out text at the center of the word, instead, he/she generally strikes out the text such that the strike-out passes through nearly all the letters present in it. For synthetically generating this pattern in the strike-outs, details such as the position of horizontal maxima of black pixels in the image are used. In many cases, students use a wave-like pattern/spiral-pattern to strikeout the text. In such cases, these patterns need to be synthetically created over the clean data. When a student strikes out the text in a multiple slanted right/left pattern, all these lines in the image are generally non-parallel. Hence, an uncertainty factor also needs to be added while synthetically creating this pattern over clean data. All these factors are considered while creating a synthetic dataset for strike-outs as shown in the sample images in Fig. 4, where we can see samples of both struck-out and non-struck-out components. A total of 1,000 images were used in the test dataset for experimentation comprising 50% struck-out component images and 50% non-struck out component images.

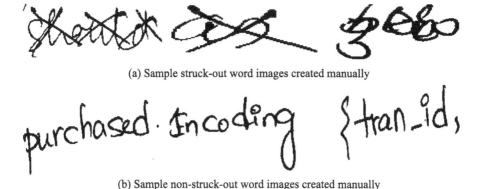

(a) Sample struck-out word images created manually

(b) Sample non-struck-out word images created manually

**Fig. 4.** Labeling struck-out and non-struck-out words manually.

For evaluating the step of text component detection, we consider standard measures such as Precision, Recall and F-Score as defined in Eqs. (3–5), where TP: True Positive, FP: False Positive, FN: False Negative.

$$\text{Precision} = \frac{TP}{TP + FP} \tag{3}$$

$$\text{Recall} = \frac{TP}{TP + FN} \tag{4}$$

$$\text{F1} - \text{score} = 2 * \frac{Precision * Recall}{Precision + Recall} \tag{5}$$

For evaluating the performance of the proposed classification, we use standard measures called the confusion matrix and Average Classification Rate which is the mean of diagonal elements of the confusion matrix. For classifying if a bounding box was correctly predicted, we consider an IoU (Intersection over Union) measure as defined in Eq. (6). It is defined as the ratio between the common area shared by the predicted bounding box and the actual bounding box present in the ground truth to the total area of both the predicted bounding box and actual bounding box combined together. In this work, we consider IoU > 0.5, which is the standard one used for struck-out component classification for counting classified and misclassified components.

$$IoU = \frac{Overlapping\ Area\ between\ the\ prediction\ and\ ground\ truth}{Union\ of\ the\ Area\ between\ the\ predicition\ and\ ground\ truth} \tag{6}$$

In order to show the effectiveness of the proposed classification, we implemented two state-of-the art methods as a comparative study, namely, Chaudhuri et al.'s method [1] which uses aspect ratio of words as features and an SVM classifier for classification of normal (non-struck-out) and struck-out words, and Adak et al.'s method [11] which uses a convolutional neural network for classification of struck-out and non-struck-out words. It is noted that the former one uses handcrafted features for classification while the latter one used CNNs. To show that handcrafted features alone and CNNs alone may not be effective for classifying multiple types of struck-out components whilst the combination is effective, we compared the proposed method with two existing methods for classification in this work.

For smoothing operation presented in Sect. 3.1 for detecting component, we used a threshold value, 5, which is determined experimentally as illustrated in Fig. 5, where for the different threshold values, the Average Classification (ACR) is calculated for 500 samples chosen randomly from our dataset. It is observed from Fig. 5 that for the threshold value, 5, the ACR is the highest and for the subsequent values, the ACR decreases. Therefore, 5 is considered as the optimal threshold value for component detection.

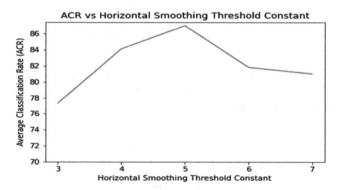

**Fig. 5.** Determining optimal threshold for smoothing operations

## 4.2   Ablation Study

To test the performance of the proposed method on each of the struck-out type of components shown in Fig. 1, we calculated the Average Classification Rate (ACR) as reported in Table 2. It is noted from Table 2 that the proposed approach scores more than 79% for all the 20 types of struck-out components. This shows that the proposed model is consistent and stable in achieving the best ACR irrespective of the type of struck-out components. Therefore, one can infer that the proposed model is robust to multiple types of struck-out components.

**Table 2.** Analyzing the performance of the proposed method for each type of struck-out components mentioned in Fig. 1.

| # | Type | ACR | # | Type | ACR | # | Type | ACR |
|---|------|-----|---|------|-----|---|------|-----|
| (i) | Single Straight | 84.0 | (viii) | Multiple Crossed Strikeout | 90.0 | (xv) | Box Multiple Left Slanted | 79.0 |
| (ii) | Single Left Slanted | 85.0 | (ix) | Box Single Straight | 81.0 | (xvi) | Box Multiple Crossed | 92.0 |
| (iii) | Single Right Slanted | 87.0 | (x) | Box Single Right Slanted | 87.0 | (xvii) | Zigzag | 89.0 |
| (iv) | Single Crossed | 88.0 | (xi) | Box Single Crossed | 85.0 | (xviii) | Wavy Slanted | 80.0 |
| (v) | Multiple Straight | 83.0 | (xii) | Box Single Left Slanted | 79.0 | (xix) | Hybrid | 84.0 |
| (vi) | Multiple Left Slanted | 88.0 | (xiii) | Box Multiple Right Slanted | 77.0 | (xx) | Blackout | 81.0 |
| (vii) | Multiple Right Slanted | 85.0 | (xiv) | Box Multiple Straight | 88.0 | | | |

## 4.3   Experiments on Component Detection

Qualitative results of the proposed struck-out and non-struck-out component detection are illustrated in Fig. 6 and Fig. 7, where one can see that the proposed model fixes bounding boxes for struck-out and non-struck-out words properly. It is observed from Fig. 6 and Fig. 7 that the image contains different types of struck-out and non-struck-out components for which the proposed method detects successfully. This shows that the proposed text component step works well irrespective of component type. In Fig. 7, a few components are missed by the component detection step because those are eliminated during binarization through Otsu threshold. Quantitative results of the proposed text component detection step reported in Table 3 confirm that the proposed component detection works well for multiple struck-out components in the image. Table 3 shows that the proposed method achieves a promising score for text component detection in terms of Precision, Recall and F-Score. Since the precision is larger than recall, one can infer that the proposed text component detection step is stable and reliable for different situations.

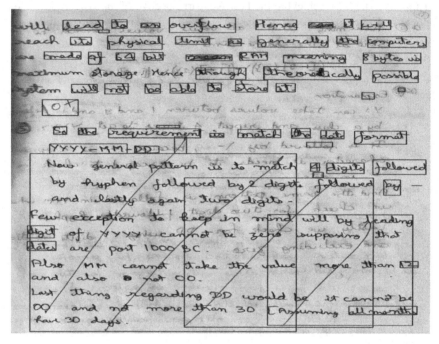

**Fig. 6.** The results of component detection by the proposed method

**Table 3.** Evaluating the component step detection of the proposed method.

| Precision | Recall | F-score |
|-----------|--------|---------|
| 0.96      | 0.93   | 0.94    |

### 4.4  Experiments on Struck-Out and Non-struck-Out Component Classification

Qualitative results of the proposed classification are shown in Fig. 8 and Fig. 9, where it is noted that our method classifies struck-out components marked by a red color rectangle and non-struck-out components are marked by a green color rectangle correctly. It is seen from Fig. 8 and Fig. 9 that the proposed classification approach works well for

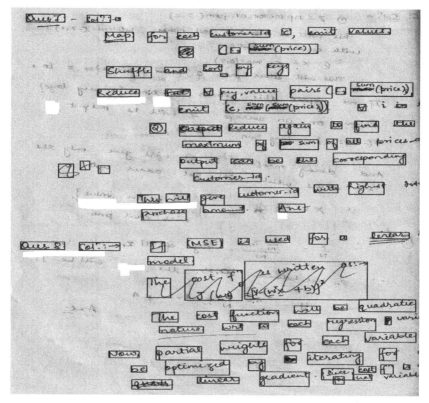

**Fig. 7.** One further sample result of component detection by the proposed method

multiple types of struck-out words. It can be verified by the quantitative results of our method reported in Table 4, that the proposed method performs best at the average classification rate compared to two existing methods. When we compare the results of existing methods, Adak et al.'s method [11] is better than Chaudhuri et al.'s [1]. This indicates that the use of a CNN is better than handcrafted features for classification of struck-out and non-struck-out components. However, the result of Chaudhuri et al.'s approach is worse than the proposed classification method in terms of ACR (Average Character Recognition). The main reason for reporting the results of the two existing methods is that the scope is limited to classification of a particular type of struck-out component. On the other hand, since the proposed classification method considers the advantages of handcrafted features and deep learning, our method is the best compared to the two existing methods.

Sometimes, when the non-struck-out words contain one or two characters with some noise as shown in the samples in Fig. 10(a), there is a high chance of misclassification. This is due to the loss of contextual information. At the same time, when struck-out components contain a mixture of non-struck-out components and when there is no clear distinction between the cross-out line and the text, as shown in Fig. 10(b), the proposed method gets confused and hence, produces erroneous results. In Fig. 10(b), we can

**Table 4.** Comparative study with the existing methods for struck-out and non-struck-out handwritten word classification. ACR: Average Classification Rate in (%).

| Classes | Proposed | | Chaudhuri et al. [1] | | Adak et al. [11] | |
|---|---|---|---|---|---|---|
| | Non-struck-out | Struck-out | Non-struck-out | Struck-out | Non-struck-out | Struck-out |
| Non-struck-out | 89.0 | 11.0 | 72.0 | 25.0 | 76.0 | 12.0 |
| Struck-out | 15.0 | 85.0 | 28.0 | 75.0 | 24.0 | 88.0 |
| ACR | **0.87** | | 73.0 | | 82.0 | |

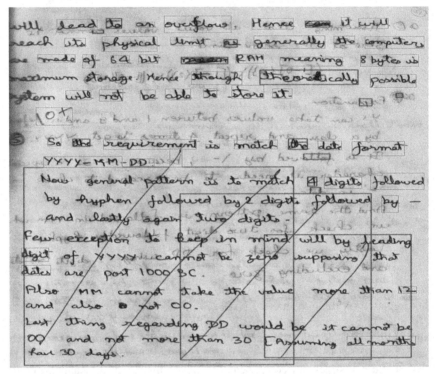

**Fig. 8.** Example of struck-out and non-struck-out component classification by the proposed method. Red boxes denote struck-out.

observe that some of the components have been classified as non-struck-out correctly and some have been misclassified incorrectly. To overcome this problem, there is a need to extract contextual information at the pixel level but not at the character level or word level. This is beyond the scope of the proposed work and can be considered in the future.

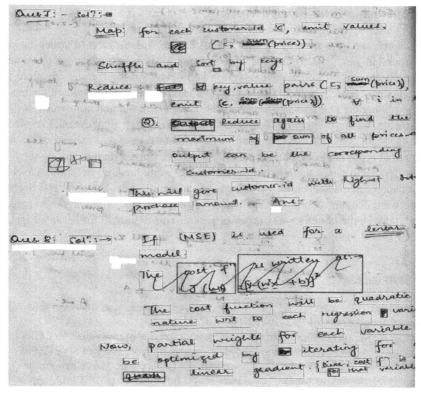

**Fig. 9.** One more example of struck-out and non-struck-out component classification by the proposed method

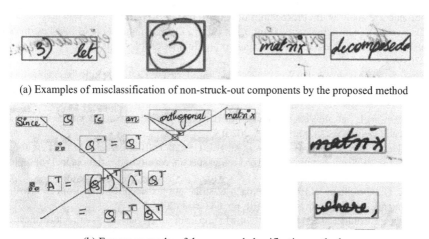

(a) Examples of misclassification of non-struck-out components by the proposed method

(b) Erroneous results of the proposed classification method.

**Fig. 10.** Examples of the limitations of the proposed method

## 5 Conclusions and Future Work

In this work, we have proposed a new model for the classification of struck-out and non-struck-out handwritten components, which combines a connected component analysis approach for text component detection and deep learning for classification. For text component detection, the proposed method explores the stroke width of characters and morphological operations. For classification of struck-out and non-struck-out text components, we have explored DenseNet in a different way. Experimental results on text component detection and classification show that the proposed method works well for multiple types of struck-out words. A comparative study with the existing methods shows that the proposed method outperforms the existing methods in terms of average classification rate. However, in some cases, as discussed in the Experimental Section, when a word contains one or two characters, the performance of the proposed method degrades due to the loss of contextual information. This requires a further investigation of the proposed method, which will be our future work.

## References

1. Chaudhuri, B.B., Adak, C.: An approach for detecting and cleaning of struck-out handwritten text. Pattern Recogn. **61**, 282–294 (2017)
2. Brink, A., Klauw, H.V.D., Schomaker, L.: Automatic removal of crossed-out handwritten text and the effect on writer verification and identification. In: Proceedings of SPIE (2008)
3. Shivakumara, P., Pal, U., Lu, T., Chakarborti, T., Blumenstein, M.: A new roadmap for evaluating descriptive handwritten answer script. In: Proceedings of ICPRAI, pp. 83–96 (2019)
4. Navya, B.J., et al.: Multi-gradient directional features for gender identification. In: Proceedings of ICPR, pp. 3657–3662 (2018)
5. Raghunundan, K.S., et al.: Fourier coefficients for fraud handwritten document identification through age analysis. In: Proceedings of ICFHR, pp. 25–30 (2016)
6. Basavaraj, V., Shivakumara, P., Guru, D.S., Pal, U., Lu, T., Blumenstein, M.: Age estimation using disconnectedness features in handwriting. In: Proceedings of ICDAR, pp. 1131–1136 (2019)
7. Nag, S., Shivakumara, P., Wu, Y., Pal, U., Lu, T.: New COLD feature based handwriting analysis for ethnicity/nationality identification. In: Proceedings of ICFHR, pp. 523–527 (2018)
8. Nandanwar, L., et al.: A new method for detecting altered text in document images. In: Proceedings of ICPRAI, pp. 93–108 (2020)
9. Kundu, S., Shivakumara, P., Grouver, A., Pal, U., Lu, T., Blumenstein, M.: A new forged handwriting detection method based on Fourier spectral density and variation. In: Proceedings of ACPR, pp. 136–150 (2019)
10. Adak, C., Chaudhuri, B.B.: An approach of strike-out text identification from handwritten documents. In: Proceedings of ICFHR, pp. 643–648 (2014)
11. Adak, C., Chaudhuri, B.B., Blumenstein, M.: Impact of struck-out text on writer identification. In: Proceedings of IJCNN, pp. 1465–1471 (2017)
12. Nisa, H., Thom, J.A., Ciesielski, V., Tennakoon, R.: A deep learning approach to handwritten text recognition in the presence of struck-out text. In: Proceedings of IVCNZ (2019)
13. Qi, Y., Huang, W.R., Li, Q., DeGange, J.L.: DeepErase: weakly supervised ink artifact removal in document text images. In: Proceedings of WACV, pp. 3511–3519 (2020)

14. Bhattacharya, N., Frinken, V., Pal, U., Roy, P.P.: Overwriting repetition and crossing-out detection in online handwritten text. In: ACPR 2015, pp. 680–684 (2015)
15. Wan, Z., Yuxiang, Z., Gong, X.Z., Yu, B.: DenseNet model with RAdam optimization algorithm for cancer image classification. In: Proceedings of ICCECE, pp. 771–775 (2021)
16. Tong, W., Chen, W., Han, W., Li, X., Wang, L.: Channel-attention-based DenseNet network for remote sensing image scene classification. IEEE Trans. AEORS **13**, 4121–4132 (2020).
17. Marti, U., Bunke, H.: The IAM-database: an english sentence database for off-line handwriting recognition. IJDAR **5**, 39–46 (2002)
18. Frinken, V., Fischer, A., Manmatha, R., Bunke, H.: A novel word spotting method based on recurrent neural networks. IEEE Trans. PAMI **34**, 211–224 (2012)

# Contextualized Knowledge Base Sense Embeddings in Word Sense Disambiguation

Mozhgan Saeidi[(✉)] [iD], Evangelos Milios, and Norbert Zeh

Dalhousie University, Halifax, Canada
mozhgan.saeidi@dal.ca, {eem,nzeh}@cs.dal.ca

**Abstract.** Contextualized sense embedding has been shown to carry useful semantic information to improve the final results of various Natural Language Processing tasks. However, it is still challenging to integrate them with the information of the knowledge base, which is one lack in current state-of-the-art representations. This integration is helpful in NLP tasks, specifically in the lexical ambiguity problem. In this paper, we present C-KASE (Contextualized-Knowledge base Aware Sense Embedding), a novel approach to producing sense embeddings for the lexical meanings within a lexical knowledge base. The novel difference of our representation is the integration of the knowledge base information and the input text. This representation lies in a space that is comparable to that of contextualized word vectors. C-KASE representations enable a simple 1-Nearest-Neighbour algorithm to perform as well as state-of-the-art models in the English Word Sense Disambiguation task. Since this embedding is specified for each individual knowledge base, it also outperforms in other similar tasks, i.e., Wikification and Named Entity Recognition. The results of comparing our method with recent state-of-the-art methods show the efficiency of our method.

**Keywords:** Sense embedding · Representation learning · Word sense disambiguation · Pre-trained language models

## 1 Introduction

Word representations have been shown to play an important role in different Natural Language Processing (NLP) tasks [21,35], especially in disambiguation tasks [37]. Embeddings based on pre-trained deep language models have attracted much interest recently as they have proved to be superior to classical embeddings for several NLP tasks, including Word Sense Disambiguation (WSD). These models, e.g., ELMO [27], BERT [6], XLNET [43], encode several pieces of linguistic information in their word representations. These representations differ from static neural word embeddings [26] in that they are dependent on the surrounding context of the word. This difference makes these vector representations especially interesting for disambiguation, where effective

© Springer Nature Switzerland AG 2021
E. H. Barney Smith and U. Pal (Eds.): ICDAR 2021 Workshops, LNCS 12917, pp. 174–186, 2021.
https://doi.org/10.1007/978-3-030-86159-9_12

contextual representations can be highly beneficial for resolving lexical ambiguity. These representations enabled sense-annotated corpora to be exploited more efficiently [14].

One important factor in text ambiguity problem is knowledge base. In WSD with Wikipedia as the knowledge base, we face the problem of link ambiguity, meaning a phrase can be usually linked to more than one Wikipedia page in which the correct link depends on the context where it occurs. For example, the word "bar" can be linked to different articles, depending on whether it was used in a business or musical context. In this study, next section, we overview different current approaches for text embedding with focusing on the contextualized sense representation. We also provide an overview of the disambiguation methods and the most used ones in the literature. Our novel contribution provides a new representation of learning using the context of the input text and the context of the knowledge base and uses the nearest neighbor heuristic algorithm to disambiguate ambiguous words. We finally compare the performance of our proposed approach with our representations with the most recent methods in the disambiguation task.

Text ambiguity problem arises from the difficulty of picking the right meaning for a word with multiple meanings. This task is easy for human, specially when one considers the surrounding words, and the human reader identifies the correct meaning of each word based on the context in which the word is used. Computational methods try to mimic this approach [9]. These methods often represent their output by linking each word occurrence to an explicit representation of the chosen sense [42]. Two approaches to tackle this problem are the machine learning-based approach and the knowledge-based approach [37]. In the machine learning-based approach, systems are trained to perform the task [35]. The knowledge-based approach requires external lexical resources such as Wikipedia, WordNet [17], a dictionary, or a thesaurus. The machine learning-based approaches mainly focus on achieving maximal precision or recall and have their drawbacks of run-time and space requirement at the time of classifier training [5]. So, knowledge-based disambiguation methods still have advantages to study. Among different knowledge-based methods, coherence-based has been more effective to explain it [7]. In the coherence-based approach, one important factor is the coherence of the whole text after disambiguation, while in other approaches, this factor might change to considering the coherence of each sentence or paragraph. It is a significant challenge to perform disambiguation accurately but also fast enough to process long text documents [15].

Knowledge bases are different in nature [3]; for example, WordNet is a lexical graph database of semantic relations (e.g., synonyms, hyponyms, and meronyms) between words. Synonyms are grouped into synsets with short definitions and usage examples. WordNet can thus be seen as a combination and extension of a dictionary and thesaurus [4]. Wikipedia is a hyperlink-based graph between encyclopedia entries.

The task of "**Entity Recognition and Disambiguation**" (**ERD**) is to identify mentions[1] of entities and link them to a relevant entry in an external knowledge base, which is also known under the names of "**Entity Linking**", "**Wikification**" or more generally "**Text Annotation**" [1]. Given an input document, a Wikifier links entities of the document to the most relevant corresponding Wikipedia pages. Automated document annotation has become an important topic due to the vast growth in Natural Language Processing (NLP) applications [40]. One benefit of document annotation is enhancing text readability and its unambiguousness by inserting connections between the text and an external knowledge base, like Wikipedia, which is the most popular among online encyclopedias [32].

The problem of Wikification is closely related to other core NLP tasks, such as Named-Entity Recognition (NER) and Word Sense Disambiguation (WSD) [39]. Wikification, in particular, is the task of associating a word in context with the most suitable meaning from the predefined sense inventory of Wikipedia. Named-Entity Recognition involves identifying certain occurrences of noun phrases as belonging to particular categories of named-entities [24]. These expressions refer to names, including person, organization, and location names, and numeric expressions including time, date, money, and percent expressions [12]. In Word Sense Disambiguation, our knowledge base is not limited to Wikipedia, which is the difference between Wikification and WSD [13]. In WSD, the knowledge base is a treasury like WordNet and Wikipedia [11]. Details and performance of each WSD method are highly dependent on the knowledge base to link to.

## 2    Related Work

In Natural Language Processing (NLP), Word Sense disambiguation (WSD) is one of the main tasks. This task is at the core of lexical semantics and has been tackled with many various approaches. We divide these approaches into two categories of knowledge-based and supervised approaches [21].

### 2.1    Knowledge-Based Approaches

Knowledge-based methods use the semantic network structure, e.g., Wikipedia [10], WordNet [17], BabelNet [23], to find the correct meaning based on its context for each input word [20]. These approaches employ algorithms on graphs to address the word ambiguity in texts [2]. Disambiguation based on Wikipedia has been demonstrated to be comparable in terms of coverage to domain-specific ontology [41] since it has broad coverage, with documents about entities in a variety of domains. The most widely used lexical knowledge base is WordNet, although it is restricted to the English lexicon, limiting its usefulness to other vocabularies. BabelNet solves this challenge by combining lexical and semantic information from various sources in numerous languages, allowing knowledge-based

---

[1] A mention can be one or more tokens.

approaches to scale across all languages it supports. Despite their potential to scale across languages, knowledge-based techniques on English fall short of supervised systems in terms of accuracy.

## 2.2 Supervised Approaches

The supervised approaches surpass the knowledge-based ones in all English data sets. These approaches use neural architectures, or SVM models, while still suffering from the need of creating large manually-curated corpora, which reduces their usability to scale over unseen words [25]. Automatic data augmentation approaches developed methods to cover more words, senses, and languages.

## 2.3 Language Modelling Representation

Most NLP tasks now use semantic representations derived from language models. There are static word embeddings and contextual embeddings. This section covers aspects of the word and contextual embeddings that are especially important to our work.

**Static Word Embeddings.** Word embeddings are distributional semantic representations usually with one of two goals: predict context words given a target word (Skip-Gram), or the inverse (CBOW) [16]. In both, the target word is at the center, and the context is considered as a fixed-length window that slides over tokenized text. These models produce dense word representations. One limit for word embeddings, as mentioned before, is meaning conflict around word types. This limitation affects the capability of these word embeddings for the ones that are sensitive to their context [34].

**Contextual Word Embeddings.** The problem mentioned as a limitation for the static word embeddings is solved in this type of embeddings. The critical difference is that the contextual embeddings are sensitive to the context. It allows the same word types to have different representations according to their context. The first work in contextual embeddings is ELMO [27], which is followed by BERT [6], as the state-of-the-art model. The critical feature of BERT, which makes it different, is the quality of its representations. Its results are task-specific fine-tuning of pre-trained neural language models. The recent representations which we analyze their effectiveness are based on these two models [28,29].

# 3 Preliminaries

Our new C-KASE representations use different resources to build their vectors. In this section, we provide information on these resources.

**Wikipedia** is the largest electronic encyclopedia freely available on the Web. Wikipedia organized its information via articles called Wikipedia pages.

Disambiguation based on Wikipedia has been demonstrated to be comparable in terms of coverage to domain-specific ontology [41] since it has broad coverage with documents about entities in a variety of domains [15]. Moreover, Wikipedia has unique advantages over the majority of other knowledge bases, which include [44]:

- The text in Wikipedia is primarily factual and available in a variety of languages.
- Articles in Wikipedia can be directly linked to the entities they describe in other knowledge bases.
- Mentions of entities in Wikipedia articles often provide a link to the relevant Wikipedia pages, thus providing labeled examples of entity mentions and associated anchor texts in various contexts, which could be used for supervised learning in WSD with Wikipedia as the knowledge base.

**BabelNet** is a multilingual semantic network, which comprises information coming from heterogeneous resources, such as WordNet, and Wikipedia [23]. It is organized into synsets, i.e., sets of synonyms that express a single concept, which, in their turn, are connected to each other by different types of relation. One of Babelnet's feature which is useful for our representation is *hypernym-hyponym* relations. In this relation, each concept is connected to other concepts via hypernym relation (for generalization) and via hyponym relation (for specification). *Semantically-related* relation is the other feature that we use that expresses a general notation of relatedness between concepts. The last feature of Babelent used in this work is *mapping to Wikipedia*, which maps its concepts to Wikipedia pages.

**WordNet** is the most widely used lexical knowledge repository for English. It can be seen as a graph, with nodes representing concepts (synsets) and edges representing semantic relationships between them. Each synset has a set of synonyms, such as the lemmas spring, fountain, and natural spring in the synset, A natural flow of groundwater.

**SemCor** is the typical manually-curated corpus for WSD, with about 220K words tagged with 25K distinct WordNet meanings, resulting in annotated contexts for around 15% of WordNet synsets.

**BERT** is a Transformer-based language model for learning contextual representations of words in a text. The contextualized representation of BERT is the key factor that has changed the performance in many NLP tasks, such as text ambiguity. In our representations, we use BERT-base-cased to generate the vectors of each sense [6].

**SBERT** is a modification of the pre-trained BERT network that uses siamese and triplet network structures to derive semantically meaningful sentence embeddings that can be compared using cosine-similarity. We use this sentence representation when generating the vector representations of sense sentences, both in the input text and in the knowledge base text.

# 4   C-KASE

This section presents C-KASE, the novel knowledge-based approach of creating sense representations of BabelNet senses. C-KASE is created by combining semantic and textual information from the first paragraph of each sense's Wikipedia page and the input document paragraph, which includes the concept. C-KASE uses the representation power of neural language models, i.e., BERT and SBERT. We divide our approach into the following steps:

## 4.1   Context Retrieval

In this step, we collect suitable contextual information from Wikipedia for each given concept in the semantic network. Similar to [36], we exploit the mapping between synsets and Wikipedia pages available in BabelNet, as well as its taxonomic structure, to collect textual information that is relevant to a target synset s. For each synset s, we collect all the connected concepts to s through hyponym, and hypernym connections of the BabelNet knowledge base. We show this set of related synsets to s by $R_s$ which is:

$$R_s = \{s^{'} | (s, s^{'}) \in E\}$$

Similar to [36], we use E as the set includes all hyponyms and hypernyms connections. In this work, for each page $p_s$, we consider the first opening paragraph of the page and compute its lexical vector by summing the SBERT vector representation of the sentences in this first paragraph. These lexical representations are later used for the similarity score finding between $p_s$ and $p_{s'}$, for each $s^{'} \in R_s$ by using the weighted overlap measure from [30], which is defined as follows:

$$WO(p_1, p_2) = \left( \sum_{w \in O} \frac{1}{r_w^{p_1} + r_w^{p_2}} \right) \left( \sum_{i=1}^{|O|} \frac{1}{2i} \right)^{-1}$$

where O is the set of overlapping dimensions of $p_1$ and $p_2$ and $r_w^{p_i}$ is the rank of the word w in the lexical vector of $p_i$. We preferred the weighted overlap over the more common cosine similarity as it has proven to perform better when comparing sparse vector representations [30]. Similar to [36], Once we have scored all the $(p_s, p_{s'})$ pairs, we create partitions of $R_s$, each comprising all the senses $s'$ connected to $s$ with the same relation $r$, where $r$ can be one among: hypernymy, and hyponymy. We then retain from each partition only the top-k scored senses according to $WO(p_s, p_{s'_i})$, which we set $k = 15$ in our experiments.

## 4.2   Word Embedding

In the second step, we use BERT for the representation of the given concepts from the input text. For each ambiguous word–which we call this word by mention– of the input, we extract the BERT representation of the mention. Using the BabelNet relations of hyponymy and hypernymy, we extract all synsets of mention from BabelNet (set E). For each one of these senses, use the link structure of BabelNet and Wikipedia, we collect all the Wikipedia pages for each sense. We use BERT representation for the second time to generate vector representation for senses. In the settings, each word is represented as a 300-dimensional vector, as the BERT dimension.

### 4.3    Sense Embedding

In this step, we build the final representation of each concept. From the previous step, we took the representation of mention, $R(m)$, and the representation of each one of its senses. We show the representations of each $k$ sense of $m$ by $R(s_i)$ which $i$ varies from 1 to $k$. Our unique representations are combining the mention representation with sense representation, concatenating the two vector representations of $R(m)$ and $R(s_i)$. If mention $m$ has $k$ senses, C-KASE generates $k$ different representations of $R(m, s_1)$, $R(m, s_2)$, ..., $R(m, s_k)$. Since the dimension representation of $R(m)$ and each $R(s_i)$ is 300, these concatenated representation dimensions are 600. Next novelty in our C-KASE representations is ranking the $k$ senses of each mention based on their relevancy degree to the context. To this aim, we concatenate representations of the first step. In the first step, we took the representation of the input text paragraph, which contains the ambiguous mention, show it by $R(PD)$ which stands for representation of the **P**aragraph of the input **D**ocument. In the first step, we also took the representation of the first paragraph of the Wikipedia page, which represents it by $R(PW)$, which stands for representation of the first **P**aragraph of the **W**ikipedia page. Finally, we concatenate these two representations as $R(PD, PW)$. The dimension of this concatenated representation is also equal to the word representation, so it makes it possible to calculate their cosine similarities. To rank the senses most related to the context, we use the cosine similarity as follows:

$$\text{Sim}(m, s_i) = Cosine(R(m, s_i), R(PD, PW)), \text{ for } i = 1, ... , k$$

This ranking provides the most similar sense to the context for each mention. This novelty makes this representation more effective than the previous contextualized-based embeddings, especially in the task of sense disambiguation.

At the end of these three steps, each sense is associated with a vector that encodes both the contextual information and knowledge base semantic information from the extracted context of Wikipedia and its gloss.

## 5    Experimental Setup

We present the settings of our evaluation of C-KASE in the English WSD task. This setup includes the benchmark, C-KASE setup for disambiguation task and state-of-the-art WSD models as our comparison systems.

**Evaluation Benchmark.** We use the English WSD test set framework which is constructed by five standard evaluation benchmark datasets[2]. It is included of

---

[2] http://lcl.uniroma1.it/wsdeval/.

Senseval-2 [8], Senseval-3 [38], SemEval-07 [31], SemEval-13 [22], SemEval-15 [19] along with ALL, i.e., the concatenation of all the test sets [33].

**C-KASE Setup.** In our experiments, we use BERT pre-trained cased model. Similar to [36], among all the configurations reported by Devlin et al. (2019), we used the sum of the last four hidden layers as contextual embeddings of the words, since they showed it has better performance. In order to be able to compare our system with supervised models, we build a supervised version of our C-KASE representations. This version combines the gloss and contextual information with the sense-annotated contexts in SemCor [18], a corpus of 40K sentences where words have been manually annotated with a WordNet meaning. We leveraged SemCor for building a representation of each sense therein. To this end, we followed [27], given a mention-sense pair $(m, s)$, we collected all the sentences $c_1, ..., c_n$ where m appears tagged with s. Then, we fed all the retrieved sentences into BERT and extracted the embeddings $BERT(c_1, m)$, ... , $BERT(c_n, m)$. The final embedding of s was built by concatenating the average of its context and sense gloss vectors and its representation coming from SemCor, i.e., the average of $BERT(c_1, m)$, ... , $BERT(c_n, m)$. We note that when a sense did not appear in SemCor, and we built its embedding by replacing the SemCor part of the vector with its sense gloss representation.

**WSD Model.** For WSD modeling, we employed a 1-nearest neighbor approach– as previous methods in the literature– to test our representations on the WSD task. For each target word m in the test set, we computed its contextual embedding by means of BERT and compared it against the embeddings of C-KASE associated with the senses of m. Hence, we took as a prediction for the target word the sense corresponding to its nearest neighbor. We note that the embeddings produced by C-KASE are created by concatenating two BERT representations, i.e., context and sense gloss (see Sect. 4.3), hence we repeated the BERT embedding of the target instance to match the number of dimensions.

**Comparison Systems.** We compared our representation against the best recent performing systems evaluated on the English WSD task. LMMS is one of these systems which generates sense embedding with full coverage of Wordnet. It uses pre-trained ELMO and BERT models, as well as the relations in a lexical knowledge base to create contextual embeddings [14]. SensEmBERT is the next system that relies on different resources for building sense vectors. These resources include Wikipedia, BabelNet, NASARI lexical vectors, and BERT. It computes context-aware representations of BabelNet senses by combining the semantic and textual information derived from multilingual resources. This model uses the BabelNet mapping between WordNet senses and Wikipedia pages which drops the need for sense-annotated corpora [36]. The next comparison system is ARES, a semi-supervised approach to produce sense embeddings for all the word senses in a language vocabulary. ARES compensates for the lack of manually annotated examples for a large portion of words' meanings. ARES is the most recent contextualized word embedding system, to our knowledge. In our comparisons, we also considered BERT as a comparison system since it is at the core of all the considered methods. BERT also has shown good performance in most NLP tasks by using pre-trained neural networks.

## 6   Results

The results of our evaluations on the WSD task are represented in this section. We show the effectiveness of C-KASE representation by comparing it with the existing state-of-the-art models on the standard WSD benchmarks. In Table 1 we report the results of C-KASE and compare it against the results obtained from other state-of-the-art approaches on all the nominal instances of the test sets in the framework of [33]. All performances are reported in terms of F1-measure, i.e., the harmonic mean of precision and recall. As we can see, C-KASE achieves the best results on the datasets when compared to other precious contextualized approaches. It indicates that C-KASE is competitive with these previous models. These results show the novel idea in the nature of creating this C-KASE representation has improved the lexical ambiguity. It is a good indicator of the dependency of the WSD task to the representation that is aware of the context and the information extracted from the reference knowledge base.

**Table 1.** F-Measure performance of WSD evaluation framework on the test sets of the unified dataset.

| Model | Senseval-2 | Senseval-3 | Semeval-7 | Semeval-13 | Semeval-15 | All |
|---|---|---|---|---|---|---|
| BERT | $77.1 \pm 0.3$ | $73.2 \pm 0.4$ | $66.1 \pm 0.3$ | $71.5 \pm 0.2$ | $74.4 \pm 0.3$ | $73.8 \pm 0.3$ |
| LMMS | $76.1 \pm 0.6$ | $75.5 \pm 0.2$ | $68.2 \pm 0.4$ | $75.2 \pm 0.3$ | $77.1 \pm 0.4$ | $75.3 \pm 0.2$ |
| SensEmBERT | $72.4 \pm 0.1$ | $69.8 \pm 0.2$ | $60.1 \pm 0.4$ | $78.8 \pm 0.1$ | $75.1 \pm 0.2$ | $72.6 \pm 0.3$ |
| ARES | $78.2 \pm 0.3$ | $77.2 \pm 0.1$ | $71.1 \pm 0.2$ | $77.2 \pm 0.2$ | $83.1 \pm 0.2$ | $77.8 \pm 0.1$ |
| C-KASE | $79.6 \pm 0.2$ | $78.5 \pm 0.2$ | $74.6 \pm 0.3$ | $79.3 \pm 0.6$ | $82.9 \pm 0.4$ | $78.9 \pm 0.1$ |

We also evaluate the effectiveness of our representation on parts of speeches. The parts of speech that we have in the dataset are nouns, verbs, adjectives, and adverbs. Table 2 shows the number of instances in each category. In our second evaluation, we examined the effect of our representation against previous ones on each word category. Table 3 represents the F-Measure performance of the 1-NN WSD of each one of the contextualized word embeddings which we considered on All datasets split by parts of speech.

**Table 2.** The Number of instances and ambiguity level of the concatenation of all five WSD datasets [33].

|           | Nouns | Verbs | Adj. | Adv | All  |
|-----------|-------|-------|------|-----|------|
| #Entities | 4300  | 1652  | 955  | 346 | 7253 |
| Ambiguity | 4.8   | 10.4  | 3.8  | 3.1 | 5.8  |

**Table 3.** F-Measure performance of the 1-NN WSD of each embedding on All dataset split by parts of speech. The dataset in this experiment is a concatenation of all five datasets, which is split by Part-of-Speech tags.

| Model      | Nouns          | Verbs          | Adjectives     | Adverbs        |
|------------|----------------|----------------|----------------|----------------|
| BERT       | $76.2 \pm 0.2$ | $62.9 \pm 0.5$ | $79.7 \pm 0.2$ | $85.5 \pm 0.5$ |
| LMMS       | $78.2 \pm 0.6$ | $64.1 \pm 0.3$ | $81.3 \pm 0.1$ | $82.9 \pm 0.3$ |
| SensEmBERT | $77.8 \pm 0.3$ | $63.4 \pm 0.5$ | $80.1 \pm 0.4$ | $86.4 \pm 0.2$ |
| ARES       | $78.7 \pm 0.1$ | $67.3 \pm 0.2$ | $82.6 \pm 0.3$ | $87.1 \pm 0.4$ |
| C-KASE     | $79.6 \pm 0.2$ | $69.6 \pm 0.1$ | $85.2 \pm 0.1$ | $89.3 \pm 0.5$ |

# 7 Conclusion

In this paper, we present C-KASE, a novel approach for creating sense embeddings considering the knowledge base and the context of the input document text. We showed that this context-rich representation is beneficial for lexical ambiguity in English. The results of experiments in the WSD task show the efficiency of C-KASE representations in comparison with other state-of-the-art methods, despite relying only on English data only. The results across other different datasets show the high quality of our embeddings and also enable the English WSD while at the same time relieving the heavy requirement of sense-annotated corpora. We further tested our embeddings on the split data into four parts of speeches. As future work, we plan to extend our approach to cover multiple languages. The other point to improve our representations in the text ambiguity task is by training the model with data including more verbs than the current one. As the results of our second experiment show, the effectiveness of the contextualized embeddings in WSD on verbs is not as good as on nous. This defect is because of the lack of instances in the dataset in each word category.

# References

1. Aghaebrahimian, A., Cieliebak, M.: Named entity disambiguation at scale. In: Schilling, F.-P., Stadelmann, T. (eds.) ANNPR 2020. LNCS (LNAI), vol. 12294, pp. 102–110. Springer, Cham (2020). https://doi.org/10.1007/978-3-030-58309-5_8, https://digitalcollection.zhaw.ch/bitstream/11475/21530/3/2020_Aghaebrahimian-Cieliebak_Named-entity-disambiguation-at-scale.pdf

2. Agirre, E., de Lacalle, O.L., Soroa, A.: Random walks for knowledge-based word sense disambiguation. Comput. Linguist. **40**(1), 57–84 (2014). https://direct.mit.edu/coli/article/40/1/57/145
3. Aleksandrova, D., Drouin, P., Lareau, F.C.C.O., Venant, A.: The multilingual automatic detection of 'e nonc é s bias 'e s in wikip é dia. ACL (2020). https://www.aclweb.org/anthology/R19-1006.pdf
4. Azad, H.K., Deepak, A.: A new approach for query expansion using wikipedia and wordnet. Inf. Sci. **492**, 147–163 (2019). https://www.sciencedirect.com/science/article/pii/S0020025519303263
5. Calvo, H., Rocha-Ramírez, A.P., Moreno-Armendáriz, M.A., Duchanoy, C.A.: Toward universal word sense disambiguation using deep neural networks. IEEE Access **7**, 60264–60275 (2019). https://ieeexplore.ieee.org/abstract/document/8706934
6. Devlin, J., Chang, M.W., Lee, K., Toutanova, K.: BERT: pre-training of deep bidirectional transformers for language understanding. In: North American Association for Computational Linguistics (NAACL) (2018). https://www.aclweb.org/anthology/N19-1423/
7. Dixit, V., Dutta, K., Singh, P.: Word sense disambiguation and its approaches. CPUH-Res. J. **1**(2), 54–58 (2015). http://www.cpuh.in/academics/pdf
8. Edmonds, P., Cotton, S.: SENSEVAL-2: overview. In: Proceedings of SENSEVAL-2 Second International Workshop on Evaluating Word Sense Disambiguation Systems, pp. 1–5. Association for Computational Linguistics, Toulouse, France, July 2001. https://www.aclweb.org/anthology/S01-1001.pdf
9. Ferreira, R.S., Pimentel, M.D.G., Cristo, M.: A wikification prediction model based on the combination of latent, dyadic, and monadic features. IST **69**(3), 380–394 (2018). https://asistdl.onlinelibrary.wiley.com/doi/pdf/10.1002/asi.23922
10. Fogarolli, A.: Word sense disambiguation based on wikipedia link structure. In: 2009 IEEE International Conference on Semantic Computing, pp. 77–82. IEEE (2009). https://ieeexplore.ieee.org/stamp/stamp.jsp
11. Kwon, S., Oh, D., Ko, Y.: Word sense disambiguation based on context selection using knowledge-based word similarity. Inf. Process. Manag. **58**(4), 102551 (2021). https://www.sciencedirect.com/science/article/pii/S0306457321000558
12. Li, B.: Named entity recognition in the style of object detection. arXiv preprint arXiv:2101.11122 (2021). https://arxiv.org/pdf/2101.11122.pdf
13. Logeswaran, L., Chang, M.W., Lee, K., Toutanova, K., Devlin, J., Lee, H.: Zero-shot entity linking by reading entity descriptions. arXiv preprint arXiv:1906.07348 (2019). https://arxiv.org/pdf/1906.07348.pdf
14. Loureiro, D., Jorge, A.: Language modelling makes sense: propagating representations through wordnet for full-coverage word sense disambiguation. In: Proceedings of the 57th Annual Meeting of the Association for Computational Linguistics, Florence, Italy, pp. 5682–5691 (2019). https://www.aclweb.org/anthology/P19-1569
15. Martinez-Rodriguez, J.L., Hogan, A., Lopez-Arevalo, I.: Information extraction meets the semantic web: a survey. Semantic Web Preprint, pp. 1–81 (2020). http://repositorio.uchile.cl/bitstream/handle/2250/174484/Information-extraction-meets-the-Semantic-Web.pdf?sequence=1
16. Mikolov, T., Chen, K., Corrado, G.S., Dean, J.: Efficient estimation of word representations in vector space. In: Proceedings of ICLR, vol. 4, pp. 321–329 (2013). https://arxiv.org/pdf/1301.3781.pdf
17. Miller, G.A., Beckwith, R., Fellbaum, C., Gross, D., Miller, K.J.: Introduction to wordnet: An on-line lexical database. Int. J. Lexicography **3**(4), 235–244 (1990). https://watermark.silverchair.com/235.pdf

18. Miller, G.A., Leacock, C., Tengi, R., Bunker, R.T.: A semantic concordance. In: Human Language Technology: Proceedings of a Workshop Held at Plainsboro, New Jersey, 21–24 March, 1993 (1993). https://www.aclweb.org/anthology/H93-1061/
19. Moro, A., Navigli, R.: SemEval-2015 task 13: multilingual all-words sense disambiguation and entity linking. In: SEM, pp. 288–297. Association for Computational Linguistics, Denver, Colorado, June 2015. https://www.aclweb.org/anthology/S15-2049.pdf
20. Moro, A., Raganato, A., Navigli, R.: Entity linking meets word sense disambiguation: a unified approach. Trans. Assoc. Comput. Linguist. **2**, 231–244 (2014). https://watermark.silverchair.com/tacl_a_00179.pdf
21. Navigli, R.: Word sense disambiguation: a survey. ACM Comput. Surv. (CSUR) **41**(2), 1–69 (2009). https://dl.acm.org/doi/abs/10.1145/1459352.1459355
22. Navigli, R., Jurgens, D., Vannella, D.: SemEval-2013 task 12: multilingual word sense disambiguation. In: SEM. Association for Computational Linguistics, Atlanta, Georgia, USA, June 2013. https://www.aclweb.org/anthology/S13-2040.pdf
23. Navigli, R., Ponzetto, S.P.: Babelnet: The automatic construction, evaluation and application of a wide-coverage multilingual semantic network. Artif. Intell. **193**, 217–250 (2012). https://www.sciencedirect.com/science/article/pii/S0004370212000793
24. Nguyen, D.B., Hoffart, J., Theobald, M., Weikum, G.: Aida-light: high-throughput named-entity disambiguation. LDOW **14**, 22–32 (2014). http://ceur-ws.org/Vol-1184/ldow2014_paper_03.pdf
25. Pasini, T., Elia, F.M., Navigli, R.: Huge automatically extracted training sets for multilingual word sense disambiguation. arXiv preprint arXiv:1805.04685 (2018). https://arxiv.org/abs/1805.04685
26. Pennington, J., Socher, R., Manning, C.D.: Glove: global vectors for word representation. In: Proceedings of the 2014 Conference on Empirical Methods in Natural Language Processing, pp. 1532–1543. EMNLP, Qatar (2014). https://www.aclweb.org/anthology/D14-1162.pdf
27. Peters, M., et al.: Deep contextualized word representations. Association for Computational Linguistics, pp. 2227–2237 (2018). https://www.aclweb.org/anthology/N18-1202
28. Peters, M.E., Logan IV, R.L., Schwartz, R., Joshi, V., Singh, S., Smith, N.A.: Knowledge enhanced contextual word representations. arXiv preprint arXiv:1909.04164 (2019). https://arxiv.org/pdf/1909.04164.pdf
29. Peters, M.E., Neumann, M., Zettlemoyer, L., Yih, W.T.: Dissecting contextual word embeddings: architecture and representation. In: Proceedings of the 2018 Conference on Empirical Methods in Natural Language Processing, pp. 1499–1509 (2018). https://www.aclweb.org/anthology/D18-1179/
30. Pilehvar, M.T., Jurgens, D., Navigli, R.: Align, disambiguate and walk: a unified approach for measuring semantic similarity. In: Proceedings of the 51st Annual Meeting of the Association for Computational Linguistics (Volume 1: Long Papers), pp. 1341–1351 (2013). https://www.aclweb.org/anthology/P13-1132.pdf
31. Pradhan, S., Loper, E., Dligach, D., Palmer, M.: SemEval-2007 task-17: English lexical sample, SRL and all words. In: Proceedings of the Fourth International Workshop on Semantic Evaluations (SemEval-2007), pp. 87–92. Association for Computational Linguistics, Prague, Czech Republic, June 2007. https://www.aclweb.org/anthology/S07-1016

32. Raganato, A., Bovi, C.D., Navigli, R.: Automatic construction and evaluation of a large semantically enriched wikipedia. In: IJCAI, pp. 2894–2900 (2016). http://wwwusers.di.uniroma1.it/~navigli/pubs/IJCAI_2016_Raganatoetal.pdf
33. Raganato, A., Camacho-Collados, J., Navigli, R.: Word sense disambiguation: a unified evaluation framework and empirical comparison. In: Proceedings of the 15th Conference of the European Chapter of the Association for Computational Linguistics: Volume 1, Long Papers, pp. 99–110 (2017). https://www.aclweb.org/anthology/E17-1010/
34. Reisinger, J., Mooney, R.: Multi-prototype vector-space models of word meaning. In: Human Language Technologies: The 2010 Annual Conference of the North American Chapter of the Association for Computational Linguistics, pp. 109–117 (2010). https://www.aclweb.org/anthology/N10-1013.pdf
35. Saeidi, M., da Sousa, S.B.S., Milios, E., Zeh, N., Berton, L.: Categorizing online harassment on twitter. In: Cellier, P., Driessens, K. (eds.) ECML PKDD 2019. CCIS, vol. 1168, pp. 283–297. Springer, Cham (2020). https://doi.org/10.1007/978-3-030-43887-6_22, https://link.springer.com/chapter/10.1007/978-3-030-43887-6_22
36. Scarlini, B., Pasini, T., Navigli, R.: Sensembert: context-enhanced sense embeddings for multilingual word sense disambiguation. In: Proceedings of the AAAI Conference on Artificial Intelligence, vol. 34, pp. 8758–8765 (2020). https://ojs.aaai.org//index.php/AAAI/article/view/6402
37. Scarlini, B., Pasini, T., Navigli, R.: With more contexts comes better performance: contextualized sense embeddings for all-round word sense disambiguation. In: Proceedings of the 2020 Conference on Empirical Methods in Natural Language Processing (EMNLP), pp. 3528–3539 (2020). https://www.aclweb.org/anthology/2020.emnlp-main.285/
38. Snyder, B., Palmer, M.: The English all-words task. In: Proceedings of SENSEVAL-3, the Third International Workshop on the Evaluation of Systems for the Semantic Analysis of Text, pp. 41–43. Association for Computational Linguistics, Barcelona, Spain, July 2004. https://www.aclweb.org/anthology/W04-0811
39. Wang, A., et al.: Superglue: a stickier benchmark for general-purpose language understanding systems. arXiv preprint arXiv:1905.00537 (2019)
40. Wang, Y., Wang, M., Fujita, H.: Word sense disambiguation: a comprehensive knowledge exploitation framework. Knowl. Based Syst. 43, 105–117 (2019). https://www.sciencedirect.com/science/article/pii/S0950705119304344
41. Weikum, G., Dong, L., Razniewski, S., Suchanek, F.: Machine knowledge: creation and curation of comprehensive knowledge bases. arXiv preprint arXiv:2009.11564 (2020). https://arxiv.org/pdf/2009.11564.pdf
42. West, R., Paranjape, A., Leskovec, J.: Mining missing hyperlinks from human navigation traces: a case study of wikipedia. In: Proceedings of the 24th International Conference on World Wide Web, pp. 1242–1252 (2015). https://dl.acm.org/doi/pdf/10.1145/2736277.2741666
43. Yang, Z., Dai, Z., Yang, Y., Carbonell, J., Salakhutdinov, R.R., Le, Q.V.: Xlnet: generalized autoregressive pretraining for language understanding. Curran Associates, Inc., vol. 32, pp. 221–229 (2019). https://proceedings.neurips.cc/paper/2019/file/dc6a7e655d7e5840e66733e9ee67cc69-Paper.pdf
44. Zhao, G., Wu, J., Wang, D., Li, T.: Entity disambiguation to wikipedia using collective ranking. Inf. Process. Manag. 52(6), 1247–1257 (2016). https://www.sciencedirect.com/science/article/pii/S0306457316301893

# ICDAR 2021 Workshop on Open Services and Tools for Document Analysis (OST)

# ICDAR-OST 2021 Preface

The ICDAR Workshop on Open Services and Tools for Document Analysis (ICDAR-OST) aims at promoting open tools, software, and open services (for processing, evaluation or visualization), as well as facilitating public dataset usage, in the domain of document image analysis research, building on the experience of our community and of other ones.

Such tools, softwares, services, formats, or datasets observe the principles of being reusable (I can use it on my data), transferable (I can use it on my premises), and reproducible (I can obtain the same results). The accepted contributions are presented during interactive pitch and demo sessions, enabling authors to advertise their work, identify potential issues and solutions in their approach, and ignite collaboration with other participants.

ICDAR-OST 2021 was a half-day online workshop comprising interactive pitch and demo sessions and a keynote speech by Marcel Gygli (Institut für Interaktive Technologien FHNW, Switzerland).

For this third edition of the workshop, we received contributions from authors in a number of different countries and each paper was reviewed by at least three experts from the field. We accepted two papers which deal with two different topics: platforms for visualizing documents and automatic generation of semistructured data.

As the Program Committee chairs and organizers, we would like to warmly thank all the authors who submitted their work to the ICDAR-OST workshop and made the choice to release the fruit of their research as broadly as possible. Finally, we also thank the ICDAR conference for sponsorship and the International Association for Pattern Recognition for their endorsement.

September 2021

Lars Vögtlin
Fouad Slimane
Oussama Zayene
Paul Märgner
Ridha Ejbali

# Organization

## General Chairs

Fouad Slimane     University of Fribourg, Switzerland
Oussama Zayene     University of Applied Sciences Western
       Switzerland, Switzerland
Lars Vögtlin     University of Fribourg, Switzerland

## Program Committee Chairs

Paul Märgner     University of Fribourg, Switzerland
Ridha Ejbali     National School of Engineers Gabes, Tunisia

## Program Committee

Vlad Atanasiu     University of Fribourg, Switzerland
Christopher Kermorvant     Teklia, France
Maroua Mehri     University of Sousse, Tunisia
Nicholas Journet     Universitè de Bordeaux, France
Marcel Gygli     Institut für Interaktive Technologien FHNW,
       Switzerland
Mourad Zaied     University of Gabes, Tunisia
Günter Mühlberger     University of Innsbruck, Austria
Bertrand Coüasnon     IRISA, INSA Rennes, France
Josep Llados     Universitat Autònoma de Barcelona, Spain
Olfa Jemai     University of Gabes, Tunisia
Barsha Mitra     BITS Pilani, Hyderabad, India
Dorra Sellami     University of Sfax, Tunisia
Pascal Monasse     LIGM, Ecole des Ponts ParisTech, France
Jihad El-Sana     Ben-Gurion University of the Negev, Israel
Salwa Said     University of Gabes, Tunisia
Nouredine Boujnah     Waterford Institute of Technology, Ireland
Tahani Bouchrika     University of Gabes, Tunisia
Mohamedade Farouk Nanne     University of Nouakchott Al Aasriya,
       Mauritania
Mohamed Lemine Salihi     University of Nouakchott Al Aasriya,
       Mauritania
Nouha Arfaoui     University of Gabes, Tunisia

# Automatic Generation of Semi-structured Documents

Djedjiga Belhadj$^{(\boxtimes)}$ ⓘ, Yolande Belaïd ⓘ, and Abdel Belaïd ⓘ

Université de Lorraine-LORIA, Campus Scientifique,
54500 Vandoeuvre-lès-Nancy, France
{yolande.belaid,abdel.belaid}@loria.fr

**Abstract.** In this paper, we present a generator of semi structured documents (SSDs). This generator can provide samples of administrative documents that are useful for learning information extraction systems. It can also take care of the document annotation operation which is generally difficult to do and time consuming. We propose a general structure for SSDs and we prove that it perfectly works on three SSD types: invoices, payslips and receipts. Both the content and the layout are managed by random variables allowing them to be varied and to obtain different samples. These documents have some sort of similarity that gives them a common global model with particularities for each of them. The generator outputs the documents on three formats: pdf, xml and tiff image. We add an evaluation step to choose an adequate dataset for the learning process and avoid the overfitting. We can easily extend the actual implementation (https://github.com/fairandsmart/facogen) to other SSD types. We use this generator results to experiment an information extraction system from SSDs.

**Keywords:** Semi-structured documents · Automatic generation · Random variables

## 1 Introduction

Document recognition using deep neural models requires having many examples relating to both structure and content. In addition, if one seeks to identify several types of information in the document, several other samples will be necessary to relate the different forms of content and their context. Unfortunately, it is not often possible to have such a mass of documents, and even if it were possible, these documents still have to be annotated correctly, which is very costly in time and money. This is why the current trend is to want to generate its own documents in an artificial way allowing to have annotated documents and therefore ready for use.

Many solutions offer different deep learning architectures to generate document layouts. The LayoutGAN approach of [5] proposes a layout generation using a generative adversarial network (GAN), while [4] generates layouts

Supported by BPI DeepTech.

E. H. Barney Smith and U. Pal (Eds.): ICDAR 2021 Workshops, LNCS 12917, pp. 191–205, 2021.
https://doi.org/10.1007/978-3-030-86159-9_13

based on graph neural networks and variational auto-encoder (VAE). The solution of [6] generates bounding boxes from graphs using different and multiple constraints. Whereas the READ framework proposed in [8], uses a REcursive Autoencoders. LayoutTransformer framework suggested in [2], generates layouts of graphical elements by building a self-attention model to capture contextual relationships between different layout elements and generates new layouts. All of these approaches demonstrate effective results in generating reasonable document layouts with almost ten elements, represented as bounding boxes in a document. However, various types of highly organized documents can significantly have a large number of elements, up to dozens or even hundreds. In addition, their training data are thousands or at least hundreds of annotated documents. This can be difficult to obtain for different types of documents, especially for the administrative semi-structured documents that are usually confidential and don't exist in public datasets. Here comes the need to have a first database of annotated documents, even for deep learning solutions and especially for confidential administrative documents.

The authors of [1] proposed an invoice generator using two techniques: one cloning sample invoices by varying the contents in existing structures, and a completely random one generating from scratch new structures and contents.

Here we try to extend this latter work to generate a set of semi-structured documents encompassing invoices, payslips and receipts. These documents present a certain resemblance which gives them a common global model with particularities for each of them.

The remaining paper is structured as follows: Sect. 2 defines the semi-structured documents; Sect. 3 shows our SSD modeling, the SSD graphical description, its layout and content variation as well as the annotation method; Sect. 4 summarizes the implemented cases: payslips, receipts and invoices; Sect. 5 exposes the evaluation method of the generated SSDs; and Sect. 6 concludes the global contribution of this work.

## 2 Semi-structured Document Modeling

### 2.1 Semi-structured Document Definition

Typically, a document falls into one of the following three categories: structured document, semi-structured document and unstructured document. A structured document has the same page layout with fields always positioned in the same places on the page. The number and type of these fields are always the same. Regular expressions or templates can be used to extract these types of information. Conversely, semi-structured documents are documents in which the location of the data and the fields containing the data vary from one document to another. For example, the delivery address on the order form may be in the top left area or in the top middle or in the bottom right of the page. The quantity of fields, the quantity of rows per table, or the quantity of items per transaction may also differ from document to document and from supplier to supplier. Certain fields

- even solid columns in the tables - may be optional, present on certain documents but absent on other documents of the same type. Content elements can be of three types: highly formatted, accompanied by a keyword, such as "Total: amount", or a few grouped lines containing a few keywords, such as "street", "postcode" for an address. Finally, unstructured documents contain information embedded in the text, requiring the course to use NLP techniques for its extraction.

## 2.2 SSD Graphical Description

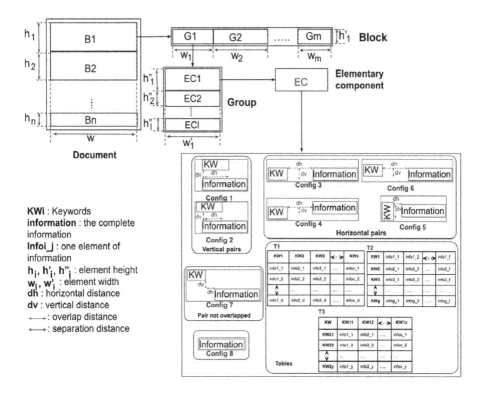

**Fig. 1.** Global SSD structure.

As illustrated in Fig. 1, a SSD page is modeled as a vertical sequence of blocks. A block B is an horizontal sequence of sub-blocks that we call groups. Each group G, in turn, is a vertical sequence of elementary components (ECs). The EC content is usually a pair (keyword (KW), information). As shown in Fig. 1, seven configurations (Config 1 to Config 7) are possible for this pair corresponding to different positions of one in relation to the other (horizontal, vertical, or both) at horizontal (dh) and vertical (dv) distances. The EC could also contain only the information without keywords (Config 8) or a table (T1, T2 or T3) built

from a list of pairs where information is a list of elementary information. The
sequence of the blocks forming the SSD page is centered in the page. The content
of the blocks, groups and ECs could be vertically justified (top, center or in the
bottom) and horizontally (right, left and in the center).

## 2.3   Layout and Content Variation in a SSD

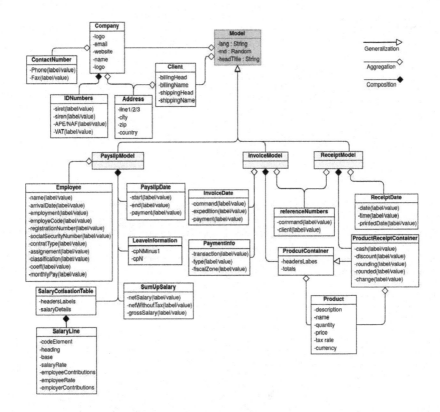

**Fig. 2.** UML representation of SSD content generation class diagram: the composition
association is used when a class is a part of another class and it cannot exist without
it (IDNumbers and Company), whereas the child class in the aggregation could exist
independently of its parent class (Address and Client)

The creation of a SSD is made in two separated stages. The first stage concerns
the content creation of the three SSD types (payslips, receipts and invoices) and
the second one relates to the organization of this content on the document, thus
managing its layout. We give in this section the details of these two steps.

**Content Variation:** As we can see in Fig. 2, a generic "Model" is created to generate the content of the three SSD types (invoices, payslips and receipts). It brings together the common properties of the three types of SSDs, namely: the company's and customer information. Most administrative documents contain these two types of information. The "Model" class contains also the document language (lang), its title (headTitle) and a random variable (rnd) useful for the content management. We then make, for each type of SSD, its own model that instantiates its specific properties ("PayslipModel", "InvoiceModel" and "ReceiptModel"). The various types of SSD can share multiple properties, such as: the invoice and receipt that share together the products information and the document references. As shown in Fig. 2, most of the classes attributes are represented by their label and value. That corresponds to the EC content in the SSD: the pair (keywords, information), where the labels are the keywords and the values are the information.

To generate a new SSD type, we can extend the actual schema by adding new classes if necessary (payments table in a bank statement document for example) and exploit the existing classes that could be found in most administrative documents (company, client, date, etc.).

*Variation Strategy:* Random variables are responsible for managing the content generation of each SSD, according to the schema of Fig. 2. The "rnd" attribute of the Model class is used to facilitate this task. In order to generate the keywords, a random variable is used to select randomly keywords from lists. Since keyword variations are rather limited, keywords lists are created and filled in the bodies of different classes.

To generate the information part, different types of random variables come to manage this part, depending on the type and nature of the information:

– If the information is a numerical value or a date: the random variable takes random numerical value in view of defined thresholds or precise formats;
– If the information is a text phrase: the random variable chooses its value randomly from a list of values, stored in the class's body if the variation is limited (for example the type of contract in French payslips could be either CDD of CDI) or loaded from a source file if the variation is bigger.

If the information is a numerical value and it is calculated based on other information values, a function is used to calculate it, such as the totals in invoices and receipts. Both keywords and information value could be empty in some cases.

The text information variations can be enriched by adding new values to the lists in the different classes, as well as the resource files.

**Layout Variation:** In this step, we create and organize containers that contain the content, generated in the first step, in the SSD. As shown in Fig. 3, three layout classes implement a generic layout interface (SSDLayout). The three classes define "buildModel" function, which is responsible of collecting and managing the SSD layout and saving the final SSD in PDF and image format as well as the SSD annotation in a XML file.

The three layout classes (PayslipLayout, InvoiceLayout and ReceiptLayout) make use of multiple elementary layout components in the "buildModel" function. These components represent generally the different blocks of each SSD type, which have complex configurations. It's the case when the blocks have multiple groups and ECs and it exists different possibilities to organize them like: "EmployeeInfoBox", "SalaryBox" in payslips and "ProductBox", "Receipt-ProductBox" in receipts and invoices. When the blocks have a simple composition (contain one group or one pair), we use two generic containers types: "HorizontalContainer" and "VerticalContainer" to create them directly in the "BuidModel" function.

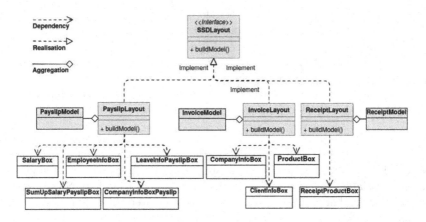

**Fig. 3.** UML class diagram of the SSD layout generation. The aggregation interprets the java class attributes, the dependency indicates the classes used in the buildModel method and the realisation represents the implementation of the SSDLayout interface

*Variation Strategy:* The SSD layout elements (like their content) are managed using different random variables which are described in the following:

- *xi_available*: decides if an optional layout component will be added or not;
- *pos_xi*: chooses the order of the blocks (if one or more blocks could be placed in different emplacement in the SSD);
- *width_random*: chooses the width and the height of the SSD from a list of proposed values;
- *xi_justif*: selects an horizontal or a vertical justification from three possible values of each justification type;
- *columns_order*: designates the columns order and width in a table from a list of fixed values;
- *font*: selects a text font.

Figure 4 shows the algorithm of generating an example of a SSD containing 4 mandatory blocks (A, B, C, D) and an optional block (E). The blocks can have two different orders: A, B, C, D, E or A, D, B, C, E. Each block contains a number of mandatory and optional groups and ECs as shown in Fig. 4. We start by choosing the SSD width and we create its global container. The first block A is generated; the detailed generation of this latter is as follows: the mandatory EC $A_1$ is generated; according to the value of the random variable A2_available, we decide if the optional block $A_2$ would be generated or not; we generate $A_3$ and then we add all the generated blocks to $G_{A1}$ group. The justification of the $G_{A1}$ is selected according to the value of the GA1_justif variable. Three values are possible: to the right(JR), to the left(JL) and in the center(JC). And finally $G_{A1}$ is added to A, which is, in turn, added to the SSD_Container. According to pos_b, the position of the block B is chosen and then, in the same way, the remaining blocks are generated as shown in the Fig. 4.

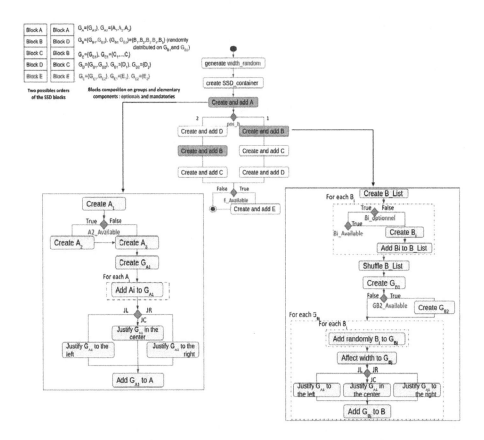

**Fig. 4.** SSD layout generation algorithm.

To add new layout variations to the three generated SSD types, we can add new values to the different lists of variables. To create a new layout of a new SSD type, we add a layout class that implements the SSD Layout interface, then, we add new blocks if necessary. In the "buildModel" method, we could manage its different layout variations.

## 2.4 SSD Annotation

The appropriate classes are assigned to the information values and an "undefined" class to both the keywords and the additional information. The "build-Model" method saves each SSD annotations in an XML file. This latter contains all the SSD words attributes: their classes, bounding boxes coordinates and text fonts. A simpleTextBox class that gathers all this properties, is used to create the layout component of the labels and values and save their properties in the XML file.

# 3   Implemented Cases

In this section, we present the three implemented cases of SSDs, namely the French payslips, the invoices and the receipts in both French and English languages. We mention the mandatory and the optional information of each SSD type and we show some generated examples.

## 3.1 Payslips

The French payslip is divided into three main parts: employer and employee information; components of the salary and employee contributions; and synthesis of important information as shown in the Fig. 5(a).

The first part could be divided into one or two blocks, it groups the company's information, the payment dates and the employee information. It may contain the leave information (in table format) as well as the national collective agreement name. The second part is represented as a table that contains the salary information and the different employee and employer contributions. The third part sum up the important information of the salary: the gross, the net, the taxes, etc. and could also contain the leave information and the national collective agreement if they don't appear in the first part. We detail in the Table 1 all the mandatory and the optional information that we can find in a payslip.

## 3.2 Receipts

A receipt is a SSD that proves a payment. This SSD document has a variable layout because of the variability of the companies and the softwares that produce their receipts (see Fig. 5(b)).

Most of the receipts may be divided into three parts: the first part contain the company's information, the receipt's references, date and time, name of the costumer, etc. The second part is reserved to the sold products table and the last part summarizes the totals values, the taxes, the payment amount, the payment mode and the footnotes. The first and third part can be composed of several

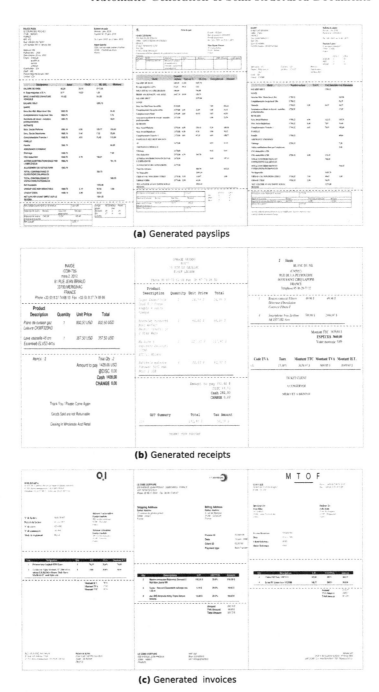

**(a)** Generated payslips

**(b)** Generated receipts

**(c)** Generated invoices

**Fig. 5.** Generated SSDs examples: a) payslips, b) receipts and c) invoices. The three SSDs are all divided into three main parts (with the red line), but the content and layout details of each part are very varied in each single SSD type (Color figure online)

**Table 1.** Mandatory and optional payslips information

|           | Employer       | Employee              | Salary component | Others               |
| --------- | -------------- | --------------------- | ---------------- | -------------------- |
| Mandatory | Name           | Name                  | Gross salary     | Period               |
|           | Address        | Address               | Net salary       | Number of hours      |
|           | APE code       | Employment            |                  | Payment date         |
|           | NAF code       | Level/Coefficient     |                  | Leave information    |
|           | Siret number   | Contributions         |                  | Collective agreement |
|           | Contributions  |                       |                  |                      |
| Optional  | Logo           | Social security number | Net before taxes | Payment mode         |
|           | Phone          | Contract type         |                  | Date                 |
|           | Fax            | Seniority             |                  |                      |
|           | RCS            | Social security ceiling |               |                      |
|           | VAT            | Affectation           |                  |                      |
|           |                | Qualification         |                  |                      |
|           |                | Coefficient/index     |                  |                      |
|           |                | Monthly salary        |                  |                      |
|           |                | Reference salary      |                  |                      |

**Table 2.** Mandatory and optional receipts information

|           | Company                    | Sold products        | Others                      |
| --------- | -------------------------- | -------------------- | --------------------------- |
| Mandatory | Name                       | Names list           | Date and time of transaction |
|           | Address                    | Prices               | Payment amount              |
|           |                            | Total amount paid    |                             |
|           |                            | Sales tax            |                             |
|           |                            | Discounts or coupons |                             |
| Optional  | Logo                       |                      | Payment mode                |
|           | Contact information        |                      | Receipts reference numbers  |
|           | SIRET and SIREN number     |                      | Organization's return policy |
|           | VAT number                 |                      | Terms of sale               |

blocks while the second one contains one block represented by a table of products. The detailed mandatory and optional information that could be found in the receipts is summarized in Table 2. Other additional information could be found in other types of receipts. These additional information represent the details of the sold products or the represented service. For example, the restaurants could add the table number, the server name, etc.

### 3.3 Invoices

Invoices are documents that are very similar to the receipts. The main difference between the two documents is: an invoice is a request for payment which is issued from the company before the payment, while a receipt is a proof of payment.

Thus, invoices contain other information (in addition to those mentioned in Table 2) related to the customer, such as the shipping and the billing addresses. On the other hand, they do not generally contain payment information (cash, change, etc.). Unlike receipts that can have different dimensions, invoices have an A4 format mostly.

The invoice is divided into 3 parts where: the first contains the company's information as well as the shipping and billing information; the second part is reserved to the products table; the last part contains the totals and the footnotes. We can consider this the examples shown in Fig. 5(c).

## 4    Generated Dataset Evaluation

We add the evaluation step to choose an adequate dataset for the learning process and avoid the overfitting. The generated dataset content must be varied as well as the layout, depending on the SSD type and its generation restrictions.

### 4.1    Content Text Diversity

We evaluate the content diversity of the generated datasets using the more recent metric that evaluates the diversity of generated texts known as SELF-BLEU metric, proposed in [10]. It measures the differences between generated phrases of the same class. A low SELF-BLEU score implies a high variety. BLEU score, as defined in [7], is calculated for every generated sentence, and the average BLEU score is then the SELF-BLEU of the dataset.

$$BLEU = BP * \exp(\sum_{n=1}^{N} w_n \log p_n) \tag{1}$$

where PB is the brevity penalty. It is computed as follows:

$$BP = \left\{ \begin{array}{cc} 1 & if\ c > r \\ \exp(\frac{1-r}{c}) & if\ c \leq r \end{array} \right\} \tag{2}$$

where c is the length of the selected phrase (the hypothesis), and r is the length of another sentence (reference), N is the maximum n-gram length and $w_n$ is a positive weight ($\frac{1}{N}$). The $p_n$ is the modified precision score, which is calculated as follows:

$$p_n = \frac{\sum_{C \in Candidates} \sum_{n\text{-}gram \in C} Count_{clip}(n\text{-}gram)}{\sum_{C' \in Candidates} \sum_{n\text{-}gram \in C'} Count(n\text{-}gram)} \tag{3}$$

*Candidates* are the generated phrases that we want to evaluate, *Count* is the number of times a *n-gram* appears in the candidate phrase and $Count_{clip} = min(Count, Max\_Ref\_Count)$, where $Max\_Ref\_Count$ is the maximum number of times a *n-gram* occurs in any reference phrase (the other generated phrases).

Table 3 shows the SELF-BLEU score on three generated datasets: receipts, invoices and payslips. Each dataset contains 1000 SSDs. SROIE dataset is issued from ICDAR competition [3] and is taken here as a reference. SROIE is a real public dataset that contain 973 receipts. We notice that the SELF-BLEU score recorded on the three generated datasets has a low value, even lower than the score recorded on SROIE which is a real dataset. This means that the three generated datasets have a high content diversity.

**Table 3.** SELF-BLEU scores of SROIE dataset and our three generated datasets

| SROIE | G_Receipts | G_Invoices | G_Payslips |
|-------|------------|------------|------------|
| 0.44  | 0.29       | 0.29       | 0.23       |

### 4.2 Layout Diversity

Most metrics used to compute the layout diversity in generated documents, calculate the distance between the blocks of the same class. This distance can be given in different ways: the Euclidean distance between the center of the blocks, the Manhattan distances of the corners, the difference between the height and the width, the overlap area between the blocks or combinations of these distances [9]. An alignment score defined by the sum of the Manhattan distance between the two corners of the blocks gives us both position and area information, while the overlapping zone depends on the position, size and ratio of the height/width of the blocks. These two measures might be reasonable block distances to use, because both include several layout information in one measure.

As a result, we adopt these two latter diversity metrics introduced in [9] to compute our layout diversity. These two metrics (Align and Over) are computed on the mandatory ECs of each SSD type. We define also another metric that we call SCR (SSD_Compositions Ratio) where SSD_composition of a document is the list of all its elementary components (ECs), whether they are mandatory or optional, ordered according to their position. SCR is a measure of layout diversity that focuses on the general SSD layout composition.

We give in what follows the details of these three adopted metrics: Align, Over and SCR respectively.

Let $B_i = (p_a, p_b)$, $B_j = (p_c, p_d)$ be two mandatory EC blocks of the same class in two SSD layouts, where $p_a$ and $p_b$ are the corner points that define the EC block $B_i$ (its bounding box). The alignment ($Align$) score is calculated as follows:

$$Align(B_i, B_j) = Dmh(p_a, p_b) + Dmh(p_c, p_d) \qquad (4)$$

where $Dmh(p_a, p_b) = |x_a - x_b| + |y_a - y_b|$ is the Manhattan distance between two points $p_a$ and $p_b$.

While the overlapping $Over$ score is computed as:

$$Over(B_i, B_j) = 1 - \frac{2 * ov(B_i, B_j)}{area(B_i) + area(B_j)} \qquad (5)$$

where $ov(B_i, B_j)$ is the overlapping area of the two EC blocks $B_i$ and $B_j$, and $area(B_i)$ is the area of the EC block $B_i$.

SCR is computed as the number of distinct *SSD_compositions* divided by the total number of *SSD_compositions* which is the same number of generated SSDs in one dataset. Two *SSD_compositions* are considered distinct if their ECs lists are not perfectly identical (comparing the number of ECs, their classes as well as their order in the list).

$$SCR = \frac{\#distinct\ SSD\_Compositions}{\#SSD} \tag{6}$$

The alignment and the overlap scores of the SSD's EC blocks are performed on SROIE dataset as well as on 1000 generated samples of each SSD type as shown in the Table 4. While, the SCR score is computed on 2000 generated SSDs of each type. SCR is not computed on SROIE dataset because the calculation is very costly in time and also we don't have enough information about all the ECs classes in the SROIE samples.

**Table 4.** Alignment, overlap and SCR scores where G_ indicates the generated dataset

| SSD type | SROIE | G_Receipts | G_Invoices | G_Payslips |
|---|---|---|---|---|
| Alignment | 0.48 | 0.06 | 0.14 | 0.06 |
| Overlap | 0.998 | 0.998 | 0.997 | 0.986 |
| SCR | – | 0.90 | 0.88 | 0.79 |

As can be seen in the Table 4, the normalized alignment score of the three generated datasets has very low values, even lower than SROIE score. The variation of EC positions in SSDs is the main source of this low values and the variation of content also affects this score. The corners of the EC bounding boxes depend on the EC content's font size and the number of characters that make it up.

The overlapping score has high values, they are almost the same as the SROIE score. Restrictions on the positions of the mandatory ECs in SSDs is the first reason of this score value in this type of documents. The content variation has almost no influence on this score, even if the corner positions of the EC bounding boxes change, but their shared area remains very important. And even for the ECs that have variable positions, the possibility of overlapping is very likely because the variation is generally in the same region of the SSD.

While the SCR score, as could be noted, has high values. It means that the generated layouts of each SSD type are quite diverse. The various optional ECs that could be found in a SSD and not in another as well as the position variation of both mandatory and optional ECS are the main explications of this high diversity score.

In conclusion, the generator provides SSDs datasets, whose content and layout diversity are quite important. This makes it possible to avoid the overfitting in the learning process of the information extraction systems. Despite this diversity, the SSDs are subjected to some layout constraints, which characterize this type of documents.

## 5 Use of the SSDs Generator in an Information Extraction System

As could be seen in the Fig. 6, the generator is responsible for providing datasets (annotated SSDs samples) for both the pre-processing step and the graph and word modeling in the learning step in the system architecture. The pre-processing step calculate a max distance (d-max) useful for the graph modeling, by applying an heuristic.

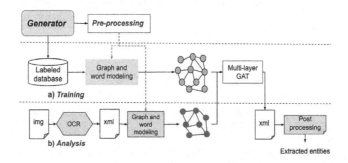

**Fig. 6.** The generator role in the information extraction system architecture

In the system training process, the main labeled database is used to model the SSDs content as a graph of sub-star graphs, using the d-max. The graph nodes represent the document words and are modeled by features vectors that include both textual (word embedding) and visual (normalized positions) features, while the edges are the neighborhood relationships. The graph is then fed to a multi-layer GAT (Graph attention network), which applies the multi-head attention mechanism. In the analysis step, a SSD image is OCRed and the result is transformed to an xml file with the same format as the labeled generated SSDs. The associated graph to this new SSD is created based on d-max. The graph is then fed to the trained Multi-layer GAT, which classifies all the nodes of the graph to their appropriate classes and saves the result in an xml file. A post-processing step is finally applied to regroup the words of the same classes and form the entities to extract.

## 6 Conclusion

In this paper, we introduced a semi structured documents (SSD) generator that generates automatically three SSDs types (payslips, receipts and invoices) with a significant content and layout variations. Based on a general structure of a SSD, we have provided random variables, to handle both aspects (content and layout), that guarantee the SSD's variation. We set some constraints to ensure that our artificial documents are almost close to real ones. Then we introduced layout and content metrics to measure the diversity of our generated SSDs. The evaluation results showed that the generated datasets of the three SSD types have a high content and layout diversity. This diversity in the generated datasets avoids the overfitting during the learning process of the information extraction system.

**Acknowledgements.** This work was carried out within the framework of the BPI DeepTech project, in partnership between the University of Lorraine (Ref. UL: GECO/2020/00331), the CNRS, the INRIA Lorraine and the company FAIR&SMART. The authors would like to thank all the partners for their fruitful collaboration.

# References

1. Blanchard, J., Belaïd, Y., Belaïd, A.: Automatic generation of a custom corpora for invoice analysis and recognition. In: 2019 International Conference on Document Analysis and Recognition Workshops (ICDARW), vol. 7, p. 1. IEEE (2019)
2. Gupta, K., Achille, A., Lazarow, J., Davis, L., Mahadevan, V., Shrivastava, A.: Layout generation and completion with self-attention. arXiv preprint arXiv:2006.14615 (2020)
3. Huang, Z., et al.: ICDAR 2019 competition on scanned receipt OCR and information extraction. In: 2019 International Conference on Document Analysis and Recognition (ICDAR), pp. 1516–1520. IEEE (2019)
4. Lee, H.-Y., et al.: Neural design network: graphic layout generation with constraints. In: Vedaldi, A., Bischof, H., Brox, T., Frahm, J.-M. (eds.) ECCV 2020. LNCS, vol. 12348, pp. 491–506. Springer, Cham (2020). https://doi.org/10.1007/978-3-030-58580-8_29
5. Li, J., Yang, J., Hertzmann, A., Zhang, J., Xu, T.: LayoutGAN: generating graphic layouts with wireframe discriminators. arXiv preprint arXiv:1901.06767 (2019)
6. Nauata, N., Chang, K.-H., Cheng, C.-Y., Mori, G., Furukawa, Y.: House-GAN: relational generative adversarial networks for graph-constrained house layout generation. In: Vedaldi, A., Bischof, H., Brox, T., Frahm, J.-M. (eds.) ECCV 2020. LNCS, vol. 12346, pp. 162–177. Springer, Cham (2020). https://doi.org/10.1007/978-3-030-58452-8_10
7. Papineni, K., Roukos, S., Ward, T., Zhu, W.J.: BLEU: a method for automatic evaluation of machine translation. In: Proceedings of the 40th Annual Meeting of the Association for Computational Linguistics, pp. 311–318 (2002)
8. Patil, A.G., Ben-Eliezer, O., Perel, O., Averbuch-Elor, H.: READ: recursive autoencoders for document layout generation. In: Proceedings of the IEEE/CVF Conference on Computer Vision and Pattern Recognition Workshops, pp. 544–545 (2020)
9. Van Beusekom, J., Keysers, D., Shafait, F., Breuel, T.M.: Distance measures for layout-based document image retrieval. In: Second International Conference on Document Image Analysis for Libraries (DIAL 2006), p. 11-pp. IEEE (2006)
10. Zhu, Y., et al.: Texygen: a benchmarking platform for text generation models. In: The 41st International ACM SIGIR Conference on Research & Development in Information Retrieval, pp. 1097–1100 (2018)

# DocVisor: A Multi-purpose Web-Based Interactive Visualizer for Document Image Analytics

Khadiravana Belagavi[ORCID], Pranav Tadimeti[ORCID],
and Ravi Kiran Sarvadevabhatla[✉][ORCID]

Centre for Visual Information Technology, International Institute of Information
Technology, Hyderabad 500032, India
ravi.kiran@iiit.ac.in
https://github.com/ihdia/docvisor

**Abstract.** The performance for many document-based problems (OCR,
Document Layout Segmentation, etc.) is typically studied in terms of a
single aggregate performance measure (Intersection-Over-Union, Character Error Rate, etc.). While useful, the aggregation is a trade-off between
instance-level analysis of predictions which may shed better light on a
particular approach's biases and performance characteristics. To enable
a systematic understanding of instance-level predictions, we introduce
DocVisor - a web-based multi-purpose visualization tool for analyzing
the data and predictions related to various document image understanding problems. DocVisor provides support for visualizing data sorted using
custom-specified performance metrics and display styles. It also supports
the visualization of intermediate outputs (e.g., attention maps, coarse
predictions) of the processing pipelines. This paper describes the appeal-
ing features of DocVisor and showcases its multi-purpose nature and
general utility. We illustrate DocVisor's functionality for four popular
document understanding tasks – document region layout segmentation,
tabular data detection, weakly-supervised document region segmentation
and optical character recognition. DocVisor is available as a documented
public repository for use by the community.

## 1 Introduction

Tasks in document image analysis are evaluated using task-specific performance
measures. For example, measures such as average Intersection-over-Union (IoU)
or the Jaccard Index, mAP@x (mean Average Precision for an IoU threshold
of x), and mean HD (Hausdorff distance) are used to evaluate semantic segmentation approaches [6,10]. Similarly, measures such as Character Error Rate
(CER) and Word Error Rate (WER) are used to evaluate Optical Character
Recognition (OCR) systems [5]. These measures are undoubtedly informative
as aggregate performance measures. However, they are prone to statistical bias
arising from the imbalanced distribution of performance measures - e.g., mean

E. H. Barney Smith and U. Pal (Eds.): ICDAR 2021 Workshops, LNCS 12917, pp. 206–219, 2021.
https://doi.org/10.1007/978-3-030-86159-9_14

averaging [11]. Therefore, it is helpful to have a mechanism that enables predictions to be visualized on a per-instance (image or region) basis after sorting them by the performance measure of interest. Doing so can be valuable in understanding factors contributing to performance or lack thereof. Sometimes, it may also be essential to visualize outputs of the intermediate stages in the systems' pipeline. For example, in an end-to-end OCR system, the prediction from a skew correction module is fed to a line segmenter module, which then produces an input to the recognizer module [15]. Similarly, in attention-based OCR systems, visualizing attention maps intermediately may help better understand the final prediction [25].

Moreover, multiple tasks may be of interest to different project groups working on the same document corpus. One group could be working on document layout segmentation, while another could be working on OCR. Instead of maintaining separate visualization mechanisms, it may be more efficient to develop and maintain a single interface that can be used concurrently by all project groups.

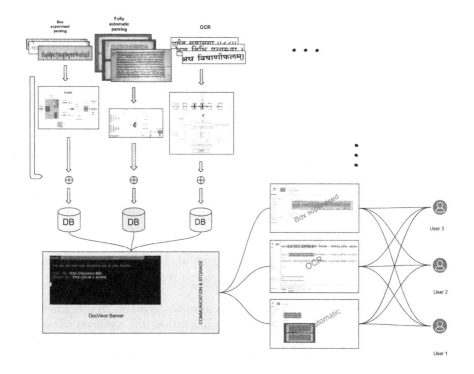

**Fig. 1.** Architecture of the DocVisor tool.

Motivated by the above requirements, we introduce DocVisor - a configurable web-based multi-purpose visualization tool for analyzing the data and predictions related to various document image understanding problems. Currently,

DocVisor supports four popular document understanding tasks and their variants – document region layout segmentation, tabular data detection, weakly-supervised document region segmentation, and optical character recognition. The tool is flexible and provides support for visualizing data sorted using custom-specified performance metrics and display styles.

**Table 1.** Salient attributes of popular visualization tools.

| Methods | Released in | Metric-sorted | Allow comparison with g.t | Multi-model comparison | Task(s) | Web-based |
|---|---|---|---|---|---|---|
| dhSegment [17] | 2018 | ✗ | ✓ | ✗ | Layout | ✓ |
| PAGEViewer[a] | 2014 | ✗ | Partially | ✗ | Layout | ✗ |
| OpenEvaluationTool [1] | 2017 | ✗ | ✓ | ✗ | Layout | ✓ |
| HInDoLA [23] | 2019 | ✗ | ✗ | ✗ | Layout | ✓ |
| **DocVisor** | 2021 | ✓ | ✓ | ✓ | Multiple | ✓ |

[a]https://www.primaresearch.org/tools/PAGEViewer.

The architecture of the DocVisor tool is depicted in Fig. 1. Model outputs for the three document-analysis tasks (Fully Automatic Region Parsing, Box Supervised Region Parsing and OCR) are pre-processed and given as an input to the DocVisor tool, which then renders an interactive multi-layout web-page. The tool and associated documentation can be accessed from https://github. com/ihdia/docvisor.

## 2    Related Work

Tools for document analysis and Optical Character Recognition (OCR) form an essential part of improving the related models. They are broadly classified into annotation tools, visualization tools, post-correction tools and evaluation tools. Prominent tools for layout parsing, image-based text search and historical manuscript transcription include HInDoLA[1], Aletheia[2], Monk[3] and Transkribus[4] respectively [4,13,20,23]. Kiessling et al. [14] propose eScriptorium for historical document analysis, which allows to segment and transcribe the manuscripts manually or automatically.

AnyOCR model is designed to be unsupervised by combining old segmentation-based methods with recent segmentation-free methods to reduce the annotation efforts [12]. It also includes error correction services like any-LayoutEdit and anyOCREdit [2]. OpenOCRCorrect[5] is designed to detect and correct OCR errors by exploiting multiple auxiliary sources [19]. It includes

---

[1] https://github.com/ihdia/.
[2] www.primaresearch.org/tools/Aletheia.
[3] http://monkweb.nl/.
[4] https://readcoop.eu/transkribus/.
[5] https://tinyurl.com/5dd7dh6a.

features like error detection, auto-corrections, and task specific color coding to reduce the cognitive load. OCR4all[6] is designed with semi-automatic workflow for historical documents OCR.

Some of these tools naturally allow analysis of annotations and related metadata as a secondary feature. One work close to ours is open evaluation tool[7] by Alberti et al. [1], which allows for evaluation of document layout models at pixel level. Another is dhSegment [17] which enables predictions and ground-truth to be serially visualized using the Tensorboard functionality of the popular machine learning library, Tensorflow.

Unlike DocVisor, these tools omit certain important features such as interactive overlays, visualization at multiple levels of granularity, metric-sorted visualization of all test set predictions, simultaneous comparison of multiple model outputs etc. The few tools which provide a reasonable degree of functionality for visualization are summarized in Table 1.

We describe the functionalities of DocVisor in subsequent sections.

## 3    Fully Automatic Region Parsing

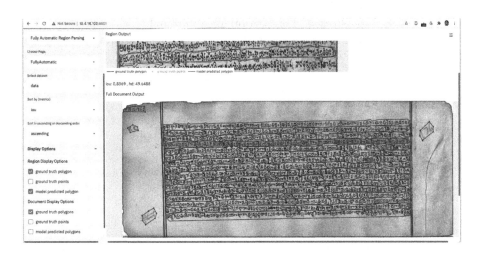

**Fig. 2.** Fully Automatic Region Parsing interface in DocVisor. Refer to Sect. 3.

Segmenting a document or tabular image into regions based on their semantic category (e.g. text lines, pictures, boundary lines, tabular rows) is an important problem [7,16,18,26]. This is especially crucial for isolating popularly sought regions such as text lines within handwritten and historical manuscripts. We consider three problem settings - historical manuscript region layout segmentation, printed document region layout detection and tabular data layout detection.

---

[6] https://github.com/OCR4all/OCR4all.

[7] https://tinyurl.com/69thzj3c.

### 3.1  Historical Manuscript Region Layout Segmentation

For our setting, we consider a state-of-the-art deep network for handwritten document layout segmentation [21]. The document images are sourced from Indiscapes, a large-scale historical manuscript dataset [18]. The dataset is annotated using the HInDoLA annotation system [23]. As part of the annotation process, each region of interest is captured as a user-drawn polygon along with a user-supplied region category label ('text line','picture','library stamp' etc.). These polygons are rendered into binary image masks and used along with the associated document images to train the deep network. The outputs from the deep network are also binary image masks. Selecting `Go to→Fully Automatic Region Parsing` and `Page→Fully Automatic` at top-left corner of DocVisor enables visualization of the document images, associated ground-truth and deep network predictions (Fig. 2). To analyze the performance at a per-region level, the region type is first selected at the top. The metric by which the selected regions are to be sorted (IoU, HD) are selected from the `Sort by (metrics)` dropdown in the left navigation bar. An important utility of the setup becomes evident when regions with IoU value = 0 are examined. These regions correspond to false negatives (i.e. they failed to be detected by the network). In the `Display Options→Region Display Options` on the bottom-left, the visualization elements of interest are checked. This section allows ground-truth or prediction to be displayed (on the top-right) either in terms of annotator provided polygon vertices or the associated polygon boundary.

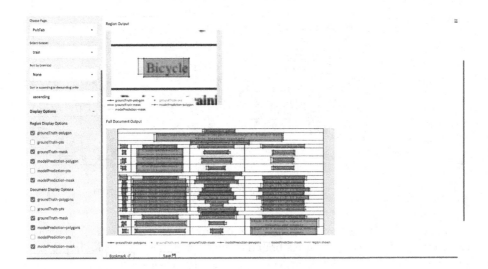

**Fig. 3.** PubTabNet [26] data visualized using DocVisor.

In the main panel on the right, the region image (sorted by the metric) is shown at the top. To enable viewing in the original document context, the region is highlighted as a shaded region in the full image. As with the region display elements, the navigation bar on left can also be used to view visualization elements related to the full document image (`Display Options`→`Document Display Options`).

Finally, two additional options are also provided for selective navigation. The `Bookmark` button at the bottom enables the user to shortlist select regions for later viewing during the session. The `Save` button enables select regions to be saved to the user's device for offline analysis.

## 3.2  Table Data Layout Segmentation

Image-based table recognition is an important problem in document analysis. In our case, we have obtained the table-based images from PubTabNet [26]. Given the prominence of this dataset, we have provided all the necessary scripts to parse the data in the format akin to PubTabNet, and convert it to a format recognized by DocVisor.

This particular instance of Fully Automatic Region Parsing can be viewed by selecting `Go To`→`Fully Automatic Region Parsing` as well as selecting `Choose Page`→`PubTab` in the sidebar to the left. Since this page is an instance of the Fully Automatic Region Parsing layout, the functionalities available are identical to the `FullyAutomatic` instance. Refer to Fig. 3 for a depiction of an example visualization.

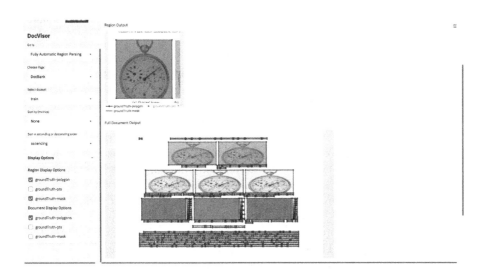

**Fig. 4.** DocBank [16] data visualized using DocVisor.

### 3.3   DocBank Dataset

DocBank [16] is a popular, large-scale dataset which consists of around 500,000 annotations, which have been obtained from a weak supervision approach. This data is also visualized as an instance of Fully Automatic Region Parsing layout. To select the DocBank page, select Go To→Fully Automatic Region Parsing and Choose Page→DocBank. Additionally, a script is provided in the DocVisor repository for parsing and converting data from the DocBank format to the DocVisor format. Much like PubTabNet, since DocBank is an instance of Fully Automatic Region Parsing layout, the features are identical to FullyAutomatic. Figure 4 shows an example visualization of DocBank data.

## 4   Box-Supervised Region Parsing

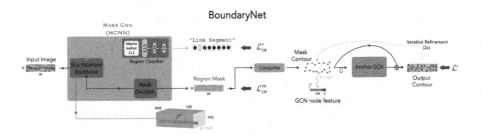

**Fig. 5.** The architectural diagram of the box-supervised deep network BoundaryNet [24].

A popular alternative to fully supervised region parsing (of the type described in previous section) is weakly supervised region parsing. In this setup, a weak supervisory input is provided (e.g. region class or the bounding box enclosing the region). We consider a state-of-the-art bounding box-supervised deep network called BoundaryNet [24] which outputs a sequence of predicted points denoting the vertices of the polygon enclosing the region (Fig. 5). The network also predicts the class label associated with the region. BoundaryNet is trained on an earlier version of the dataset [18] mentioned in the context of previous section.

As Fig. 5 shows, BoundaryNet comprises of two component modules. The first module, Mask-CNN, produces an initial estimate of the region boundary mask. This estimate is processed by a contourization procedure to obtain the corresponding polygon. This polygon estimate is refined by the Anchor-GCN module to produce the final output as a sequence of predicted points. Note that the output is not a mask unlike the full image setup in previous section.

In our setup, the set of intermediate and final outputs from BoundaryNet can be viewed in DocVisor by making Box-supervised Region Parsing selection at top-left corner. The various menu and visualization choices resemble the ones present for Fully Supervised Region Parsing as described in previous section (Fig. 6).

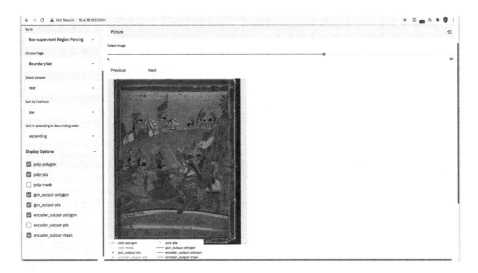

**Fig. 6.** Box-Supervised Region Parsing interface in DocVisor. A region image of type 'Picture' is shown in the example. Refer to Sect. 4.

# 5   Optical Character Recognition (OCR)

The OCR task involves recognizing text from images and is an intensely studied problem in the domain of document image understanding [3]. Although OCR systems work on various levels of granularities, we focus on the line-image variant. In this variant, the input is a single line of a document page, and the output is corresponding text.

We consider the line-level OCR developed by Agam et al. [5]. This OCR has been developed for historical documents in Sanskrit, a low-resource Indic language. We use DocVisor for analyzing two OCR models - one trained on printed texts and another on handwritten manuscripts. The accuracy of an OCR is typically characterized by Word Error Rate (WER) and Character Error Rate (CER); the smaller these quantities, the better the performance. The final numbers reported in OCR literature are typically obtained by averaging over WER and CER across line images from the test set. However, we use WER and CER scores at line-level as the metric for sorting the image set (train, validation, or test) in DocVisor.

The DocVisor interface enables selection between WER and CER metrics for sorting. The slider bar enables random access across the sorted image sequence. For each line image, the predicted text, corresponding ground-truth text, WER, and CER are shown (see Fig. 7). To enable aligned comparison with the image, we provide an option for changing the font size in the left-side navigation bar (under `Settings`→`Render`). For quick comprehension of differences between ground-truth and predicted text, we numerically index and color-code matching and mismatched segments (see `Visualize diffs` in Fig. 7).

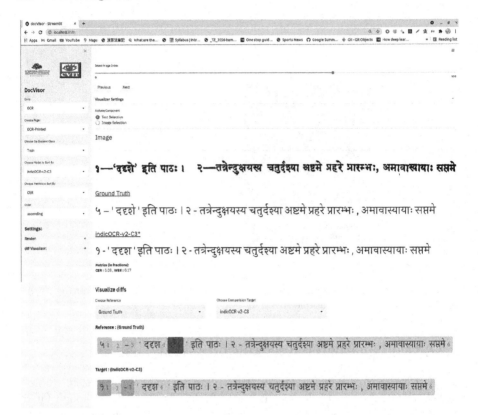

**Fig. 7.** OCR interface in DocVisor depicting line image, predicted text and ground-truth text. The metrics (CER, WER) and the color-coded mismatches and matches can also be seen. Refer to Sect. 5.

An interesting aspect of the OCR mentioned above lies in its 'attention' component [5]. In the process of learning to map a line image to its corresponding text, an intermediate stage produces an alignment matrix via the attention component mentioned before. The alignment matrix can be used to map a line image segment to an estimate of the corresponding matching sub-string in the predicted text and vice-versa. In the DocVisor UI, we enable the mapping via a selection process (Fig. 8). For instance, selecting Visualizer Settings→Image Selection enables selecting a segment of the line image. Upon completing the selection, the corresponding portion of the predicted text is highlighted. A similar mechanism for highlighting the image segment corresponding to a selection within the predicted text is achieved by switching to Visualizer Settings→Text Selection. From an explainability point of view, this feature is useful for understanding the image to text mapping decision process of the deep network, particularly for incorrectly predicted text portions. DocVisor also allows the user to select the color used for highlighting the image through a color picker for added flexibility.

**Fig. 8.** OCR interface in DocVisor depicting the image-text alignment by leveraging the attention mechanism from the OCR architecture of Agam et al. [5]. Refer to Sect. 5. Notice that there are multiple (2) attention OCR models along with non-OCR models that can be loaded in the OCR layout that allows the user to compare line level results of multiple models simultaneously.

While analysing the results of a model on a particular dataset, sometimes it is important to compare the predictions of two different models or even two identical models trained with different hyper-parameter settings (e.g., learning rates) or different user-defined configurations. The DocVisor tool interface allows for visualization and comparison of multiple OCR outputs for the same image or dataset. The user can load multiple attention models along with outputs of other non-attention models (models that do not produce an alignment matrix) to compare results. The user can select the model by which the dataset is to be sorted using the metrics of interest. The user can choose to select and visualize the difference between output texts with the Ground-Truth or between outputs of any two models.

In our setting, we use the DocVisor tool to compare the results of C1 and C3 configurations of the IndicOCR model as defined by Agam et al. [5] along with results of popular state of the art non-attention models such as Google-OCR [8], Tessaract [22], and IndSenz [9]. Figure 8 shows an example of this setting.

```
{
    "metaData":{
        "pageLayout": "Fully Automatic Region Parsing",
        "pageName": "FullyAutomatic",
        "dataPaths": {
            "Train": "/home/user/Desktop/DocVisor/jsonData/fullyAutomatic/train/train_data.json",
            "Validation": "/home/user/Desktop/DocVisor/jsonData/fullyAutomatic/val/val_data.json",
            "Test": "/home/user/Desktop/DocVisor/jsonData/fullyAutomatic/test/test_data.json"
        },
        "outputMasks": {"groundTruth":1,"modelPrediction":1}
    }
}
```

(a)

```
{
    "metaData":{
        "pageLayout": "Box-supervised Region Parsing",
        "pageName": "BoundaryNet",
        "dataPaths": {
            "Train": "/home/user/Desktop/DocVisor/jsonData/boxSupervised/train/train_data.json",
            "Validation": "/home/user/Desktop/DocVisor/jsonData/boxSupervised/val/val_data.json",
            "Test": "/home/user/Desktop/DocVisor/jsonData/boxSupervised/test/test_data.json"
        },
        "outputMasks":{"poly":1,"gcn_output":0,"encoder_output":1}
    }
}
```

(b)

```
{
    "metaData": {
        "pageLayout": "OCR",
        "pageName": "OCR-Printed",
        "dataPaths": {
            "Train": "/home/user/Desktop/DocVisor/jsonData/ocr/train/train_data.json",
            "Validation": "/home/user/Desktop/DocVisor/jsonData/ocr/val/val_data.json",
            "Test": "/home/user/Desktop/DocVisor/jsonData/ocr/test/test_data.json"
        }
    }
}
```

(c)

**Fig. 9.** Example metadata files to launch the DocVisor tool. (a) Metadata file for Fully Automatic Layout. (b) Metadata file for Box Supervised Region Parsing Layout and (c) Metadata file for OCR layout. Refer to Sect. 6.

## 6  Configuration

To launch DocVisor, the user needs to provide the meta-information required to set up the tool through metadata files. Essentially, the metadata files act as configuration files in which the user not only provides the relevant information regarding the data to be visualized, but also certain settings the user can tweak for their own use cases. Figure 9 shows the structure of example metadata files for the three different layouts of DocVisor.

The user can load DocVisor with one or more of DocVisor's layouts. Additionally, a single layout can have multiple instances in the app: for example, a user may be interested in viewing the results of a model on two different datasets. In such cases, the user provides metadata files which would each contain the meta-information for the corresponding layout instance.

Figure 10 depicts standard data-file formats for each of the three layouts provided in the DocVisor tool. Each layout can be loaded provided the data files have the mandatory fields. Another advantage of DocVisor lies in its ability to import standard, pre-existing configuration formats for tasks which omits the need to write custom configuration files.

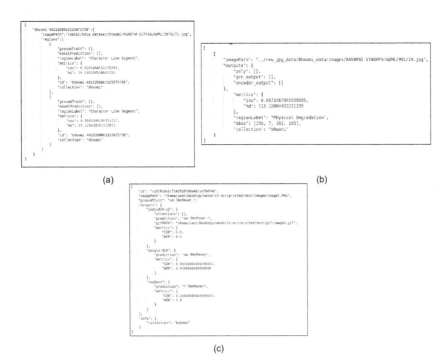

(a)                                                      (b)

(c)

**Fig. 10.** Example data files for (a) Fully Automatic Layout (b) Box Supervised Region Parsing (c) OCR Layout. Refer to Sect. 6.

For more details of the data file format, refer to the tool's documentation at https://github.com/ihdia/docvisor.

# 7   Future Work

The existing work on DocVisor can be extended to include formats for some popular image segmentation tasks such as hOCR, PAGE format, LatexMath etc., making it much easier for users to integrate their data into DocVisor.

Additionally, support can be included for approaches with a large amount of data, where it may not be feasible to include all data into a single file.

# 8   Conclusion

We have presented an overview of DocVisor - our web-based multi-purpose visualization tool for analyzing the data and predictions related to various document image understanding problems. DocVisor currently supports four popular document understanding tasks and their variants – document region layout segmentation, tabular data detection, weakly-supervised document region segmentation, and optical character recognition. Given the lack of general-purpose data and prediction analysis tools in the document image community, we expect our tool with its interactive and flexible display options to be a valuable contribution, readily available for reuse and modification given its open-source, public availability. Given the tool's capability for displaying point, polygon, and mask overlays, we also expect the tool to be helpful to researchers working on the broader computer vision problems of layout prediction.

# References

1. Alberti, M., Bouillon, M., Ingold, R., Liwicki, M.: Open evaluation tool for layout analysis of document images. In: 2017 14th IAPR International Conference on Document Analysis and Recognition (ICDAR), vol. 4, pp. 43–47. IEEE (2017)
2. Bukhari, S.S., Kadi, A., Jouneh, M.A., Mir, F.M., Dengel, A.: anyOCR: an open-source OCR system for historical archives. In: 2017 14th IAPR International Conference on Document Analysis and Recognition (ICDAR), vol. 1, pp. 305–310. IEEE (2017)
3. Cheriet, M., Kharma, N., Liu, C.L., Suen, C.: Character Recognition Systems: A Guide for Students and Practitioners. John Wiley & Sons, Hoboken (2007)
4. Clausner, C., Pletschacher, S., Antonacopoulos, A.: Aletheia - An advanced document layout and text ground-truthing system for production environments. In: 2011 International Conference on Document Analysis and Recognition, pp. 48–52. IEEE (2011)
5. Dwivedi, A., Saluja, R., Kiran Sarvadevabhatla, R.: An OCR for classical indic documents containing arbitrarily long words. In: The IEEE/CVF Conference on Computer Vision and Pattern Recognition (CVPR) Workshops, June 2020
6. Everingham, M., Van Gool, L., Williams, C.K., Winn, J., Zisserman, A.: The PASCAL visual object classes (VOC) challenge. Int. J. Comput. Vis. **88**(2), 303–338 (2010)
7. Gatos, B., et al.: Ground-truth production in the tranScriptorium project. In: 2014 11th IAPR International Workshop on Document Analysis Systems, pp. 237–241. IEEE (2014)
8. Google: Convert pdf and photo files to text (2020). https://support.google.com/drive/answer/176692?hl=en. Accessed 26 March 2020
9. Hellwig, O.: Indsenz OCR (2020). http://www.indsenz.com/. Accessed on 26 March 2020
10. Huttenlocher, D.P., Klanderman, G.A., Rucklidge, W.J.: Comparing images using the hausdorff distance. IEEE Trans. Pattern Anal. Mach. Intell. **15**(9), 850–863 (1993)
11. James, G., Witten, D., Hastie, T., Tibshirani, R.: An Introduction to Statistical Learning, vol. 112. Springer, New York (2013). https://doi.org/10.1007/978-1-4614-7138-7
12. Jenckel, M., Bukhari, S.S., Dengel, A.: anyOCR: a sequence learning based OCR system for unlabeled historical documents. In: 2016 23rd International Conference on Pattern Recognition (ICPR), pp. 4035–4040. IEEE (2016)
13. Kahle, P., Colutto, S., Hackl, G., Mühlberger, G.: Transkribus-a service platform for transcription, recognition and retrieval of historical documents. In: 2017 14th IAPR International Conference on Document Analysis and Recognition (ICDAR), vol. 4, pp. 19–24. IEEE (2017)

14. Kiessling, B., Tissot, R., Stokes, P., Ezra, D.S.B.: escriptorium: an open source platform for historical document analysis. In: 2019 International Conference on Document Analysis and Recognition Workshops (ICDARW), vol. 2, p. 19. IEEE (2019)

15. Kumar, M.P., Kiran, S.R., Nayani, A., Jawahar, C., Narayanan, P.: Tools for developing OCRS for Indian scripts. In: 2003 Conference on Computer Vision and Pattern Recognition Workshop, vol. 3, p. 33. IEEE (2003)

16. Li, M., Xu, Y., Cui, L., Huang, S., Wei, F., Li, Z., Zhou, M.: DocBank: a benchmark dataset for document layout analysis (2020)

17. Oliveira, S.A., Seguin, B., Kaplan, F.: dhsegment: a generic deep-learning approach for document segmentation. In: 2018 16th International Conference on Frontiers in Handwriting Recognition (ICFHR), pp. 7–12. IEEE (2018)

18. Prusty, A., Aitha, S., Trivedi, A., Sarvadevabhatla, R.K.: Indiscapes: instance segmentation networks for layout parsing of historical indic manuscripts. In: 2019 International Conference on Document Analysis and Recognition (ICDAR), pp. 999–1006. IEEE (2019)

19. Saluja, R., Adiga, D., Ramakrishnan, G., Chaudhuri, P., Carman, M.: A framework for document specific error detection and corrections in Indic OCR. In: 2017 14th IAPR International Conference on Document Analysis and Recognition (ICDAR), vol. 4, pp. 25–30. IEEE (2017)

20. Schomaker, L.: Design considerations for a large-scale image-based text search engine in historical manuscript collections. IT-Inf. Technol. **58**(2), 80–88 (2016)

21. Sharan, S.P., Aitha, S., Amandeep, K., Trivedi, A., Augustine, A., Sarvadevabhatla, R.K.: Palmira: a deep deformable network for instance segmentation of dense and uneven layouts in handwritten manuscripts. In: International Conference on Document Analysis Recognition, ICDAR 2021 (2021)

22. Smith, R.: Tesseract-OCR (2020). https://github.com/tesseract-ocr/. Accessed 26 Mar 2020

23. Trivedi, A., Sarvadevabhatla, R.K.: HInDoLA: A Unified Cloud-based Platform for Annotation, Visualization and Machine Learning-based Layout Analysis of Historical Manuscripts. In: 2nd International Workshop on Open Services and Tools for Document Analysis, OST@ICDAR 2019, Sydney, Australia, September 22–25, 2019. pp. 31–35. IEEE (2019). https://doi.org/10.1109/ICDARW.2019.10035, https://doi.org/10.1109/ICDARW.2019.10035

24. Trivedi, A., Sarvadevabhatla, R.K.: BoundaryNet: an attentive deep network with fast marching distance maps for semi-automatic layout annotation. In: International Conference on Document Analysis Recognition, ICDAR 2021 (2021)

25. Wojna, Z., et al.: Attention-based extraction of structured information from street view imagery. In: 2017 14th IAPR International Conference on Document Analysis and Recognition (ICDAR), vol. 1, pp. 844–850. IEEE (2017)

26. Zhong, X., ShafieiBavani, E., Yepes, A.J.: Image-based table recognition: data, model, and evaluation. arXiv preprint arXiv:1911.10683 (2019)

# ICDAR 2021 Workshop on Industrial Applications of Document Analysis and Recognition (WIADAR)

# WIADAR 2021 Preface

There is a lot of innovative work being conducted in industry, in public administration, and in universities in partnership with industry or public administration. Usually, the aim of applied research for industrial innovation is to breakthrough two main dimensions: performance (highest success with lowest critical error, resistance on specific constraints) and user scope (application to a new domain, ability to adapt the application to another domain with a minimized effort). Hence, the nuance is that scientists target the methodology, whereas industrial fellows target the application.

Most applied research reuses, combines, tunes, and trains existing methods with the advantage of accessing large corpuses of data. Industrial researchers adapt or create methods with scientific innovations, but also with heuristics, workarounds, and whatever can pragmatically be used to reach the target for an acceptable scope. Furthermore, we recognize that when industrial researchers create a really novel approach, the business strategy that includes industrial competition often prevents the disclosure of the information. Hence, that applied research cannot be submitted to ICDAR.

All these works are of interest to the ICDAR audience, because experiments with large real-life datasets provides pragmatic feedback on technologies. They really demonstrate the state-of-the-art limitations and infer a ranking of technology performance, and both workarounds and optimizations may point out some new perspectives for future research. In addition, such experiments highlight end-user problematics and needs, i.e. they may infer academic perspectives.

The Workshop on Industrial Applications of Document Analysis and Recognition (WIDAR) topics are the same as for ICDAR, including applications of document analysis, physical and logical layout analysis, character and text recognition, interactive document analysis, cognitive issues, and semantic information extraction.

For the second edition of WIDAR we received seven proposals. Each paper was carefully evaluated by at least three reviewers, including two Program Committee (PC) members and one workshop co-chair. Our PC members were from industry or academia with a strong experience in industrial projects. Following the review process, five proposals with very positive evaluations were accepted. All of the accepted papers were related to industrial frameworks or embedded in industrial applications. These papers were presented during the poster sessions of the main conference.

The WIADAR papers and posters were an effort to make information about innovative applications available to the general ICDAR attendee. It allowed those researchers and engineers to complete ICDAR discussions even if an application is not scientifically original or not clearly described. Moreover, this workshop was an opportunity to build more interaction between industry and academia. For this reason, we believe the second WIADAR was a success.

September 2021

Elisa H. Barney Smith
Vincent Poulain d'Andecy
Hiroshi Tanaka

# Organization

## Organizing Chairs

Elisa H. Barney Smith       Boise State University, USA
Vincent Poulain d'Andecy    Yooz, France
Hiroshi Tanaka        Fujitsu, Japan

## Program Committee

Sameer Antani       National Institutes of Health, USA
Bertrand Coüasnon     IRISA, France
Volkmar Frinken      Onai, USA
Christopher Kermorvant   Teklia, France
Liuan Wang        Fujitsu R&D Center, China

# Object Detection Based Handwriting Localization

Yuli Wu[1], Yucheng Hu[2], and Suting Miao[3]

[1] Rheinisch-Westfälische Technische Hochschule Aachen, Aachen, Germany
[2] Nanjing Normal University, Nanjing, China
[3] SAP Innovation Center Network (ICN) Nanjing, Nanjing, China
phoebe.miao@sap.com

**Abstract.** We present an object detection based approach to localize handwritten regions from documents, which initially aims to enhance the anonymization during the data transmission. The concatenated fusion of original and preprocessed images containing both printed texts and handwritten notes or signatures are fed into the convolutional neural network, where the bounding boxes are learned to detect the handwriting. Afterwards, the handwritten regions can be processed (*e.g.* replaced with redacted signatures) to conceal the *personally identifiable information* (PII). This processing pipeline based on the deep learning network Cascade R-CNN works at 10 fps on a GPU during the inference, which ensures the enhanced anonymization with minimal computational overheads. Furthermore, the impressive generalizability has been empirically showcased: the trained model based on the English-dominant dataset works well on the fictitious unseen invoices, even in Chinese. The proposed approach is also expected to facilitate other tasks such as handwriting recognition and signature verification.

**Keywords:** Handwriting localization · Object detection · Regional convolutional neural network · Anonymization enhancement

## 1 Introduction

Handwriting localization plays an important role in the following scenarios: First, the handwritten regions in the documents may contain sensitive information, which must be anonymized before transmission. Second, handwriting localization can be naturally served as the first stage to achieve *handwriting-to-text* recognition. Third, signatures to be verified must be extracted via localization from their surrounding texts or lines in the documents.

This work is initially motivated by the demanding case of anonymization enhancement. Access to data is vital to undertake enterprise today. One of the most common data types would be the invoices: with digitalized invoices and all kinds of powerful AI-driven technologies, the companies would be able to analyze

---

This work was done during their (Y. Wu and Y. Hu) internship at SAP ICN Nanjing.

© Springer Nature Switzerland AG 2021
E. H. Barney Smith and U. Pal (Eds.): ICDAR 2021 Workshops, LNCS 12917, pp. 225–239, 2021.
https://doi.org/10.1007/978-3-030-86159-9_15

customers' behaviors and extract business intelligence automatically, offering utmost help to refine strategies and make decisions. However, the *personally identifiable information* (PII) must be anonymized beforehand, as it is not worth the risk of any privacy exposure.

Ostensibly, tabular texts are of overwhelming majority in the business processing. In fact, the documents containing handwritten notes or signatures, such as invoices, also play an important role. The goal of this work is to localize the handwritten regions from the full-page invoice images for anonymization enhancement. The detected handwriting shall be anonymized afterwards, while the detailed implementation of which is beyond the scope of this work. As the expected results of the whole processing, handwriting should be excluded from the anonymized invoices, where it is assumed all handwritten regions would contain PII. One trivial way to realize this would be replacing the handwritten boxes with redacted signatures or notes.

Tesseract [24], the de facto paradigm of *optical character recognition* (OCR) engines, is nowadays widely used in the industry to extract textual information from images. OCR engines are competent to deal with the *optimal* data, which is referred to as the image data of documents, where all items of interest are regularly printed texts under the context of this work. In contrast, the real-world documents are usually the ones containing not only the regularly printed texts, but also some irregular patterns, such as handwritten notes, signatures, logos etc., which might also be desired.

In this work, we adopt object detection approaches with deep learning networks to localize handwritten regions in the document data based on the SAP's Data Anonymization Challenge[1]. The feasibility and effectiveness of such algorithms have been empirically shown on those scenarios, where the *objects* (handwriting) and the *backgrounds* (printed texts) are extremely similar. Besides, the improvement from Faster R-CNN [23] to Cascade R-CNN [1] can be effortlessly reproduced. In addition, the new baseline of the handwriting localization as the subtask from the SAP's Data Anonymization Challenge (see Footnote 1) has been released. Last but not least, the proposed deep learning approach with Cascade R-CNN [1] has demonstrated impressive generalizability. The trained model based on the English-dominant dataset works well on the fictitious unseen invoices, even for those in Chinese as toy examples. Empirically, it is believed that the deep learning model has learned the *irregularity* of the images.

Since the detailed types of handwritten regions, such as signatures or notes, are not discriminated during the experiments, we term *detection* and *localization* in this work interchangeably. The one-class detection merely consists of the localization regression task without classification. Despite the simplicity of the task description, it is still challenging to distinguish the handwritten notes from the printed texts, as they are similar regarding the contextual information. Furthermore, the detected bounding boxes, which should contain PII, are expected to be more accurate, compared to the general object detection tasks, *i.e.*, the primary evaluation score $AP^{FP}$ (average precision with penalty of false positive, see Sect. 4.4) is thresholded with the IoU of 80%.

---

[1] https://www.herox.com/SAPAI/.

## 2    Related Work

The input images are usually in the format of the cropped handwritten regions in the signature verification competitions [15,19]. Likewise, some handwritten text recognition datasets provide the option of the images labeled with divided lines [16]. This work is expected to bridge the gap between these researches and industrial applications through handwriting localization. Besides, text detection in natural scene images is close to our work. One significant difference between these two tasks is the *target objects*: All texts should be detected in scene text detection task (*e.g.* [18]), while only the handwritten texts in this work. The other difference is the background: The background in scene text detection task is the natural view. In this work, the background is the printed texts and tables on the blank document. Also based on Faster R-CNN [23], Zhong et al. [26] uses LocNet [6] to improve the accuracy in scene text detection, whereas we use Cascade R-CNN [1], the cascade version of Faster R-CNN.

There are two main categories of methods to localize the handwritten regions in the documents. The OCR based approaches recognize and then exclude printed texts. As a result, the unrecognizable parts are believed to be the handwriting. In contrast, the object detection based approaches regard this as a localization task, where the handwriting is the target and all other items (such as printed texts, logos, tables, etc.) are considered as the background. Thanks to the datasets and detection challenges on common objects (*e.g.* [5,13]), a considerable number of novel algorithms about object detection have been productively proposed in the recent years, *e.g.* Faster R-CNN [23], YOLO [21], SSD [14], RetinaNet [12], Cascade R-CNN [1], etc.

Three different approaches submitted to the Data Anonymization Challenge (see Footnote 1) are also briefly introduced in the following sections, including an OCR based approach and two deep learning based approaches (one with YOLOv3 [22], one with Google's paid cloud service).

**OCR Based Approaches.** In this section, an example proposal from the challenge (see Footnote 1) is demonstrated. First, the images are sequentially preprocessed, including removing the horizontal and vertical lines, median filtering (to remove salt-and-pepper noises), thresholding and morphological filtering (*e.g.* dilation and erosion). The handwritten parts are then discriminated from the printed ones with respect to the manually chosen features like the heights and widths of the text boxes, text contents and confidence scores recognized by OCR. In the experiments, this approach brings in the results on a par with those using deep learning approaches. However, the robustness and the generalizability of the deep learning approaches are believed to be advantageous.

**Object Detection Based Approaches with Deep Learning.** Since object detection is an intensively researched area in the field of computer vision, it is natural to directly apply the deep learning algorithms to the handwriting localization task. With the deep learning engine ImageAI [17], the networks like YOLOv3 [22] can be trained in an end-to-end manner. Moreover, some deep learning services like Google's Cloud AutoML Vision API take it further, managing the training process even without specifically assigning an algorithm.

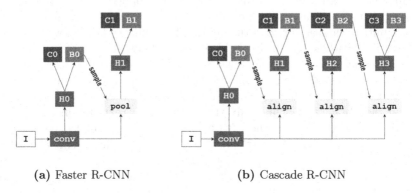

(a) Faster R-CNN                    (b) Cascade R-CNN

**Fig. 1.** Network Architectures of Faster R-CNN [23] and Cascade R-CNN [1]. Figures are adapted from [1].

## 3 Method

### 3.1 Faster R-CNN

Faster R-CNN [23] consists of two modules: the Region Proposal Network (RPN) that proposes rectangular regions containing the desired objects, and the Fast R-CNN detector [7] that predicts the classes and the locations.

The processing pipeline is demonstrated based on Fig. 1a. The input images (I) are first fed into a convolutional neural network (conv), where the shared features are extracted for both RPN and Fast R-CNN detector. Given the shared convolutional feature map of a size $w \times h \times d$ and the number of the anchors $k$ for each location in the feature map, the RPN head (H0) transforms it into two proposal features of $w \times h \times 2k$ (C0) and $w \times h \times 4k$ (B0) with one $e.g.$ $3 \times 3$ convolutional layer followed by two sibling $1 \times 1$ convolutional layers.

Now, $w \times h \times k$ proposals have been generated, each in the form of 6 representative values: 2 objectness scores and 4 coordinate offsets. The higher-scored proposals from B0 are selected as the inputs of the Fast R-CNN detector, together with the shared convolutional feature map. The transform of the proposals' coordinates between the original images and the feature maps is calculated via $e.g.$ RoIPool [7] (pool) or RoIAlign [9] (align). The pooled or aligned $region\ of\ interest$ (RoI) feature map of some fixed size is flattened then projected onto a feature vector via the RoI head (H1). Finally, two vectors of classes (C1) and locations (B1) are obtained by fully connected layers upon the feature vector.

There are two places where multi-task loss functions are calculated: RPN (C0 and B0) and Fast R-CNN detector (C1 and B1). First, log loss is used for both classification tasks (specifically, sigmoid activation function plus binary cross entropy loss for C0 and softmax activation function plus cross entropy loss for C1). Second, smooth L1 loss [7] is used for both bounding box regression tasks (B0 and B1), which is defined as: $\text{smooth}_{L1}(x) = 0.5x^2$ if $|x| < 1$ and $\text{smooth}_{L1}(x) = |x| - 0.5$ otherwise.

## 3.2  Cascade R-CNN

Figure 1 (adapted from [1]) depicts the differences of these two framework architectures. First, a cascade network is used to train the regressors and classifiers in a multi-stage manner. A four-stage version is illustrated in Fig. 1b, including one RPN stage and three detection stages. Second, the IoU threshold is different for each detection stage, which is increasingly set to {0.5, 0.6, 0.7}. Note that the mentioned IoU threshold does *not* refer to the one in the RPN or the one when calculating $mAP$ (mean Average Precision [13]). It is used to define the positive or negative candidates during the mini-batch sampling (sample in Fig. 1).

Thanks to the cascade architecture with progressively increasing IoU thresholds for sampling, Cascade R-CNN can accomplish object detection of *high quality*, which is exactly desired in the handwriting localization task to enhance anonymization.

## 3.3  Other Techniques

**Canny Edge Detection.** The gradient intensity based Canny edge detector [2] generates preliminary edges for the following processes. The detected edges might be truncated, *e.g.* under different optical circumstances. Thus, further processes of refinement or extraction are normally applied to the edges detected by Canny.

**Hough Transform.** Through the transform of line parameterizations, straight lines can be efficiently detected by the voting based Hough Transform [4]. The images are usually first processed by edge detectors like Canny, followed by thresholding. Next, each edge pixel $(x, y)$ in the binary images is represented by $k$ evenly rotating lines through it in the Hesse normal form: $r = x \cos(\theta) + y \cos(\theta)$. In the so-called *accumulator space* of $(r, \theta)$, each edge pixel would have $k$ votes. The peaks of the accumulator space are thus the desired lines. In this work, lines detected by Hough Transform are removed in the preprocessing step to generate clearer input images for the deep learning network.

**Tesseract OCR Engine.** As an open source paradigm OCR engine, Tesseract [24] has been widely used to recognize textual information in the industry. In this work, a Python wrapper for the tesseract-ocr API has been used (https://github.com/sirfz/tesserocr) to detect and eliminate the printed texts in the preprocessing step.

# 4  Experiments

## 4.1  Dataset

The dataset used in this work is the scanned full-page low-quality invoices in the tobacco industry from the 1990s (http://legacy.library.ucsf.edu/), which was once served in a document classification challenge [8]. Based on the invoice (or invoice-like) images from the same dataset, the labels and the bounding boxes of names and handwritten notes are manually annotated for the Data Anonymization Challenge (see Footnote 1).

In total, we have access to 998 gray-scale images with ground-truth labels, which are randomly split to 600, 198 and 200 images as training, validation and testing set, respectively (denoted below as `600train+198val+200test`). The hidden evaluation set consists of 400 images. In this case the training set covers 800 images, the validation set remains unchanged and the testing set covers 400 unseen images (denoted below as `800train+198val+400test`). The sizes of the images are varied around $700 \times 1000$, which are resized to $768 \times 768$.

## 4.2 Preprocessing

Despite the powerful capability of extracting features *automatically* being one of the benefits when using deep learning algorithms, it is believed that the elementarily preprocessed inputs or their fusions might improve the performance. Intuitively, if some non-handwritten parts could be omitted in the preprocessing step, the following handwriting localization task could be facilitated. Based on this assumption, texts recognized by the OCR engine (`tesseract-ocr` [24]) of high confidence and the straight lines detected by Hough transform [4] are excluded. In the experiments, the threshold confidence for the OCR engine is set to 0.7. The preprocessed images without highly confident textual information or straight lines of tables are denoted as ''`pre`'' in the following, while the original ones as ''`o`''.

Besides, the documents usually consist of a white background (of the highest intensity values, *e.g.* 1) and black texts (of the lowest intensity values, *e.g.* 0). As the background is dominant in terms of the number of pixels, it is natural to negate the images to obtain the inputs of sparse tensors, which is believed to make the learning progress more effectively. Given an image ranged in $[0, 1]$, the negated image is calculated as the original image element-wise subtracted by 1. With this in mind, the negated original and preprocessed images are denoted as ''`o-`'' and ''`pre-`'', respectively.

In addition, the inputs of the deep learning networks can be usually of an arbitrary number of channels. The original and preprocessed images are concatenated to create fused inputs, where the preprocessed layer can be interpreted as an *attention* mechanism, which highlights the most likely regions being the target objects. The concatenated inputs are denoted as *e.g.* ''`o/pre`'', the two dimensions of which are original and preprocessed images. The influences of the different inputs are investigated in Sect. 4.6.

## 4.3 Training with Deep Learning Networks

All experiments were running on a single RTX 2080 Ti GPU. The implementation of the deep learning networks are adopted by the open source toolbox from OpenMMLab for object detection: MMDetection [3].

The default optimizer is *Stochastic Gradient Descent* (SGD [11]) with a learning rate of 0.001, a momentum of 0.9, and a weight decay of 0.0001. Since the training set of 600 images is relatively small, the default number of epochs is set to a relatively large one (200) to make full use of the computational capacity if the training lasts overnight. During the experiments, it is observed that 200 epochs are appropriate (Fig. 4). The `train/val/test` sets are defined as in Sect. 4.1. Model weights after each epoch with the best results of `val` set are chosen to make predictions on `test` set. The different preprocessing steps are introduced in Sect. 4.2 and they are compared with Faster R-CNN [23] and Cascade R-CNN [1] in details. Next, additional two deep learning networks, RetinaNet [12] and YOLOv3 [22], have been tested with the preprocessing step which yields the best result on Cascade R-CNN. Detailed experimental results can be found in Sect. 4.6.

### 4.4    Evaluation Scores

**IoU.** Intersection over Union of two bounding boxes $p$ and $g$ is defined as below:

$$\text{IoU}(p, g) = \frac{\text{Area}\{\, p \cap g \,\}}{\text{Area}\{\, p \cup g \,\}} \tag{1}$$

**GIoU.** Global IoU of two lists of bounding boxes $P = \{p_1, p_2, ...\}$ and $G = \{g_1, g_2, ...\}$ is defined as below:

$$\text{GIoU}(P, G) = \frac{\text{Area}\{\, (p_1 \cap p_2 \cap ...) \cap (g_1 \cap g_2 \cap ...) \,\}}{\text{Area}\{\, (p_1 \cap p_2 \cap ...) \cup (g_1 \cap g_2 \cap ...) \,\}} \tag{2}$$

**AP$^{\text{FP}}$.** Average Precision with penalty of False Positive is the original evaluation score for the handwriting detection used in the Data Anonymization Challenge (see Footnote 1), which is defined as follows:

$$AP^{FP} = \begin{cases} \dfrac{|\mathcal{M}^G|}{|\mathcal{G}|} \cdot 0.75^{|\mathcal{P}| - |\mathcal{M}^P|}, & \text{if } |\mathcal{G}| \neq 0; \\ 0.75^{|\mathcal{P}| - |\mathcal{M}^P|}, & \text{otherwise.} \end{cases} \tag{3}$$

In Eq. 3, $\mathcal{P}$, $\mathcal{G}$, $\mathcal{M}^G$, $\mathcal{M}^P$ denote the sets of predicted, ground-truth, matched *w.r.t.* ground-truth and matched *w.r.t.* predicted bounding boxes, respectively, and $|\cdot|$ denotes the number of the bounding boxes in this set.

The criterion to call some predicted bounding box $p_i \in \mathcal{P}$ a match *w.r.t.* the ground-truth bounding box $g \in \mathcal{G}$ is, when the IoU between $p_i$ and $g$ is greater than a threshold $T$, *i.e.*:

$$\mathcal{M}^G = \{\, g \in \mathcal{G} \mid \exists\, p_i \in \mathcal{P} \,(\text{IoU}(p_i, g) > T)\}. \tag{4}$$

Analogously,

$$\mathcal{M}^P = \{\, p \in \mathcal{P} \mid \exists\, g_i \in \mathcal{G} \,(\text{IoU}(p, g_i) > T)\}. \tag{5}$$

It differs from the popular evaluation score AP (Average Precision) for object detection in COCO [13], where the false positive (FP) has not been particularly punished. Moreover, considering the potential application on the anonymization enhancement, the IoU threshold $T$ in the Data Anonymization Challenge (see Footnote 1) is set to 0.8, which is more strict than the common single threshold of 0.5 in COCO. In this work, the results are evaluated both with $T = 0.8$ and $T = 0.5$, which are denoted with $AP_{80}^{FP}$ and $AP_{50}^{FP}$, respectively. In the Data Anonymization Challenge, there is a mechanism of Bad-Quality: if an image is marked as Bad-Quality, its $AP_{80}^{FP}$ is assigned with 35%. We therefore record another two evaluation scores $AP_{80}^{FP}*$ and $AP_{80}^{FP}+$, where $*$ denotes the Bad-Quality mechanism is used when calculating $AP_{80}^{FP}$ and $+$ denotes the images marked as Bad-Quality are excluded when calculating $AP_{80}^{FP}$.

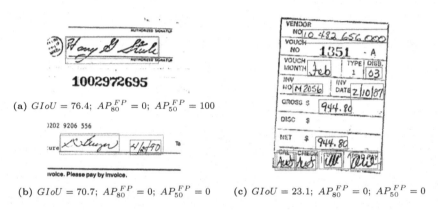

(a) $GIoU = 76.4$; $AP_{80}^{FP} = 0$; $AP_{50}^{FP} = 100$

(b) $GIoU = 70.7$; $AP_{80}^{FP} = 0$; $AP_{50}^{FP} = 0$     (c) $GIoU = 23.1$; $AP_{80}^{FP} = 0$; $AP_{50}^{FP} = 0$

**Fig. 2.** Comparison of different evaluation scores (in %). *Green*: ground-truth box; *Red*: predicted box. See Sect. 4.4 for the definitions of $GIoU$, $AP_{80}^{FP}$, $AP_{50}^{FP}$. (a) shows the IoU threshold of 80% might be too strict; (b) shows the $AP$ family might not effectively indicate the decent quality of the prediction if the ground-truth contains multiple adjacent boxes and a larger one is detected, while $GIoU$ could; (c) shows an unacceptable prediction where most of sensitive data are exposed, in which case the evaluation score of $AP$ family (with a high threshold) is desired. (Color figure online)

In addition, the overall evaluation score is calculated by averaging all image level scores, which applies to $GIoU$ and $AP^{FP}$. These 3 evaluation scores are illustrated with visual results of ground-truth and predicted bounding boxes in Fig. 2. It is shown that there is no single evaluation score considered as *silver bullet*, especially if the ground-truth annotations are not perfect. In this dataset, *imperfect* is referred to as the fact that the neighboring handwritten regions are sometimes annotated as a single large box, sometimes as multiple separate small boxes. With this in mind, the results are evaluated with the following five scores: $AP_{80}^{FP}$, $AP_{80}^{FP}*$, $AP_{80}^{FP}+$, $AP_{50}^{FP}$ and $GIoU$. Note that the first two evaluation scores ($AP_{80}^{FP}$ and $AP_{80}^{FP}*$) are eligible in the SAP's Data Anonymization Challenge (see Footnote 1).

## 4.5  Postprocessing

The deep learning classifier outputs a confidence score for each corresponding class. This confidence score can be thresholded as a hyperparameter to control the false positive rate in the postprocessing step. It has been observed during the experiments that the best results ($w.r.t.$ $AP_{80}^{FP}$) can be achieved, if this confidence is thresholded as 0.8, which is thus chosen as default.

In addition, to reduce the overlapped bounding boxes, some postprocessing steps are applied in the end, which follows the simple criterion: one large box is preferable to multiple small ones. First, all the predicted bounding boxes from each image are sorted in ascending order by their areas. Second, starting from the smallest one, each box is checked if the intersection area over the smaller box area is greater than a threshold (chosen as 0.9). If it is the case, the smaller box is omitted. This postprocessing is applied by default and we observed a minimal improvement, namely around 0.3% in terms of $AP_{80}^{FP}$.

## 4.6  Results

In this section, the experimental results are presented regarding the different preprocessing steps (Table 1), the influences of the mechanism of `Bad-Quality` (Table 1, 3), the comparison of various deep learning networks (Table 2) and the results released on the leaderboard of Data Anonymization Challenge (see Footnote 1) (Table 3). The improved performance from Fast R-CNN and Cascade R-CNN is specifically demonstrated (Fig. 3, 4). Furthermore, the examples of predicted handwritten regions from the `val` set are visualized in Fig. 5, together with the ground-truth bounding boxes.

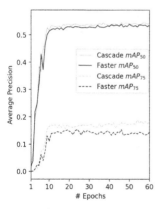

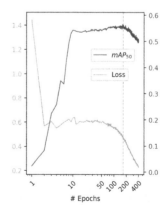

**Fig. 3.** $mAP_{50}$ and $mAP_{75}$ (`val`) of Cascade and Faster R-CNN. The cascade version surpasses Faster R-CNN by a larger margin under the more strict criterion.

**Fig. 4.** Overall loss (bottom; left axis) and $mAP_{50}$ (top; right axis) during the training. The dashed vertical line indicates the boundary of overfitting after around 170 epochs.

**Preprocessing.** To begin with, different types of preprocessed images as the inputs of the deep learning networks are investigated based on Faster R-CNN and Cascade R-CNN. As shown in Table 1, the preprocessed images alone can bring in worse results, compared to the original ones, while the concatenated preprocessed and original images as inputs (''o/o-/pre-'') have achieved the best result *w.r.t.* $AP_{80}^{FP}$, which are therefore used as default in the following experiments (Table 2, 3). It is believed that the fused preprocessed images could be served as an *attention* mechanism, which highlights the regions more likely containing the desired objects. Moreover, it is empirically shown in Table 1 that sparse inputs have not always yielded better results, if compared the negated inputs with the original ones.

**Table 1.** Handwriting detection results (in %) with different preprocessing steps as inputs. Images which contain more than 3 detected boxes are marked as `Bad-Quality`. Dataset: `600train+198val+200test`. Input abbreviations see Sect. 4.2. Column-wise best results are made bold.

| Network | Input | $AP_{80}^{FP}$ | $AP_{80}^{FP}*$ | $AP_{80}^{FP}+$ | $AP_{50}^{FP}$ | $GIoU$ |
|---|---|---|---|---|---|---|
| Faster R-CNN | o | 34.2 | 45.4 | 59.2 | 65.5 | 64.6 |
| Faster R-CNN | o- | 35.1 | 45.7 | 58.2 | 64.9 | 65.1 |
| Faster R-CNN | pre | 31.3 | 43.1 | 55.9 | 54.0 | 56.3 |
| Faster R-CNN | pre- | 29.6 | 42.4 | 54.0 | 54.6 | 55.3 |
| Faster R-CNN | o/pre | 34.6 | 43.9 | 56.3 | 64.1 | 64.4 |
| Faster R-CNN | o-/pre- | 35.1 | 44.5 | 53.3 | **66.7** | 65.4 |
| Faster R-CNN | o/o-/pre | 37.2 | 45.6 | 59.6 | 63.3 | 64.9 |
| Faster R-CNN | o/o-/pre- | 35.1 | 45.0 | 57.4 | 62.4 | 64.3 |
| Cascade R-CNN | o | 37.7 | 45.7 | 56.6 | 65.6 | 66.4 |
| Cascade R-CNN | o- | 34.7 | 42.9 | 49.2 | 65.1 | 66.5 |
| Cascade R-CNN | pre | 31.9 | 41.6 | 47.5 | 55.4 | 56.7 |
| Cascade R-CNN | pre- | 32.2 | 42.0 | 47.6 | 57.0 | 58.7 |
| Cascade R-CNN | o/pre | 36.3 | 44.3 | 56.0 | 64.4 | 66.4 |
| Cascade R-CNN | o-/pre- | 35.0 | 44.7 | 56.6 | 64.1 | 64.3 |
| Cascade R-CNN | o/o-/pre | 37.2 | **46.9** | **60.0** | 64.4 | 66.3 |
| Cascade R-CNN | o/o-/pre- | **38.3** | 46.0 | 56.5 | 65.0 | **66.8** |

**Mechanism of `Bad-Quality`.** As introduced in Sect. 4.4, the mechanism of `Bad-Quality` is provided in the challenge (see Footnote 1), which has been made full use of with the evaluation scores $AP_{80}^{FP}*$ and $AP_{80}^{FP}+$. It is observed based on the `val` set that the performance tends to be worse with the increasing number of the detected bounding boxes. With this in mind, all images with more than 3 detected bounding boxes are marked as `Bad-Quality`. The purpose of this mechanism is to raise the abnormal or complicated cases and turn to the

**Table 2.** Handwriting detection results (in %) using different deep learning networks. Inference speed is tested on a single RTX 2080 Ti GPU. Images which contain more than 3 detected boxes are marked as `Bad-Quality`. Dataset: `800train+198val+400test`. Column-wise best results are made bold.

| Network | Inference | $AP_{80}^{FP}$ | $AP_{80}^{FP}*$ | $AP_{80}^{FP}+$ | $AP_{50}^{FP}$ | $GIoU$ |
|---|---|---|---|---|---|---|
| YOLOv3 | **42** fps | 36.6 | 43.2 | 47.4 | 61.0 | 62.2 |
| RetinaNet | 11 fps | 27.9 | 40.8 | 44.4 | 51.7 | 54.9 |
| Faster R-CNN | 11 fps | 37.1 | 45.3 | 57.2 | 62.1 | 66.6 |
| Cascade R-CNN | 10 fps | **41.8** | **47.5** | **57.5** | **66.9** | **68.2** |

**Table 3.** Comparison with the leaderboard. *OCR*: Tesseract with manual engineering. *Service*: Google's Cloud API. * and † denote YOLOv3 results from the leaderboard and ours, respectively. BQ: if `Bad-Quality` is used.

| Method | $AP_{80}^{FP}$ | BQ |
|---|---|---|
| YOLOv3* | 26.3 | ✗ |
| OCR | 37.5 | ✗ |
| Service | 42.5 | ✗ |
| YOLOv3† | 36.6 | ✗ |
| YOLOv3† | 43.2 | ✓ |
| Cascade | 41.8 | ✗ |
| Cascade | 47.5 | ✓ |

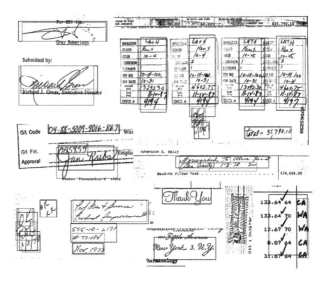

**Fig. 5.** Visual results of cropped handwritten regions from val set (Cascade R-CNN with ``o/o-/pre-''). *Green*: ground-truth box; *Red*: predicted box. (Color figure online)

manual process, which is practical in the industry. However, if the number of images marked as `Bad-Quality` is too large to maintain the productive advantage of machines, the evaluation scores $AP_{80}^{FP}*$ and $AP_{80}^{FP}+$ might bring in pseudo good results. Therefore, the number of images marked as `Bad-Quality` is loosely limited up to 50% of all images. Conclusively, the mechanism of `Bad-Quality` is believed to be a flexible trick to deal with the hard cases.

**Faster and Cascade R-CNN.** Not surprisingly, it is shown in Table 1 that Cascade R-CNN outperforms Faster R-CNN in general except for $AP_{50}^{FP}$, as Cascade R-CNN focuses on the object detection of higher quality and might perform worse than Faster R-CNN in terms of $AP_{50}^{FP}$. As shown in Fig. 3, the cascade version surpassed Faster R-CNN by a larger margin under the more strict criterion, when compared $mAP_{75}$ to $mAP_{50}$ (using COCO's evaluation scores [13] for brevity) in the val set. In addition, as depicted in Fig. 4, the training progress has been overfitted after around 170 epochs, from where the loss values and $mAP_{50}$ start decreasing. Thus, the chosen 200 epochs during all the experiments are appropriate.

**Comparison with Other Approaches.** Some of the popular deep learning networks for object detection are compared in Table 2, including YOLOv3 [22], RetinaNet [12], Faster R-CNN [23] and Cascade R-CNN [1]. They are implemented with [3]. Darknet-53 [20] is used as the backbone for YOLOv3 and ResNeXt-101 [25] for the other three. The dataset used in this experiment, 800train+198val+400test (see Sect. 4.1), is identical with the one used in the leaderboard. The images are preprocessed to the form of ''o/o-/pre-'' as described in Sect. 4.2. The results show that Cascade R-CNN outperforms other networks, with the trade-off regarding the inference speed on a single RTX 2080 Ti GPU due to the extra computational overhead though. As showcased in Table 3, the results of our approaches are compared with those submitted to the leaderboard. The first three rows are the results from the leaderboard, and the following four rows are our approaches. Our best achieved result (with Cascade R-CNN and BQ) has surpasses previous submissions on the leaderboard. Without considering the mechanism of BQ, however, the paid Google AutoML Vision API is slightly more advantageous (by 0.7%). Besides, it is noticeable that our YOLOv3 result (implemented by [3]) has outperformed the version submitted to the leaderboard (implemented by [17]) by more than 10% in terms of $AP_{80}^{FP}$.

### 4.7 Generalizability

English is the vast majority of the languages used in the dataset. Other languages such as Dutch or German are also included. However, the deep learning network is not expected to recognize the discrepancy of different languages. It is natural to categorize the languages using Latin alphabets indiscriminately. In this section, it is tested if the trained model works on the redacted real-world images in *foreign* languages.

Figure 6 illustrates two toy examples of fictitious and unseen invoices to evaluate the generalizability of the trained model. The model used to localize the handwritten regions is Cascade R-CNN. It is noteworthy that the language in the left image in Fig. 6 is Chinese, which can be considered as a foreign language in the dataset. Analogous to the German invoice demonstrated in the right one, the handwritten regions of both images are accurately detected as desired. The generalizability of the R-CNN family has also been observed by [26] during the text detection in natural scene images.

It is believed to be beneficial in the industry, if the model can be trained once and applicable to various cases. Additionally, it has also raised the common question of what the deep learning network has learned. In this case, it is supposed that the *irregularity* might be learned to discriminate the printed and handwritten texts.

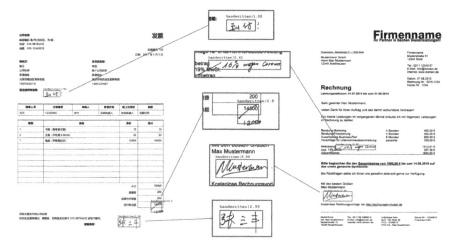

**Fig. 6.** Test of generalizability on toy examples with fictitious and unseen invoices. The handwritten regions are accurately localized from the images in Chinese and German.

# 5  Discussion

## 5.1  Conclusion

In this work, we present an object detection based approach to localize the handwritten regions, which is effective, fast and applicable to the unseen languages. First, the influences of the preprocessing steps are investigated. It has been empirically found that the fused concatenation of original and preprocessed images as the inputs can achieve the best performance. Second, different deep learning networks are compared. It is noticeable that the improvement from Faster R-CNN to Cascade R-CNN can be reproduced and the *high quality* characteristic of the cascade version suits the problem of the handwriting localization well. The results of our approaches can be served as a baseline of deep learning approaches in the handwriting localization problem. At last, the generalizability of the deep learning approach is impressive. The learned model is capable to successfully detect the handwritten regions on the real-world unseen images, even for those in the unseen language of Chinese. We believe it is of great interest both for the future research and for the industrial applications.

## 5.2  Outlook

As showcased in Fig. 5, some printed cursive texts are also detected as handwriting. It remains challenging to distinguish such nuanced discrepancies. Furthermore, apart from the object detection approaches, other proposals in the field of computer vision can also be adopted to differ the handwritten texts from the printed ones. One example is the anomaly detection, where the printed texts can be considered as the *normal* instances, since they are more regularly shaped. Thanks to the algorithms like *variational autoencoder* (VAE) [10], it is also promising to accomplish such tasks in a semi-supervised or even unsupervised manner. The other benefit of using the algorithms like VAE is that the learned intermediate representations can also be exploited to synthesize the artificial signatures, further enhancing the anonymization without eliminating the existence of such entities.

# References

1. Cai, Z., Vasconcelos, N.: Cascade R-CNN: high quality object detection and instance segmentation. IEEE Trans. Pattern Anal. Mach. Intell. (2019)
2. Canny, J.: A computational approach to edge detection. IEEE Trans. Pattern Anal. Mach. Intell. **PAMI-8**(6), 679–698 (1986). https://doi.org/10.1109/TPAMI.1986. 4767851
3. Chen, K., et al.: MMDetection: open MMLab detection toolbox and benchmark. arXiv preprint arXiv:1906.07155 (2019)
4. Duda, R.O., Hart, P.E.: Use of the hough transformation to detect lines and curves in pictures. Commun. ACM **15**(1), 11–15 (1972)
5. Everingham, M., Eslami, S.M.A., Van Gool, L., Williams, C.K.I., Winn, J., Zisserman, A.: The pascal visual object classes challenge: a retrospective. Int. J. Comput. Vision **111**(1), 98–136 (2015)
6. Gidaris, S., Komodakis, N.: LocNet: improving localization accuracy for object detection. In: Proceedings of the IEEE Conference on Computer Vision and Pattern Recognition, pp. 789–798 (2016)
7. Girshick, R.: Fast R-CNN. In: Proceedings of the IEEE International Conference on Computer Vision, pp. 1440–1448 (2015)
8. Harley, A.W., Ufkes, A., Derpanis, K.G.: Evaluation of deep convolutional nets for document image classification and retrieval. In: 2015 13th International Conference on Document Analysis and Recognition (ICDAR), pp. 991–995. IEEE (2015)
9. He, K., Gkioxari, G., Dollár, P., Girshick, R.: Mask R-CNN. In: Proceedings of the IEEE International Conference on Computer Vision, pp. 2961–2969 (2017)
10. Kingma, D.P., Welling, M.: Auto-encoding variational bayes. arXiv preprint arXiv:1312.6114 (2013)
11. LeCun, Y., et al.: Backpropagation applied to handwritten zip code recognition. Neural Comput. **1**(4), 541–551 (1989)
12. Lin, T.Y., Goyal, P., Girshick, R., He, K., Dollár, P.: Focal loss for dense object detection. In: Proceedings of the IEEE International Conference on Computer Vision, pp. 2980–2988 (2017)
13. Lin, T.-Y., et al.: Microsoft COCO: common objects in context. In: Fleet, D., Pajdla, T., Schiele, B., Tuytelaars, T. (eds.) ECCV 2014. LNCS, vol. 8693, pp. 740–755. Springer, Cham (2014). https://doi.org/10.1007/978-3-319-10602-1_48
14. Liu, W., et al.: SSD: single shot MultiBox detector. In: Leibe, B., Matas, J., Sebe, N., Welling, M. (eds.) ECCV 2016. LNCS, vol. 9905, pp. 21–37. Springer, Cham (2016). https://doi.org/10.1007/978-3-319-46448-0_2
15. Malik, M.I., Liwicki, M., Alewijnse, L., Ohyama, W., Blumenstein, M., Found, B.: ICDAR 2013 competitions on signature verification and writer identification for on-and offline skilled forgeries (SIGWICOMP 2013). In: 2013 12th International Conference on Document Analysis and Recognition, pp. 1477–1483. IEEE (2013)

16. Marti, U.V., Bunke, H.: The IAM-database: an English sentence database for offline handwriting recognition. Int. J. Doc. Anal. Recogn. **5**(1), 39–46 (2002)
17. Moses, Olafenwa, J.: ImageAI, an open source python library built to empower developers to build applications and systems with self-contained computer vision capabilities, March 2018. https://github.com/OlafenwaMoses/ImageAI
18. Nayef, N., et al.: ICDAR 2019 robust reading challenge on multi-lingual scene text detection and recognition—RRC-MLT-2019. In: 2019 International Conference on Document Analysis and Recognition (ICDAR), pp. 1582–1587. IEEE (2019)
19. Ortega-Garcia, J., et al.: MCYT baseline corpus: a bimodal biometric database. IEE Proc. Vis. Image Signal Process. **150**(6), 395–401 (2003)
20. Redmon, J.: DarkNet: open source neural networks in C (2013–2016). http://pjreddie.com/darknet/
21. Redmon, J., Divvala, S., Girshick, R., Farhadi, A.: You only look once: unified, real-time object detection. In: Proceedings of the IEEE Conference on Computer Vision and Pattern Recognition, pp. 779–788 (2016)
22. Redmon, J., Farhadi, A.: Yolov3: an incremental improvement. arXiv preprint arXiv:1804.02767 (2018)
23. Ren, S., He, K., Girshick, R., Sun, J.: Faster R-CNN: towards real-time object detection with region proposal networks. IEEE Trans. Pattern Anal. Mach. Intell. **39**(6), 1137–1149 (2016)
24. Smith, R.: An overview of the tesseract OCR engine. In: Ninth international conference on document analysis and recognition (ICDAR 2007), vol. 2, pp. 629–633. IEEE (2007)
25. Xie, S., Girshick, R., Dollár, P., Tu, Z., He, K.: Aggregated residual transformations for deep neural networks. arXiv preprint arXiv:1611.05431 (2016)
26. Zhong, Z., Sun, L., Huo, Q.: Improved localization accuracy by LocNet for faster R-CNN based text detection. In: 2017 14th IAPR International Conference on Document Analysis and Recognition (ICDAR), vol. 1, pp. 923–928. IEEE (2017)

# Toward an Incremental Classification Process of Document Stream Using a Cascade of Systems

Joris Voerman[1,2(✉)], Ibrahim Souleiman Mahamoud[2], Aurélie Joseph[2], Mickael Coustaty[1], Vincent Poulain d'Andecy[2], and Jean-Marc Ogier[1]

[1] La Rochelle Université, L3i, Avenue Michel Crépeau, 17042 La Rochelle, France
{joris.voerman,mickael.coustaty,jean-marc.ogier}@univ-lr.fr
[2] Yooz, 1 Rue Fleming, 17000 La Rochelle, France
{joris.voerman,ibrahim.mahamoud,aurelie.joseph,
vincent.dandecy}@getyooz.com

**Abstract.** In the context of imbalanced classification, deep neural networks suffer from the lack of samples provided by low represented classes. They can't train enough their weights with a statistically reliable set. All solutions in the state of the art that could offer better performance for those classes, sacrifice in return a huge part of their precision on bigger classes. In this paper, we propose a solution to this problem by introducing a system cascade concept that could integrate deep neural network. This system is designed to keep as mush as possible the original network performance while it reinforces the classification of the minor classes by the addition of stages with more specialised systems. This cascade offers the possibility to integrate few-shot learning or incremental architecture following the deep neural network without major restrictions on system internal architecture. Our method keeps intact or slightly improves the performances of a deep neural network (used as first stage) in conventional cases and improves performances in strongly imbalanced cases by around +8% accuracy.

**Keywords:** Documents classification · Imbalanced classification · Deep learning · Image processing · NLP

## 1 Introduction

Private companies and public administrations have to manage a huge quantity of documents every day. These documents come from internal processes and from exchanges with external entities like subcontractors or the public administration. Processing so many documents requires a lot of human resources without an automatic system. In addition, these documents are generally linked to the administrative part or to the company's core activities. Such documents are consequently of primary importance as they generally validate an action or a decision inside and/or outside the company. The management of these documents

© Springer Nature Switzerland AG 2021
E. H. Barney Smith and U. Pal (Eds.): ICDAR 2021 Workshops, LNCS 12917, pp. 240–254, 2021.
https://doi.org/10.1007/978-3-030-86159-9_16

becomes a challenge between speed and precision. Indeed, an error could have a heavy cost by causing a wrong action or decision. Consequently, any automatic system that could be used in this context needs to have a high precision.

Many companies, like ABBYY, KOFAX, PARASCRIPT, or YOOZ, propose some document classification solutions known as a Digital Mailroom [5,25]. Those systems relies on a combination of deep learning methods and expert systems, where experts are needed to make those systems operational. Expert systems offer high performance in almost every situation, but with a very high maintenance cost to keep them up-to-date. They are then gradually replaced by machine learning methods where the performance is linked to the availability of large labeled datasets. These machine learning methods haven't entirely replaced expert systems, because in imbalanced cases the most reliable machine learning methods like deep learning lose a significant part of their performance and can't face to the lack of samples available to train them [30]. In the state of the art, this problem can only be solved by retraining the model each time enough samples have been gathered to properly train a new class.

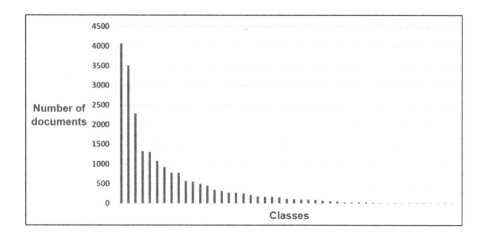

**Fig. 1.** Distribution of samples among all classes in our private database (a real document stream)

In order to formalize the document workflow, we use a document stream model proposed in [5]. A document stream is a sequence of very heterogeneous documents that appear irregularly over time. The stream is composed of numerous classes unequally distributed inside the stream with a group of core classes that is the majority of the stream and a lot of minor classes less represented. Figure 1 illustrates the distribution of documents among classes in a real document stream. Many of the minor classes have only few documents that do not allow a good training. Even if the number of documents per class increases as the content of a document stream evolves, new classes spontaneously appear time-to-time and others disappear. So a training set generated from a document stream has two main properties: it is incomplete and imbalanced. In other words, it does not contain a sample for each class and the distribution of samples between classes is not equal.

The classification of documents issuing from a document stream is already studied in the literature, and the most recent and efficient approaches rely on machine learning and deep learning methods. Those methods have to deal with two main issues: the first one is the imbalance distribution of documents between classes. This leads to classes with a very few samples (sometimes only 1) to train the network while a few classes gather more than 70% of the total stream. In addition, the network has to be trained again each time a new class appears in the stream (which is far from being simple to detect). The second issue is precisely this addition of new classes, or samples with a really different template in an existing class. This second point has also been studied and some technology using incremental learning [3] or few-shot machine learning methods [31] have been proposed. However, even if they are able to learn new classes, the global performances on major classes (*i.e.* with the largest number of documents) did not compete with the elder methods.

Finally, the last challenge related to our context is the document classification itself. Indeed, this task is not simple and many articles have been published recently to adapt methods from image and language processing to the document classification. In fact, documents have the particularity to have two main modalities whose the importance varies according to the context: the image and the text. Here, with a majority of administrative and companies documents the text is the most important modality. Recently, a multi-modal approach of deep learning [2] appears to combine these two modalities. In the next section, we will review the state of the art around this theme.

## 2   Related Work

In the literature the document classification is traditionally divided into two approaches: text and image processing. For each of these approaches, many highly accurate deep learning architectures have been developed during the last years. The majority of newest deep learning methods for text classification use two steps. The first step is the generation of a word embedding because words are not suitable input for a neural network. The embedding is used to change a word into a numerical vector that could represent as much as possible the contextual word meaning. Multiple methods like Word2Vec [20], Fasttext [12] or BERT [6] have been developed to generate word embedding by training them on huge textual dataset like Wikipedia articles. The second step is a deep neural network trained on the targeted dataset. Many methods use a recurrent architecture like biRNN (bidirectional recurrent) [26] or RCNN (recurrent convolutional) [16] to take in account the sequence information of the sentence. These methods use specific recurrent neuron architectures with GRU [4] and LSTM [10]. Not recurrent methods also work, like textual CNN [13,33]. CouldScan [21] is a good example of these methods apply to the industrial context.

The majority of image processing methods used for classification are deep convolutional pixel-based neural network (CNN) with multiple architecture: VGG [27], NasNet [34], InceptionResNet [29], DenseNet [11], etc. Initially, those methods are designed for image classification, but they offer suitable performance on document classification. There are also some methods designed specifically for document classification [9]. Recently, a multi-modal strategy emerges and seems to take advantage of previous approaches. The state of the art proposes multiple combinations of previous deep neural network for a multi-modal classification [2]. Architectures are mainly two networks with one for text and the second for image that combine their output.

The one-shot and few-shot learning strategies [31] are also relevant in our case. The few/one-shot learning is a challenge in image processing which is defined as train with only a few numbers of samples per classes. Firstly, there is Bayesian based approach methods [7,17]. Secondly, some methods try to adapt the neural network architecture to few-shot learning like Neural Turing Machine and other Memory Augmented Network [23], Siamese Network [14] and Proto-typical Network [28]. The incremental training also competes for this task, but the multiple tries to adapt neural networks for incremental learning [15,22,24] do not show great results. Older methods like incremental SVM [18], K-means [1] and Growing Neural Gas (IGNG) [3] seem to be more reliable for now. The literature shows some good samples of industrial applications based on these methods like INTELLIX [25] or INDUS [5].

At first sight, zero-shot learning [32] seems to be an interesting option to solve the problem of new class classification. However, zero-shot learning strategies are mainly designed for image classification and use transfer learning from a textual description or all other prior knowledge. In our case, we have no prior knowledge about new classes so these methods are unsuitable.

## 3 The Cascade of Systems

### 3.1 Main Idea

In order to solve the issue of low represented classes classification, we propose to introduce a cascade of systems. The objective is to keep the precision of deep learning network for the main classes (i.e. which represent the most frequent documents of the stream), while using more specific architectures designed for the least represented classes.

Our cascade follows a divide and conquer strategy, where the decision proposed by a system will be taken into account if and only if its confidence is high and this decision is reliable from a global point of view. More formally, the first stages of the cascade will deal with the most frequent classes, while the lowest stages will rely on other systems trained as a specialist to reinforce the classification of outliers and small classes. The cascade aims at promoting a high precision of each cascade stage. This also implies the use of a rejection system with a high threshold to send all rejected elements to the next stage of the cascade. The more we advance in the cascade, the more the systems become specialized in rejected cases. This process is then repeated as many times as we need or want.

This strategy offers some advantages. Firstly, we have the possibility to combine the network with a system that could balance a deep learning network weakness, like an incremental method. The next stage system could equally use another modality than the previous one. This seems to offer better performance when modalities are complementary.

## 3.2   Training Architecture

The new problem introduced by the cascade is the training of the next stage system. Keeping the original training set to train the stage $n + 1$ will mainly duplicate the results without significant benefit. The $n + 1$ system will reject the same elements as stage $n$. To train stage $n + 1$ effectively, we need to remove from the new training set all elements easily classified by the stage $n$ and keep all possible rejected cases. We define this new training set in Eq. 1, where:

- $D_n$ is the training dataset of stage $n$ with $D_n \subset \{d_1, d_2, ...d_i\}$
- Each sample $d_i$ has a class $c_j$
- $D_{n+1}$ is the training dataset used by stage $n + 1$ (*i.e.* that have been rejected at stage $n$)
- $D'_n$ is all the documents with a very high confidence level and with a trustworthy class representation, according to stage $n$ (will be detailed in the next section)

$$D_n = D'_n \cup D_{n+1} \tag{1}$$

The evolution of the training set from one stage to the next one acts as a focusing step for the next stage and imposes the system to better discriminate the documents rejected in the previous stages. In our architecture, each training phase is done successively as illustrated in Fig. 2. We operate this set division between each system training phase and then update the training set for stage $n + 1$.

## 3.3   Set Division

With this architecture, the global performance of the next stages will depend on the division of the training set that becomes one of the main parts of the cascade. $D_{n+1}$ needs to model as much as possible all the elements that will be rejected by the previous stage. This includes complex classes with a high intra-class variance, which overlaps with the closest classes (in the feature representation space), and that have not enough samples to train properly the model. This also includes outliers that have a lower confidence rate than other samples from the same class.

The selection process, which in practice corresponds to the confidence we grant to the system at stage $n$, is defined by a parametric selection function presented in Eq. 2. This selection function uses four different features to assess how

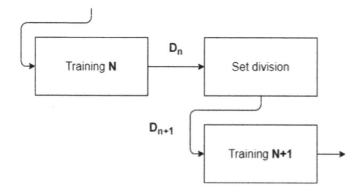

**Fig. 2.** Training architecture

the system performs at this stage on the proposed training set. These features gather information at the document level and at the class level.

We define $D_{n+1}$ as $\forall d_i \in D_n$ with class $c_j$:

$$d_i \in D_{n+1}, \text{ if and only if, } \underbrace{\alpha A}_{\text{doc level}} + \underbrace{\beta B + \gamma C + \delta D}_{\text{class level}} < T \qquad (2)$$

where:

- $\alpha, \beta, \gamma, \delta$ are weighting parameters, with $\alpha + \beta + \gamma + \delta = 1$.
- $A$ is the confidence rate computed by the stage $n$ system $N_n$ for the corresponding class $c_j$. As deep neural networks generally propose high confidence rate, we use a normalization function to enlarge the confidence level.
  In accordance with the confidence result distribution given by the system we advise to use a min/max or an exponential normalisation. The objectives is to keep only the documents with the lowest score.
- $B$ is the accuracy obtained by the stage $n$ system $N_n$ on class $c_j$. This feature is used to keep more documents from classes with a low accuracy.
- $C$ is a ratio between the inter-class and the intra-class variance.
  Variances are computed with an euclidean distance between document's embeddings $E_n$ trained by $N_n$ and their class centroids. $Ed_n(d_i)$ is the distance between embedding of $d_i$ and its class centroid, $Ed_n(c_j)$ is the same for class level.
  $\overline{Ed_n(c_j)}$ is the means of distances for class $c_j$, same for $\overline{Ed_n(D_n)}$ with the dataset. $x_j$ is the number of samples within the class $c_j$ and $x_n$ number of class in $D_n$. As $C \in [0 : +\infty[$, we recommend to restrain values in $[0 : 2]$ then

normalize to keep all features in $[0:1]$.

$$
\begin{cases}
C = \dfrac{var_{inter}(c_j)}{var_{intra}(c_j)} \\[2ex]
\forall c_y \in D_n \\[1ex]
var_{inter}(c_j) = \dfrac{1}{x_n} \sum_{y=1}^{x_n} (Ed_n(c_y) - \overline{Ed_n(D_n)})^2 \\[2ex]
\forall d_y \in c_j \\[1ex]
var_{intra}(c_j) = \dfrac{1}{x_j} \sum_{y=1}^{x_j} (Ed_n(d_y) - \overline{Ed_n(c_j)})^2 \\[2ex]
\overline{Ed_n(c_j)} = \dfrac{1}{x_j} \sum_{y=1}^{x_j} Ed_n(d_y)
\end{cases}
\tag{3}
$$

- $D$ is the representation of class $c_j$ inside $D_n$ with $size(c_j)$ the number of document in class $c_j$ and $max(size(c))$ the number of document in the largest class. This feature is used to ensure that the lowest represented classes remain in $D_{n+1}$.

$$
D = \frac{size(c_j)}{max(size(c))}
\tag{4}
$$

- $T$ is a threshold parameter

The division of the training set is done once the stage $n$ system training phase is completed. Consequently, the features are computed when the whole system is fully trained. The global proposed architecture is summarized in Fig. 3.

The selection function is applied to the training set and the validation set in order to ensure a fair validation process for the stage $n$ system and to be sure that the training step of the stage $n+1$ system will focus on the remaining samples. The last step of our architecture is related to the reduction of the training and validation size. The balance between them will be broken in the majority of cases because the validation set become proportionally greater than the training set. We adjust them class by class by randomly choosing enough samples to adjust the balance between set. This need to be done class by class, because each class is not reduced with the same rate.

### 3.4 Decision Process

The decision process starts once the training phase is ended for all the stages of the cascade. The decision process uses a rejection system to separate the less reliable answer from the others. The rejection is applied on the confidence rate returned by the system (in the one-hot vector case, the confidence rate is the highest score of the vector). As mentioned earlier in this paper, our document stream classification problem entails a really high precision rate to avoid mistakes. To this end, we chose to set a high value for the threshold. All rejected

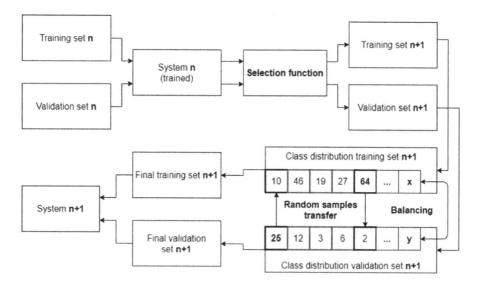

**Fig. 3.** Set division architecture

results are then sent to the next stage of the cascade that will apply the same process. If no decision is taken when the document goes through the last stage of the cascade, the document is finally labeled with a "reject label".

The last adjustment of our architecture concerns the computation of the final confidence rate in an imbalanced context. Indeed, the confidence rate returned for a class trained with two samples will not have the same value than the confidence rate returned for a class with five thousand samples. In addition, all cascade stages will potentially not be trained on all classes. To take this into account, we propose to weight this confidence rate. The main idea is to use the validation set to assess the reliability of our cascade stage for each class and adapt the confidence rate proportionally to the accuracy score of the class. This system will increase the rejection rate on classes that we estimate with the validation set to have the lowest accuracy and so the highest risk to give a wrong answer. The classes with the lowest accuracy are theoretically the same as those that were sent to the stage $n + 1$ training set by the selection function. The new weighted confidence score of a document $d_i$ (denoted as $N'_n(d_i)$) is computed by Eq. 5. It is composed of the former confidence score $N_n(d_i)$ and a penalty score. This penalty is based on the stage $n$ accuracy $Acc_n(c_j)$ on validation set for the predicted class $c_j$. An additional $r$ parameter tends to limit the class weight effect into specific bound. This parameter prevents the system to reject only more documents of low reliable classes and not whole classes.

$$N'_n(d_i) = \underbrace{N_n(d_i)}_{\text{Confidence score of } d_i} - \underbrace{r(1 - Acc_n(c_j))}_{\text{Weighted penalty Wb}} \qquad (5)$$

The Fig. 4 illustrates the proposed decision process architecture with the new weighted scores system. The system will generally reduce a bit the stage $n$ accuracy and increase the precision, mainly on low represented classes. The Next stages of the cascade will compensate the loss of accuracy as we will demonstrate in the next section.

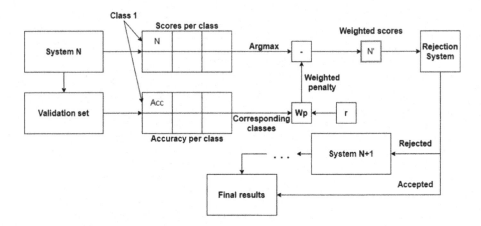

**Fig. 4.** Cascade decision with weighted rejection

# 4 Experiments

## 4.1 Architectures Used

All experiments done in this paper use a two stage cascade adapted for our context (private dataset). For the first stage, we used deep neural networks for their abilities to classify classes with a large number of labeled data. For the second stage, we propose more specific architectures able to deal with few samples per class during the learning phase. The selection has been done with one method per modality. To validate our ideas, we use two different networks in the first stage. A VGG16 [27] network as an image-based classification model and a textual-based bidirectional recurrent network (biRNN) [26] with CamemBERT [19] as embedding model. The biRNN is reinforced with an attention model to improve the trained features as proposed in [8]. These two methods correspond to state of the art deep neural network models for their respective modality classification task.

For the second stage, we propose to compare a few-shot learning method called prototypical network (ProtoNet [28]) known as one of the best few shot learning method, and an incremental learning approach (growing gas based method) called A2ING [3] and developed for document classification. The first method is based on visual features while the second one uses raw text from OCR as input. Those systems are selected to offer a maximum of variety in modality and state of the art architectures designed for documents, image or text classification in the experimentation results.

Finally, and in order to allow a fair comparison of our proposed cascade architecture performance, we also used two other methods in our baseline. An image-based holistic CNN designed for document classification (HCNN [9]) and a non-recurrent textual network (called TCNN). The TCNN network is a combination of architecture from [13] and [33].

## 4.2   Datasets

To evaluate our system we used two datasets, one private and one public. The first is a private document database provided by Yooz company to assess the performance on a real-case document stream. We will call it the "YoozDB" set. It gathers real documents coming from Yooz's customers, with an access to scanned image of document and text extracted by ABBYY-FineReader (version 14.0.107.232). It is composed of 23 577 documents unequally distributed in 47 classes, with 15 491 documents (65.71%) for training, 2203 (9.34%) for validation and 5883 (24.95%) for test. Classes are imbalanced with between 1 to 4500 documents for each class (the documents distribution per class is proposed in Fig. 1). The classes includes multiple variation of invoice, business order, dues, tax notice, bank details, contract, mail, cheque, ID card, etc. The main drawback of this dataset comes out from this test set which is as the same distribution as its training set with some classes composed of only one (or very few) document. This leads to a lack of samples for the very low represented classes that do not offer statistically reliable results for those classes and reduce the effect on global performance.

The second dataset we used in our comparison is the well-known RVL-CDIP dataset [9]. Composed of 400 000 industrial documents gatherers from tobacco companies equally distributed in 16 classes: letter, memo, email, file-folder, form, handwritten, invoice, advertisement, budget, news, article, presentation, scientific publication, questionnaire, resume, scientific report and specification. Some of them have almost no text and overall this dataset is built as an image dataset and so disadvantages the text based methods. RVL is divided in a training set of 320 000 documents (20 000 per classes), a validation and a test set of 40 000 documents each (2500 per classes).

## 4.3   Experimentation Methodology

To evaluate the performance of our cascade architecture and all the methods used in our comparison, we use three classical information retrieval metrics: the accuracy, the precision and the rejection rate. We use accuracy to compare our results to the other methods of the state of the art. The precision is important for our context and finally, the rejection rate aims at displaying the relevance of our proposed rejection system.

All results on YoozDB are computed in a k-fold cross-validation process (in our case, k = 10 with seven folds for train, one for validation and two for test) to deal as much as possible with low amounts of documents. Folds are randomly generated class by class to take into account the imbalanced distribution between classes.

In addition, we apply specific strategies to evaluate system adaptation to imbalanced and incomplete context of document stream. Those strategies follow the same methodology as that proposed in our previous work [30]. Briefly, we generate two new sets called Imbalanced and Realistic from RVL-CDIP. The first evaluates the resilience to imbalanced class distribution and so modifies the original set to simulate the imbalanced distribution of YoozDB. The modification is done by gathering all classes in four groups of four random classes and each group keeps only a part of their samples (5%, 10%, 50% and 100%). The second, add a group of classes with only one sample to represent classes that will appear later in the document stream. The distribution between groups becomes four classes with only one document and three for each of the imbalanced groups as previously. All these modifications are done only on the train and validation sets, but not on the test set to keep the results as reliable as possible.

### 4.4 Results

All experimental results on a balanced RVL-CDIP and YoozDB sets are presented in Table 1. We can observe that all the results given by cascade's methods are at least equivalent to basic methods. One exception appears on the biRNN-ProtoNet architecture on RVL-CDIP that offer better performances. This upgrade is linked to the fact that modalities complementing each other between the biRNN (text) and the ProtoNet (visual). The gain is mainly on file-folder and handwritten classes that are very difficult for a text classifier due to the lack of text. The results on YoozDB are not very significant here because, as explained previously, the test set is imbalanced and incomplete. In fact, the under-representation of minor classes in the test set limits or nullifies the impact of these classes on the global result. In addition, it is impossible to evaluate the real internal diversity with only two samples of the same document template and at least the results are not statistically reliable on these classes. Consequently, the results on this dataset come mainly from a good classification of major classes with a less importance of imbalanced, and global performance remains close to balanced dataset. The result of A2ING combines on RVL-CDIP are not displayed, because this method is not adapted to RVL documents that lack of words for a majority of cases. Finally, the multi-modal potential seems to be limited. The other methods from the state of the art have better gain for the combination of text and image modality in document classification task [2]. The results given by the table for VGG16 and biRNN alone can be used as raw performance of the first stage $(n)$ of our cascades (respectively to their combination). If you want, it is possible to calculate the second stage $(n + 1)$ only performances from combining result and corresponding raw method.

On the second table (Table 2) we have all results of methods computed on RVL-CDIP modified to simulate an imbalanced set with the protocol introduce in Sect. 4.3. For the imbalanced case, the cascade system offers a +11% accuracy

**Table 1.** Balanced case results

| Dataset | RVL-CDIP | | | YoozDB | | |
|---|---|---|---|---|---|---|
| Method/Measure | Accuracy | Precision | Reject rate | Accuracy | Precision | Reject rate |
| HCNN | **80.76%** | **95.42%** | 15.36% | 84.57% | 95.67% | 11.61% |
| TCNN | 65.52% | 93.72% | 30.09% | 88.36% | 98.97% | 10.72% |
| VGG16 | 75.14% | 94.41% | 20.41% | 84.70% | 95.47% | 11.28% |
| VGG16 - ProtoNet | 75.79% | 94.04% | 19.41% | 84.21% | 95.61% | 11.92% |
| VGG16 - A2ING | – | – | – | 82.05% | 96.62% | 15.08% |
| biRNN | 77.95% | 88.58% | 12.01% | 93.57% | 99.22% | 5.69% |
| biRNN - ProtoNet | 80.72% | 88.99% | **9.30%** | **93.77%** | 98.95% | **5.23%** |
| biRNN - A2ING | – | – | – | 91.45% | **99.30%** | 7.90 % |

for the VGG16 and +6.5% accuracy for biRNN against around −1% precision. It means that the cascade recovers a third of documents rejected by VGG16, and close to the half for biRNN, with almost the same precision in an imbalanced context. With absolute values, the cascade system recovers more than 5 000 documents with the same precision as the network alone. This means 5 000 documents less to process manually of 40 000. This gain comes from the better adaptation of 5-shot learning methods like ProtoNet to train classes with fewer samples and the rebalancing done by the cascade. Indeed, an important part of major class samples has not been transferred to the second stage training set, so minor classes become proportionally more represented. For example on VGG16, the cascade with ProtoNet recover 43.52%, 17.16%, 29.16% and 13% of rejected document for the classes that was reduced to 5% of training document as you can see with Table 3. The results for realistic case are less impressive, even if the accuracy earns +8%, because the precision is more reduced by the second stage. In fact, the Prototypical Network is less effective in the one-shot learning situation.

**Table 2.** Imbalanced case results

| Dataset | Imbalanced | | | Realistic | | |
|---|---|---|---|---|---|---|
| Method/Measure | Accuracy | Precision | Reject rate | Accuracy | Precision | Reject rate |
| HCNN | 67.97% | 89.14% | 23.75% | 55.70% | 77.55% | 28.18% |
| TCNN | 51.98% | **91.63%** | 43.27% | 36.88% | **78.50%** | 53.02% |
| VGG16 | 59.17% | 88.90% | 33.44% | 51.28% | 77.97% | 34.24% |
| VGG16 - ProtoNet | 70.19% | 87.67% | 19.94% | **58.89%** | 73.28% | 19.64% |
| biRNN | 68.37% | 79.73% | 14.25% | 50.71% | 67.56% | 24.94% |
| biRNN - ProtoNet | **74.90%** | 79.00% | **5.19%** | 58.05% | 64.30% | **9.72%** |

**Table 3.** Differences per classes between VGG and VGG-ProtoNet cascade on imbalanced RVL-CDIP

| Groups | 100% | | | | 50% | | | |
|---|---|---|---|---|---|---|---|---|
| Classes | Advert | File F | Handwr | Sc report | Budget | Email | Invoice | Resume |
| Precision | 0.06% | −5.97% | −5.80% | 0.69% | −1.45% | −0.25% | −0.06% | −0.23% |
| Reject rate | −4.60% | −7.00% | −2.12% | −5.00% | −6.52% | −1.72% | −3.20% | −3.44% |
| Groups | 10% | | | | 5% | | | |
| Classes | Form | Letter | Presen | Questi | Memo | News A | Sc Public. | Speci |
| Precision | −8.16% | −1.98% | −5.48% | −2.02% | −1.27% | −4.05% | −3.51% | −1.10% |
| Reject rate | −18.40% | −17.76% | −22.88% | −20.56% | −13.00% | −29.16% | −43.52% | −17.16% |

## 5   Conclusion

### 5.1   Overview

We propose a new architecture to combine successively in a cascade multiple neural networks to reinforce them in an imbalanced context. A two stages combination between a deep network and a 5-shot learning method offers between +11% and +8% accuracy in imbalanced context and keeps equivalent performance in balanced cases. This cascade method offers multiple possibilities of combination and do not limit too much the internal architecture of combined systems. It needs an embedding representation of the document and a class confidence score to compute the next stage training set. The results of the cascade system in imbalanced context are promising but need a better adaptation for one-shot cases.

### 5.2   Perspectives

In perspectives, we will try to combine our cascade with a multi-modal architecture and evaluate the potential of a cascade with more than two stages. Equally, we want to explore further the potential of a neural network cascade with an incremental system and build a better synergy between them to create an incremental deep learning system with high performances. In addition, we consider designing a methodology to assess the quality of the next stages training set and so the selection function. All parameters used for our experiments have been selected empirically for now. We will find a way to compute them automatically in future research.

## References

1. Aaron, B., Tamir, D.E., Rishe, N.D., Kandel, A.: Dynamic incremental k-means clustering. In: 2014 International Conference on Computational Science and Computational Intelligence, vol. 1, pp. 308–313. IEEE (2014)
2. Bakkali, S., Ming, Z., Coustaty, M., Rusinol, M.: Visual and textual deep feature fusion for document image classification. In: Proceedings of the IEEE/CVF Conference on Computer Vision and Pattern Recognition Workshops, pp. 562–563 (2020)

3. Bouguelia, M.R., Belaïd, Y., Belaïd, A.: A stream-based semi-supervised active learning approach for document classification. In: 2013 12th International Conference on Document Analysis and Recognition, pp. 611–615. IEEE (2013)
4. Cho, K., et al.: Learning phrase representations using RNN encoder-decoder for statistical machine translation. arXiv preprint arXiv:1406.1078 (2014)
5. d'Andecy, V.P., Joseph, A., Ogier, J.M.: IndUS: incremental document understanding system focus on document classification. In: 2018 13th IAPR International Workshop on Document Analysis Systems (DAS), pp. 239–244. IEEE (2018)
6. Devlin, J., Chang, M.W., Lee, K., Toutanova, K.: BERT: pre-training of deep bidirectional transformers for language understanding. arXiv preprint arXiv:1810.04805 (2018)
7. Fei-Fei, L., Fergus, R., Perona, P.: One-shot learning of object categories. IEEE Trans. Pattern Anal. Mach. Intell. **28**(4), 594–611 (2006)
8. Górriz, M., Antony, J., McGuinness, K., Giró-i Nieto, X., O'Connor, N.E.: Assessing knee OA severity with CNN attention-based end-to-end architectures. arXiv preprint arXiv:1908.08856 (2019)
9. Harley, A.W., Ufkes, A., Derpanis, K.G.: Evaluation of deep convolutional nets for document image classification and retrieval. In: 2015 13th International Conference on Document Analysis and Recognition (ICDAR), pp. 991–995. IEEE (2015)
10. Hochreiter, S., Schmidhuber, J.: Long short-term memory. Neural Comput. **9**(8), 1735–1780 (1997)
11. Huang, G., Liu, Z., Van Der Maaten, L., Weinberger, K.Q.: Densely connected convolutional networks. In: Proceedings of the IEEE Conference on Computer Vision and Pattern Recognition, pp. 4700–4708 (2017)
12. Joulin, A., Grave, E., Bojanowski, P., Douze, M., Jégou, H., Mikolov, T.: Fasttext. zip: Compressing text classification models. arXiv preprint arXiv:1612.03651 (2016)
13. Kim, Y.: Convolutional neural networks for sentence classification. arXiv preprint arXiv:1408.5882 (2014)
14. Koch, G., Zemel, R., Salakhutdinov, R.: Siamese neural networks for one-shot image recognition. In: ICML Deep Learning Workshop, Lille, vol. 2 (2015)
15. Kochurov, M., Garipov, T., Podoprikhin, D., Molchanov, D., Ashukha, A., Vetrov, D.: Bayesian incremental learning for deep neural networks. arXiv preprint arXiv:1802.07329 (2018)
16. Lai, S., Xu, L., Liu, K., Zhao, J.: Recurrent convolutional neural networks for text classification. In: Twenty-Ninth AAAI Conference on Artificial Intelligence (2015)
17. Lake, B.M., Salakhutdinov, R., Tenenbaum, J.B.: Human-level concept learning through probabilistic program induction. Science **350**(6266), 1332–1338 (2015)
18. Laskov, P., Gehl, C., Krüger, S., Müller, K.R.: Incremental support vector learning: analysis, implementation and applications. J. Mach. Learn. Res. **7**, 1909–1936 (2006)
19. Martin, L., et al.: CamemBERT: a tasty French language model. arXiv preprint arXiv:1911.03894 (2019)
20. Mikolov, T., Chen, K., Corrado, G., Dean, J.: Efficient estimation of word representations in vector space. arXiv preprint arXiv:1301.3781 (2013)
21. Palm, R.B., Winther, O., Laws, F.: CloudScan-a configuration-free invoice analysis system using recurrent neural networks. In: 2017 14th IAPR International Conference on Document Analysis and Recognition (ICDAR), vol. 1, pp. 406–413. IEEE (2017)
22. Rosenfeld, A., Tsotsos, J.K.: Incremental learning through deep adaptation. IEEE Trans. Pattern Anal. Machine Intell. **42**, 651–663 (2018)

23. Santoro, A., Bartunov, S., Botvinick, M., Wierstra, D., Lillicrap, T.: One-shot learning with memory-augmented neural networks. arXiv preprint arXiv:1605.06065 (2016)
24. Sarwar, S.S., Ankit, A., Roy, K.: Incremental learning in deep convolutional neural networks using partial network sharing. IEEE Access **8**, 4615–4628 (2019)
25. Schuster, D., et al.: Intellix-end-user trained information extraction for document archiving. In: 2013 12th International Conference on Document Analysis and Recognition, pp. 101–105. IEEE (2013)
26. Schuster, M., Paliwal, K.K.: Bidirectional recurrent neural networks. IEEE Trans. Signal Process. **45**(11), 2673–2681 (1997)
27. Simonyan, K., Zisserman, A.: Very deep convolutional networks for large-scale image recognition. arXiv preprint arXiv:1409.1556 (2014)
28. Snell, J., Swersky, K., Zemel, R.S.: Prototypical networks for few-shot learning. arXiv preprint arXiv:1703.05175 (2017)
29. Szegedy, C., Ioffe, S., Vanhoucke, V., Alemi, A.A.: Inception-v4, inception-ResNet and the impact of residual connections on learning. In: Thirty-First AAAI Conference on Artificial Intelligence (2017)
30. Voerman, J., Joseph, A., Coustaty, M., Poulain d'Andecy, V., Ogier, J.-M.: Evaluation of neural network classification systems on document stream. In: Bai, X., Karatzas, D., Lopresti, D. (eds.) DAS 2020. LNCS, vol. 12116, pp. 262–276. Springer, Cham (2020). https://doi.org/10.1007/978-3-030-57058-3_19
31. Wang, Y., Yao, Q., Kwok, J.T., Ni, L.M.: Generalizing from a few examples: a survey on few-shot learning. ACM Comput. Surv. (CSUR) **53**(3), 1–34 (2020)
32. Xian, Y., Schiele, B., Akata, Z.: Zero-shot learning-the good, the bad and the ugly. In: Proceedings of the IEEE Conference on Computer Vision and Pattern Recognition, pp. 4582–4591 (2017)
33. Zhang, X., Zhao, J., LeCun, Y.: Character-level convolutional networks for text classification. In: Advances in Neural Information Processing Systems, pp. 649–657 (2015)
34. Zoph, B., Vasudevan, V., Shlens, J., Le, Q.V.: Learning transferable architectures for scalable image recognition. In: Proceedings of the IEEE Conference on Computer Vision and Pattern Recognition, pp. 8697–8710 (2018)

# Automating Web GUI Compatibility Testing Using X-BROT: Prototyping and Field Trial

Hiroshi Tanaka[✉]

Fujitsu Limited, 4-1-1 Kamikodanaka, Nakahara-ku, Kawasaki 211-8588, Japan
htnk@fujitsu.com

**Abstract.** The main purpose of this paper is to introduce the application area of GUI test automation to researchers in the document analysis and recognition (DAR) field. Continuous integration (CI) represents the main stream in the field of web application development and reducing the burden of testing has become an important issue. In particular, GUI testing relies on human labor in many cases and test automation in this area is the least advanced. Because web content is essentially a type of "document," DAR technology is useful for promoting automation in this field. However, there have been few presentations on testing technology at conferences in the DAR field. This may be because researchers in the DAR field do not know much about which types of technologies are required in the field of software testing. We developed X-BROT, which is a tool for automatically determining the compatibility of web applications, and attempted to automate GUI testing. X-BROT can detect degradation by comparing the behaviors of web applications between versions under development. Compatibility verification between web documents is realized using DAR technology. X-BROT has shown nearly 100% accuracy for detecting differences between browsers [1] and it seems to be sufficient for automating testing, but many issues have been pointed out in the field of web application development. This paper describes this feedback and discusses new features that have been developed in response, as well as cases where DAR technology is required for GUI test automation.

**Keywords:** Web browser · HTML · DOM · Clique · Relative position · OCR · Selenium · WebDriver · CI

## 1 Introduction

In the field of software engineering (software industry), as continuous integration (CI) has become more widely used, the automation of web GUI testing, which relies on human labor, has become an increasingly important issue. In general, CI automatically executes tests in the background every time a software version is released. Therefore, to reduce the number of required man-hours, a fully automated function for determining the pass/fail of tests is required.

Software behavior testing involves executing typical usage procedures (test cases) and verifying the correctness of the produced results. In the case of CI, it can be assumed that the version prior to a revision has been tested and is working properly, so if there are

© Springer Nature Switzerland AG 2021
E. H. Barney Smith and U. Pal (Eds.): ICDAR 2021 Workshops, LNCS 12917, pp. 255–267, 2021.
https://doi.org/10.1007/978-3-030-86159-9_17

differences between the web pages before and after revisions, they can be considered as possible bugs. In other words, the execution results of the previous version can be regarded as expected values.

Because web applications can behave differently in different browsers, cross-browser testing (XBT) is required to compare behavior across browsers [1]. There are two types of compatibility tests for XBT called functional compatibility test (FCT) and visual compatibility test (VCT). FCT verifies the compatibility of the behavior of webpages, whereas VCT verifies the compatibility of the visual representations of webpages. To automate VCT, document analysis and recognition (DAR) techniques such as image processing and document structure analysis are particularly useful. However, there are few presentations on testing techniques at conferences in the DAR field. This may be because the web GUI testing is not well known as an application field of DAR technology.

VCT requires the ability to detect differences between web pages. The simplest way to achieve this goal is to compare images captured by browser screens in pixel-by-pixel manner using functions such as OpenCV. There are other image comparison techniques such as Browserbite [2], which divides images based on Harris Corners, and the technique presented in [3], which iteratively divides page images and compares each partial image. Another approach to detecting page incompatibilities is DOM-based comparison, where the DOM is a tree structure from which the content on a web page is extracted by each browser using the Selenium WebDriver [4]. Because each DOM element contains attribute values (position, size, borders, etc.), visual incompatibilities between browsers can be detected by referring to DOM information [5–9].

We developed X-BROT, which is a tool for DOM-based web page compatibility testing using Selenium, and incorporated it into a scenario testing tool (STT) (Figs. 1 and 2) that was applied to CI-based development sites. Because CI requires comparing different versions of web applications, X-BROT has the ability to identify DOM elements changes between different DOM trees. X-BROT has achieved sufficient practical verification accuracy, but several practical problems were reported from the development site. In this paper, we first present an overview of the previously reported X-BROT system. We then discuss the issues identified in the field of development and explain the newly developed functions as solutions to these issues. Finally, we discuss how DAR technology can be used to apply web GUI test automation to CI.

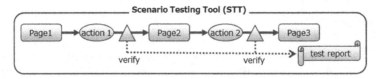

**Fig. 1.** Typical structure of a scenario-based test.

## 2 X-BROT: Web Compatibility Testing Tool

In this section, we discussed the web GUI compatibility testing tool X-BROT, which we developed in a previous study [1]. First, we describe the overall structure and processing

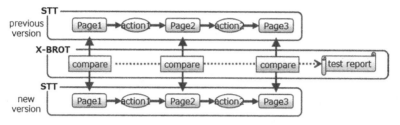

**Fig. 2.** Incorporating X-BROT into a scenario-based test.

modules, followed by the comparison algorithm. Additionally, we describe the results of a simple evaluation experiment to assess the accuracy of web page comparisons.

## 2.1 Overall Structure

First, it should be noted that there are two ways to use X-BROT for inter-browser comparison and inter-version comparison. The former detects differences on the same web page displayed in different browsers and the latter detects the differences between the previous version and new version of a web page to check for degradation. In this paper, we mainly focus the latter function.

The structure of X-BROT is illustrated in Fig. 3. The FCT module accesses the browser through WebDriver and obtains DOM data. When X-BROT is used for inter-browser comparison, multiple browsers must perform the same operation synchronously, so the simultaneous operation module operates each browser and obtains web page data from each browser. Because this paper focuses on the case of inter-version comparison, it is the display content of a single browser that is saved for a single web operation. As shown in Fig. 2, the STT operates twice, once for the previous version and once for the new version, so the web page is saved for each operation. The VCT module reads the stored webpage data, compares them, and generates a difference detection report. Here, either layout comparison or content comparison can be performed to compare web pages.

X-BROT has three main functions: element identification, layout comparison, and content comparison. Element identification is a function that identifies the same DOM elements to compare different versions of a DOM. This is similar to a "diff" command, which maps components that match before and after a web page has been modified. Layout comparison compares the positions of DOM elements on a browser screen. Subtle changes in position are difficult for humans to detect, so automating display comparisons is effective at reducing workloads and improving accuracy. Content comparison compares the contents of each DOM element and detects differences between versions using functions for directly comparing the attribute values of DOM elements and comparing character codes through the optical character recognition (OCR) of text images displayed in DOM elements. The details of each process are described below.

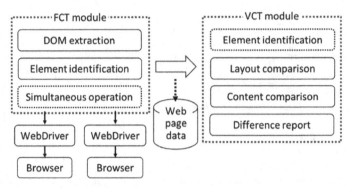

**Fig. 3.** Structure of X-BROT.

## 2.2 Element Identification

The element identification function compares DOM structures between web pages and identifies the corresponding DOM elements (Fig. 4). Typically, a given DOM element has a consistent "xpath" and "id," so identification is easy. However, in the case of inter-version comparisons, because the web page has been modified, the xpath may change based on changes in the DOM structure, the id and attribute values may change, or a corresponding DOM element may not exist (Fig. 5). Therefore, the element identification function must be able to identify a target DOM element, even if its xpath or id changes.

If the DOM structure and attribute values are unreliable, it is necessary to evaluate changes in structures and values comprehensively to find the best mapping. If the change in a DOM structure is local and slight, similar xpaths can be mapped based on the edit distance between xpaths [5–7], but if changes in the DOM structure are large, then a correct mapping cannot be obtained. Therefore, we use tree edit distance [10, 11], which evaluates edit operations (insertions, deletions, and replacements) at the tree level to find the correspondence that minimizes the changes in both structures and attribute values.

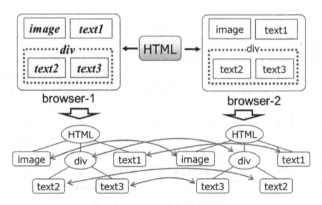

**Fig. 4.** Element identification.

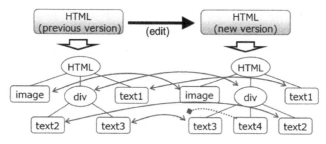

**Fig. 5.** Element identification (inter-version comparison).

## 2.3 Layout Comparison

The layout comparison function compares the relative positions of DOM elements identified as identical elements and detects differences in their layouts on a display screen (Fig. 6). In our layout comparison technique, we select arbitrary pairs of DOM elements from a set of DOM elements displayed on the same screen and consider the differences in their relative positions as differences in layout. For example, in Fig. 6, the pair of "text2" and "text3" are in the same relative position, but their positional relationships with the other elements are different. In this case, it is easy to understand that "text2" and "text3" are considered as a group and that this group has moved downward. In other words, we classify items that do not change their positional relationships with each other into groups and detect positional differences between minor groups.

To detect minor groups, we use graph analysis (Fig. 7). We generate a complete graph with all DOM elements as nodes and remove the links whose relative positions have changed. We then use the remaining links to find maximal cliques [12, 13]. The largest clique obtained is considered as the major clique and the rest are minor cliques. As shown in Fig. 8, the DOM elements belonging to minor cliques are displayed as differences in the results of layout comparisons.

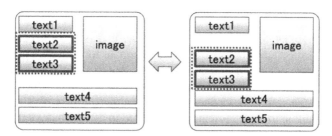

**Fig. 6.** Difference area detection.

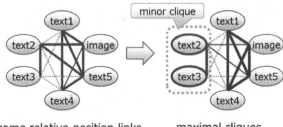

**Fig. 7.** Maximal clique extraction.

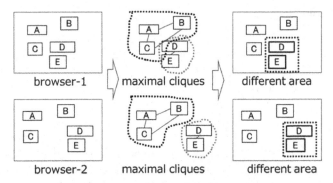

**Fig. 8.** Difference area detection based on cliques.

## 2.4 Content Comparison

Content comparison is a function that compares the contents of corresponding DOM elements. The comparison functions include "attribute data comparison," which compares the attribute values stored by each DOM element, and "image comparison," which compares the images displayed in the DOM element area.

For attribute data comparison, for example, coordinate values (x and y coordinates, height, and width) of DOM elements, internal text, scroll bars, and other attribute values are compared. Text comparison is particularly important for web GUI testing because internal text is often contained in item data values that are displayed to users in web applications. There are many types of attributes that make it difficult to perform comprehensive analysis. Therefore, we defined a set of common attributes that are used often (Table 1). An addition to these attributes, we decided to use attributes defined by users in HTML sources for comparison.

Image comparison involves the comparison of partial images displayed in the regions corresponding to each DOM element. Because pixel-level comparisons are commonly used [3], we specifically focus on the comparison of text images. Figure 9 presents an example of garbled text displayed in a font. There are two problems: one is the character code conversion problem and the other is the line feed position problem. In particular, the line feed position problem affects the size of the DOM element and leads to layout collapse. To detect the occurrence of such garbled characters, we use OCR to recognize

**Table 1.** Common attributes

| < tag name> | alt | height | font-family |
|---|---|---|---|
| <parent node> | src | display | color |
| nodeText | type | position | test-align |
| border | value | margin | hidden |
| scrollbar | disabled | padding | visibility |
| id | checked | overflow | opacity |
| class | selected | font-size | |
| href | width | font-weight | |

text images and compare recognition results. To this end, we must prevent as many OCR errors as possible. Therefore, we adopted the following three approaches.

The first approach involves hiding images other than the target DOM element. This improves the accuracy of character recognition by using the hidden attribute to hide irrelevant areas so that images of other DOM elements do not have a negative impact when capturing text images. The second approach is to enlarge the character image displayed by a font when capturing text images to increase the resolution of the target image for OCR (Fig. 10). The third approach is correction using contextual post-processing. Each DOM element has an "innerText" attribute and in many cases, this text is displayed on the screen. However, because there are cases in which text other than the innerText is displayed, the value of the innerText can be used as reference information for context processing to correct recognition results (Fig. 11).

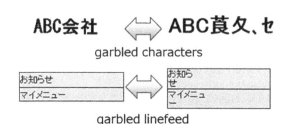

garbled characters

garbled linefeed

**Fig. 9.** Garbled text.

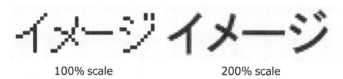

100% scale                    200% scale

**Fig. 10.** Expanded text image.

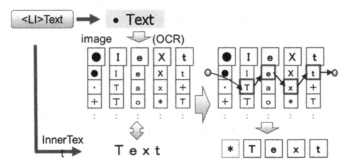

**Fig. 11.** Lexical post-processing based on HTML innerText elements.

## 2.5 Evaluation

We evaluated the performance of the developed tool through inter-browser comparisons by using the behavior of IE11 as a reference. Difference detection was performed by displaying the same web page in IE9, IE10, and Firefox, and the accuracy of detecting differences was determined through visual inspections. The URLs used in for evaluation are listed in Table 2. Accuracy evaluation was performed based on layout comparison and text image comparison using OCR (Tables 3 and 4). Here, the true positive (TP), false positive (FP), and false negatives (FN) occurrences are counted.

**Table 2.** Web site URLs used to evaluate the developed tool

| |
| --- |
| http://www.smbc.co.jp/ |
| http://www.mizuhobank.co.jp/index.html |
| http://www.bk.mufg.jp/ |
| http://t21.nikkei.co.jp/ |
| http://t21.nikkei.co.jp/public/guide/article/index.html |
| http://jp.fujitsu.com/ |
| http://www.au.kddi.com/ |
| https://www.nttdocomo.co.jp/ |
| http://www.softbank.jp/ |
| http://www.jreast.co.jp/ |
| http://www.westjr.co.jp/ |
| http://www.jr-central.co.jp/ |
| http://www.tokyo-solamachi.jp/ |
| http://www.motomachi.or.jp/ |
| http://www.ytv.co.jp/index.html |

For layout comparison, all 33 differences were detected and no false positives occurred. For text image comparison, most differences (97.0%) were detected, but many false positives occurred in the "IE11-FF" comparison, resulting in a significant decrease in precision. The main causes of false positives were OCR errors (7 cases) and line feed detection errors (10 cases).

As described previously, the X-BROT tool we developed was determined to be very accurate at detecting differences between browsers. Although our evaluations were performed based on inter-browser comparisons, the same principles of comparison can be applied to inter-version comparison if DOM elements are successfully identified, so a similar level of accuracy can be expected.

**Table 3.** Incompatibility detection (layout comparison)

|           | TP | FP | FN | Recall | Precision |
|-----------|----|----|----|--------|-----------|
| IE11-IE9  | 18 | 0  | 0  | 100%   | 100%      |
| IE11-IE10 | 7  | 0  | 0  | 100%   | 100%      |
| IE11-FF   | 8  | 0  | 0  | 100%   | 100%      |
| Total     | 33 | 0  | 0  | 100%   | 100%      |

**Table 4.** Incompatibility detection (text image comparison)

|           | TP | FP | FN | Recall | Precision |
|-----------|----|----|----|--------|-----------|
| IE11-IE9  | 20 | 0  | 1  | 95.2%  | 100%      |
| IE11-IE10 | 10 | 1  | 0  | 100%   | 90.9%     |
| IE11-FF   | 34 | 16 | 1  | 97.1%  | 57.6%     |
| Total     | 64 | 17 | 2  | 97.0%  | 79.0%     |

# 3   Field Trial: Feedback and Solutions

In this section, we discuss the problems identified as a result of testing X-BROT in the development field and the solutions to these problems. As described in the previous section, the accuracy of X-BROT for detecting browser differences is sufficiently high, making it easy to apply for practical use. However, unexpected problems may occur during trials. One reason for this is that the data used for evaluation may not sufficiently reflect actual usage conditions. To make our evaluation environment more similar to the real world, field trials at an early stage were necessary.

Our field trial was conducted in our web application development department using form software under development. The developers conducted tests using X-BROT, listed issues encountered when using X-BROT for practical purposes, and developed improvements to solve the identified issues. In this section, we present various examples.

## 3.1   Multi-frame Page

The first problem identified is the inability to handle web pages with multiple frames. A normal web page is written in a single HTML file, but if a frame is created using the "frame" or "iframe" tags, then an external HTML file can be imported into a page (Fig. 12). Because the inside of a frame has a different coordinate system and DOM tree structure, and because the DOM elements in different frames cannot be mapped, web pages with frames cannot be compared to each other.

This problem initially went unnoticed because there were no samples containing frames in the evaluation URLs, but feedback from professionals in the field made us aware of the need for frame support. We solved this problem by adopting a two-step process: 1) creating a frame-only tree structure to determine the correspondence between frames (Fig. 13) and 2) comparing the DOMs for each frame to detect differences.

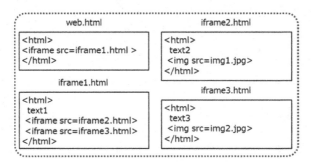

**Fig. 12.** A web page consisting of multiple HTML files.

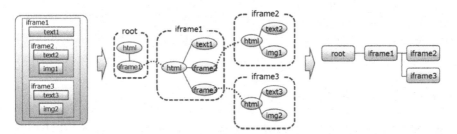

**Fig. 13.** A web page with multiple frames.

## 3.2 Non-comparison Areas

The next issue was a request to remove DOM elements that changed their contents every time they were executed from comparisons. For example, in areas that count the number of accesses to a web page or display the date and time, differences are always detected because the corresponding data change every time the page is opened, even if the web application does not change. It was pointed out that this is not a form of degradation and that the professionals in the field do not want to detect it as a difference. As shown in Fig. 2, the STT considers differences from previous versions as candidates for degradation, so this request was made to reduce unnecessary differences as much as possible.

This problem was solved by implementing a function to set areas in which detected differences should be ignored ("non-comparison areas") (Fig. 14). To specify non-comparison areas, it is sufficient to compare the results of displaying the same web page twice before executing a test and registering the detected difference areas as non-comparison areas. This function is difficult to notice without conducting field trials, but it is intuitive when pointed out.

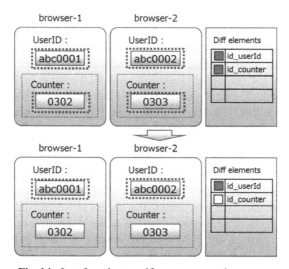

**Fig. 14.** Interface that specifies non-comparison areas.

### 3.3 Huge Web Pages

It was pointed out that the comparison process stopped with an error when the target web page was too large. This is because there was insufficient processing memory available when there were more than 10,000 DOM elements in a single web page. In our X-BROT implementation, a dynamic programming table is created when determining the correspondences of DOM elements, but the length of the array required to compare more than 10,000 elements exceeds 100 million and the memory required exceeds 4 GB (implemented in C++). Although this amount of memory can be compressed several times through programming choices, large web pages are still problematic.

Through our investigation, we determined that this large number of DOM elements is mainly caused by the large numbers of nodes under "Table" tags. Therefore, a potential solution is to group all the child nodes of Table tags together to reduce the number of DOM elements significantly.

### 3.4 Discussion

In this section, we presented examples of the findings obtained when X-BROT was applied in the software development field. As discussed in the previous section, the accuracy of X-BROT at detecting differences is close to 100% and it seems feasible to put it into practical use immediately. However, a number of problems were identified that were not noticed in our laboratory analysis. The issues described in this section can be addressed by investigating the real scenarios of software development and by improving our evaluation database, but there are likely other issues that have not been noticed and other possibilities for improvement. Overall, feedback from field trials is important for discovering new technological issues.

## 4  Conclusion

We developed X-BROT, which is a tool for automatically determining the compatibility of web applications, and attempted to automate web GUI testing. In this paper, we discussed the details of the X-BROT tool reported in [1] and described issues and solutions that emerged when X-BROT was tested in the field of software development. X-BROT has very high accuracy for detecting differences between browsers, but there are lingering issues that must be addressed to improve the functionality of the tool for practical use, such as the ability to handle web pages that contain frames and set areas that should not be compared. We were able to confirm the importance of identifying issues related to practical use through field trials.

Although most of the issues described in this paper can be resolved by extending the functionality of our tools, there are some technical issues that remain unsolved. One such issue is the problem of mapping DOM elements when a web page is heavily modified. As discussed in Sect. 2, we use tree edit distance to identify DOM elements. This is a method of mapping elements in a partially modified DOM structure, but it is unable to find an appropriate mapping when the structure of a DOM changes significantly. Research on identifying DOM elements across different DOMs is part of the research on automatic test case repair for web test automation and various methods have been proposed [14–16]. By using this related research as a reference, we believe it should be possible to develop a method that can stably obtain a mapping of DOM elements, even when comparing web pages that have changed significantly.

Additionally, as a future task, we will conduct a more detailed survey of specific scenarios in which X-BROT is used and enhance our database for evaluation. Particularly for inter-version comparisons, it is very difficult to enrich evaluation data because it is necessary to prepare new and old versions during the process of developing a web application. We believe it is necessary to collect data tailored to the actual conditions of development by surveying how web content is edited.

## References

1. Tanaka, H.: X-BROT: prototyping of compatibility testing tool for web application based on document analysis technology. In: International Conference on Document Analysis and Recognition Workshops (ICDARW), (WIADAR 2019), pp. 18–21 (2019)
2. Saar, T., Dumas, M., Kaljuve, M., Semenenko, N.: Browserbite: cross-browser testing via image processing. Softw. Pract. Exp. **46**(11), 1459–1477 (2016). arXiv:1503.03378v1
3. Lu, P., Fan, W., Sun, J., Tanaka, H., Naoi, S.: Webpage cross-browser test from image level. Proceedings of IEEE ICME **2017**, 349–354 (July 2017)

4. Selenium Project. https://www.selenium.dev/
5. Choudhary, S.R., Versee, H., Orso, A.: WEBDIFF: automated identification of cross-browser issues in web applications. In: IEEE International Conference on Software Maintenance (ICSM 2010), pp. 1–10 (2010)
6. Choudhary, S.R., Prasad, M.R., Orso, A.: Crosscheck: combining crawling and differencing to better detect cross-browser incompatibilities in web applications. In: IEEE International Conference on Software Testing, Verification and Validation (ICSTVV 2012), pp. 171–180 (2012)
7. Choudhary, S.R., Prasad, M., Orso, A.: X-PERT: accurate identification of cross-browser issues in web applications. In: International Conference on Software Engineering (ICSE 2013), pp. 702–711, May 2013
8. Dallmeier, V., Pohl, B., Burger, M., Mirold, M., Zeller, A.: WebMate: web application test generation in the real world. In: 7th International Conference on Software Testing, Verification and Validation Workshops (ICSTW 2014), pp. 413–418, March 2014
9. Mesbah, A., Prasad, M.R.: Automated cross-browser compatibility testing. In: International Conference on Software Engineering (ICSE 2011), pp. 561–570, May 2011
10. Masuda, S., Mori, I., Tanaka, E.: Algorithms for finding one of the largest common subgraphs of two trees. IEICE Trans. **J77-A**(3), 460–470, March 1994 (in Japanese)
11. Wang, J.T.L., Shapiro, B.A., Shasha, D., Zhang, K., Currey, K.M.: An algorithm for finding the largest approximately common substructures of two trees. IEEE Trans. Pattern Anal. Mach. Intell. **20**(8), 889–895 (1998)
12. Östergård, P.R.J.: A fast algorithm for the maximum clique problem. Discrete Appl. Math. **120**(1–3), 197–207 (2002) . https://core.ac.uk/download/pdf/82199474.pdf
13. Tomita, E., Matsuzaki, S., Nagao, A., Ito, H., Wakatsuki, M.: A much faster algorithm for finding a maximum clique with computational experiments. IPSJ J. Inf. Process. **25**, 667–677 (2017) . https://www.jstage.jst.go.jp/article/ipsjjip/25/0/25_667/_pdf
14. Leotta, M., Stocco, A., Ricca, F., Tonella, P.: Using multi-locators to increase the robustness of web test cases. In: Proceedings of 8th IEEE International Conference on Software Testing, Verification and Validation (ICST 2015), pp. 1–10, April 2015
15. Eladawy, H.M., Mohamed, A.E., Salem, S.: A new algorithm for repairing web-locators using optimization techniques. In: 13th International Conference on Computer Engineering and Systems (ICCES 2018), pp. 327–331, December 2018
16. Leotta, M., Stocco, A., Ricca, F., Tonella, P.: Robula+: an algorithm for generating robust XPath locators for web testing. J. Softw. Evol. Process **28**(3), 177–204 (2016)

# A Deep Learning Digitisation Framework to Mark up Corrosion Circuits in Piping and Instrumentation Diagrams

Luis Toral[1], Carlos Francisco Moreno-García[1]([✉]) [iD], Eyad Elyan[1] [iD],
and Shahram Memon[2]

[1] Robert Gordon University, Aberdeen AB10 7QB, Scotland, UK
c.moreno-garcia@rgu.ac.uk
[2] Archimech Limited, Aberdeen AB15 9RJ, Scotland, UK

**Abstract.** Corrosion circuit mark up in engineering drawings is one of the most crucial tasks performed by engineers. This process is currently done manually, which can result in errors and misinterpretations depending on the person assigned for the task. In this paper, we present a semi-automated framework which allows users to upload an undigitised Piping and Instrumentation Diagram, i.e. without any metadata, so that two key shapes, namely pipe specifications and connection points, can be localised using deep learning. Afterwards, a heuristic process is applied to obtain the text, orient it and read it with minimal error rates. Finally, a user interface allows the engineer to mark up the corrosion sections based on these findings. Experimental validation shows promising accuracy rates on finding the two shapes of interest and enhance the functionality of optical character recognition when reading the text of interest.

**Keywords:** Digitisation · Corrosion detection · Piping and instrumentation diagrams · Convolutional neural networks

## 1 Introduction

Experienced corrosion and material engineers have the task of defining corrosion circuits within a system based on construction materials, operating conditions and active damage mechanisms [1]. This is part of a recommended practice developed by the American Petroleum Institute (API), which outlines the basic elements to maintain a credible risk-based inspection (RBI) programme[1]. Once the circuits have been defined, the Condition Monitoring Locations (CMLs) and the Thickness Monitoring Locations (TMLs) are installed and documented on a type of engineering drawings known as a Piping and Instrumentation Diagram (P&ID). This process involves a manual mark up which becomes time-consuming and error prone, as shown in Fig. 1.

---

[1] https://store.nace.org/corrosion-looping-for-down-stream-petroleum-plants-an-enigma-for-rbi-engineers-a-perspective-from.

© Springer Nature Switzerland AG 2021
E. H. Barney Smith and U. Pal (Eds.): ICDAR 2021 Workshops, LNCS 12917, pp. 268–276, 2021.
https://doi.org/10.1007/978-3-030-86159-9_18

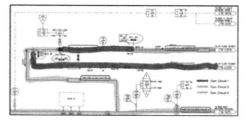

**Fig. 1.** Extract of a P&ID representing three corrosion circuits (colour sections), pipe specification (rectangles) and connection points (ellipses) (original resolution: 2048 × 080 pixels).

To define a corrosion circuit, engineers need to identify two key elements within the piping system on the P&IDs. The first one is the pipe specification (pipe spec), which is a character string formed by seven sections divided by hyphens. The second one is the connection point. This is a pair of text lines and arrows pointing towards a division line. In a specific P&ID, both the pipe spec and the connection point can be oriented in different directions (see Fig. 2a and 2b). The more complex a piping system is, the larger the amount of information displayed in the drawing. At first, our aim is to detect the text in both pipe specs and connection points, to then read it and identify the limits of the corrosion sections.

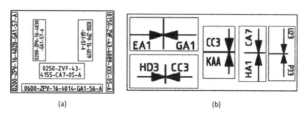

**Fig. 2.** Different types of shapes and orientations for (a) Pipe specifications and (b) Connection points detections.

In this paper, we propose a framework to develop a semi-automatic novel tool that allows the user to mark up the corrosion circuits in P&IDs based on automatically locating pipe specs and the connection points. The paper is organised as follows: Sect. 2 presents the related work, Sect. 3 the methodology, Sect. 4 the experiments and results, and Sect. 5 concludes and presents future directions.

## 2  Related Work

Text detection is one of the cornerstones in document image analysis, as it can lead to the location and identification of the depicted shapes. Later on, it can also help on the mapping of the structural representation, as the labels usually contain relevant information such as the direction of flow, sectioning, and other useful data [2].

Many attempts have been presented in literature to digitise P&IDs, mostly following two lines of work. The first one involves the use of heuristics to detect certain well-known shapes, such as geometrical symbols, arrows, connectors, tables and even text [3–6]. The second and most recent one relies on deep learning techniques in which the algorithms are trained to recognise shapes based on the collection and tagging of numerous samples [7–10]. Both approaches have advantages and disadvantages depending on the use case. The first family of methods is better suited when the characteristics of the P&ID follow a certain standard; however, if this is not the case, then the latter option can be used.

In the previous edition of this workshop, we presented a paper addressing the challenges and future directions of P&ID digitisation [11]. In that work, we addressed three main challenges: image quality, class imbalance and information contextualisation. Why being able to digitise drawings with reduced quality is still a work in progress, in this new challenge we focus on the two latter. On the one hand, there is a need to locate symbols and text strings which do not appear often in these drawing. Therefore, we must resort to consider heuristic and automatic solutions to properly locate the text strings and symbols which depict pipe specs and connection points respectively. On the other hand, by correctly identifying these pointers, we are able to allow the user to manually mark up the corrosion sections, bringing us one step closer to identifying the structures depicted within the engineering diagram [12].

## 3   Methodology

The digitisation workflow of our method is comprised of five steps:

I.   Train two different deep learning models to detect pipe specs and connection points using the YOLO v5 framework and store the detection coordinates.
II.  Create a binarised image for each detection and look over for connected components based on pixel connectivity to locate the region of interest (ROI) containing the text.
III. Get the components statistics from the ROIs and apply a heuristic method to align the detections horizontally.
IV.  Apply the Tesseract[2] Optical Character Recognition (OCR) engine with a custom configuration.
V.   Link the codes found in both pipe specs and connection points with their respective locations on the drawing.

### 3.1   Connection Points and Pipe Specs Detection

A pre-processing step was applied to the dataset to standardise the P&IDs and convert them into grayscale images to reduce noise. Subsequently, two different models were built to detect the pipe specs and connection points separately (see Fig. 3). The convolutional neural network selected was YOLOv5x[3] with the default configuration. Each model was trained at 3000 epochs with a batch size of 8. The marker tool runs the trained models on *PyTorch* to generate a cropped image of every single detection and store the positional coordinate relative to the P&ID.

---

[2] https://github.com/tesseract-ocr/tesseract.
[3] https://github.com/ultralytics/yolov5/releases.

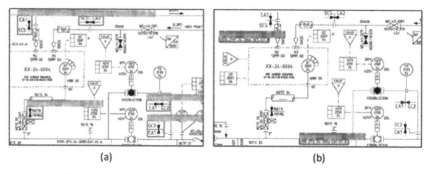

**Fig. 3.** Detection boxes (a) Connection points and (b) Pipe specs.

## 3.2 Text Bloc Localisation

The next step is to localise the text block regions. While the text in pipe specs tends to have a consistent shape, the text in the connection points can vary considerably in size and orientation. Depending on the amount of information shown on the P&ID, textual and non-textual elements can overlap and appear on the detection boxes. Thus, to extract the ROI, a noise removal technique was introduced. Firstly, a binarisation process is applied to all the cropped images to reduce noise. Secondly, a connected component labelling (CCL) technique is used to get the shape and size information of the elements in the image. Finally, the components that fulfil a heuristic threshold criterion are kept (see Figs. 4a and 4b for more details).

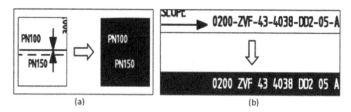

**Fig. 4.** Location ROI for (a) Connection points and (b) Pipe specifications ROI.

## 3.3 Text Alignment

One of the limitations of the Tesseract OCR engine is that it can only interpret the text when horizontally aligned. Although this engine has a built-in configuration to assess the orientation of the text in an image, the number of characters in the detections is not enough to implement this feature. Hence, an additional method is applied to adjust those detections that are misaligned.

This method consists of a two-step process. First, the marker tool identifies the nonaligned detections, and then it determines the direction to rotate them. Given the height and width of the cropped image, we can classify which detections are vertically oriented. While this is a general rule for pipe specs, it does not apply to connection points (see Fig. 2b). Thus, a conditional criterion is added.

Figure 5 shows the rotation conditions for connection points. For each ROI, the *x*-coordinate of the components are calculated and checked for collinearity among themselves. If the group standard deviation falls behind a heuristically learned threshold, the image is vertically oriented. Subsequently, if the ROI is on the left side of the image, a clockwise rotation is applied. Conversely, if it is located on the right side, the flip direction is counter-clockwise.

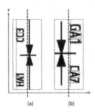

**Fig. 5.** (a) Clock-wise and (b) Counter clockwise rotation criteria.

### 3.4  Text Recognition

The text strings inside the detected areas contain the code which delimits the corrosion circuits; therefore, the next stage is to read them as accurately as possible. The ROI in the image can contain text in a single or a double line (see Fig. 2b). Connection points are composed of two text blocs with three characters, whereas pipe specs contain a long code string. These aspects can affect accuracy performance. Hence, different experiments were tested on Tesseract OCR to set a custom configuration which delivered the highest accuracy. Given the size of the detections being significantly small compared to the original image, different filters were tested to remove the noise caused by this distortion. The results of these configurations will be discussed in detail in Sect. 4.2.

### 3.5  Linkage

The last step is to link the codes extracted in the connection points and pipe specs with their respective positional coordinates relative to the P&ID. The pair codes on the connection points are processed and stored as three-string characters (see Fig. 6). For pipe specs, since the code is right next to the fourth hyphen, this is extracted by iterating over the string of characters. Finally, both codes and coordinates are stored in a dictionary.

## 4  Experiments and Results

The private dataset consists of 85 P&IDs provided by our industrial partner Archimech Limited. A total of 75 images were labelled, 70% were used for training, 20% for validation, and 10% for the testing, having 1653 and 537 annotations for pipe specs and connection points, respectively. The remaining 10 images from our dataset were used to test the end-to-end performance of the marker tool.

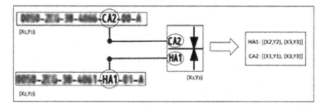

**Fig. 6.** The linkage between pipe specs and connection points codes.

## 4.1 Detection

Several experiments were deployed modifying the hyperparameters. The optimal results were attained by setting the input image size to 2048 × 1080 pixels, a batch size of 8, and 3000 epochs. The confidence threshold used to run the detection was 0.4 for both models. In the end, we have achieved an accuracy of 96.23% for detecting detect pipe specs and 92.68% for connection points (see Table 1).

**Table 1.** Stages and their accuracies.

| Deep learning model | Accuracy (%) |
| --- | --- |
| Pipe specs localisation | 96.23 |
| Connection points localisation | 92.68 |
| Pipe specs text recognition | 98.72 |
| Connection points text recognition | 93.55 |
| End-to-end-performance | 82.37 |

## 4.2 Text Recognition

The text recognition accuracy for pipe specs is 98.72% and 93.55% for connection points. Due to the variation in shape and the overlapping of text with graphics, it becomes more challenging to recognize the codes on the connection points. Different experiments were tested in order to improve the text recognition accuracy. The optimal performance was achieved by applying a median blur filter to the detections and setting a custom page segmentation model in the Tesseract-OCR engine. In Tables 2 and 3, we show how this method can successfully improve the accuracy text recognition for both connection points and pipe specs, respectively.

## 4.3 Final Output

The marker tool allows uploading either a single or a set of multiple P&IDs at once. After running the application, a list of all the codes detected in the drawing is deployed. The user can then select the code to visualize on the drawing and the colour of the marker.

**Table 2.** Connection points text recognition accuracy improvement.

| Connection points | Text localisation | Tesseract OCR | Filter + Tesseract OCR-custom config |
|---|---|---|---|
| HA1 \| CA7 | HA1 CA7 | HA` CA? | HA1 CA7 |
| HA1 HD3 | HA1 HD3 | Te | HA1 HD3 |

**Table 3.** Pipe specs text recognition accuracy improvement.

| Pipe specs | Text localisation | Tesseract OCR | Filter + Tesseract OCR-custom config |
|---|---|---|---|
| 0250-ZWC-40-4000-DA4-13-A | 0250 ZWC 40 4000 DA4 13 A | 0250 ZWC 40 4000 DAG 13 A | 0250 ZWC 40 4000 DA4 13 A |
| 0200-ZVF-43-4038-DD2-05-A | 0200 ZVF 43 4038 DD2 05 A | 0200 ZVF 43 4038 D02 05 A | 0200 ZVF 43 4038 DD2 05 A |
| 0050-ZAI-63-4000-CC2-01-A | 0050 ZA 63 4000 CC2 01 A | 0050 ZA 63 4000 CC2 01 A | 0050 ZAI 63 4000 CC2 01 A |

Finally, the user can mark up the corrosion circuits and save the document as a jpg file. Figure 7 shows an example of a P&ID section with three different corrosion circuits marked with our novel tool. The company works with many stakeholders which have even more P&IDs to digitise.

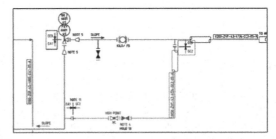

**Fig. 7.** Extract of a P&ID with three corrosion circuits marked using the proposed framework (original resolution: 2048 × 1080 pixels).

## 5   Conclusion and Future Work

In this paper, we presented an end-to-end framework which allows Oil & Gas engineers to load undigitised P&IDs and locate the corrosion circuits with minimal intervention and error. We used two models trained with a state-of-the-art deep learning technique (i.e. YOLOv5) to find the two shapes of interest, namely the pipe specs and the connection points. Once these are located, there are additional post-processing steps that have to be performed prior to presenting these symbols to the user. In the case of the connection points, the text needs to be found, oriented properly and read with total accuracy. This allows the system to identify which are the symbols of interest which will be shown to the engineer so that the corrosion circuits can be marked up.

In future work, we would like to explore the scalability of our system to perform appropriately in P&IDs generated in other standards and qualities. Moreover, we aim to test more novel deep learning frameworks which allow us to increase the accuracy of the detection and OCR tasks by using techniques that work with limited character sets [13]. Finally, we aim to automate the last step so that the engineer is shown the corrosion circuits automatically.

**Acknowledgements.** We would like to thank Innovate UK for funding this research under the Innovation Voucher scheme.

# References

1. Mohammed, M.: Corrosion looping for down stream petroleum plants: An Enigma for RBI Engineers A Perspective from the Review of Mechanical Integrity Systems. AMPP. Corrosion and Monitoring Control (2016)
2. Moreno-García, C.F., Elyan, E., Jayne, C.: New trends on digitisation of complex engineering drawings. Neural Comput. Appl. **31**(6), 1695–1712 (2019)
3. Howie, C., Kunz, J., Binford, T., Chen, T., Law, K.H.: Computer interpretation of process and instrumentation drawings. Adv. Eng. Softw. **29**, 563–570 (1998)
4. Arroyo, E., Hoernicke, M., Rodríguez, P., Fay, A.: Automatic derivation of qualitative plant simulation models from legacy piping and instrumentation diagrams. Comput. Chem. Eng. **92**, 112–132 (2016)
5. Moreno-García, C.F., Elyan, E., Jayne, C.: Heuristics-based detection to improve text/graphics segmentation in complex engineering drawings. In: Engineering Applications of Neural Networks (EANN), pp. 87–98 (2017)
6. Sinha, A., Bayer, J., Bukhari, S.S.: Table localization and field value extraction in piping and instrumentation diagram images. In: Graphics Recognition Methods and Applications (GREC), pp. 26–31 (2019)
7. Mani, S., Haddad, M.A., Constantini, D., Douhard, W., Li, Q., Poirier, L.: Automatic digitization of engineering diagrams using deep learning and graph search. In: Computer Vision and Pattern Recognition (CVPR) (2020)
8. Yun, D.Y., Seo, S.K., Zahid, U.: Deep neural network for automatic image recognition of engineering diagrams. Appl. Sci. **10**, 1–16 (2020)
9. Sierla, S., Azangoo, M., Fay, A., Vyatkin, V., Papakonstantinou, N.: Integrating 2D and 3D digital plant information towards automatic generation of digital twins. In: 2020 IEEE 29th International Symposium on Industrial Electronics (ISIE), pp. 460–467 (2020)
10. Elyan, E., Jamieson, L., Ali-Gombe, A.: Deep learning for symbols detection and classification in engineering drawings. Neural Netw. **129**, 91–102 (2020)

11. Moreno-García, C.F., Elyan, E.: Digitisation of assets from the oil & gas industry: challenges and opportunities. In: International Conference on Document Analysis and Recognition (ICDARW), pp. 16–19 (2019)
12. Rica, E., Moreno-García, C.F., Álvarez, S., Serratosa, F.: Reducing human effort in engineering drawing validation. Comput. Ind. **117**, 103198 (2020)
13. Majid, N., Barney Smith, E.H.: Performance comparison of scanner and camera-acquired data for Bangla offline handwriting recognition. In: International Conference on Document Analysis and Recognition Workshops (ICDARW), pp. 31–36 (2019)

# Playful Interactive Environment for Learning to Spell at Elementary School

Sofiane Medjram[1](✉), Véronique Eglin[1], Stephane Bres[1], Adrien Piffaretti[2], and Jobert Timothée[3]

[1] Université de Lyon, INSA Lyon, LIRIS, UMR5205, 69621 Lyon, France
{veronique.eglin,stephane.bres}@insa-lyon.fr
[2] 20 rue du Professeur Benoît Lauras, 42000 St-Étienne, France
adrien@superextralab.com
[3] 22 Avenue Benoit Frachon, 38 400 Saint Martin D'Hères, France
timothee.jobert@iskn.co

**Abstract.** This paper proposes a new playful approach to improve children's writing skills in primary school. This approach is currently being developed in the Study project ([1]Study Project, French Region Auvergne-Rhône-Alpes R&D Booster Program, 2020-2022), that proposes a global learning solution through the use of an innovative interactive device, combining advanced technologies for the writing acquisition (Advanced Magnetic Interaction tech from AMI/ISKN [2](ISKN Repaper Tablet,https://www.iskn.co/fr), and the conception of a gamified environment for children (Super Extra Lab [3](SuperExtraLab, https://www.superextralab.com/) distributed by Extrapage, publisher of the solution) [4](Extrapage, https://www.extrapage.io/fr/pages/index.html) The proposition associates traditional school materials (textbook, paper, pencil...) with very recent technologies and didactic innovations. While the state-of-art is usually satisfied with solutions strictly focusing on lexical spelling, the Study project aims to contribute to a broader control of handwriting skills, i.e., the acquisition of efficient handwriting micromotor skills and also strong grammatical skills. To achieve this, the approach relies on a personalized, regular and enriched copy and dictation training program. It establishes conditions that encourage children to be involved in an activity that is often unpleasant, and is inspired by the mechanics of video games. In the paper, we present an overview of the project, the construction of the general framework in relation with the design of a game interface for learning and the first recognition results based on a dedicated MDSLTM-CTC trained on a GAN-generated dataset that cover different writing styles.

**Keywords:** Child handwriting recognition · Dictation · French language · ISKN Repaper · Extrapage publisher · SCOLEDIT dataset · Study R&D project · MDLSTM technologies · GAN-generated handwriting styles · Human computer interaction · Deep-learning

© Springer Nature Switzerland AG 2021
E. H. Barney Smith and U. Pal (Eds.): ICDAR 2021 Workshops, LNCS 12917, pp. 277–290, 2021.
https://doi.org/10.1007/978-3-030-86159-9_19

# 1    Introduction

## 1.1    Context of Study Project

Learning to read and write a mother tongue is an indispensable task for children in school. It will build the basis of their learning and their integration into different modules of higher grades, as well as it will allow them to express themselves and communicate properly.

Nowadays, learning to write in France still uses classical methods where the child learns to write by dictation. This method, although it has shown its reliability, remains difficult for teachers because they have to check all the pupils' copies, adapting this check to the different pupils' profiles, and then return their spelling errors. This intertwining of dictation, the child's handwriting and the teacher's correction feedback can become degrading because the student's state of concentration when writing and receiving the correction is different. So, in order to help the teacher in this task, we thought about the integration of modern interaction technologies and recent advances in deep-learning based handwriting recognition. Thanks to the ISKN interaction device that preserves the naturel handwriting aspect [7], trainees can receive feedback and spelling correction on the fly during dictation, each by his or her level of expertise. We can also notice that the teacher's task become lighter and more precise in terms of monitoring the children's progresses. Using ISKN Repaper the child writes on a sheet of paper attached to the tablet like a slate and on the fly, the system retrieves the handwritten patterns of what has been written. We also propose a deep learning recognition model for child handwriting recognition which will allow spelling error detection and synchronous interactions with the learner.

## 1.2    Destination Market and Product Industrialisation

Without mentioning the french speaking market (with 51.5 millions learners worldwide), the national target market is already very significant since it is made up of 47000 french elementary schools. In addition, we can also count about 30000 kindergarten classes with approximately 700000 trainees. The french market of Technologies for Education (EdTech) is split up with more than 350 companies, and 86% of them contains less than twenty employees. These are concentrated in the three main segments: ongoing education (58%), higher education (34%), and schools (42%). The segment in which we operate, educational resources, represents 42 million euros and should double by 2022. At the same time, EdTech market in the field of educational innovations is also expanding worldwide (in China and United States). Especially, AMI/ISKN's experience in China has made the Study project very successful and will encourage the consortium to further developments in foreign languages (especially Spanish and Chinese). In this context, Study offers a complete solution to the crucial issue of spelling acquisition and handwriting quality, where the concurrence only approaches those questions from the more reductive angle of lexical spelling skills, focusing on the learning of writing motor skills.

The project extends and enriches the work of AMI/iskn (advanced magnetic interaction), which, with tori[TM] (a learning platform co-developed with the Japanese group Bandai Namco)[1] and Repaper (a pen and paper tablet, already used by a Chinese company on its territory for learning calligraphy in schools), has already developed several products at the frontiers of the education market. Study project also supports the economic model of SuperExtraLab society (for the software application and design distributed by the Extrapage publisher[2]) by ensuring the marketing of a joint product. This work is also prepared in collaboration with a team of linguists, speech therapists and primary school teachers who allow us to access a unique database of handwritten texts from children learning to spell [13]. At the end of the value chain, the project is also associated to a publisher (Extrapage) and further distributors specialized in education supplying the equipment (graphic tablet) and the service (gamified learning software), see Fig. 1.

**Fig. 1.** Different software and hardware components for an integrative solution to improve children's writing.

Because Study project aims to enhance child handwriting for French language, it mainly depends on the performance of handwriting recognition model. Indeed, a bad handwriting recognition will result in an incorrect words alignment and many inaccuracies in the feedback to trainees. In the following sections, we propose to develop technical and scientific aspects relative to HTR (Handwriting Text Recognition) and the interactive scenario as it will be proposed to young trainees through the design of a gamified application.

---

[1] tori[TM], https://www.tori.com/fr/.
[2] Extrapage, https://www.extrapage.io/.

## 2    Advances in Handwriting Recognition

### 2.1    Software Advances and Recent Deep Neural Networks Techniques

Nowadays, handwriting recognition still represents a challenging problem. The high variance in handwriting styles for the same person/child or across people and the specific difficulty to read children's writing pose significant hurdles in transcribing it to machine readable text [10]. Handwriting Recognition methods are divided into online and offline approaches. Online methods consist of tracking a digital pen and receiving information about position and stroke while text is written. Offline methods transcribe text contained in an image into digital text. Offline methods are mostly used when the temporal dimension in the writing execution is not required, [11]. Recently, like in other application domains, deep learning approaches improved tremendously handwriting recognition accuracy [1,6,9]. In [6] off-line recognition is handled by CRNN networks, processing the image through several convolution layers CNN to produce deep spatial features of the image, and then passing them to a recurrent network RNN to produce deep contextual features. After being collapsed to a one dimensional sequence by summing over the height axis, the sequence is processed through a softmax layer with a Connectionist Temporal Classification (CTC) loss handling the right alignment between the input and the output sequence.

In [1], an attention based model for end-to-end handwriting recognition has been proposed. Attention mechanisms are often used in seq2seq model in machine translation and speech recognition task, [2,8]. In handwriting recognition tasks, attention mechanism allows control of text transcription without segmenting the image into lines as a pre-processing step. It can process an entire page as a whole, while giving predictions for every single line and words in it [10], see Fig. 2. In fact, although models in [1,6] achieving pretty good results in offline handwriting recognition, they are slow in training due to the LSTM layers involved. To overcome this drawback, Transformer have been introduced as efficient alternative solution to LSTM mechanism. In [9], the authors proposed a non-recurrent model for handwriting recognition based on transformers. It uses attention in spatial and contextual stages which allow to built character representation models while training language model at the same time. Using both recent deep-learning based handwriting recognition technologies adapted with transfer learning to specific handwriting pupils, and the ISKN Repaper interactive tablet, our proposition aims at establishing a comprehensive system improving children's handwriting in their mother French language.

### 2.2    Recent Application Programming Interface in Free Access

Different children's handwriting recognition API, that are now available in open-source have been tested and we measured their performances in the context of our industrial project of gamification. Both *Microsoft Ink Recognizer API*[3] and

---

[3] Microsoft Ink Recognizer API: https://www.iskn.co/fr.

*Cognitive Services Vision* provide a cloud-based REST API to analyze and recognize digital ink content. Unlike services that use Optical Character Recognition (OCR), the API requires digital ink stroke data as the input. Digital ink strokes are time-ordered sets of 2D points (X,Y coordinates) that represent the motion of the input tools such as digital pens or fingers. It recognizes both shapes and handwritten content from the input and returns all recognized features (containing position related to time) in JSON files. According to a series of tests, this API leads to very weak results for French text recognition on adult handwritings (less than 70% of correct character recognition for a set of adult handwriting styles). Based on the same technology, Google Cloud Vision (GCV)[4] provides more reliable results than its competitor:on the number of tests performed on children handwriting texts, the API managed to recognize children's handwriting in a fair way (about a 60% average rate for text recognition). Finally, on the recent MyScript Interactive Ink API[5], results outperform the others, even on children handwriting styles, but the input requirements are based on a drawing technology (called Digital Ink Recognition) which is not in the lineage with the way Study Project intends to acquire and process the data.

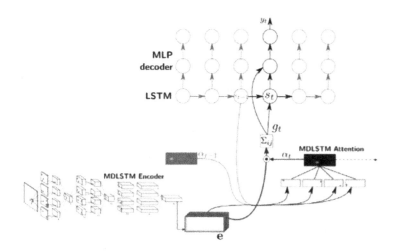

**Fig. 2.** Attention-based model for end-to-end handwriting recognition - MDLSTM-RNN for handwriting recognition [1]

# 3   Comprehensive Scenario for Tablet-Trainee Interactions

## 3.1   The Interactive Environment

**ISKN Repaper Tablet**. Using a complex interaction medium, we increase the difficulty to produce efficient online answers to the trainee. The ISKN Repaper

---

[4] Google Cloud Vision: https://cloud.google.com/vision.

[5] MyScript: https://www.myscript.com/interactive-ink/.

Tablet has been designed to respect the traditional way of writing on a sheet of paper with a standard pen. This latter consists of a tablet coupled with a pen: the tablet acts as a slate where a sheet of A5 paper is attached. The pen is the standard pen that we usually use (pencil, felt-tip or ballpoint pen) connected to the tablet by a magnetic ring, without battery and without electronics [7].

**A Playful Dictation-Based Scenario.** Dictation is a method used in French schools for learning the language and it is often done by the teacher in front of the students. In the Study project, dictation can be done in two ways: face-to-face or using an audio recording. Unlike face-to-face dictation alone, audio recordings allow the teacher to manage difference in levels between students and to create a new environment of concentration for each of them. The presence of the teacher and students in the classroom will maintain the necessary communication aspect, see Fig. 3 for the complete interaction cycle.

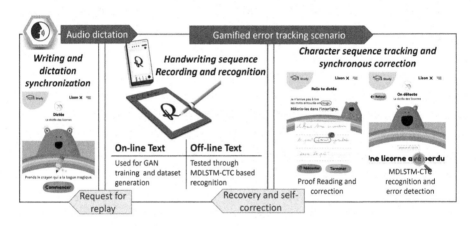

**Fig. 3.** Interactions cycle (from left to right) between trainee and tablet for an automatic and real-time handwriting spelling errors detection and correction.

**Children's Handwriting Complexity and Sparse Data.** Children's handwriting presents several variations and ambiguities, see Fig. 4. For a given child, a word, repeated several times, presents a large panel of intra-class variability. From one child to another, the variability becomes more complex, including: the writing style, thickness and curves, the alignment of words, the spelling accuracy. In addition, written production at this age is difficult to achieve. It is only possible to work with limited amounts of text and vocabulary.

### 3.2   Handwriting Recognition Pipeline

To interact with a child through a graphical interface and guide him in his language learning, an intelligent model is implemented. As mentioned in Sect. 2, deep learning methods are the most efficient and suitable models for our project. For the recognition architecture, we propose to use an adapted MDLSTM (HTR

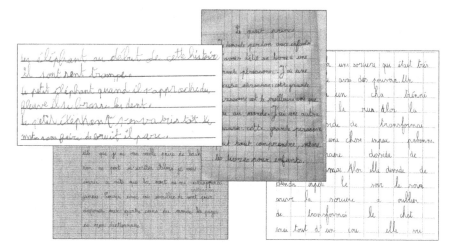

**Fig. 4.** Intra-class handwriting variations for trainees of primary classes, [13]

model) [12] that receives, as an input, an image of text and returns, as an output, a transcription and a reliability score. The model transforms the input image into a feature vector after a series of convolution operations, which are then passed as an input to a bidirectional recurrent neural network. Operations series of the recurrent network are then decoded by a decoder in order to produce a clean transcription.

The proposed architecture in [12] is adapted to our needs considering primarily two central key steps:

1. Data preparation and data augmentation for the training steps,
2. Correct and misspelling words alignment.

Let's notice that the project is designed for the French language. Data preparation consists of collecting French word images with their transcription with respect to IAM Dataset format described in [4]. For the first key step, due to the lack of annotated children data (with the current available ScolEdit[6] dataset), we investigated the preparation of a synthetic multi-writer dataset with GAN [5], which enable the generation of handwriting samples from online handwriting features. The standard architecture of the model used is illustrated in Fig. 2 stacking multiple layers of alternating convolution and multidirectional MDLSTM. A softmax layer with a Connectionist Temporal Classification (CTC) loss is finally used to handle the alignment between the input and the output sequence.

For the second key step, let's notice that children are very likely to make spelling errors and also to produce illegible sequences of characters. To localize errors or illegible characters/words, transcription results are (periodically) compared with the expected sequences of the dictation. A transcription that produces an incomplete, incorrect or illegible sequence (false correspondences according

---

[6] ScoleEdit, http://scoledit.org/scoledition/corpus.php.

to the Levenshtein distance) leads to an alert that is delivered through a customized answer adapted to the trainee via the application's interface (Fig. 3). We will not detail this part in the paper.

### 3.3   Software Environment of the STUDY Application

This section presents the interactions existing between the different software modules of the Study application, including the data circulation from the writing acquisition to its recognition, see Fig. 5.

The handwriting data are acquired by the Repaper-Unity device through the motion and pressure tracker of the magnetic ring (consisting of a magnet and a silicone). The information is collected in .json/xml files that sample the handwriting in a temporal sequence (50 position points x,y are captured per second). Concomitantly with the production of dynamic points, the ISKN-Repaper device produces bitmap images that are sent to the local HTR module.

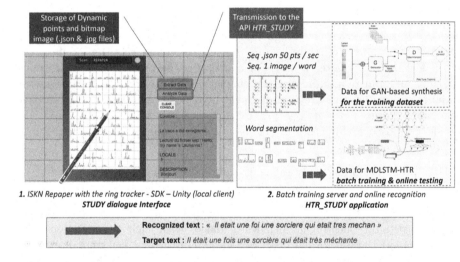

**Fig. 5.** Process order sequences : from the acquisition module via online Repaper-Unity application to the HTR module via local MD-LSTM recognition

The processing pipeline is composed of two major parts either for data acquisition and recognition. Due the lack of annotated data of handwriting children, we thought about exploring the power of GAN to learn and imitate children handwriting after series of acquisition form dictation. Extracted from the acquisitions coming from a specific child, online handwriting features are stored to feed the GAN model that will learn to imitates his style. Once the GAN is trained, we can generate augmented data (synthetic handwritings of the chosen style) to enhance offline handwriting recognition performance.

For the standard process of handwriting recognition, the student listens to the dictation text and writes corresponding words on one or several lines on a A5 paper sheet. After a defined time of waiting, the system of offline handwriting recognition processes the input and segments written words with respect to language and text lines. For each segmented word, a proposition with alignment is returned to the student and showed in its playful interface.

After describing how both parts of the pipeline works, it is convenient to illustrates minimum and recommended material configuration. The main hardware of the pipeline is the ISKN Repaper tablet. It can be coupled with a smartphone or tablet as an interactive interface of the playful application installed on it. For data processing (for GAN and MDLSTM-HTR models), the best choice is to use an integrated solution in the same mobile application. However, in order to reduce the processing load dedicated to smartphones or tablets, we can use external servers to process data as well. The playful mobile application runs under Unity 3D and data process models uses Python and Tensorflow.

## 4   Experimentations

### 4.1   Results Obtained on Synthetic Writings and IAM Dataset

The evaluation criteria for this application are twofold: evaluation of the playful learning scenario for children and performance evaluation of the integrated handwriting recognition system.

The procedure of collecting annotated data (text with transcription) requires schools administrative protocols and important amount of time. In order to study recognition module architecture feasibility, we generated data from a Generative Adversarial Network (GAN), producing synthetic text from different writing styles of adults [5], later improved by Alonso et al. [3].

Three major training experiments are proposed. The first one consists of studying the model trained with one style of handwriting generated by a GAN and tested on the same style. The second one considers three styles of handwriting generated by the GAN, two of which are used for training, and one for test. The third experiment consists of training the model with two styles of handwriting generated by a GAN and two styles of handwriting from the IAM Dataset. In addition, for each experiment considering synthetic data, the model has been tested with repeated sentences produced with a same style. We chose a sentence composed of six words, repeated five times to simulate intra-class variability, see Fig. 6. For each experience, we trained an MD-LSTM architecture which is an adaptation of the HTR model proposed in [12].

The architecture consists of processing the input image by a five-layer CNN each layer performs a convolution operation, which consist of applying a filter kernel of size 55 in the first two layers and 33 in the last three layers to the input image. Then, the feature sequence obtained after convolutions contains 256 features. A multi-directional MDLSTM with 4 LSTM hidden layers propagates relevant information through this sequence and provides more robust training-characteristics than a classical RNN. The output sequence is mapped to a matrix

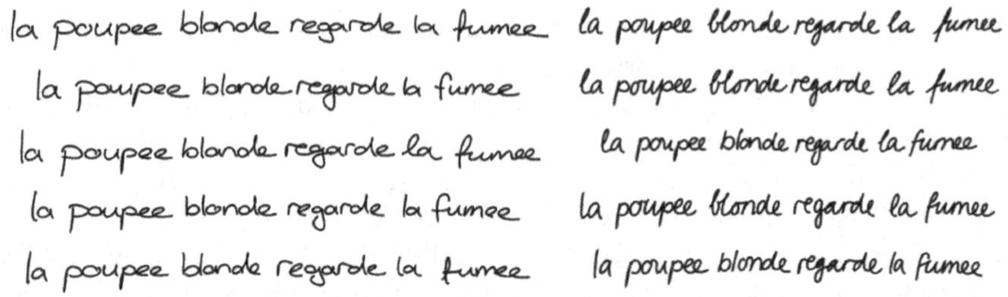

**Fig. 6.** Repeated sentence generated over two styles of GAN used for testing model against intra-class variability

of size 3280. At the end, CTC operations are required to train the MDLSTM during the training phase with pairs of images and GT texts. They operate on MD-LSTM output matrix through a loss function, evaluating the difference between the expected (ground-truth) text and the output (at most 32 characters long). For the inference part, the CTC decodes the output matrix into the final text.

For the first experience, we trained the HTR model [12], which is a modified version of the MDLSTM architecture, on 1566 word images from scratch (from Synthetic dataset). We divided the data into 95% training, 5% validation, we set the mini-batch size into 64 images and we apply Word Beam Search (WBS) algorithm as decoder, using a prefix tree that is created from a lexicon of words to constrain the words in the recognized text. The results that we obtained were very encouraging : WER (word error rate) of 9.38% and a CER (character error rate) of 2.58%.

For the second experiment, we trained the HTR model on 3077 word images from scratch. We divided the data into 95% training, 5% validation. We set the mini-batch size to 64 images and we apply WBS algorithm as decoder. Here again, we obtained very encouraging results: WER (word error rate) of 15.58% and CER (character error rate) of 17.41%.

The third style of GAN has been used for the tests. Within this set, we chose 26 random images of words for testing. The model succeeded in recognizing correctly 20 over the 26 word images, and on the 6 images not correctly recognized, one or two characters were missing. As for the first experiment, we also tested the robustness of the model towards a repeated sentence using three different styles of GAN, two styles used for training and one style reserved for testing. As a result, the model succeeded in recognizing 4/5 of the variability inter-class.

In the last experiment, we trained, from scratch, the HTR model on 3181 word images composed by data previously used in experiment two, plus 52 new images from two different styles of IAM Data-set. We divided the data into 95% training, 5% validation and we set the mini-batch size into 64 images and we apply WBS algorithm as decoder. The result obtained were very encouraging also, with a WER (word error rate) of 29.09% and a CER (character error rate) of 35.38%.

**Table 1.** Experimentation results of HTR model [12] on one style of GAN, two styles of GAN and two styles of GAN and IAM.

| Experimentations and results | | | | |
|---|---|---|---|---|
| | Human level | First exp | Second exp | Third exp |
| CER | 1% | 2.58% | 15.58% | 29.09% |
| WER | 1% | 9.38% | 17.41% | 35.38% |

From the Table 1 results, we can notice that increasing the variability (intra and inter-writer) causes a drop in performance compared to human recognition. In order to enhance them, important amount and balance of annotated data are highly recommended in addition to a customized deep-learning tuning strategy. Although actual experiments results are still far from human level, the model is very encouraging, especially because it quickly converges with very little data.

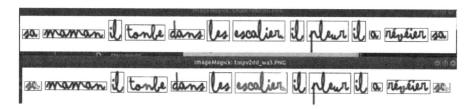

**Fig. 7.** Words alignment results according to three scenarios of spelling errors correction.

The last step of our system is the visual error tracking to the child interface. For errors detection and alignment, we used a simple French lexicon acting as a spellchecker. If a spelling error exists, we compare the number of characters between prediction and lexicon. At this stage, for the three error scenarios that we defined, we used color codes to indicate the type of detection: red for character recognition error, cyan for insertion of incorrect character and purple for missing characters. Regarding image word characters alignment, we proposed two approaches. The first approach is slow but precise: it consists of making a retro-prediction from transcription of the model to character images using sliding window crops. The second approach, the simplest and fastest, consists of cutting the image into vertical slices according to the number of characters obtained in the transcription, see Fig. 5.

## 4.2 Results on a Sample of Real Child Handwriting

Previous experiments were fully based on GAN data of several handwriting adult styles, whereas the Study project targets child handwriting. Because of the limitation of available children annotated data (children's data acquisition being still in progress), we are limited in our experiments with the SCOLEDIT dataset. However, the system trained on synthetic children writings produced by the GAN shows a promising efficiency when tested on real child writings, coming from scans of an elementary French school (50 random clean words well written by five different children) [13].

From those scans, we first extract handwriting words by a classical morphologial segmentation, then we passed them through the trained model inference for prediction. Due to changing style of handwriting, we used beamsearch decoder for more precision.

The results obtained on isolated words produced by young trainees are quite encouraging, see Table 2. The HTR-model trained on a couple of synthetic styles achieves promising preliminary results on five trainees handwriting styles, see Table 2. For the misspredicted words, several character (CER) are however well recognized: this paves the way to further research requiring adaptations on the training stage, with some combinations of learning datasets composed of potentially more abundant children synthetic data and limited real writings.

**Table 2.** Average WER and CER on a reduced dataset for five young trainees handwriting [13].

|      | Child 1 | Child 2 | Child 3 | Child 4 | Child 5 |
|------|---------|---------|---------|---------|---------|
| CER  | 22.5%   | 32.5%   | 19%     | 32.8%   | 32%     |
| WER  | 23%     | 50%     | 30.1%   | 40%     | 50%     |

## 5   Conclusion

In this paper, we present a new framework providing a comprehensive learning solution for children developing their writing skills in French language through an innovative device (Advanced Magnetic Interaction technologies of the AMI/ISKN Society, and the Extrapage software environment from Super Extra Lab) and traditional school material (textbook, paper, pencil...). We also focus on how we can recognize handwriting and localize spelling errors with high precision, by combining ISKN REPAPER tablet, SuperExtraLab's playful interface and a deep-learning offline handwriting recognition model. By integrating this framework with an educational dictation program, children will interact smoothly with their spelling errors, just after the writing period and most probably, at the period of their best concentration. The teacher will manage easily the progress of their pupils and adapt adequate solutions to their different profiles. We have illustrated, across several experiments on adult GAN styles, that a dedicated MD-LSTM architecture has a quite promising capacity to adapt to handwriting intra-variability and new writing. The results of these experiments are encouraging for future works. This ability to adapt to new scripts suggests that the model also allows adaptation to Latin languages other than French. The overall end-to-end scenario for recognizing children's handwriting dictations is technically sound and will require further investigations to improve the transfer learning from real data augmented with children GAN-synthetized writings. Study project concerns an end-to-end chain of handwriting recognition in a context of dictation. It is not the first industrial experience in the international educational sphere but it represents a first hardware and software integrative solution for training in spelling and writing at primary school with a first collaboration between two french start'ups, one dedicated to harware environment, and the other to the develoment of the gamified learning application.

# References

1. Bluche, T., Louradour, J., Messina. R.: Scan, attend and read: end-to-end handwritten paragraph recognition with MDLSTM attention. In: Proceedings of the International Conference on Document Analysis and Recognition (ICDAR), vol. 1, pp. 1050–1055 (2017). ISSN:15205363. https://doi.org/10.1109/ICDAR.2017.174
2. Chorowski, J., Bahdanau, D., Serdyuk, D., Cho, K., Bengio, Y.: Attention-based models for speech recognition. Adv. Neural Inf. Proces. Syst. **2015**, 577–585 (2015). ISSN 10495258
3. Eloi, A., Bastien, M., Ronaldo, M.: Adversarial generation of handwritten text images conditioned on sequences. In: Proceedings of the International Conference on Document Analysis and Recognition (ICDAR), vol. 1, pp. 481–486 (2019). ISSN 15205363. https://doi.org/10.1109/ICDAR.2019.00083
4. FKI: Research group on computer vision and artificial intelligence — computer vision and artificial intelligence (2002). https://fki.tic.heia-fr.ch/databases/iam-handwriting-database
5. Graves, A.: Generating Sequences With Recurrent Neural Networks. pp. 1–43 (2013). http://arxiv.org/abs/1308.0850
6. Graves, A., Fernández, S., Schmidhuber, J.: Multi-dimensional recurrent neural networks. Lecture Notes in Computer Science (including subseries Lecture Notes in Artificial Intelligence and Lecture Notes in Bioinformatics), 4668 LNCS (PART 1), pp. 549–558 (2007). ISSN 16113349. https://doi.org/10.1007/978-3-540-74690-4_56
7. ISKN: Tablette graphique Repaper: tablette papier crayon originale — iskn, 2020. URL https://www.iskn.co/fr
8. Jia, Y.: Attention mechanism in machine translation. J. Phys. Conf. Ser: **1314**(1), 012186 (2019). ISSN:17426596. https://doi.org/10.1088/1742-6596/1314/1/012186
9. Kang, L., Riba, P., Rusiñol, M., Fornés, A.: Mauricio villegas. pay attention to what you read: non-recurrent handwritten text-line recognition. arXiv, (2020). ISSN 23318422
10. Matcha, A.C.N.: How to easily do handwriting recognition using deep learning. (2020). https://nanonets.com/blog/handwritten-character-recognition/
11. Pham, V., Bluche, T., Kermorvant, C., Louradour, J.: Dropout improves recurrent neural networks for handwriting recognition. In: Proceedings of International Conference on Frontiers in Handwriting Recognition (ICFHR), pp. 285–290. December 2014. ISSN 21676453. https://doi.org/10.1109/ICFHR.2014.55

12. Scheidl, H., Fiel, S., Sablatnig, R.: Word beam search: a connectionist temporal classification decoding algorithm. In: 2018 16th International Conference on Frontiers in Handwriting Recognition (ICFHR) (2018)
13. Université Grenoble Alpes. SCOLEDIT - Etat du Corpus (2013). http://scoledit. org/scoledition/corpus.php

# ICDAR 2021 Workshop on
# Computational Paleography (IWCP)

# IWCP 2021 Preface

Computational paleography is an emerging field investigating new computational approaches for analyzing ancient documents. Paleography, understood as the study of ancient writing systems (scripts and their components) as well as their material (characteristics of the physical inscribed objects), can benefit greatly from recent technological advances in computer vision and instrumental analytics. Computational paleography, being truly interdisciplinary, creates opportunities for experts from different research fields to meet, discuss, and exchange ideas. Collaborations between manuscript specialists in the humanities rarely overcome the chronological and geographical boundaries of each discipline. However, when it comes to applying optical, chemical, or computational analysis, these boundaries are often no longer relevant. Computer scientists are keen to confront their methodologies with actual research questions based on solid data. Natural scientists open new perspectives by focusing on the physical properties of the written artefacts. In many cases, only a collaboration between experts from the three communities can yield significant results.

In this workshop, we aimed to bring together specialists from the different research fields analyzing handwritten scripts in ancient artefacts. It mainly targeted computer scientists, natural scientists, and humanists involved in the study of ancient scripts. By fostering discussion between the three communities, it facilitated future interdisciplinary collaborations that tackle actual research questions on ancient manuscripts.

The first edition was held in hybrid form on September 7, 2021, in Lausanne, Switzerland, in conjunction with ICDAR 2021. We had two invited speakers: Peter Stokes from Université PSL, France, and Sebastian Bosch from Universität Hamburg, Germany. The Program Committee was selected to reflect the interdisciplinary nature of the field.

For this first edition, we welcomed two kinds of contributions: short papers and abstracts. We received a total of 11 submissions. Each short paper was reviewed by two members of the Program Committee via EasyChair and five out of six were accepted. A single blind review was used for the short paper submissions, but the authors were welcome to anonymize their submissions. The five abstract submissions were evaluated and accepted by the organizers, and were published separately by the organizers in a dedicated website.

September 2021

Isabelle Marthot-Santaniello
Hussein Adnan Mohammed

# Organization

## General Chairs

Isabelle Marthot-Santaniello      Universität Basel, Switzerland
Hussein Mohammed      Universität Hamburg, Germany

## Program Committee

| | |
|---|---|
| Andreas Fischer | University of Applied Sciences and Arts Western Switzerland, Switzerland |
| Dominique Stutzmann | IRHT CNRS, France |
| Giovanni Ciotti | Universität Hamburg, Germany |
| Imran Siddiqi | Bahria University, Pakistan |
| Jihad El-Sana | Ben-Gurion University of the Negev, Israel |
| Marie Beurton-Aimar | LaBRI, University of Bordeaux, France |
| Maruf Dhali | University of Groningen, The Netherlands |
| Nachum Dershowitz | Tel Aviv University, Israel |
| Peter Stokes | Université PSL, France |
| Sebastian Bosch | Universität Hamburg,Germany |
| Volker Märger | Universität Hamburg, Germany |

# A Computational Approach of Armenian Paleography

Chahan Vidal-Gorène[1]([✉]) [iD] and Aliénor Decours-Perez[2]

[1] École Nationale des Chartes - Université Paris, Sciences & Lettres,
65 rue de Richelieu, 75002 Paris, France
chahan.vidal-gorene@chartes.psl.eu
[2] Calfa, MIE Bastille, 50 rue des Tournelles, 75003 Paris, France
alienor.decours@calfa.fr
http://www.chartes.psl.eu
https://calfa.fr

**Abstract.** Armenian Paleography is a relatively recent field of study, which emerged in the late 19th century. The typologies are fairly well established and paleographers agree on the distinction between four main types of writing in describing Armenian manuscripts. Although these types characterize clearly different and lengthy periods of production, they are less appropriate for more complex tasks, such as precise dating of a document. A neural network composed of a stack of convolutional layers confirms the relevance of the classification, but also highlights considerable disparity within each type. We propose a precise description of the specificities of Armenian letters in order to explore some other possible classifications. Even though the outcomes need to be further developed, the intermediate evaluations show a 8.07% gain with an extended classification.

**Keywords:** Armenian manuscripts · Computational paleography · Handwritings classification

## 1 Introduction

The Armenian language belongs to the Indo-European family [5] and is documented from the 5th century, when the monk Maštoc' created the alphabet in 405 AD. The original Armenian alphabet has 36 letters sorted in the Greek alphabetical order and according to the gradual introduction of characters specific to Armenian phonetics. Numbers are written using letters, as in Greek (from 1 to 9,000). Two letters were added in the Middle Ages and expanded the alphabet to 38 letters. There exists a well established classification made by Armenian paleographers. Four major types of writing have been identified and used to describe Armenian manuscripts in the major catalogues, namely the *erkat'agir*

Supported by the École Nationale des Chartes-PSL and the Calouste Gulbenkian Foundation. We especially thanks libraries for providing us manuscripts for this research.

E. H. Barney Smith and U. Pal (Eds.): ICDAR 2021 Workshops, LNCS 12917, pp. 295–305, 2021.
https://doi.org/10.1007/978-3-030-86159-9_20

script, "majuscule" or uncial [9]; the *bolorgir* script, word modeled on the latin *rotunda*, that refers to a minuscule script [9]; the *nōtrgir* script that refers to a notarial script, also named late minuscule [9]; and finally the *šłagir* script or ligatured cursive script [9]. It is very frequent to find manuscripts that combine several types: sometimes as a means structuring a document (e.g. with one script restricted to the title) but mostly within the same text, in "transition" periods between two different types of writing.

The names of the scripts are themselves attested in colophons of manuscripts, although their meaning is still open to debate. As soon as 1895, the Mekhitarist monk Yakovbos Tašean [10] introduced more detailed descriptions of the scripts: even though he kept the same terminology, he introduced formal nuances (e.g. "large", "medium", "thick" or "regular"), morphological details (e.g. "rounded" or "slanted") and chronological observations (e.g. "intermediate" or "transitional"); his extended terminology was also completed *a posteriori* with geographical origins (e.g. "eastern"). At times, these characterizations are used in catalogs without consistency or consensus on the terminology, due to their ambiguity and obvious subjectivity. However, they are seldom used and the description of manuscripts is very often limited to the distinction between the four main types of writing, without any further detail that could be helpful. Even though the relevance of the four types is uncontested, the profusion of names mirrors their great heterogeneity. It reveals the need for a more efficient classification, more appropriate to meet the challenges of cataloguing Armenian manuscripts, especially issues of dating. 40% of the manuscripts are not dated, due to the loss of their colophon [7], and the current classification does not provide enough criteria to date the documents on a paleographical basis.

In this article, we will begin to explore how to refine this classification through the measurement of the intra-class similarity of the four main types of writing. To that effect, we proceed to an automatic classification monitored by a CNN, following a proven approach tested on Latin scripts (see *infra* Sect. 3). The article will first shed new light on the paleographic study of Armenian writings and their specificities in the context of automatic processing. Then, it will assess the relevance of several groupings achieved on the basis of paleographical criteria. The CNN, whose intermediate outcomes we outline, has been trained on a dataset created for the purposes of this research.

## 2   Notions of Armenian Paleography

The classification created by Armenian copyists and later adopted by paleographers enables us to represent very distinct types of writings, with their own ductus and temporality.

The *erkat'agir* script is monumental. The letters are comprised of very thick downstrokes and thin connecting upstrokes, if not non-existant, which complicates reading and create ambiguities. As for Latin and Greek manuscripts of the first centuries, there are no opposition between majuscule and minuscule letters, yet, but a distinction is drawn between body text and initials.

These are capital letters, also found in lapidary inscriptions. It is a type of writing with no ligatures, the letters are in most cases distinctly separate. The script can be upright (*boloragic' erkat'agir*, see E1 in Table 1, shapes from manuscript W538) or slanted (*ułłagic' erkat'agir*, see E2 in Table 1, shapes from manuscript W539), the difference lies in the inclination of minims and the curvature of upstrokes. In the oldest manuscripts using this script all letters have the same height – between the baseline and the cap line (bilinear system) – and the same width, with very two exceptions: $p'$ and $k'$, which show ascenders and descenders reaching sometimes the previous or the following lines. With time, letters comprised of a horizontal stroke on the baseline tended to overreach and extend below the next letter. The letters are well drawn and particularly easy to read in the upright form, whereas in the smaller slanted form, the size reduction of letters and connecting strokes (sometimes barely suggested) impairs legibility. With time, the use of this type of writing decreases and is limited to highly sophisticated bibles (e.g. the W539 manuscript copied by T'oros Roslin in 1262), to rubrics and the first lines of the latest manuscripts.

The *bolorgir* script (see B in Table 1, shapes from manuscript W540) introduces a distinction between majuscule and minuscule letters (bicameral system), subsequently adopted by the *nōtrgir* and *šłagir* scripts. It is a book script. Sometimes copyists used the *erkat'agir* script for majuscules, sometimes new shapes were created. The stark contrast between downstrokes and upstrokes of *erkat'agir* letters diminishes. However, the minims are still significantly thicker than the crossbars,

**Table 1.** Armenian printed and handwritten alphabet

| E1 | E2 | B | N | Š | Maj. | Min. | Greek | Trans. |
|---|---|---|---|---|---|---|---|---|
|  |  |  |  |  | Ա | ա | $\alpha$ | a |
|  |  |  |  |  | Բ | բ | $\beta$ | b |
|  |  |  |  |  | Գ | գ | $\gamma$ | g |
|  |  |  |  |  | Դ | դ | $\delta$ | d |
|  |  |  |  |  | Ե | ե | $\epsilon$ | e |
|  |  |  |  |  | Զ | զ | $\zeta$ | z |
|  |  |  |  |  | Է | է | $\eta$ | ē |
|  |  |  |  |  | Ը | ը | - | ə |
|  |  |  |  |  | Թ | թ | $\theta$ | t' |
|  |  |  |  |  | Ժ | ժ | - | ž |
|  |  |  |  |  | Ի | ի | $\iota$ | i |
|  |  |  |  |  | Լ | լ | - | l |
|  |  |  |  |  | Խ | խ | - | x |
|  |  |  |  |  | Ծ | ծ | - | c |
|  |  |  |  |  | Կ | կ | $\kappa$ | k |
|  |  |  |  |  | Հ | հ | - | h |
|  |  |  |  |  | Ձ | ձ | - | j |
|  |  |  |  |  | Ղ | ղ | $\lambda$ | ł |
|  |  |  |  |  | Ճ | ճ | - | č |
|  |  |  |  |  | Մ | մ | $\mu$ | m |
|  |  |  |  |  | Յ | յ | - | y |
|  |  |  |  |  | Ն | ն | $\nu$ | n |
|  |  |  |  |  | Շ | շ | $\xi$ | š |
|  |  |  |  |  | Ո | ո | o | o |
|  |  |  |  |  | Չ | չ | - | č' |
|  |  |  |  |  | Պ | պ | $\pi$ | p |
|  |  |  |  |  | Ջ | ջ | - | ǰ |
|  |  |  |  |  | Ռ | ռ | $\rho$ | ṙ |
|  |  |  |  |  | Ս | ս | $\sigma$ | s |
|  |  |  |  |  | Վ | վ | - | v |
|  |  |  |  |  | Տ | տ | $\tau$ | t |
|  |  |  |  |  | Ր | ր | - | r |
|  |  |  |  |  | Ց | ց | - | c' |
|  |  |  |  |  | Ւ | ւ | $\upsilon$ | w |
|  |  |  |  |  | Փ | փ | $\phi$ | p' |
|  |  |  |  |  | Ք | ք | $\chi$ | k' |
| - | - |  |  |  | Օ | օ | - | ō |
| - | - |  |  |  | Ֆ | ֆ | - | f |

producing a very uniform overall appearance. Henceforth, it was a quadrilinear writing system, with tilted and slanted letters, comprised of ascenders and descenders. The letters in themselves are easy to decipher and do not raise particular problems, but ligatures appear. *Ew* and *mn* are the most common (see Table 3, *bolorgir*, miscellaneous). Copyists sometimes innovated and introduced new ligatures of their own making (e.g. a pair of letters with ascenders joined at the top).

The *nōtrgir* script (see N in Table 1, shapes from manuscript M1495) differs from *bolorgir* by a significant change in letter morphology, so that paleographers identify it as a new ductus [9]. The acceleration of the copying process of *bolorgir* was carried further: the script is cursive, fine, and minims are now often limited to simple dots, whereas ascenders and descenders are excessively long compared to the body of the letter (e.g. see Table 3). *Bolorgir* majuscules are preserved. New ligatures appear and the horizontal stretching of some letters results in disproportions (e.g. see Table 3). Depending on the diligence of the copyist, numerous shapes can be ambiguous: for instance, *z*, *ł* and *š* don't always differ from one another. It is a script type difficult to read and that introduces numerous recognition difficulties (innovations, ambiguities, etc.).

Lastly, the *šłagir* script (see S in Table 1, shapes from manuscript MAF68) finalized the additions of *nōtrgir*. They share the same appearance with thinner strokes for *šłagir*, leading to the frequent confusion between the two types. This script favors non formalized ligatures and abbreviations (see Table 3, *šłagir*, miscellaneous).

The *erkat'agir* script, used for inscriptions, is prevalent, at least in the written production that has reached us, until the 12th century (then limited to rubrics, titles, etc.). From that date, the *bolorgir* script was widely adopted and its use extends until the 19th century. The *nōtrgir* and *šłagir* scripts appeared respectively in the 15th and 19th century [1,9]. These periods of prevalence do not exclude earlier or later evidence [1]. It is worth noting that since the first use of printing for Armenian (first book printed in Venice in 1512), *erkat'agir* has been used for majuscule letters, *bolorgir* for plain roman minuscule and *nōtrgir* for italic minuscule. This system continued in the digital age, except for more unusual fonts.

*Abbreviations and Ideograms:* We find a huge amount of abbreviations in Armenian manuscripts, originally restricted to nomina sacra as in Latin or in Greek. Abbreviations are created through contraction using only the first and last letters of the word, and adding a horizontal stroke above (*badiw* and sometimes a double accent). Very early on, abbreviations multiplied and took on various forms: still created through contraction, they may include some other consonants in the abbreviated word (see Table 3, *nōtrgir*). The letters of the abbreviated word are sometimes ligated. In *nōtrgir* manuscripts, we can observe a very high density of abbreviations. Starting with *bolorgir* manuscripts, we also find ideograms that transform a whole word into a figure. These ideograms can be declined, adding the inflection to the figure, adding complexity for other task such as HTR [12]. There is a very wide range of ideograms (more than 500 words, with sometimes three different ideograms for a same word) that cover a very large lexicon. Abbreviations and ideograms can be a reliable criterion to date a manuscript or to identify a script.

# 3   A Dataset for Armenian Script Identification

There is no dataset for the analysis and classification of Armenian manuscripts, to this day. Only one dataset is proposed by Calfa, focused on the description of medieval Armenian handwritten characters and annotated with the Calfa Vision platform [12], that enables different customized levels of annotation on handwritten documents, here at the character level as with DigiPal (now Archetype) [2]. We have therefore adopted a methodology similar to that used for the creation of the datasets CLAMM 2016 [4] and 2017 [3]. We have built an internal dataset of 2,300 images (D-ARM), from 190 manuscripts: 500 images come from the Calfa database to include older witnesses, and 1,800 have been newly added (a mix of thesis corpus and new images). The dataset covers a period from the 10th to the 20th century and approximately 230 hands (Table 2). Most of the images are copyrighted and used within the scope of partnerships we have with libraries.

**Table 2.** Manuscripts provenance

| Provider | Manuscripts | Scripts | Images |
|---|---|---|---|
| Gallica (Bibliothèque Nationale de France (BnF), Paris) https://gallica.bnf.fr | 80 | 1–4 | 1,386 |
| Matenadaran (Erevan) | 26 | 1–4 | 112 |
| BVMM (Musee Armenien de France-IRHT, Paris) https://bvmm.irht.cnrs.fr | 23 | 1–3 | 368 |
| Congregation of Mekhitarist Fathers (Vienna) | 20 | 1 | 49 |
| StaatBibliothek zur Berlin (Berlin) https://digital.staatsbibliothek-berlin.de | 18 | 2–4 | 195 |
| Congregation of Mekhitarist Fathers (Venice) | 11 | 2–3 | 49 |
| Walters Art Museum (Baltimore) https://www.thedigitalwalters.org | 7 | 1–2 | 110 |
| Armenian Patriarcate of Jerusalem (Jerusalem) | 4 | 2–3 | 23 |
| British Library (London) https://www.bl.uk/manuscripts/ | 1 | 2 | 8 |
| Total | 190 | | 2,300 |

We have primarily collected the samples on non religious manuscripts, to limit the number of manuscripts copied with great care and consistency. Manuscripts come from 9 libraries: the Bibliothèque Nationale de France (80), the Matenadaran (26), the Musée Arménien de France-IRHT (23), the Congregation of Mekhitarist Fathers of Venice (11) and Vienna (20), the StaatsBibliothek zu Berlin (18), the Walters Art Museum (7), the Armenian Patriarcate of Jerusalem

(4) and the British Library (1). That corresponds on average to 12 pages per manuscript, non-consecutive and mixing different sections of manuscripts (cover pages, title-pages, colophons, etc.) thus limiting redundancy. The images are tagged according to 4 classes (1 = *erkat'agir*, 2 = *bolorgir*, 3 = *nōtrgir*, 4 = *šłagir*), following the traditional typology. We have strictly replicated the descriptions featured in the catalogs, except in the case of manifest error.

The classes are not uniformly distributed (13% for class 1; 39% for class 2; 38% for class 3 and 8% for class 4) – which is consistent with the distribution of these scripts within the digital libraries – subsequently we created 3 sub-datasets to have an equal distribution (156 images for each class). Each sub-dataset (D-ARM$_i$) is composed of training data (624 images) and test data (1,546 images), completely separated and randomly generated. Each sub-dataset is evaluated independently from the others for a cross-validation of the results. The size of the test data allows for a very wide variety of unseen variations of a script.

**Table 3.** Overview of the intra-class and inter-class heterogeneity within the dataset (non exhaustive list). *Sources: mss Walters Art Museum (W537, W538, W539, W540), Bibliothèque Virtuelle des Manuscrits Médiévaux (BVMM) – IRHT-CNRS (MAF52, MAF67, MAF80), Bibliothèque nationale de France-Gallica (P52, P55, P79, P118, P119, P132, P135, P178, P193, P206, P207, P208, P218, P220, P236, P242, P304, P317, P324), Staatsbibliothek zu Berlin - PK (BER Petermann I 143, BER or. oct. 282, BER or. oct. 341)*

| Word | Erkat'agir | Bolorgir | Nōtrgir | Šłagir |
|---|---|---|---|---|
| ŭnpш | | | | |
| ŭngш | | | | |
| ŭnuш | | | | |
| (j)unшɋ | | | | |
| Miscellaneous (h, ŭ, ш, ɥ, ɋ and ե) | | | | |

Images offer different resolutions (from 300 to 600 DPI) and consist of grey-level images in TIFF format. Most of the BnF images consist in digitized microfilms. For each image, automatic semantic segmentation to localize text-regions has been performed on Calfa Vision and manually proofread. Neither the images have been straightened, nor enhanced, that constitute a bias in our experiment.

A first look at the dataset highlights the great diversity in the morphology of the letters included (see Table 3). The framed examples show notably the mix of shapes, that is representative of transitional writing. *Erkat'agir* displays numerous various shapes and shares sometimes with the *bolorgir* script some letters such as *a*. Many more similarities exist between the three other bicameral scripts. One of the more selective criteria, obvious in Table 3, is the higher proportion of ligatures in the most cursive shapes. We also notice words that don't follow the bilinear or quadrilinear system of their class (see *supra* Sect. 2). The *šłagir* script appears to be less codified, like a modern cursive script, than a formally defined type of writing, as it is classically the case in the ancient manuscripts. The table doesn't display all the variations present in our dataset or in Armenian.

# 4 Experiments and Discussion

*Architecture and Training Parameters:* The classification task was operated by a neural network comprised of several convolution layers [4]. This is a simplification of the VGG16 architecture [8] used by M. Kestemont for DeepScript [6]. The model takes the form of a stack of convolutional layers with a ReLU activation function, each with a $3 \times 3$ perceptive field and an increasing number of filters ($2 \times 64 > 3 \times 128 > 2 \times 256$), followed by a MaxPooling with a size of $2 \times 2$, two fully-connected dense layers with a dimensionality of 1048, and a softmax layer. This streamlined version is more effective on our dataset than the deeper version proposed by DeepScript, which we reproduce for the main parameters. We use an adaptive learning rate starting from 0.001, the RMSprop optimizer proposed in Keras and a batch of 100 images. 90% of our training dataset is used for the training process and 10% for validation.

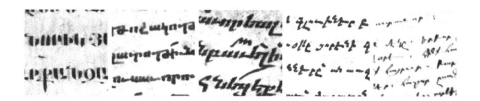

**Fig. 1.** Examples of generated patches for class 1 (ms W549), class 2 (mss MAF50 and MAF52), class 3 (ms MAF68), and class 4 (fragment from Venice)

*Data Augmentation:* Due to the small size of the dataset and the great disparity of image resolutions, we also follow the strategy adopted by DeepScript regarding data augmentation and disruption [6]. We especially apply resizing depending on the original resolution of the images, in order to smooth out discrepancies in image resolutions and to limit the influence of the size of the script or of text density (see Fig. 1). In total, the network will be faced with 62,400 patches each new epoch.

*Results:* The three evaluations provide convergent results, with a clear improvement for the D-ARM-2 (see Fig. 2). The classification established by paleographers is clearly verified. A clear distinction between the four types of writing can be observed despite the very different datasets for classes 2 and 3. The global accuracy is 82.97%. This outcome is convergent with the accuracy achieved by DeepScript on Latin Script [4], that was 76.49%, but with three times as many classes to predict. The original architecture of DeepScript, that is with a convolution layer of 256 additional filters, here, achieves an average accuracy of 76.86%.

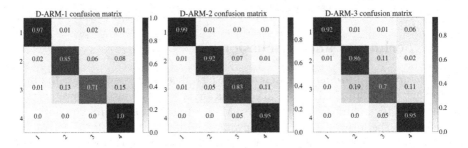

**Fig. 2.** Confusion matrix for D-ARM-1, D-ARM-2 and D-ARM-3

*Generalities:* Classes 1 and 4 appear to be the most discriminating, and class 3, conversely, presents the most difficulties. If the distinction between class 1 and every other appears consistent regarding the morphology and ductus of letters, that are starkly different; from a paleographical point of view, there is no obvious reason for such a variance in class 3, in particular with class 2. Following Table 3, we could expect more confusion between classes 3 and 4. The latter surprise may be explained in part by the difficulty of labelling *šłagir*, often mistakenly catalogued as *nōtrgir* – the model then correcting the mistake.

Class 1 is accurately identified. Some very marginal errors can be observed for datasets D-ARM-1 and D-ARM-3, the mistakes are located on a set of patches in marginal notes, often in less formal writing, or in very damaged areas (data extracted from manuscript fragments used as binding material).

Class 2 seems to be confused with class 3, and less frequently with classes 1 and 4. The confusion with class 1 coincides with the existence of a mixed script (see Fig. 1 and Table 3). Patches with *erkat'agir* titles or majuscules (see *supra* Sect. 2) are not misclassified. More generally, *bolorgir* is not cursive and comprises thicker strokes without being monumental either. The use of a thinner writing tool, similar to the pen used for *nōtrgir* and *šłagir*, could be the source of the error (see Table 3).

Class 3 has the worst classification score. It is the least uniform class: it often shares its ductus and morphologies with class 2 (e.g. letters *y*, *c* and *l*, Table 1), even though the script is more cursive and has its own ductus for other letters (e.g. *j* and *t*, Table 1). With 19% of confusion for D-ARM-3, we can reasonably rule out the coincidence factor for a patch lacking different scripts.

Concerning class 4, the network is undermined by errors of cataloguing, but class 4, which incorporates widely varying but very cursive scripts, is still well recognized under all scenarii.

*Cross-Entropy Loss:* Although the accuracy provides information that is easily interpreted by paleographers, it is worth noting that the loss function, calculated at each epoch (train loss and validation loss) remains high (0.3 at epoch 30 on average), even when increasing the complexity of the model or the number of data. The training and validation sets are widely different and each epoch has his own patches randomly generated. This resilience of the loss score may be explained by the significant data heterogeneity in each class.

*Which New Classes to Consider?* Class 1 has already at least two distinct subclasses (see *supra* Sect. 2) that we keep. The paleographical studies on the letter *A* [11] show some evolutions of the ductus that could be the basis for new types. The so called mixed-script between *erkat'agir* and *bolorgir* is also a strong candidate. In our dataset, there is a mix of *boloragic' erkat'agir* (E1) and *ułłagic' erkat'agir* (E2), kept for creating two new classes. As shown in Fig. 1, we obtain a lot of variations inside each class. We start from misclassified images and this variety to experiment with two new classifications. The first one is composed of 10 classes: (1) *boloragic'erkat'agir*, (2) *ułłagic' erkat'agir*, (3) a rounded and compact *erkat'agir*, (4) the mixed-script between *erkat'agir* and *bolorgir*, (5) *bolorgir*, (6) a rounded *bolorgir*, (7) a mixed-script between *bolorgir* and *nōtrgir*, (8) *nōtrgir*, (9) a more cursive *nōtrgir*, and (10) *šłagir*. The lack of data could strongly penalize these classes, so we also tried an other simpler distribution: (1) *boloragic' erkat'agir*, (2) *ułłagic' erkat'agir*, (3) the mixed-script between *erkat'agir* and *bolorgir*, (4) *bolorgir*, (5) a mixed-script between *bolorgir* and *nōtrgir*, (6) *nōtrgir*, and (7) *šłagir*.

Experiment 1 presents a tendency, but inconclusive (see Fig. 3). The lack of data for the classes 3, 6, 7 and 9 creates additional noise and the final accuracy achieved (86.6%) is impacted. Experiment 2 is more relevant with a robust model (91.04%) and a loss under 0.25 at epoch 30 (see Fig. 3). Class 2 suffers from a shortfall of quality data (from damaged fragments), but class 5 seems also more reliable and ensures a better intra-class inertia for *nōtrgir*. The new classification gain is 8.07% with three additional classes, even if the tendency should be confirmed in future work. Aside from the morphology of the letters or the study of their ductus, there are other relevant criteria to identify the type of writing or, more generally, to classify Armenian manuscripts. Their layout is a good indicator even if the correlation between text layout and script is not self-evident. We can simply observe that text density increases when the minuscule scripts appear. Often, the medium allows to discriminate *erkat'agir* from the other scripts. Abbreviations, ideograms or specifics punctuation marks could prove to be an interesting avenue.

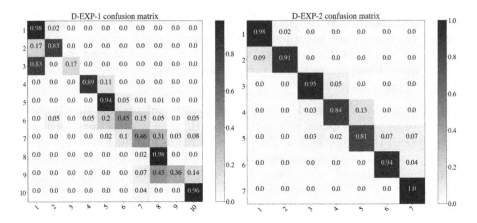

**Fig. 3.** Confusion matrix for experiments 1 and 2

# 5   Conclusion

Armenian paleography constitute a new field of study within the scope of Digital Humanities, with only very recent developments for HTR. Manuscripts datation notably is an open issue, where a new classification may be needed. The exploratory approach we have undertaken highlights simultaneously the homogeneity of main Armenian scripts and their considerable intra-class heterogeneity. A large variance is evidenced for the scripts that became prevalent after the 12th century, which are also the most widely used. We have provided some directions for possible classifications without a full assessment due to the lack of data. If the existing classification is appropriate, a potential extension appears to be relevant. Future work will focus on the constitution and diffusion of a representative and open dataset, for which different classification scenarii will be considered. The supervised approach that we have adopted may also be compared with an unsupervised clustering. Going forward, it will be useful to discern which are the points of interest of the neural network in order to visualize its criteria for classification and to put them into perspective with a paleographical approach. Lastly, another aspect that we intend to study in the future is the relevance of a new paleographical-based classification for the improvement of HTR models for Armenian.

# References

1. Ačaṙean, H.: Hayoc' gir (= Armenian letters). Erevan (1928)
2. Brookes, S., Stokes, P.A., Watson, M., De Matos, D.M.: The DigiPal project for European scripts and decorations. Essays Stud. **68**, 25–59 (2015)
3. Cloppet, F., Eglin, V., Helias-Baron, M., Kieu, C., Vincent, N., Stutzmann, D.: ICDAR 2017 competition on the classification of medieval handwritings in Latin script. In: 2017 14th IAPR International Conference on Document Analysis and Recognition (ICDAR), vol. 1, pp. 1371–1376. IEEE (2017)

4. Cloppet, F., Eglin, V., Stutzmann, D., Vincent, N., et al.: ICFHR 2016 competition on the classification of medieval handwritings in Latin script. In: 2016 15th International Conference on Frontiers in Handwriting Recognition (ICFHR), pp. 590–595. IEEE (2016)
5. Hubschmann, H.: Uber die Stellung des Armenischen im Kreise der indogermanischen Sprachen. Weimar (1875)
6. Kestemont, M., Christlein, V., Stutzmann, D.: Artificial paleography: computational approaches to identifying script types in medieval manuscripts. Speculum **92**(S1), S86–S109 (2017)
7. Mahé, J.P.: Une approche récente de la paléographie arménienne. Revue des Études Arméniennes **30**, 433–438 (2005–2007)
8. Simonyan, K., Zisserman, A.: Very deep convolutional networks for large-scale image recognition. arXiv preprint arXiv:1409.1556 (2014)
9. Stone, M.E., Kouymjian, D., Lehmann, H.: Album of Armenian Paleography. Aarhus University Press, Aarhus (2002)
10. Tašean, Y.: C'uc'ak hayerēn jeṙagrac' Matenadaranin Mxit'areanc' i Vienna (= Catalog der Armenischen Handschriften in der Mechitharisten Bibliothek zu Wien) (1895)
11. Vidal-Gorène, C.: Notes de paléographie arménienne à propos de la lettre ayb. Revue des Études Arméniennes **39**, 143–168 (2020)
12. Vidal-Gorène, C., Dupin, B., Decours-Perez, A., Thomas, R.: A modular and automated annotation platform for handwritings: evaluation on under-resourced languages. In: Lladós, J., et al. (eds.) ICDAR 2021. LNCS, vol. 12823. Springer, Cham (2021). https://doi.org/10.1007/978-3-030-86334-0_33

# Handling Heavily Abbreviated Manuscripts: HTR Engines vs Text Normalisation Approaches

Jean-Baptiste Camps$^{(\boxtimes)}$ [iD], Chahan Vidal-Gorène[iD], and Marguerite Vernet

École Nationale des Chartes – Université Paris, Sciences and Lettres,
65 rue de Richelieu, 75002 Paris, France
{jean-baptiste.camps,chahan.vidal-gorene,
marguerite.vernet}@chartes.psl.eu
http://www.chartes.psl.eu/

**Abstract.** Although abbreviations are fairly common in handwritten sources, particularly in medieval and modern Western manuscripts, previous research dealing with computational approaches to their expansion is scarce. Yet abbreviations present particular challenges to computational approaches such as handwritten text recognition and natural language processing tasks. Often, pre-processing ultimately aims to lead from a digitised image of the source to a normalised text, which includes expansion of the abbreviations. We explore different setups to obtain such a normalised text, either directly, by training HTR engines on normalised (i.e., expanded, disabbreviated) text, or by decomposing the process into discrete steps, each making use of specialist models for recognition, word segmentation and normalisation. The case studies considered here are drawn from the medieval Latin tradition.

**Keywords:** Abbreviations · Handwritten text recognition · Medieval western manuscripts

## 1 Introduction

### 1.1 Abbreviations in Western Medieval Manuscripts

In medieval Latin manuscripts, abbreviations are fairly common and follow a practice that was established, by and large, during the first centuries A.D., reserved for a time to administrative and everyday written production and then extended to literary manuscripts [10,11]. They mostly derive from two antique conventional systems: *notae antiquae*, on the one hand, that proceed by suspension, superscript letters or tachygraphic signs, and mostly affect grammatical morphemes such as inflections, adverbs, prepositions, pronouns and the forms of the verb *esse*; Christian *nomina sacra*, on the other, abbreviations of holy names, by contraction [1]. Extended by Irish monks, with the addition of new (insular) signs such as ÷ (*est*) and then standardised and generalised by the

© Springer Nature Switzerland AG 2021
E. H. Barney Smith and U. Pal (Eds.): ICDAR 2021 Workshops, LNCS 12917, pp. 306–316, 2021.
https://doi.org/10.1007/978-3-030-86159-9_21

Carolingian *renovatio*, this system forms the basis for abbreviation practices in medieval Latin manuscripts, but also for the abbreviations of many vernaculars. They were notably adapted to Old French by Anglo-Norman scribes [10]. Inherited from this history are abbreviations that can be categorised as

**tachygraphic sign** e.g. Tironian ⁊ (*et*) or ꝯ (*cum, con-, com-,*...).

**superscript letter** e.g. superscript *a* for *ua* or *ra* in q̃ (*qua*), t̊ns (*trans*).

**suspension** e.g. ẽ (*est*).

**contraction** e.g. D̃S (*Deus*).

In the 12<sup>th</sup> and 13<sup>th</sup> centuries, the intellectual flourishing and the development of schools and universities caused a heavy demand for written artefacts. The copying of manuscripts expanded beyond the sole framework of monastic *scriptoria* and spread to the city in professional workshops and lay *scriptoria*. The development of a larger literate milieu of students and masters, and the growth of book production led to modifications in intellectual practices and ultimately in the processes of reading and writing. A switch, at least in scholastic milieus, from slow syllabic reading to faster, expert modes of reading, based on the global perception of each word, led to a very significant increase in word-level abbreviations, and, specifically, abbreviations by contraction [10]. They display much variety and include:

**simple contraction** using letters from the original word (with a marker to make the presence of the abbreviation explicit) that can be relatively **unambiguous**, e.g. eccl̄a (*ecclesia*), rōe (*ratione*) or **ambiguous**, e.g. i̊ for *ita, illa* or *infra*, or even sometimes *prima* or *una*, depending on the context.

**composite contraction** combining other conventional devices with the contraction itself, for instance p̊ (*persona*).

In Latin manuscripts, we already encounter a great versatility of signs and many homographic abbreviations (or alternative expansions), a situation that is made even more uncertain in vernacular manuscripts, due to spelling variation.

## 1.2   Expanding Abbreviations

Expanding abbreviations is not a trivial task because there is no unambiguous character-, syllable- or word-level mapping between abbreviation and expansion: the same abbreviation can correspond to several expanded forms, and an expanded form can have several abbreviations. In addition, the same sign can fulfil alternative functions on different levels. Attempts to model the relationship between abbreviations and expanded forms exist on a theoretical level in linguistic research [16], but the situation remains complex. In graph theory terms, the binary relation between the set of abbreviations, and the set of expanded forms can be characterised as a many-to-many relation and not as a function. On an individual level, a sign can have

**one character expansion** e.g., ⁊ → *et* (word) or -*et* (word syllable) in Latin; while on the contrary ⁊ → *et, e, ed* (Old French).

**multiple character expansion** 9 → *cum* (word) and *cum-*, *con-*, *com-* (prefix) in both Latin and Old French.

The same is also true at the level of the sequence of signs, that can have

**one expanded form** e.g., rõe → *ratione* (Latin).
**many expanded forms** ĩ → *ita, illa, infra, prima, una...* (Latin).

The same sign can enter into different relations both in isolation and as part of groups. A good example of this plasticity is given by the common abbreviative mark known as 'titulus' or 'tittle': alone, it can be used to stand for a nasal consonant (*-m-* or *-n-*), while it is also the most common marker to indicate that a word is globally abbreviated, having, in that case *no explicit character value* per itself, for instance in the aforementioned rõe example. The actual incidence of abbreviation polyvalence varies in time, between languages and language variants, as well as per types of documents or texts, and ultimately, scribes.

### 1.3   Computational Approaches

Handling abbreviations is a general problem with manuscripts, especially medieval manuscripts, but we find relatively few studies dealing with computational approaches to their expansion. Romero et al. report on recording both diplomatic transcriptions (with abbreviations) and normalised (expanded) transcriptions of Dutch Medieval manuscripts, through the use of XML/TEI, but give only results for the first version [19].

The problem of homograph abbreviations and the versatility of signs seems to call for a representation of context. In practice, two main kinds of approaches have been used: HTR systems trained on normalised data on one hand; treating abbreviation expansion as a text normalisation task on the other.

HTR approaches can include some representation of token context, because state-of-the-art HTR systems usually take into account the full text line. From a pragmatic perspective, this makes the creation of ground truth easier, because it facilitates the reuse of existing transcriptions and has been investigated for this very reason, yet tended to show relatively high character error rate (CER), where 'deletions' (including letters that should have been added as part of abbreviation expansion) represent more than half of the errors [2,25].

Alternatively, normalisation can be treated as a separate (posterior) normalisation task, based on the output of the HTR phase. The literature concerning historical text normalisation is considerably larger, and includes approaches based on substitution lists, rules, as well as distance-based, statistical approaches (in particular, character-based neural machine translation) and more recently neural models [4]. To include a modelling of context, normalisation systems can reuse deep-learning architectures originally intended for neural machine translation [4,9] or lemmatisation [6,15].

In this paper, we plan to explore two approaches to expand abbreviations in Latin manuscripts, with and without post-processing. Evaluations are carried out with a small dataset in order to highlight benefits of each approach within the scope of an under-resourced language.

## 2    MS BnF lat. 14525

For this work, we used MS BnF lat. 14525, the subject of an ongoing master's thesis [22]. It belonged to the library of Saint-Victor of Paris, a canonical Abbey that played a central role in the intellectual life of Paris, especially during the 12th century, and was situated at the intersection between the monastic and Parisian worlds.

BnF lat. 14525 was produced for the library of Saint-Victor in the first half of the 13th century. It is not made up of a single codicological unit, but was completed and improved about ten years later. This manuscript of 305 folios brings together various texts from different origins (Victorian, Cistercian, Parisian schools) and includes spiritual treatises, works on practical theology, numerous sermons, Constitutions of the Fourth Lateran Council or synodal constitutions, and even particular material and spiritual privileges. Despite this heterogeneity, it is a very useful item for a better understanding of Saint-Victor during this time. About ten different hands wrote it, though the handwriting is quite similar. We are dealing here with a script akin to scholastic writing, with a few broken stems, a reduced module, as well as numerous abbreviations.

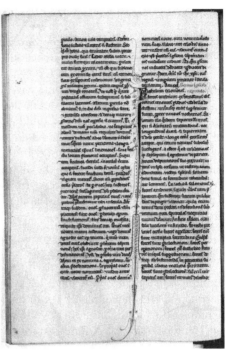

**Fig. 1.** Folio 45v, MS BnF lat. 14525

We focus here on one of these hands, which occupies a quarter of the manuscript, i.e. 79 fols, copied on two columns of 42 ruled lines (the first of which also carries writing), most likely between 1215 and 1225, on parchment of relatively good quality despite some defects. The copy is neat with few errors, and corrections by expunctuation or crossing out, as well as some interlinear or marginal additions. Hyphens and dotted 'i' s are also present, although irregularly. The writing itself is very regular and skilled, and the use of abbreviations is quite essential. The density of abbreviations, measured as the ratio of words with at least one abbreviation, is around 47%, not out of the ordinary for this type of manuscript, but much larger than what is found for instance for contemporary Old French epic manuscripts, with figures in the 10 to 20% range [5]. We find the systematic abbreviation of *et* in two forms: & and ꝯ. The copyist uses different tittles: ꝰ mostly *er*, while ˗ stands for nasals or signals suspensions or contractions. ꝰ is used for *ur* and ꝰ for *us*. Superscript letters are also abundantly used, in particular *a*, *o* and *i*. This frequent use of abbreviations is typical of the kind of scholastic writing whose use was widespread from the mid-13$^{th}$ century in university circles.

Images of the manuscript are available on Gallica as a grayscale digitisation of the microfilm [3].

## 3   Experiments and Results

The paper aims to compare two approaches to decipher Latin abbreviations, and HTR experiments have been carried out with two neural architectures. The first one (HTR-CB) is proposed by Kraken [13]. The results produced by this architecture serve as a baseline for pure character-level recognition of this manuscript and are provided to other modules in the defined pipeline (see *infra* 3.3). The second architecture (HTR-WB) is an adaptation of that proposed on Calfa Vision [24], originally developed for the reading of medieval Armenian manuscripts and the management of abbreviations and ideograms specific to this language and which cannot be recognised at the character level [23]. Character recognition is preceded by a word-based system, to which we first provide an exhaustive list of abbreviations encountered in the manuscript. The results of this architecture serve as a point of comparison.

### 3.1   Ground Truth Creation

Training and testing data have been annotated with a layout analysis and a baseline model, and manually proofread via eScriptorium [12,14]. The dataset is composed of 1,861 lines of text (1,524 reserved for training, 168 for validation and 169 for testing). We built a total of four datasets:

1. **D-exp**, consisting of a transcription with full expansion of the abbreviations. We consider two variants, **D-exp1** with inter-word spaces restored (separation of words according to Latin grammar and not the spacing present in the manuscript), and **D-exp2** without spaces. These datasets are respectively composed of 34 and 33 classes. The number of classes has been limited to include enough samples in each. This can lead to strong intra-class inertia, due in particular to the unsystematic grouping of upper- and lower-case letters for the less endowed classes.
2. **D-abb**, composed of transcriptions registering the abbreviation system used in this manuscript (see *supra* 1.1). We also consider two variants, **D-abb1** with inter-word spaces, and **D-abb2** without spaces. These datasets are respectively composed of 60 and 59 classes. In detail, we have 36 classes representing alphanumeric characters and punctuation marks, 10 classes specific to Latin paleography to represent certain abbreviations according to the same scheme as the ORIFLAMMS project (e.g. ƀ and ꝗ; see *infra*), and 13 classes made up of combining signs (written above or below one or more letters). These last 13 classes can be difficult to identify for an HTR engine.

Moreover, additional data was used for some of the trainings:

1. **Oriflamms** diplomatic and allographetic transcriptions provided by D. Stutzmann and team [7,18,20,21].
2. **PL** 216 volumes of normalised editions from Migne's *Patrologia latina* [17].

### 3.2 HTR on Abbreviated and Expanded Data

Common parameters have been chosen for training steps of HTR-CB and HTR-WB. Line images provided as input are resized to 64px in height and have varying widths. No Unicode normalisation or data augmentation is applied. A repolygonisation has been performed to equalise polygons of first and last rows of each text columns [24]. Due to the small size of the dataset and to avoid overfitting, training steps are carried out with a dynamic learning rate starting at 0.001 – to which a coefficient of 0.75 is applied every 10 epochs –, and we use a batch of one image for each iteration. We have limited training to 30 epochs. First results are summarised in Table 1.

**Table 1.** Evaluation of character-based and word-based HTR system on datasets with and without expanded abbreviations.

|  | CER (%) | | | |
|---|---|---|---|---|
|  | D-exp1 | D-exp2 | D-abb1 | D-abb2 |
| HTR-CB | 10.59 | 9.69 | 4.89 | 4.55 |
| HTR-WB | 3.76 | 2.96 | 5.57 | 4.83 |

The two architectures give equivalent results on the two variants of the D-abb dataset. At identical initial parameters, we do not observe any significant difference, except that HTR-CB converges twice as fast as HTR-WB. A dynamic learning rate brings a real benefit for both of two architectures (stagnation of the CER until epoch 10 then gradual reduction). Epoch 30 is never reached. 20% of the errors of HTR-CB are focused on combining signs, but it generally recognises combined letters well. It also mistakes close classes as P and p, or q and q, but the lack of data can be a good explanation of this phenomenon.

There is a clear benefit to using a word-based approach for the management of abbreviations directly within the HTR process. If it seems quite logical that the absence of spaces, an ambiguous notion in manuscripts, can really benefit to text recognition, there is however only a marginal gain between D-abb1 and D-abb2. On the other hand, the HTR-WB model takes advantage of the lack of spaces. Most errors are focused on small and independent abbreviations, generally limited to one single character (e.g. m̊ > modo), but also on long abbreviated words (e.g. **micdia** > misericordia) for which we do not have enough samples in the training set (e.g. only one sample of misericordia). We give in Table 2 an example of predictions. For the rest of the paper, we consider the HTR-CB output, as shown in Table 2.

**Table 2.** Example of predictions with and without abbreviations.

| | |
|---|---|
| | *p̄omo d̄n̄i nos̄ oppon̄e n̄ curam̄. Ti* |
| GT-exp1 | pro domo domini nos opponere non curamus. ti |
| **HTR-WB** | prodomo domini nos oponere non curamus. ti |
| GT-abb2 | pdomodñinosoppoñeñcuram̊.ti |
| **HTR-CB** | pdomodñinosoppoñeñcuram̊.tiē |

### 3.3 Text Normalisation Approach

The modular text normalisation approach uses several consecutive steps, with specialised tools each necessitating a training of its own [6], as follows (Fig. 2):

1. **HTR** (*see above*).
2. **word segmentation** using a deep-learning word segmentor, Boudams [8].
3. **Abbreviation expansion and word normalisation** using a deep learning based word annotator, Pie [15].

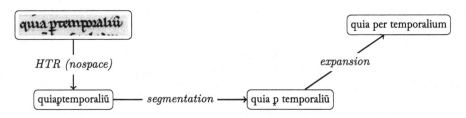

**Fig. 2.** Workflow for the full modular approach [6]. The digital image is segmented in lines that are submitted to recognition, and the recognised text then goes through word segmentation and normalisation (abbreviation expansion).

*Word Segmentation.* A neural word segmentor, Boudams, was trained using the best known configuration: convolutional encoder without position embeddings (Embedding 256, KernelSize 5, Dropout 0.25, Layers 10) [8]. Three datasets were used: for abbreviated text, D-abb (with word separation and line hyphenations normalised) alone, and with the addition of Oriflamms data; for normalised version, D-exp (with word separation and line endings normalised) with the addition of PL data. Three models were trained for each setting, and the best of the three (based on F-statistic) was selected. Results are shown in Table 3.

**Table 3.** Results for training of the word segmentor Boudams

| dataset | type | n-lines | | | test scores | | |
|---|---|---|---|---|---|---|---|
| | | train | dev | test | F-s. | Prec. | Recl. |
| D-abb | abbr. | 1 962 | 196 | 200 | 95.1 | 96.1 | 94.1 |
| D-abb+Oriflamms | abbr. | 17 139 | 196 | 200 | 97.3 | 97.4 | 97.1 |
| D-exp+PL | norm. | 3 891 929 | 196 | 200 | 98.8 | 98.6 | 98.9 |

*Word Normalisation.* For abbreviation expansion, the neural tagger Pie was trained [15], following a setup already used for Old French [6]. It was trained on an aligned version of the previous D-abb and D-exp (with word separation and line hyphenations normalised) alone, and with the addition of the Oriflamms data. Results are shown in Table 4.

**Table 4.** Results for the training of Pie for abbreviation expansion

|  | Test accuracy (%) | | | | |
|---|---|---|---|---|---|
| dataset | all | known | unkn. | ambig. | unkn. targ. |
| D-abb/D-exp | 94.04 | 95.37 | 92.42 | 89.58 | 91.23 |
| D-abb/D-exp+Oriflamms | 97.02 | 98.65 | 95.02 | 97.92 | 92.86 |

### 3.4 Results

The full pipeline was evaluated globally on the normalised transcription of one folio of the manuscript, with expanded abbreviations, normalised word segmentation and line hyphenations. The metrics used were character error rate (CER) and word error rate (WER) in the Python `fastwer` implementation [26]. Results are shown in Table 5.

**Table 5.** Results of the application of the full pipeline for a selection of the most accurate setups, with or without additional data.

| SETUPS | | | | | | | SCORES | |
|---|---|---|---|---|---|---|---|---|
| HTR | | Segment. | | Norm. | | Test | | |
| Data | Soft. | Data | Soft. | Data | Soft. | CER | WER | |
| 1a | D-exp1 | CB (HTR-CB) | | | | | 15.63 | 82.38 |
| 1b | D-exp1 | WB (HTR-WB) | | | | | **7.46** | **54.68** |
| 2a | D-exp2 | CB (HTR-CB) | D-exp1+PL | Boudams | | | 12.65 | 50.83 |
| 2b | D-exp2 | WB (HTR-WB) | D-exp1+PL | Boudams | | | **6.89** | **34.01** |
| 3a | D-abb2 | CB (HTR-CB) | D-abb1 | Boudams | D-abb/exp | Pie | 10.89 | 41.28 |
| 3b | D-abb2 | CB (HTR-CB) | +Orifl. | Boudams | +Orifl. | Pie | **8.60** | **31.81** |

The simple use of an HTR engine, trained on a normalised transcription with normalised spaces and expanded abbreviations actually provides a strong baseline. This is particularly true of the word-based HTR, and in terms of character error rate. Yet, they are overperformed by the more refined setups, either in terms of character error rate or word error rate. The best scores (Table 5) are obtained, for the character error rate, by the HTR-WB trained on normalised transcriptions whose output is then resegmented by a word segmentor trained on the *Patrologia latina*; for the word error rate, by the HTR-CB trained on abbreviated transcriptions, whose output is then resegmented and normalised with models trained using the additional Oriflamms data.

**Table 6.** Facsimile and ground truth with sample predictions of the three best performing setups (word with content or segmentation errors **in bold**), from the beginning of fol. 45v (see Fig. 1).

| Facsimile | GT |
|---|---|
| | periculo . amore uite temporalis . a defensione iusticie *et* u*eritatis* n*on* flectitur . sed qu*is* est hodie . qui animam suam ponat pro ouibus suis : cadit asella uicini . et |

| 1b | 2b |
|---|---|
| piculo. amore uite temporalis. **adefen-** **sione** iusticie et ueritatis non flectitur. **sedquis** est hodie. qui animam suam **ponatpro** ouibus suis : cadit asella uicini. et | piculo . amore uite **temprus** . a **de sen-** sione iusticie et ueritatis non flectitur . sed quis est hodie . qui animam suam ponat pro ouibus suis : cadit asella uicini . et |

| 3b | |
|---|---|
| pculo . **amoreure temporauis** ; a **def-** **sasione** iusticie et ueritatis non flectitur . **sedquas st** hodie . qui animam suam ponat pro ouibus suis : **cadita sella** uicini . et | |

# 4   Discussion and Further Research

Results seem to confirm the importance of the word level rather than the character level, with word-based overperforming character-based HTR, and with character-based HTR results being improved by adding tools dealing with word-level tokens and context, in particular for a small dataset. In this regard, artificial intelligence can be compared to human intelligence and seems to confirm the practice of global reading and global perception of the words rather than individual letters, reflected in the use of abbreviations in Latin (especially scholastic) manuscripts. Future research should investigate differences with vernacular manuscripts, for instance literature in Old French, where reading and the use of abbreviations are supposed to have remained mostly syllabic.

Another conclusion of this paper is that, given the nature of spacing in medieval manuscripts, the word error rate regarding normalised words can be greatly reduced by using a dedicated word segmentor, for which (normalised) training data can be easily collected (since all that is needed are normalised editions). In our experiments, this proved to be the best performing setup in terms of character error-rate.

Thirdly, a fully modular approach combining HTR on abbreviated data with a word segmentor and a text normalisation tool is the best performing in terms of word error rate. It would be possible to improve further the results of this approach by improving first the word-based approach – that still suffer of a lack of data to manage with the word level –, and then the segmentation and normalisation steps with additional data, yet the limit to this approach is the availability of adequate training material, i.e., editions recording abbreviations, that are much harder to come by than normalised editions.

Another future line of research should pursue the comparison with the vernacular, where the ambivalence of abbreviations, due to the variety of written norms and alternative spellings, should make the automated production of a normalised text more difficult. It should also investigate the impact of data augmentation techniques, particularly easy for the word segmentation training that could, for instance, include random character substitutions as to emulate HTR effect [8].

**Datasets Availability.** Datasets produced for this paper and evaluation scripts are available at DOI 10.5281/zenodo.5071963.

**Acknowledgements.** We thank the École nationale des chartes and the DIM STCN for the computing power and GPU server used for training, as well as INRIA and Calfa. We also thank Marc H. Smith for his keen review of our draft. Any remaining mistakes are only attributable to us.

# References

1. Bischoff, B.: Paläographie des römischen Altertums und des abendländischen Mittelalters. Grundlagen der Germanistik, 4th edn. E. Schmidt, Berlin (2009)
2. Bluche, T., et al.: Preparatory KWS experiments for large-scale indexing of a vast medieval manuscript collection in the Himanis project. ICDAR **1**, 311–316 (2017)
3. BnF: Petrus Pictaviensis, Tractatus de confessione (...). Latin 14525. In: Gallica. BnF (1997). https://gallica.bnf.fr/ark:/12148/btv1b9080806r/
4. Bollmann, M.: A Large-Scale Comparison of Historical Text Normalization Systems. NAACL-HLT pp. 3885–3898. arXiv: 1904.02036 (2019). https://doi.org/10.18653/v1/N19-1389
5. Camps, J.B.: La 'Chanson d'Otinel': édition complète du corpus manuscrit et prolégomènes à l'édition critique. thèse de doctorat, dir. Dominique Boutet, Paris-Sorbonne, Paris (2016). https://doi.org/10.5281/zenodo.1116735
6. Camps, J.B., Clérice, T., Pinche, A.: Stylometry for Noisy Medieval Data: Evaluating Paul Meyer's Hagiographic Hypothesis, December 2020. arXiv:2012.03845 (2020). http://arxiv.org/abs/2012.03845
7. Ceccherini, I.: Manuscrits datés (notices complètes). In: Stutzmann, D. (ed.) Github, Paris (2017). https://github.com/oriflamms/Dated-and-Datable-Manuscripts_LIRIS
8. Clérice, T.: evaluating deep learning methods for word segmentation of Scripta continua texts in Old French and Latin. J. Data Min. Digit. Humanities (2020). https://doi.org/10.46298/jdmdh.5581
9. Gabay, S., Barrault, L.: Traduction automatique pour la normalisation du français du XVIIe siècle. In: Benzitoun, C., et al. (eds.) TALN 27, vol. 2, pp. 213–222. Nancy (2020). https://hal.archives-ouvertes.fr/hal-02784770
10. Hasenohr, G.: Abréviations et frontières de mots. Langue française **119**, 24–29 (1998). https://doi.org/10.3406/lfr.1998.6257

11. Hasenohr, G.: Écrire en latin, écrire en roman: réflexions sur la pratique des abréviations dans les manuscrits français des XII[e] et XIII[e] siècles. In: Banniard, M. (ed.) Langages et peuples d'Europe: cristallisation des identités romanes et germaniques (VII[e]-XI[e] siècle), pp. 79–110. Toulouse (2002)
12. Kiessling, B.: A modular region and text line layout analysis system. In: ICFHR, pp. 313–318 (2020). https://doi.org/10.1109/ICFHR2020.2020.00064
13. Kiessling, B., Miller, M.T., Maxim, G., Savant, S.B., et al.: Important new developments in arabographic optical character recognition (OCR). Al-ᶜUṣūr al-Wusṭā **25**, 1–13 (2017)
14. Kiessling, B., Tissot, R., Stokes, P., Stökl Ben Ezra, D.: eScriptorium: an open source platform for historical document analysis. In: ICDARW, vol. 2, pp. 19–24 (2019)
15. Manjavacas, E., Kádár, A., Kestemont, M.: Improving lemmatization of nonstandard languages with joint learning. arXiv preprint arXiv:1903.06939 (2019)
16. Mazziotta, N.: Traiter les abréviations du français médiéval: théorie de l'écriture et pratiques d'encodage. Corpus **7**, 1517 (2008). http://corpus.revues.org/1517
17. Migne, J.P. (ed.): Patrologiae cursus completus ... Series Latina. Apud Garnieri Fratres, editores et J.-P. Migne successores, Parisiis (1844)
18. Muzerelle, D., Bozżolo, C., Coq, D., Ornato, E.: Psautiers IMS. In: D. Stutzmann, D. (ed.) Github, Paris (2018). https://github.com/oriflamms/PsautierIMS
19. Romero, V., Toselli, A.H., Vidal, E., Sánchez, J.A., Alonso, C., Marqués, L.: Modern vs. diplomatic transcripts for historical handwritten text recognition. In: ICIAP, pp. 103–114 (2019)
20. Stutzmann, D.: Psautiers: Transcriptions de différents manuscrits. Github, Paris (2018). https://github.com/oriflamms/PsautierIMS
21. Stutzmann, D.: Recueil des actes de l'abbaye de Fontenay. TELMA, Github, Paris (2018). https://github.com/oriflamms/Fontenay
22. Vernet, M.: Un Manuscrit victorin au service de la pastorale du XIIIe siècle. Master's thesis, Université PSL, Paris (2021)
23. Vidal-Gorène, C., Decours-Perez, A.: A computational approach of Armenian paleography. In: Accepted for IWCP Workshop of ICDAR 2021 (2021)
24. Vidal-Gorène, C., Dupin, B., Decours-Perez, A., Riccioli, T.: A modular and automated annotation platform for handwritings: evaluation on under-resourced languages. In: Accepted for ICDAR 2021 Conference (2021) by In: J. Lladós et al. (eds.) ICDAR 2021, LNCS 12823. Springer (2021). https://doi.org/10.1007/978-3-030-86334-0_33
25. Villegas, M., Toselli, A.H., Romero, V., Vidal, E.: Exploiting existing modern transcripts for historical handwritten text recognition. In: ICFHR, pp. 66–71 (2016)
26. Wang, C.: Fastwer (2020). https://github.com/kahne/fastwer, v0.1.3

# Exploiting Insertion Symbols for Marginal Additions in the Recognition Process to Establish Reading Order

Daniel Stökl Ben Ezra[1,2]([⊠]) [ID], Bronson Brown-DeVost[1,2][ID],
and Pawel Jablonski[1,2]

[1] École pratique des hautes études (EPHE), PSL University, Paris, France
{daniel.stoekl,bronson.browndevost,pawel.jablonski}@ephe.psl.eu
[2] Archéologie et philologie d'Orient et d'Occident (AOrOc), UMR 8546,
CNRS, Université PSL (ENS, EPHE), Paris, France

**Abstract.** In modern and medieval manuscripts, a frequent phenomenon is additions or corrections that are marginal or interlinear with regard to the main text-blocks. With the recent success of Long-Short-Term-Memory Neural Networks (LSTM) in Handwritten-Text-Recognition (HTR) systems, many have chosen lines as the primary structural unit. Due to this approach, establishing the reading order for such additions becomes a non-trivial problem, because they must be inserted between words inside line-units at undefined locations. Even a perfect reading order detection system ordering all lines of a text in the correct order would not be able to deal with inline insertions. The present paper proposes to include markers for the insertion points in the recognition training process, those indicators can then teach the recognition models themselves to detect scribal insertion markers for marginal or interlinear additions.

**Keywords:** Symbols · Medieval manuscripts · HTR · Reading order · Hebrew

## 1 Introduction

### 1.1 The Problem

Among the special glyphs found in many Medieval Hebrew manuscripts are insertion markers that indicate the location where marginal additions, writing adjacent to the main text blocks, are to be read. While such additions can have many semantic functions, e.g. additions to or corrections of the main text, commentaries, paraphrases, translations, and section indicators, the present paper focuses on those notes which have been written to correct the main text and are therefore central to the task of establishing the reading order of the main text. Even if the main text follows a simple Manhattan-style layout with rectangular

© Springer Nature Switzerland AG 2021
E. H. Barney Smith and U. Pal (Eds.): ICDAR 2021 Workshops, LNCS 12917, pp. 317–324, 2021.
https://doi.org/10.1007/978-3-030-86159-9_22

columns and straight lines, marginal (and interlinear) additions greatly compli-
cate establishment of the reading order. This problem is particularly challenging
for the current wave of successful LSTM based HTR systems, whose basic object
structure is often lines, since the corrections have to be inserted between some
words or characters inside such a line-object. Finding a solution to this problem
is crucial due to the frequency of such additions in manuscripts and the impli-
cations for the reconstruction of the reading order of the text itself. Beyond the
obvious approach to exploit basic topological heuristics, this paper will present
a possible path to automatically detecting at least some of the insertion points
of such additions as part of the LSTM training. In other words, the task this
paper addresses is the automatic detection of the insertion spot of the addition
inside the character sequence of the main text line. It is limited to manuscripts
with a single main text and not, e.g. a text surrounded by commentaries.

Reading order is generally a difficult and unsolved problem, which concerns
an enormous number of such items [1–6]. Newspapers and comic books can have
highly complex reading orders [1,3,6]. For Medieval manuscripts, reading order
is particularly difficult because the fourth dimension, time, plays a greater role
there than it does in modern print. The problem starts on the conceptual level.
Even in the case of a single text on each manuscript page and not, e.g. a text and
its commentary or translation, one should better speak of reading orders in the
plural. Even the first "original" scribe might have modified his original reading
order through the insertion of marginal and linear additions or through changes
in the word order. A greater number of scribes usually entails a greater number
of modifications to the original reading order, and each intervention represents
a valid reading order that may be of interest to researchers. The problem in this
particular corpus is therefore of a finer granularity than it is with regard to more
commonly researched newspapers or comic books that deal with complete lines.

A frequent task in philology is the creation of a scholarly edition of a text
that compares all manuscripts (and possibly secondary witnesses by citations).
Such an edition is usually structured in a book-chapter-verse hierarchy. Without
interlinear and marginal additions it is trivial to convert a manuscript-page-
zone-line as stored in a PageXML to a book-chapter-verse hierarchy. It is the
insertions inside lines of interlinear and marginal additions that make such a
transition considerably more difficult. We know of no system currently able to
deal with this problem in an automatic way.

Topological heuristics are one approach to the reading order detection problem [1,5,6]. Another possible technique uses trainable reading order [2–4].
Assuming all lines have been correctly detected, generally accepted rules for reading orders in Western and Middle Eastern literary texts in Manhattan style layout are top-to-bottom, following by the writing direction (leftwards/rightwards,
depending on whether it is a Right-to-Left (RTL) or a Left-to-Right (LTR)
script).[1]

---

[1] While Poetic texts can be written stichographically with half verses appearing as
if written in columns, we assume that lines in Poetic texts are correctly detected if
they cover a complete such line connecting both 'columns'.

The insertion spot of interlinear additions is usually close to the x position of the first letter of the first word of the addition. Due to human imprecision the precise spot may be one word before or after. Transcription-glyph alignment algorithms can convert this into positions in the transcription [7].

In Hebrew manuscripts, scribes indicated the intended insertion spot for a marginal addition usually by adding a small marker in the line, usually at the top of it and by placing the addition close to the intended insertion spot on the right or left side of the column at the same height.

Calculating these positions is usually trivial even though the association may be ambiguous with regard to two lines if the scribe was not precise enough.

## 1.2  Background

On the User-Interface (UI) and HTR-engine level, we use the eScriptorium document analysis UI [8,9] interacting with the kraken HTR engine for segmentation of lines and regions including their semantic classification as well as for transcription [10–12]. In addition to having an ergonomic interface for manual transcription of LTR and RTL scripts, baseline and region annotation (including a semantic ontology), as well as trainable segmentation and transcription, eScriptorium has a powerful application programming interface (API) that allows interaction with outside tools.

Our test corpus consists of literary Hebrew manuscripts from the first half of the second millennium. Our current central target corpus consists of 17 manuscripts of Tannaitic Rabbinic compositions (Mishnah, Tosefta, Mekhilta deRabbi Yishmael, Sifra, Sifre Numbers and Sifre Deuteronomy) with a total of more than 6000 pages, usually with one or two columns as well as marginal and interlinear additions whose number and precise location varies greatly. The manuscripts are written in Ashkenazy (3), Byzantine (1), Italian (8), Oriental (1) and Sephardi (4) styles and date from the 10th to 15th centuries. Yet, it is one thing to train a segmenter capable of detecting all lines including the lines of interlinear and marginal additions with a high accuracy and another thing to know where to insert the text of such additions. For the final product we want to propose not only transcriptions that follow the layout of each manuscript, but also a text oriented transcription that includes the reading order rather than a list of unconnected regions containing lists of lines.

## 2  Discussion

For the precise location of a marginal addition inside the line, scribes of Hebrew manuscripts frequently use special marks, circles, 'v'-shapes or other small glyphs. During the process of preparing the necessary ground truth data for transcription training and training attempts, we noted that we can actually train the recognition model to learn such insertion points for at least some of the interlinear and marginal additions and to mark them directly in the transcription. Intercolumnal additions for which the association with the left or right column is ambiguous, the detection of the insertion mark is crucial.

In a first experiment, we represented the insertion marks by a single glyph in the ground truth transcription used for the recognition training. Our test manuscript was ms Kaufmann A50 from the Library of the Hungarian Academy of the Sciences with the Mishnah from 11$^{\text{th}}$ or 12$^{\text{th}}$ century Italy. The training material consisted of 29 pages in 2 columns, each of which usually has 26–27 lines, altogether 1541 lines with 41488 characters in the transcription ground truth. Apart from the 58 main text columns, there are 53 additional zones, 50 of which are marginal additions. For 39 marginal additions, the scribes indicated insertion spots in the main columns, which were marked with the glyph '→' in the transcription ground truth. As the markers for the insertion spots are too rare to be easily learned we used a three-step training process. On top of (1) a generalized basic model for Medieval Hebrew manuscripts, we trained (2) a manuscript specific model on 26 pages (1380 lines). As a third step we trained on the manuscript specific recognition model a model tailored for the recognition of insertion spots by restricting only the 39 lines that did include insertion markers from the training material.

**Fig. 1.** Correctly recognized v-shaped insertion markers (indicated by '→' in the transcription) in ms Kaufmann A50 of the Library of the Hungarian Academy of the Sciences. The relevant manuscript line is indicated by the highlighted polygon. Below each manuscript image, the automatic transcription is shown (no manual correction).

The test corpus consists of 10 different pages from the same manuscript that contained 9 marginal additions with insertion marks. The best model, after only one epoch, recognized 8 out of 9 insertion marks. Figure 1 shows three true positives while Fig. 2 illustrates the one false negative. The system wrongly detected insertion marks at 9 additional places where none was present (see Fig. 3 for three examples of these false positives).

While the precision may seem low, the important positive result is the high re-call of 88.89%, because it gives a character precise insertion spot for the vast majority of marginal additions inside the text flow of the lines, i.e. we can reconstruct the reading order on a word level. The low precision can be easily

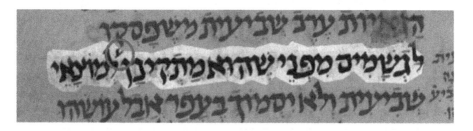

לגשמים מפני שהוא מתקינן למוצאי

**Fig. 2.** Undetected insertion marker (false negative) in ms Kaufmann A50 of the Library of the Hungarian Academy of the Sciences. Maybe the ink, darker than the other insertion markers is a reason for the miss. (automatic transcription, no manual correction).

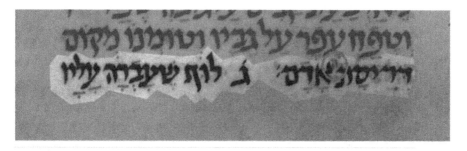

דריסת ← אדם : ג' לוף שעברה עליו

**Fig. 3.** Wrongly claimed insertion marker (false positive) in ms Kaufmann A50 of the Library of the Hungarian Academy of the Sciences. The system mistook an additional supra-linear character (the letter "he" (ה)) as insertion marker (automatic transcription, no manual correction).

over-come by a simple topological test that detects the false positives by checking whether there are any marginal regions in vicinity for this or that wrongly
detected insertion spot. The probable reasons for 8 out of 9 false positives were:

a) Confusion with single interlinear added letters (4×) that look similar to insertion marks (see Fig. 3).
b) Confusion with small ornaments (1×), high-dots (1×) or abbreviation marks
   (1×)
c) Confusion with remains from descender of previous line (1×).

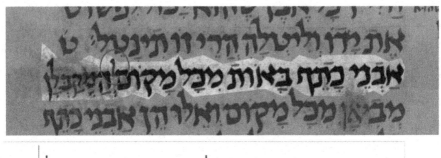

**Fig. 4.** Correctly identified interlinear "waw" (ו) in parentheses before the last word
(on the left) in ms Kaufmann A50 of the Library of the Hungarian Academy of the
Sciences

The system also learned to detect some of the single letter interlinear additions if they occur frequently enough in the training data (see Fig. 4). We also
ran a series of preliminary tests on another manuscript, ms. Parma 3259 de Rossi
139 with the Sifra.

Figure 5 shows three examples of insertion spots that were detected automatically by the recognizer. In the training material, all insertion spots for additions
longer than a single letter were marked with opening-plus-closing parentheses
and the recognizer was able to learn this to a certain degree.

The top example concerns a word inserted as interlinear addition without any
further scribal mark. The middle and the lower example show the scribe using
a circle and a v-shaped symbol to mark the insertion spots. Both interlinear
and marginal additions were marked up with the same combination '()' in the
training material and not with an arrow as in the previous experiment.

The training material for ms Parma consisted of 20 pages with a total of
1441 lines and 62 occurrences of the insertion marks '()'. This was then applied
to the remaining 252 pages with a total of 11165 lines. The combination '()' was
detected 48 times.

**Fig. 5.** Three automatic transcribed lines (without manual correction) in which the insertion points for an interlinear (top) or marginal additions (middle and bottom) have been automatically detected by the recognizer. Ms. Parma 3259, de Rossi 139 (Sifra).

# 3 Conclusion

We have shown that including mark-up for symbols indicating insertion spots of marginal additions in the transcriptions for the ground truth training process of the recognizer can cause the system to learn them well. It is a simple procedure, but as far as we know it has never been suggested before and it may be of great help in the complex process of reading order reconstruction in historical manuscripts.

**Acknowledgments.** Images from manuscript Kaufmann A50 of the Library of the Hungarian Academy of the Sciences, Budapest, by permission CC-BY-NC-SA.

# References

1. Clausner, C., Pletschacher, S., Antonacopoulos, A.: The significance of reading order in document recognition and its evaluation. In: 12th International Conference on Document Analysis and Recognition (ICDAR), Washington DC, 2013, pp. 688–692, IEEE (2013)
2. Quirós, L., Vidal, E.: Learning to sort handwritten text lines in reading order through estimated binary order relations. In: 25th International Conference on Pattern Recognition, ICPR 2020, Virtual Event/Milan, Italy, January 10–15, 2021, pp. 7661–7668, IEEE (2020)
3. Prasad, A., Déjean, H., Meunier, J.-L.: Versatile layout understanding via conjugate graph. In: 15th International Conference on Document Analysis and Recognition, (ICDAR) Sydney, Australia, September 20–25, 2019, pp. 287–294, IEEE (2019)
4. Malerba, D., Ceci, M., Berardi, M.: Machine learning for reading order detection in document image understanding. In: Marinai, S., Fujisawa, H. (eds.), Machine Learning in Document Analysis and Recognition, Studies in Computational Intelligence, vol. 90, pp. 45–69, Springer, Berlin (2008). https://doi.org/10.1007/978-3-540-76280-5_3
5. Ferilli, S., Pazienza, A.: An Abstract Argumentation-based strategy for reading order detection. In: Proceedings of 1st AI*IA Workshop on Intelligent Techniques At Libraries and Archives colocated with XIV Conference of the Italian Association for Artificial Intelligence, IT@LIA@AI*IA 2015, Ferrara, Italy, September 22, 2015, CEUR Workshop Proceedings 1509, http://ceur-ws.org/Vol-1509/ITALIA2015_paper_1.pdf
6. Kovanen, S., Aizawa, K.: A layered method for determining manga text bubble reading order. In: 2015 IEEE International Conference on Image Processing (ICIP), Quebec City, QC, 2015, pp. 4283–4287. https://doi.org/10.1109/ICIP.2015.7351614
7. Stökl Ben Ezra, D., Brown-DeVost, B., Dershowitz, N., Pechorin, A., Kiessling, B.: Transcription alignment for highly fragmentary historical manuscripts: the dead sea scrolls. In: 17th International Conference on Frontiers and Handwriting Recognition (ICFHR), Dortmund (2020), pp. 361–366, IEEE (2020)
8. Kiessling, B., Tissot, R., Stökl Ben Ezra, D., Stokes, P.: eScriptorium: an open source platform for historical document analysis, In: Open Software Technologies (OST@ICDAR) 2019, pp. 19–24, IEEE (2019)
9. Stokes, P., Kiessling, B., Tissot, R., Stökl Ben Ezra, D.: EScripta: a new digital platform for the study of historical texts and writing. In: Digital Humanities, Utrecht 2019 (DH 2019)
10. Kiessling, B.: Kraken – a universal text recognizer for the humanities. In: Digital Humanities, Utrecht 2019 (DH 2019)
11. Kiessling, B., Stökl Ben Ezra, D., Miller M.: BADAM: a public dataset for baseline detection in arabic-script manuscripts. In: HIP@ICDAR 2019, Sydney (2019), pp. 13–18, ACM (2019)
12. Kiessling, B.: A modular region and text line layout analysis system. In: 17th International Conference on Frontiers in Handwriting Recognition (ICFHR), Dortmund (2020), pp. 313–318, IEEE (2020)

# Neural Representation Learning
# for Scribal Hands of Linear B

Nikita Srivatsan[1]([✉]), Jason Vega[3], Christina Skelton[2],
and Taylor Berg-Kirkpatrick[3]

[1] Language Technologies Institute, Carnegie Mellon University, Pittsburgh, USA
nsrivats@cmu.edu
[2] Center for Hellenic Studies, Harvard University, Cambridge, USA
cskelton@ucla.edu
[3] Computer Science and Engineering, University of California, San Diego, USA
jvega@ucsd.edu, tberg@eng.ucsd.edu

**Abstract.** In this work, we present an investigation into the use of neural feature extraction in performing scribal hand analysis of the Linear B writing system. While prior work has demonstrated the usefulness of strategies such as phylogenetic systematics in tracing Linear B's history, these approaches have relied on manually extracted features which can be very time consuming to define by hand. Instead we propose learning features using a fully unsupervised neural network that does not require any human annotation. Specifically our model assigns each glyph written by the same scribal hand a shared vector embedding to represent that author's stylistic patterns, and each glyph representing the same syllabic sign a shared vector embedding to represent the identifying shape of that character. Thus the properties of each image in our dataset are represented as the combination of a scribe embedding and a sign embedding. We train this model using both a reconstructive loss governed by a decoder that seeks to reproduce glyphs from their corresponding embeddings, and a discriminative loss which measures the model's ability to predict whether or not an embedding corresponds to a given image. Among the key contributions of this work we (1) present a new dataset of Linear B glyphs, annotated by scribal hand and sign type, (2) propose a neural model for disentangling properties of scribal hands from glyph shape, and (3) quantitatively evaluate the learned embeddings on find-place prediction and similarity to manually extracted features, showing improvements over simpler baseline methods.

**Keywords:** Linear B · Paleography · Representation learning

## 1 Introduction

Neural methods have seen much success in extracting high level information from visual representations of language. Many tasks from OCR, to handwriting recognition, to font manifold learning have benefited greatly from relying on

© Springer Nature Switzerland AG 2021
E. H. Barney Smith and U. Pal (Eds.): ICDAR 2021 Workshops, LNCS 12917, pp. 325–338, 2021.
https://doi.org/10.1007/978-3-030-86159-9_23

architectures imported from computer vision to learn complex features in an automated, end-to-end trainable manner. However much of the emphasis of this field of work has been on well-supported modern languages, with significantly less attention paid towards low-resource, and in particular, historical scripts. While self-supervised learning has seen success in these large scale settings, in this work we demonstrate its utility in a low data regime for an applied analysis task. Specifically, we propose a novel neural framework for script analysis and present results on scribal hand representation learning for Linear B.

Linear B is a writing system that was used on Crete and the Greek mainland ca. 1400-1200 BCE. It was used to write Mycenaean, the earliest dialect of Greek, and was written on leaf or page-shaped clay tablets for accounting purposes. Linear B is a syllabary, with approximately 88 syllabic signs as well as a number of ideograms, which are used to indicate the commodity represented by adjacent numerals. The sites which have produced the most material to date are Knossos, Pylos, Thebes, and Mycenae (see Palmer [15] for more detail) but for the purposes of this paper, we focus specifically on Knossos and Pylos from which we collect a dataset of images of specific instances of glyphs written by 74 different scribal hands. Many signs exhibit slight variations depending on which scribe wrote them. Therefore, being able to uncover patterns in how different scribes may write the same character can inform us about the scribes' potential connections with one another, and to an extent even the ways in which the writing system evolved over time. This is a process that has previously been performed by hand [18], but to which we believe neural methods can provide new insights.

Rather than building representations of each scribe's writing style based on the presence of sign variations as determined by a human annotator, we propose a novel neural model that disentangles features of glyphs relating to sign shape and scribal idiosyncrasies directly from the raw images. Our model accomplishes this by modeling each image using two separate real-valued vector embeddings— we share one of these embeddings across glyphs written by the same scribe, and the other across glyphs depicting the same sign, thereby encouraging them to capture relevant patterns. These embeddings are learned via three networks. The first is a decoder that attempts to reconstruct images of glyphs from their corresponding embeddings using a series of transpose convolutional layers. We penalize the reconstructed output based on its mean squared error (MSE) vs. the original. In addition, we also use two discriminator networks, which attempt to predict whether a given image-embedding pair actually correspond to one another or not, both for scribe and sign embeddings respectively. These losses are backpropagated into those original vectors to encourage them to capture distinguishing properties that we care about.

This model has many potential downstream applications, as paleography has long been an integral part of the study of the Linear B texts [14]. For a handful of examples from prior work, Bennett's study of the Pylos tablets [1] revealed the existence of a number of different scribal hands, all identified on the basis of handwriting, since scribes did not sign their work. This research was carried out even before the decipherment of Linear B. Later, Driessen [5] used the archaic sign forms on Linear B material from the Room of the Chariot Tablets at Knossos, in conjunction with archaeological evidence, to argue that these tablets pre-dated the remainder of the Linear B tablets from Knossos. Skelton [18] and later Skelton and Firth [20] carried out a phylogenetic analysis of Linear B based on discrete paleographical characteristics of the sign forms, and were able to use this analysis to help date the remainder of the Linear B tablets from Knossos, whose relative and absolute dates had already been in question. Our hope is that leveraging the large capacity of neural architectures can eventually lead to further strides in these areas.

Additionally, qualitative assessments of Linear B paleography have long been used to date tablets whose history or findplaces are uncertain. This has been vital in the case of the Knossos tablets, which were excavated before archaeological dating methods were well-developed [7], and the tablets from the Pylos megaron, which seem to date to an earlier time period than the rest of the palace [19]. While qualitative techniques have been fruitful, a key motivating question in our work is to what extent we can use quantitative methods to improve on these results.

Going back further, Linear B was adapted from the earlier writing system Linear A, which remains undeciphered. There is a hundred-year gap between the last attestation of Linear A and the first attestation of Linear B. With a more thorough understanding of the evolution of these two writing systems, can we reconstruct the sign forms of the earliest Linear B, and elucidate the circumstances behind its invention? With this in mind, our approach explicitly incorporates a decoder network capable of reconstructing glyphs from learned vector representations, that we hope can eventually be useful for such investigation.

In total, three writing systems related to Linear B remain undeciphered (Cretan Hieroglypic, Linear A, and Cypro-Minoan), and a fourth (the Cypriot Syllabary) was used to write an unknown language. One motivating question is whether a better understanding of the evolution of sign forms would help contribute additional context that could make a decipherment possible. In particular, scholars are not yet in agreement over the sign list for Cypro-Minoan, and how many distinct languages and writing systems it may represent [6]. Understanding variation in sign forms is foundational for this work [14], and it makes sense to test our methods on a related known writing system (i.e. Linear B) before attempting a writing system where so much has yet to be established.

Overall these are just some of the applications that we hope the steps presented here can be directed towards.

## 2   Prior Work

One of our primary points of comparison is the work of Skelton [18], who manually defined various types of systematic stylistic variations in the way signs were written by different scribal hands, and used these features to analyze similarity between scribes via phylogenetic systematics. In a sense this work can serve as our human baseline, giving insight into what we might expect reasonable

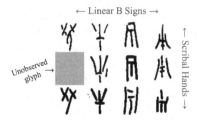

**Fig. 1.** Binarized examples from our collected dataset. Note the stylistic differences particularly in the left and right most signs, and that we do not have an observation for the left most sign in the middle scribe's hand.

output from our system to look like, and helping us gauge performance in an unsupervised domain without a clear notion of ground truth.

While neural methods have not previously been used in analyzing this particular writing system, there is established literature on applying them to typography more broadly. For example, Srivatsan et al. [21] used a similar shared embedding structure to learn a manifold over font styles and perform font reconstruction. There has also been extensive work on writer identification from handwritten samples [2,4,17,25], as well as writer retrieval [3]. However these approaches rely on supervised discriminative training, in contrast to ours which is not explicitly trained on predicting authorship, but rather learning features.

## 3    Methods

In this section we describe the layout of our model and its training procedure. At a high level, our goal is to train it on our dataset of images of glyphs, and in doing so learn vector embeddings for each scribe that encode properties of their stylistic idiosyncrasies—for example which particular variations of each sign they preferred to use—while ignoring more surface level information such as the overall visual silhouettes of the signs we happen to observe written by them. To do this, we break the roles of modeling character shape and scribal stylistic tendencies into separate model parameters, as we now describe.

Our model, depicted in Fig. 2, assigns each sign and each scribal hand a vector embedding to capture the structural and stylistic properties of glyphs in our data. These embeddings are shared by glyphs with the same sign or scribe, effectively factorizing our dataset into a set of vectors, one for each row and column. The loss is broken into two parts, a reconstruction loss and a discriminative loss. The reconstruction loss is computed by feeding a pair of sign and scribe embeddings to a decoder, which then outputs a reconstruction of the original glyph that those embeddings correspond to. A simple MSE is computed between this and the gold image. The discriminative cross-entropy loss is computed by the discriminator, a binary classifier which given an image of a glyph and a scribe embedding must predict whether or not the two match (i.e. did that scribe write that glyph). We train this on equally balanced positive and negative pairs. A similar discriminator

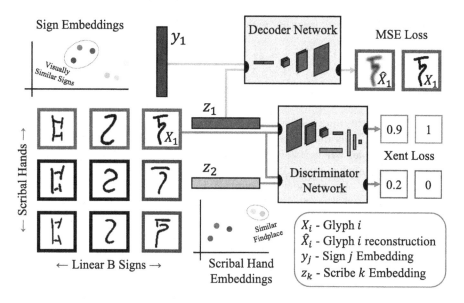

**Fig. 2.** Diagram of the model—each row is assigned a scribe embedding, and each column is assigned a sign embedding, which feed into a decoder and discriminator to collectively compute the loss function.

is applied to the sign embeddings. In our experiments, we ablate these three losses and examine the effects on performance.

More formally, suppose we have a corpus of images of glyphs $X$, where $X_i \in [0,1]^{64 \times 64}$ depicts one of $J$ characters and is authored by one of $K$ scribes. We define a set of vectors $Y$, where each $Y_j \in \mathbb{R}^d$ represents features of sign $j$. Similarly we define a set of vectors $Z$, where each $Z_k \in \mathbb{R}^d$ represents features of scribe $k$. Thus, a given $X_i$ depicting sign $j$ and authored by scribe $k$ is represented at a high level in terms of its corresponding $Y_j$ and $Z_k$, each of which is shared by other images corresponding to the same character or scribal hand respectively.

For a given $X_i$, the decoder network takes in the corresponding pair of vectors $Y_j$ and $Z_k$, and then passes them through a series of transpose convolutional layers to produce a reconstructed image $\hat{X}_i$. This reconstruction incurs a loss on the model based on its MSE compared to the original image.

Next, the scribal hand discriminator with parameters $\phi$ passes $X_i$ through several convolutional layers to obtain a feature embedding which is then concatenated with the corresponding $Z_k$ and passes through a series of fully connected layers, ultimately outputting a predicted $p(Z_k|X_i; \phi)$ which scores the likelihood of the scribal hand represented by $Z_k$ being the true author of $X_i$. Similarly, a sign discriminator with parameters $\psi$ predicts $p(Y_j|X_i; \psi)$, i.e. the likelihood of $X_i$ depicting the sign represented by $Y_j$. These discriminators are shown both one true match and one false match for an $X_i$ at every training step, giving us a basic cross-entropy loss.

These components all together yield the following loss function for our model:

$$L(X_i, Y_j, Z_k) = \|X_i - \hat{X}_i\|_2$$
$$+\lambda_1 * (\log p(Y_j|X_i; \phi) + \log(1 - p(Y_{j'}|X_i; \phi)))$$
$$+\lambda_2 * (\log p(Z_k|X_i; \psi) + \log(1 - p(Z_{k'}|X_i; \psi)))$$

where $j'$ and $k'$ are randomly selected such that $j \neq j'$ and $k \neq k'$.

The format of this model lends itself out of the box to various downstream tasks. For example, when a new clay tablet is discovered, it may be useful to determine if the author was a known scribal hand or someone completely new. The discriminator of our model can be used for exactly such as a purpose, as it can be directly queried for the likelihood of a new glyph matching each existing scribe. Even if this method is not exact enough to provide a definitive answer, we can at least use the predictions to get a rough sense of who the most similar known scribes are, despite never having seen that tablet during training.

Also, while our model treats the scribe and sign embeddings as parameters to be directly learned, one could easily imagine an extension in which embeddings are inferred from the images using an encoder network. Taking this further, we could even treat those embeddings as latent variables by adding a probabilistic prior, such as in Variational Autoencoders [9]. This would let us compute embeddings without having trained on that scribe's writing, taking the idea from the previous paragraph one step further, and letting us perform reconstructions of what we expect a full set of signs by that new scribe to look like.

## 3.1 Architecture

We now describe the architecture of our model in more detail. Let $F_i$ represent a fully connected layer of size $i$, $T$ represent a $2 \times 2$ transpose convolutional layer, $C$ be a $3 \times 3$ convolution, $I$ be an instance normalization layer [23], $R$ be a ReLU, $M$ be a $2 \times 2$ blur pool [24]. The decoder network is then $F_{1024*4*4} \rightarrow 4 \times (T \rightarrow 2 \times (C \rightarrow I \rightarrow R))$, where each convolutional layer $C$ reduces the number of filters by half. The discriminator's encoder network is then $C \rightarrow 4 \times (3 \times (C \rightarrow R) \rightarrow M) \rightarrow F_{16}$, where each convolutional layer $C$ increases the number of filters by double, except for the first which increases it to 64. Having encoded the input image, the discriminator concatenates the output of its encoder with the input scribe/sign embedding, and then passes the result through 7 fully connected layers to obtain the final output probability.

## 4    Data

### 4.1    Collection

There has been previous work on transcribing the tablets of Knossos by Olivier [11] and Pylos by Palaima [13]. These authors present their transcriptions in form of a rough table for each hand containing every example of each

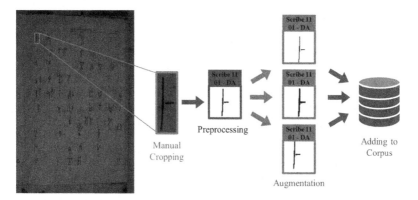

**Fig. 3.** Visualization of the data pipeline. Glyphs are (1) manually cropped from photos of pages and labeled by sign and scribe, (2) preprocessed to 64 × 64 binary images, and (3) augmented to produce multiple variants, before being added to the overall corpus.

sign written by them. While these are an important contribution and certainly parseable by a human reader, they are not of a format readily acceptable by computer vision methods. Since our model considers the atomic datapoint to be an individual glyph, we must first crop out every glyph from each table associated with a scribal hand. This is a nontrivial task as the existing tables are not strictly grid aligned, and many glyphs overlap with one another, which requires us to manually paint out the encroaching neighbors for many images. Ultimately we were able to convert the existing transcriptions into a digital corpus of images of individual glyphs, labeled both by scribal hand as well as sign type. This process was carried out by two graduate students under the guidance of a domain expert. Before being fed into our model, images were also binarized via Otsu's thresholding [12].

The digitization of this dataset into a standardized, well-formatted corpus is of much potential scientific value. While this process was laborious, it opens the doors not just to our own work, but also to any other future research hoping to apply machine learning methods to this data. Of course had this been a high-resource language such as English, we would have been able to make use of the multitude of available OCR toolkits to automate this preprocessing step. As such, one possible direction for future work is to train a system for glyph detection and segmentation, for which the corpus we have already created can serve as a source of supervision. This would be useful in more efficiently cropping similarly transcribed data, whether in Linear B or other related languages.

## 4.2   Augmentation

Due to the limited size of this corpus, there is a serious risk of models overfitting to spurious artifacts in the images rather than properties of the higher level structure of the glyphs. To mitigate this, we perform a data augmentation procedure to artificially inflate the number of glyphs by generating multiple variants

Horizontal
Translation

Vertical
Translation

Dilation
and Erosion

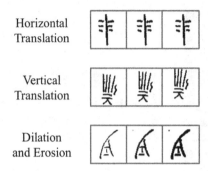

**Fig. 4.** Examples of the three types of data augmentation we perform. Each original image is augmented in all possible ways, resulting in an effective 27× increase in the size of our corpus.

of each original image that differ in ways we wish for our model to ignore. These are designed to mimic forms of variation that naturally occur in our data—for example, glyphs may not always be perfectly centered within the image due to human error during cropping, and strokes may appear thicker or thinner in some images based on the physical size of the sign being drawn relative to the stylus' width, or even as a result of inconsistencies in binarization during preprocessing. Following these observations, the specific types of augmentation we use include translation by 5 pixels in either direction horizontally and/or vertically, as well as eroding or dilating with a 2 × 2 pixel wide kernel (examples shown in Fig. 4). All together we produce 27 variations of each image. Doing this theoretically encourages our model to learn features that are invariant to these types of visual differences, and therefore more likely to capture the coarser stylistic properties we wish to study.

### 4.3   Statistics

Our dataset represents 88 Linear B sign types, written by 74 scribal hands (of these 48 are from Knossos and 26 are from Pylos). For each scribal hand, we observe an average of 56 glyph images, some which may represent the same sign. Conversely each sign appears an average of 47 times across all scribal hands. This gives us a total of 4, 134 images in our corpus, which after performing data augmentation increases to 111, 618.

## 5   Experiments

Having trained our model, we need to be able to quantitatively evaluate the extent to which it has learned features reflective of the sorts of stylistic patterns we care about, something which is nontrivial given the lack of any supervised ground truth success criteria. There are two metrics we use to accomplish this.

First, given that we expect scribes of the same findplace to share stylistic commonalities, we train a findplace classifier based on our scribal hand embeddings to probe whether our model's embeddings contain information that is informative for this downstream task. Secondly, since we have human generated scribal hand representations in the form of Skelton's manual features [18], we can measure the correlation between them and our model's learned manifold to see how well our embeddings align with human judgements of relevant features. We now describe these evaluation setups in more detail.

## 5.1   Findplace Classification

The first way we evaluate our model is by probing the learned embeddings to see how predictive they are of findplace, a metadata attribute not observed during training. Since our goal is to capture salient information about broad similarities between scribal hands, one coarse measure of our success at this is to inspect whether our model learns similar embeddings for scribes of the same findplace, as they would likely share similar stylistic properties. One reason for this underlying belief is that such scribes would have been contemporaries of each other, as can be infered from the fact that existing Linear B tablets were written "days, weeks, or at most months" before the destruction of the palaces [14]. For any scribes working together at the same place and time, we observe similar writing styles, likely as a result of the way in which they would have been trained. The number of scribes attested at any one palatial site is quite low, which would seem to rule out a sort of formal scribal school as attested in contemporary Near Eastern sources, and instead suggests that scribal training took place as more of a sort of apprenticeship, or as a profession passed from parents to children, as is attested for various sorts of craftsmen in the Linear B archives [14]. Thus, we expect that scribes found at the same findplace would exhibit similar writing styles, and therefore measuring how predictive our model's embeddings are of findplace serves as a reasonable proxy for how well they encode stylistic patterns.

To this end, we train a supervised logistic regression classifier (one per model) to predict the corresponding findplaces from the learned embeddings, and evaluate its performance on a held out set. In principal, the more faithfully a model's embeddings capture distinguishing properties of the scribal hands, the more accurately the classifier can predict a scribe's findplace based solely on those neural features. We perform $k = 5$ fold cross validation, and report average $F_1$ score over the 5 train/test splits. Our final numbers reflect the best score over 15 random initializations of the classifier parameters for each split.

Findplaces for the Knossos tablets were described in Melena and Firth [10], and for Pylos in Palaima [13]. However, there are some findplaces from which we only have one or two scribal hands in our corpus. As supervised classification is not practically possible in such cases, we exclude any findplaces corresponding to 2 or fewer scribal hands from our probing experiments. This leaves us with 63 scribal hands, labeled by 8 different findplaces to evaluate on.

## 5.2   QVEC

While findplace prediction is a useful downstream metric, we also wish to measure how well the neural features our model learns correlate with those extracted by expert humans. In order to do this, we use QVEC [22]—a metric originally developed for comparing neural word vectors with manual linguistic features—to score the similarity between our learned scribal hand embeddings and the manual features of Skelton [18]. At a high level, QVEC works by finding the optimal one-to-many alignment between the dimensions of a neural embedding space and the dimensions of a manual feature representation, and then for each of the aligned neural and manual dimensions, summing their Pearson correlation coefficient with respect to the vocabulary (or in our case the scribal hands). Therefore, this metric provides an intuitive way for us to quantitatively evaluate the extent to which our learned neural features correlate with those extracted by human experts. As a side note, while the original QVEC algorithm assumes a one-to-many alignment between neural dimensions and manual features, we implement it as a many-to-one, owing to the fact that in our setting we have significantly more manual features than dimensions in our learned embeddings, implying that each neural dimension is potentially responsible for capturing information about multiple manual features rather than vice versa as is the case in the original setting of Tsvetkov et al. [22].

## 5.3   Baselines

Our primary baseline is a naive autoencoder, that assumes a single separate embedding for every image in the dataset, therefore not sharing embeddings across signs or scribal hands. Its encoder layout is the same as that of our discriminator, and its decoder is identical to that of our main model so as to reduce differences due to architecture capacity. In order to obtain an embedding for a scribal hand using this autoencoder, we simply average the embeddings of all glyphs produced by that scribe. We report several ablations of our model, indicating the presence of the reconstruction loss, scribal hand discriminator loss, and sign discriminator loss by +Recon, +Scribe, and +Sign respectively. Finally, we provide the $F_1$ score for a naive model that always predicts the most common class to indicate the lowest possible performance.

## 5.4   Training Details

We use sign and scribe embeddings of size 16 each for all experiments. Our model is trained with the Adam [8] optimization algorithm at a learning rate of $10^{-4}$ and a batch size of 25. It is implemented in Pytorch [16] version 1.8.1 and trains on a single NVIDIA 2080ti in roughly one day. The classifier used for evaluation is trained via SGD at a learning rate of $10^{-3}$ and a batch size of 15.

**Table 1.** (Left) Best cross validated $F_1$ achieved by logistic regression findplace classifier on embeddings produced by ablations of our model, compared to an autoencoder, and naive most common label baseline. (Right) QVEC [22] score between models' embeddings and manual features of Skelton [18].

| Model | Findplace classifier $F_1$ | QVEC similarity to manual |
|---|---|---|
| Most common | 0.154 | - |
| Autoencoder | 0.234 | 54.1 |
| −Recon, +Scribe, +Sign | 0.198 | 53.2 |
| +Recon, −Scribe, −Sign | 0.234 | 51.1 |
| +Recon, +Scribe, −Sign | 0.206 | 52.9 |
| +Recon, +Scribe, +Sign | **0.292** | **57.0** |

# 6  Results

Table 1 shows results of a findplace classifier trained on our system's output vs. that of an autoencoder, with the score for a most common baseline as a lower bound. We see that our full model with all three losses achieves the highest score, with reduced performance as they are successively removed. The worst score comes from the model trained on purely discriminative losses with no decoder, suggesting that the reconstruction loss provides important bias to prevent overfitting given the limited dataset size. This also makes sense given the strong performance by the vanilla autoencoder.

For reference, Skelton [18] provides manual feature representations for 20 of the scribal hands in our corpus (17 of which are from the 6 findplaces shared by at least one other scribe in that set). If we perform a similar probing experiment with the hand crafted features for those 17 scribes, we get an $F_1$ score of 0.37. While this result is useful as an indicator of where the upper bound on this task may be, it is very important to note that this experiment is only possible on a small subset of the scribes we evaluate our model on and is therefore not directly comparable to the results in Table 1. A more direct comparison to the neural features of our model would require extracting manual features for the remaining 54 scribal hands, which is beyond the scope of this work.

Table 1 also shows the QVEC similarity between each of the models' learned embeddings and the manual features of Skelton [18]. Once again, our system ranks highest, with the autoencoder also performing more strongly than our ablations. Interestingly, we see here that our discriminative losses appear more important than the reconstructive loss, suggesting that by encouraging the model to focus on features that are highly distinguishing between scribes or signs we may end up recovering more of the same information as humans would.

Figure 5 shows example reconstructions from our decoder. These reconstructions are based solely on the scribe and sign representations for the corresponding row and column. We see that the model is able to realistically reconstruct the glyphs, showing that it is very much capable of learning shared representations for sign shape and scribal hand styles, which can then be recombined to render back the original images. This would imply that our embedding space is able to adequately fit the corpus, and perhaps requires some additional inductive bias in order to encourage those features to be more high level.

There are also additional downstream tasks we have not yet evaluated on that may be worth attempting with our system. The most obvious is that of phylogeny reconstruction. Now that we have trained scribal hand embeddings, it remains to be seen how similar a phylogenetic reconstruction based on these features would resemble that found by Skelton [18]. This may give us a better idea as to whether the embeddings we have computed are informative as to the broader hierarchical patterns in the evolution of Linear B as a writing system.

**Fig. 5.** Example reconstructions from our model for various scribal hands. The model manages to capture subtle stylistic differences, but we see a failure mode in the middle row right column where the output appears as a superposition of two images.

## 7   Conclusion

In this paper we introduced a novel approach to learning representations of scribal hands in Linear B using a new neural model which we described. To that end, we digitized a corpus of glyphs, creating a more standardized dataset more readily amenable to machine learning methods going forwards. We benchmarked the performance of this approach compared to simple baselines and ablations by quantitatively measuring how predictive the learned embeddings were of held out metadata regarding their findplaces. In addition, we directly evaluated the similarity of our scribal hand representations to those from existing manual features. We believe this work leaves many open avenues for future research, both with regard to the model design itself, as well as to further tasks it can be applied towards. For example, one of the key applications of manual features in the past has been phylogenetic analysis, something which our neural features could potentially augment for further improvements. This type of downstream work can further be used in service of dating Linear B tablets, understanding the patterns in the writing system's evolution, building OCR systems, or even analyzing other related but as of yet undeciphered scripts.

# References

1. Bennett Jr, E.L.: The Minoan linear script from Pylos. University of Cincinnati, Cincinnati (1947)
2. Bulacu, M., Schomaker, L.: Text-independent writer identification and verification using textural and allographic features. IEEE Trans. Pattern Anal. Mach. Intell. **29**(4), 701–717 (2007)
3. Christlein, V., Nicolaou, A., Seuret, M., Stutzmann, D., Maier, A.: Icdar 2019 competition on image retrieval for historical handwritten documents. In: 2019 International Conference on Document Analysis and Recognition (ICDAR), pp. 1505–1509. IEEE (2019)
4. Djeddi, C., Al-Maadeed, S., Siddiqi, I., Abdeljalil, G., He, S., Akbari, Y.: Icfhr 2018 competition on multi-script writer identification. In: 2018 16th International Conference on Frontiers in Handwriting Recognition (ICFHR), pp. 506–510. IEEE (2018)
5. Driessen, J.: The Scribes of the Room of the Chariot Tablets at Knossos: Interdisciplinary Approach to the Study of a Linear B Deposit, Ediciones Universidad de Salamanca (2000)
6. Ferrara, S.: Cypro-Minoan Inscriptions: Volume 1: Analysis, vol. 1. Oxford University Press, Oxford (2012)
7. Firth, R.J., Skelton, C., et al.: A Study of the Scribal Hands of Knossos Based on Phylogenetic Methods and Find-Place Analysis pp. 159–188, Ediciones Universidad de Salamanc (2016)
8. Kingma, D.P., Ba, J.: Adam: A method for stochastic optimization. ICLR (2015)
9. Kingma, D.P., Welling, M.: Auto-encoding variational bayes. ICLR (2014)
10. Melena, J.L., Firth, R.J.: The Knossos Tablets. INSTAP Academic Press (Institute for Aegean Prehistory), Philadelphia (2019)
11. Olivier, J.P.: Les scribes de cnossos: essai de classement des archives d'un palais mycénien (1965)
12. Otsu, N.: A threshold selection method from gray-level histograms. IEEE Trans. Syst. Man Cybern. **9**(1), 62–66 (1979)
13. Palaima, T.: The Scribes of pylos. Incunabula graeca, Edizioni dell'Ateneo, Roma (1988)
14. Palaima, T.G.: Scribes, Scribal Hands and Palaeography. pp. 33–136. Peeters Louvain-la-Neuve, Walpole (2011)
15. Palmer, R.: How to Begin? an Introduction to Linear b Conventions and Resources. pp. 25–68. Peeters Louvain-la-Neuve, Walpole (2008)
16. Paszke, A., et al., Automatic differentiation in PyTorch. In: NIPS Autodiff Workshop (2017)
17. Siddiqi, I., Vincent, N.: Text independent writer recognition using redundant writing patterns with contour-based orientation and curvature features. Patt. Recogn. **43**(11), 3853–3865 (2010)
18. Skelton, C.: Methods of using phylogenetic systematics to reconstruct the history of the linear b script. Archaeometry **50**(1), 158–176 (2008)
19. Skelton, C.: A look at early mycenaean textile administration in the pylos megaron tablets. Kadmos **50**(1), 101–121 (2011)
20. Skelton, C., Firth, R.J., et al.: A Study of the Scribal Hands of Knossos Based on Phylogenetic Methods and Find-Place Analysis. Part III Dating the Knossos Tablets Using Phylogenetic Methods pp. 215–228. Ediciones Universidad de Salamanca (2016)

21. Srivatsan, N., Barron, J., Klein, D., Berg-Kirkpatrick, T.: A deep factorization of style and structure in fonts. In: Proceedings of the 2019 Conference on Empirical Methods in Natural Language Processing and the 9th International Joint Conference on Natural Language Processing (EMNLP-IJCNLP). pp. 2195–2205. Association for Computational Linguistics, Hong Kong, China, November 2019. https://doi.org/10.18653/v1/D19-1225, https://www.aclweb.org/anthology/D19-1225
22. Tsvetkov, Y., Faruqui, M., Ling, W., Lample, G., Dyer, C.: Evaluation of word vector representations by subspace alignment. In: Proceedings of the 2015 Conference on Empirical Methods in Natural Language Processing. pp. 2049–2054 (2015)
23. Ulyanov, D., Vedaldi, A., Lempitsky, V.: Instance normalization: the missing ingredient for fast stylization. arXiv preprint arXiv:1607.08022 (2016)
24. Zhang, R.: Making convolutional networks shift-invariant again. In: International Conference on Machine Learning, pp. 7324–7334. PMLR (2019)
25. Zhang, X.Y., Xie, G.S., Liu, C.L., Bengio, Y.: End-to-end online writer identification with recurrent neural network. IEEE Trans. Hum.-Mach. Syst. **47**(2), 285–292 (2016)

# READ for Solving Manuscript Riddles: A Preliminary Study of the Manuscripts of the 3rd ṣaṭka of the *Jayadrathayāmala*

Olga Serbaeva[1][(✉)] [iD] and Stephen White[2]

[1] University of Zurich, Zürich, Switzerland
`olga.serbaeva@aoi.uzh.ch`
[2] Ca'Foscari University of Venice, Venezia, Italy
`stephen.white@unive.it`

**Abstract.** This is a part of an in-depth study of a set of the manuscripts related to the *Jayadrathayāmala*. Taking JY.3.9 as a test-chapter, a comparative paleography analysis of the 11 manuscripts was made within READ software framework. The workflow within READ minimized the effort to make a few important discoveries (manuscripts containing more than one script, identification of the manuscripts potentially written by the same person) as well as to create an overview of the shift from Nāgarī to Newārī and, finally, to Devanāgarī scripts within the history of manuscript transmission of a single chapter. Exploratory statistical analysis in R of the syllable frequency in each manuscript, based on the paleography analysis export from READ, helped to establish that there are potentially two lines of manuscript transmission of the JY.3.9.

**Keywords:** Sanskrit manuscripts · Indic scripts · Paleography · Virtual research environment

## 1 Introduction

### 1.1 Brief History of the Manuscript Transmission of the *Jayadrathayā*mala

The *Jayadrathayāmala* (further JY) is a text compiled in Northern India around the end of the 9th–10th century, and it was cited on multiple occasions by Kashmiri polymath Abhinavagupta in his *Tantrāloka*. However, none of the manuscripts from Kashmir seem to have reached the academia. The manuscript transmission is almost exclusively Nepalese: the Nepal-German Manuscript Preservation project (NGMPP) lists about 30 *Jayadrathayāmala*-related codices. Prof. Dr. Diwakar Acharya, University of Oxford[1], suggested that the oldest existing manuscript of this text comprising all four books, palm-leaf JY_B here, was written in Vārāṇasī around the time of the reign of the king Jayachandra (1170–1194). Theoretically, all Nepalese transmissions of the JY could stem from this single palm-leaf.

---

[1] Pers. comm. 05.04.2021.

© Springer Nature Switzerland AG 2021
E. H. Barney Smith and U. Pal (Eds.): ICDAR 2021 Workshops, LNCS 12917, pp. 339–348, 2021.
https://doi.org/10.1007/978-3-030-86159-9_24

## 1.2   Aim and Methods of the Present Study

The aim of the present study is to verify if all Nepalese transmissions indeed could have come from this single source, or if the paleography and statistical facts would allow any other representation of the history of transmission of JY. A secondary aim of this phase of the study is to test, whether Research Environment for Ancient Documents (READ's) paleography features can be helpful for resolving the above named issue.

The preliminary results of this phase are based on a single chapter of the 3rd book, namely, JY.3.9, dealing with an invocation and the worship of the goddess Pratyaṅgirā. This is a short chapter, comprising some 73 verses, and it is situated rather towards the middle of the text, i.e. the script shall be much less calligraphic as it is on the initial folios of a book. The chosen chapter survived in 8 manuscripts of the JY_A, B, C, D, E, G, I, K and in 3 manuscripts of the text called the *Jayadrathayāmalamantroddhāraṭippanī*, (further JYMUṬ_A, B, O). It is an independent compilation of JY materials, which often includes the whole chapters.

The overall methodology can be divided into two procedures READ[2] Paleography analysis and R statistical analysis. We use READ to segment and group the individual characters of the text, as well as hand writing characterization through side by side visual inspection of each set of characters across the different transmissions using READ's generic tagging/classification feature. We use the export of these results as the input to the R-based statistical analysis. Having mapped the syllable frequency within each manuscript, we shall cluster the manuscripts by their proximity in relation to their particular syllabic portrait. This should point to the number of independent lines of transmission.

## 2   Procedure

### 2.1   READ as a Virtual Research Environment (VRE) for Working with Manuscript Materials

Producing a curated digital edition is at the center of READ's design along with a data model in which heavily interlinked data used to support the variety of workflows for the different language types. READ has projects in (1.) syllabic Indic languages, (2.) alphabetic Latin, Greek, Hebrew, (3.) logo-syllabic Mayan. The workflow consists of adding a manuscript image (via upload or url), marking segments on the manuscript image, adding a transcript, linking the transcript and the marked segments together, and visually inspecting and classifying characters using the paleography tool.

READ is centered around the concept of a transcription (interpretation) encoded to match the original markings on the artifact. Markings are annotated by segments (boundaries). The linkage between segments and encoding of each letter/syllable/glyph of the document is preserved throughout the workflow (Fig. 1).

---

[2] READsoftware 2021: https://github.com/readsoftware/read/wiki.

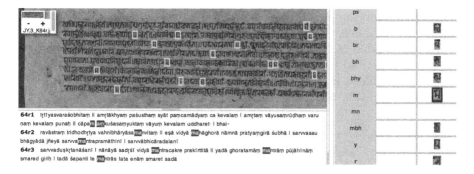

**Fig. 1.** Direct links between Paleography report (right), Document (top) and Edition (bottom) for the syllable "ma" of JY.3.9_K selected. Manuscript image: Digitale Sammlungen, Staatsbibliothek zu Berlin

## 2.2   Workflow in READ

Having obtained the manuscript images, one should find out if there are any missing folios, wrongly numbered pages, and work out the manuscript structure by assigning coherent filenames to the images of each side of the folio.

For importing transcriptions, READ has a data/constraint driven parser to validate the transcription before creating the highly link data model. One should ensure the transcription adheres to the defined encoding constraints for the language variation used (here Sanskrit, for example, transcription 1 will validate, while transcription 2 - will not.). During import READ will identify deviations which should be adjusted in the READ constraint data or changed in the transcription before import. This constitutes the one of three transcription coherency verification flowlets built into READ.

| Word on the surface | Translation | Transcr. 1 | Transcr. 2 |
|---|---|---|---|
| अमृत | "nectar" | amṛtaṃ | amṛtam |

**Fig. 2.** Various transcription standards of the same word, of which one is in the data model, and the other is not.

Having imported the document and uploaded the image(s), segmentation of every glyph (letter/syllable and punctuation mark) necessary for the research is completed by a semi-automated linking process. For manual segmentation, one should expect to spend about 10 min for marking up 500 glyphs.

The second stage of built in verification happens during the process of linking. When hovering over letter/syllable/word in the transcription with the mouse, the corresponding glyph(s) highlight on the image. Having located any misalignments, one should correct the edition or relink.

The third, and the final of level of verification comes when one works with the Paleography table. It is a grid where each cell represents the set of a single type of glyph found in the document. The exact number of rows and columns in the table is defined by the type of language. For example, for Greek one will only see the alphabet in one column, for Sanskrit a consonant vs vowel matrix will be generated (Fig. 3).

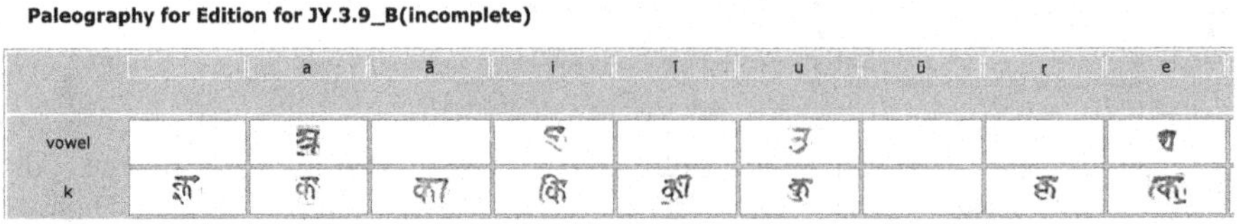

**Paleography for Edition for JY.3.9_B(incomplete)**

| | | a | ā | i | ī | u | ū | r̥ | e |
|---|---|---|---|---|---|---|---|---|---|
| vowel | | अ | | ऍ | | ऒ | | | य |
| k | कॏ | कॆ | कॉ | कि | क्य | कु | | के | क्वः |

**Fig. 3.** Paleography report: partial grid showing consonants as rows and vowels as columns, JY.3.9_B. Digitally modified segments of the manuscript JY_B, copy obtained via NGMPP.

READ shows one example of the glyphs for a given cell, clicking on it will open a presentation of all linked glyphs. In this side-by-side presentation one will immediately see any intruders (clicking on the intruder identifies its location in the document). Thanks to this triple coherency check, close to 100% correctness of reading can be achieved. This level of precision as well as preserved links between segments and glyphs constitute a solid basis for further analysis (Fig. 4).

**Fig. 4.** Paleography for Newārī "a" of BT-1 type by order, JY.3.9_A. Digitally modified segments of the manuscript JY_A, copy obtained via NGMPP.

In this side-by-side presentation, all segments can be tagged with one of several values in each of up to 3 categories (work in progress to extend this to 6 categories), a segment belongs to "unknown" until it is classified. All categories are independent from one another, thus one has sufficient number (10) of values per category (3) to mark any desired glyph-particularities.

## 3   Selected Discoveries

### 3.1   Manuscripts with Multiple Scripts/Letter Variants

While checking each letter in the paleography tables, some important manuscript characteristics became clear. JY_B and JYMUṬ_B within a single chapter and often within a single word mix two different scripts. For JYMUṬ_B there is a switch from Devanāgarī to probably one that is native to the scribe, namely, Newārī script (distribution of the two is about 50-50).

As for JY_B, going through the paleography report glyph by glyph made it clear that under the same syllables (consonant plus "e" or "o") there were two very different variations of glyphs (Fig. 5). The actual script is yet to be determined, but it has close similarities with the North-Indian variety of Nāgarī. The documents written in Nāgarī between the 10th and the 12th centuries, besides having similar shapes of glyphs, demonstrate the same variations concerning the writing of the consonant combinations with "e" and "o"[3]. This fascinating script inconsistency can point to a combination of the two different sub-types of the Nāgarī script attested within a single manuscript. If the same incoherence is found in the other 3 ṣaṭkas of the palm-leaf manuscript of JY_B), it would argue for the idea of Diwakar Acharya that all palm-leaf manuscripts of JY indeed constitute a set belonging to the same period and probably written by the same scribe.

**Fig. 5.** Variation of "me" in JY.3.9_B made apparent in READ. In the close-up vowel sign "e" is highlighted in red for each variant. Digitally modified segments of the manuscript JY_B, copy obtained via NGMPP.

### 3.2 Identification of the Same or Very Similar Handwriting

By simply opening up two paleography reports for each combinations of two of the selected 11 manuscripts side-by-side in READ, it became apparent that JY_D and JYMUṬ_O, separated in time by some 22 years according to their dating, are most probably written by the same person. The initial idea in READ was further confirmed by the similarity of the syllabic portrait of the two in the complete paleography set comprising about 690 different syllables (Fig. 6).

### 3.3 Overview of the JY Scripts from 12th to the 20th Centuries

Exporting the encoded and categorised segments from READ has allowed the creation of a compiled overview table, where the manuscripts are arranged

---

[3] Based on Gupta [n.d.]. See also Fig. 5 here.

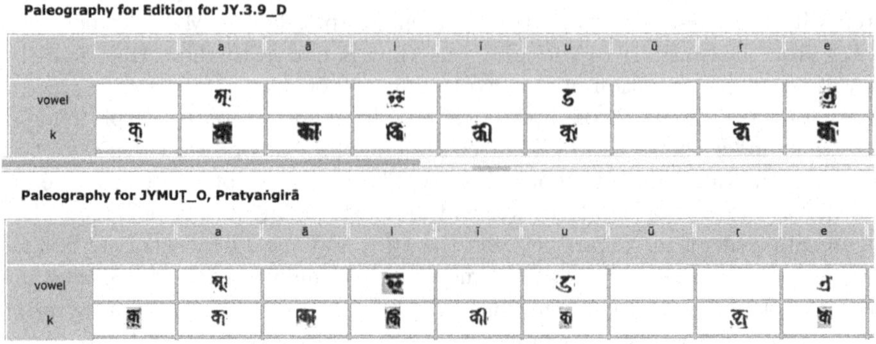

**Paleography for Edition for JY.3.9_D**

| | a | ā | i | ī | u | ū | r̥ | e |
|---|---|---|---|---|---|---|---|---|
| vowel | ग़ः | | ऌ | | ड़ | | | ऴ |
| k | क़ | क़ | क़ | ङि | क़ी | ग़ु | | ऴ | ऴा |

**Paleography for JYMUṬ_O, Pratyaṅgirā**

| | a | ā | i | ī | u | ū | r̥ | e |
|---|---|---|---|---|---|---|---|---|
| vowel | ऌ | | ऌ | | ड़ | | | ऴ |
| k | ऌ | क़ | ऌ | क़ | क़ी | ऴ | | ऴ | ऴो |

**Fig. 6.** Paleography of JY_D and JYMUṬ_O side-by-side. Digitally modified segments of the manuscripts, copies obtained via NGMPP.

according to their dates (when known). The table also represents the main differences between (1.) Nāgarī, (2.) Newārī and (3.) Devanāgarī scripts in their historical transition from 1 to 3 based on the 11 manuscript variants of the same chapter of the JY (Fig. 7).

## 3.4 Exploratory Statistical Analysis in R of the Data Produced by READ Paleography Report

It is still impossible to cluster the manuscripts based on the varieties of handwriting, but we are in the process of designing the methodology that uses Computer Vision (the present data set (JY.3.9) will serve as training data) to process all 201 chapters of the JY. However, even the paleography report constitutes a solid basis for the statistical analysis in R code adapted for the project.[4] For example, having taken just the file names of each exported segment (of the type bha_bt1_63r4.10.png, where the syllable "bha", classified under "bt1", appears on the folio "63r", line "4", syllable number "10"), we have first generated a syllabic portrait of each manuscript, i.e. how many times a syllable appears in a given manuscript. From that the mean frequency was calculated, and based on a difference from that mean, we have found that a set of just 59 syllables (punctuation excluded) was sufficient to describe the major differences between the 11 manuscripts for the same chapter. These include the preferences for single and duplicated letters (*sarva* versus *sarvva*), for rendering the nasals (Pratyaṅgirā versus Pratyaṃgirā), and the final *sandhi*, i.e. the letters at the end of the word.[5]

---

[4] For clustering function dfm_weight, library quanteda in R. Visualisation - pheatmap. We thank Dmitry Serbaev, Omsk, Russia, for his expert advise concerning R.

[5] For the historical changes in Sanskrit orthography in a close comparison to Middle-Indic languages see Edgerton 1946:199, 202–203.

**Fig. 7.** Selected digitally modified segments (all copies obtained via NGMPP, except for JY_K, belonging to Staatsbibliothek zu Berlin) from the 11 chosen manuscripts demonstrating the instability of scripts in JY_B (switch from Nāgarī to Newārī) and JYMUṬ_A and B (from Newārī to Devanāgarī) as well as the outline of change from Nāgarī to Newārī and to Devanāgarī script in the process of transmission of the JY.3. JYMUṬ_A, contains Newārī characters only occasionally.

One could argue that orthography could be accidental and therefore cannot be used to identify the manuscript relationships. However, the present case is a particular one: we are talking about 11 manuscripts of the same chapter written in verses, which means that the structure is particularly stable, and these

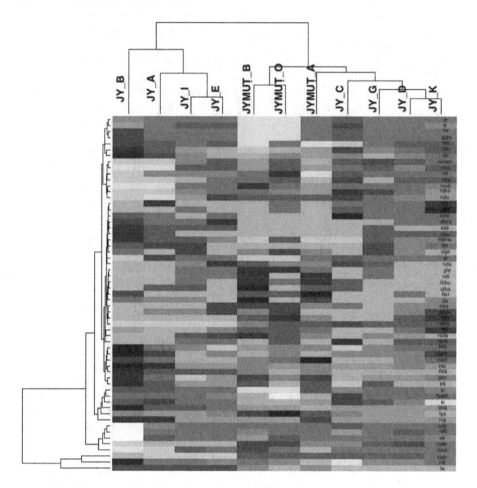

**Fig. 8.** Above heatmap is the visualisation of the clusters of the manuscripts by the closeness of their syllabic portrait. The "differentiating" syllables (listed on the right) of the 11 manuscripts are shown by the number of occurrences, where colour intensity indicates minimum (white) to maximum (dark read).

manuscripts are the same for about 98% of their material. Thus, the tiny differences found were rechecked for stability and significance, i.e. singular scribal mistakes were separated from the reoccurring, statistically significant patterns that were made apparent during the clustering process. All this is represented here as heatmap (Fig. 8).

The difference in the manuscript cluster visualisation is represented by height. One line is stemming from JY_B based on syllabic portrait similarity and including JY_A, E, I, of which only the last two are very closely related; the other is rather related to JYMUṬ transmission variant, and including the remaining 7 manuscripts, of which JYMUṬ_B and JYMUṬ_O and JY_K and JY_D are particularly close to each other. This is a preliminary result based on a single chapter, however, it already shows that this chapter has at least two independent lines of transmission, one rooted in JY_B, but another one is coming from an independent manuscript, from which both JYMUTs and JY_C, G, D, K stem. Most likely it was very close to JY_C.

# 4   Conclusion

We have discovered important variations in the scripts of two manuscripts, for JY_B it confirms the dating around the 12th century suggested by Diwakar Acharya, however, further work is needed to ascertain the precise varieties of the Nāgarī script; the JYMUT_B attests the tendency to change the script from Newārī to Devanāgarī at the beginning of the 18th century. Paleography analysis has allowed the establishment that JY_D and JYMUT_O were likely written by the same scribe. We have also compiled a table representing the evolution of the Newārī script from 1593 to 1723. The syllabic portrait of each manuscript, established with precision, has produced an exploratory clustering, which highlights the changes in orthography. The clustering results tend to support the idea that there are at least 2 different lines in the manuscript transmission, and we can suggest that another variation of JY, not rooted in JY_B, existed in Nepal which gave rise to the JYMUT variant of JY.3.9. The fact that JY_B stands very far from the rest of the manuscripts both in time and by its syllabic portrait, also suggests that there existed more manuscripts in between, which, we hope, will surface in the future.

READ software paleography tools proved to be an important asset for this research, and its data model that maintaining the link between the manuscript surface, transcription, and various forms of interpretation, allows the scholar to see the higher levels of associations and to verify the coherence of the data. READ handles highly linked Big Data, which is visualised and made available in an interactive bite-size format for human scholars. This is a good example of what a VRE should be.

Besides clarifying the questions of the manuscript transmission, this IT-DH collaboration has also identified priorities for new features to develop in READ.

The next steps of the current study include the expanding to all chapters of JY using the current/possibly modified methodology to extend the analysis of the transmission of JY. The analysis will be enriched by using data produced through other READ features (lexicography, syntax, compound analysis) with Computational Linguistics approaches.

# References

1. Benoit, K., et al. Quanteda: an R package for the quantitative analysis of textual data. J. Open Sour. Softw. **3**(30), 774 (2018). https://doi.org/10.21105/joss.00774, https://quanteda.io
2. Edgerton, F.: Meter, phonology, and orthography in Buddhist Hybrid Sanskrit. J. Am. Orient. Soc. **66**(3), 197–206 (1946). https://www.jstor.org/stable/595566
3. Gupta, P. L.: Medieval Indian Alphabets: [Nagari, North India]. All Indian Educational Supply Company, Delhi (n.d., pre 2012)
4. [Jayadrathayāmala, ṣaṭka 3]: A NGMPP Microfilm A152/9; B NGMPP Microfilm B26/9; C NGMPP Microfilm A1312/25; D NGMPP Microfilm B122/3; E NGMPP Microfilm C72/1, G NGMPP Microfilm B121/13, I NGMPP Microfilm C47/03
5. [Jayadrathayāmala, ṣaṭka 3]: K. https://digital.staatsbibliothek-berlin.de/werkansicht?PPN=PPN898781620&PHYSID=PHYS_0004. Accessed 23 May 2021
6. [Jayadrathayāmalamantroddhāraṭippaṇī]: A NGMPP Microfilm B122/07; B NGMPP Microfilm A1267/3; O NGMPP Microfilm A152/8
7. NGMPP (Nepal-German Manuscript Preservation Project) Catalogue, https://catalogue.ngmcp.uni-hamburg.de/content/search/ngmcpdocument.xed. Accessed 22 May 2021

# ICDAR 2021 Workshop on Document Images and Language (DIL)

# DIL 2021 Preface

Document image analysis and recognition has always been the focus of the ICDAR conference. Rich information can be exploited in document images and text to understand the document. In the past, researchers in this community usually turned to computer vision technology alone to solve the problems of text detection and recognition, layout analysis, and graphic structure recognition.

In order to understand the messages captured in the multimodal nature of documents, various techniques from natural language processing (NLP), such as named entity recognition and linking, visual questions answering, and text classification, can be combined with the traditional methods of optical character recognition (OCR), layout analysis, and logical labeling that are well established in the ICDAR community. This workshop, Document Images and Language (DIL), follows the tradition of ICDAR workshops and addresses the research between computer vision and NLP including the recent developments in the field of machine learning. The aim is to provide a forum to attract researchers from both communities to share their knowledge and expertise or colleagues who are already working at the edge of both fields in order to bring substantial improvement to document image understanding.

The first edition of DIL was held virtually on September 6, 2021, at the ICDAR conference. We invited three great speakers: Lei Cui from Microsoft Research Asia, Lianwen Jin from the South China University of Technology, and Heiko Maus from the German Research Center for Artificial Intelligence (DFKI). They have diverse backgrounds and research experiences, and provided unique insights on the topics of the workshop.

The first edition of DIL was held virtually on Sept 6th at ICDAR 2021 and was organized by Andreas Dengel (DFKI & University of Kaiserslautern, Germany), Cheng-Lin Liu (Institute of Automation of Chinese Academy of Sciences, China), David Doermann (University of Buffalo, USA), Errui Ding (Baidu Inc., China), Hua Wu (Baidu Inc., China), Jingtuo Liu (Baidu Inc., China).

When it comes to paper contributions, we received 11 submissions of which we accepted nine, including eight long papers and one short paper. We followed a single blind review process and each submission was reviewed by three reviewers. The accepted papers cover the topics of OCR, information extraction with visual and textual features, and language models for document understanding.

Altogether, we believe that the first edition of DIL was a success, with high-quality talks and contributed papers. We are glad to see that this topic attracted so much interest in the community and hope the research on multimodal approaches in document image understanding will become a trend in the future.

September 2021

Andreas Dengel
Cheng-Lin Liu
David Doermann
Errui Ding
Hua Wu
Jingtuo Liu

# Organization

## Organizers

Andreas Dengel      DFKI & University of Kaiserslautern, Germany
Cheng-Lin Liu      CASIA, China
David Doermann      University of Buffalo, USA
Errui Ding      Baidu Inc., China
Hua Wu      Baidu Inc., China
Jingtuo Liu      Baidu Inc., China

## Program Committee

Chengquan Zhang      Baidu Inc., China
Dominique Mercier      German Research Center for Artificial
     Intelligence, Germany
Fei Qi      Xidian University, China
Lianwen Jin      South China University of Technology, China
Lingyong Yan      Chinese Academy of Sciences, China
Longlong Ma      Chinese Academy of Sciences, China
Marco Sticker      German Research Center for Artificial
     Intelligence, Germany
Syed Yahseen Raza Tizvi      German Research Center for Artificial
     Intelligence, Germany
Xinyan Xiao      Baidu Inc., China
Yu Zhou      Chinese Academy of Sciences, China
Zhenyu Huang      Sichuan University, China

## Additional Reviewers

Chengyang Fang
Fangchao Liu
Pengfei Wang
Xugong Qin
Jianjian Cao
Minghui Liao
Dely Yu
Xinye Yang

# A Span Extraction Approach for Information Extraction on Visually-Rich Documents

Tuan-Anh D. Nguyen[1(✉)], Hieu M. Vu[1], Nguyen Hong Son[1],
and Minh-Tien Nguyen[1,2]

[1] Cinnamon AI, 10th floor, Geleximco Building, 36 Hoang Cau,
Dong Da, Hanoi, Vietnam
{tadashi,ian,levi,ryan.nguyen}@cinnamon.is
[2] Hung Yen University of Technology and Education, Hung Yen, Vietnam
tiennm@utehy.edu.vn

**Abstract.** Information extraction (IE) for visually-rich documents (VRDs) has achieved SOTA performance recently thanks to the adaptation of Transformer-based language models, which shows the great potential of pre-training methods. In this paper, we present a new approach to improve the capability of language model pre-training on VRDs. Firstly, we introduce a new query-based IE model that employs span extraction instead of using the common sequence labeling approach. Secondly, to extend the span extraction formulation, we propose a new training task focusing on modelling the relationships among semantic entities within a document. This task enables target spans to be extracted recursively and can be used to pre-train the model or as an IE downstream task. Evaluation on three datasets of popular business documents (*invoices, receipts*) shows that our proposed method achieves significant improvements compared to existing models. The method also provides a mechanism for knowledge accumulation from multiple downstream IE tasks.

**Keywords:** Information extraction · Span extraction · Pre-trained model · Visually-rich document

## 1 Introduction

Information Extraction (IE) for visually-rich documents (VRDs) is different from plain text documents in many aspects. VRDs contain sparser textual content, have strictly defined visual structures, and complex layouts which do not present in plain text documents. Due to these characteristics, existing works on IE for VRDs mostly employ Computer Vision [1] and/or graph-based [6,9,11] techniques. Recently, BERT provides more perspectives on VRDs from the Natural Language Processing point of view [3,4,12–15].

LayoutLM [13] emerges as a simple yet potential approach, where the BERT architecture was extended by the addition of 2D positional embedding and visual

© Springer Nature Switzerland AG 2021
E. H. Barney Smith and U. Pal (Eds.): ICDAR 2021 Workshops, LNCS 12917, pp. 353–363, 2021.
https://doi.org/10.1007/978-3-030-86159-9_25

embedding to its input embedding [2]. The model was trained by the Masked Visual Language Modelling (MVLM) objective [13], which is also an adaptation of positional embedding and visual embedding to the Masked Language Modelling objective [2] of BERT. More recently, LayoutLMv2 [12] was introduced with more focus on better leveraging 2D positional and visual information. There is also the addition of two new pre-training objectives that were designed for modeling image-text correlation.

While LayoutLM is very efficient to model multi-modal data like visually-rich documents, it bears a few drawbacks regarding practicality. First, pre-training a language model requires a large amount of data, and unlike plain text documents, visually-rich document data is not readily available in such large quantities, especially for low-resource languages (e.g. Japanese). Second, previous methods mostly utilize sequence labeling for IE, but we argue that this method might not work well for short entities due to the imbalance of classes and the training loss function [5,10]. Third, sequence labeling requires a fixed and predefined set of entity tags/classes for each dataset, which hinders the application of applying the same IE model to multiple datasets.

To address the aforementioned problems, we propose a new method employing span extraction instead of using sequence labeling. The method combines QA-based span extraction and multi-value extraction using a novel recursive relation predicting scheme. We also introduce a pre-training procedure that extends the LayoutLM to both English and non-English datasets by using semantic entities. Since the query is embedded independently with the context, this new method can be applied across multiple datasets, thus enable the seamless accumulation of knowledge through one common formulation.

## 2    Information Extraction as Span Extraction

The most common approach for extracting information from visually-rich documents is sequence labeling which assigns a label for each token of the document [3,4,12–15]. Even though this solution showed promising results, we argue that the sequence labeling might not work well when some entity types have only a small amount of samples or when the target entities are short spans of text in dense documents. The possible reason is the imbalance of labels when training IE models. Moreover, sequence labeling requires an explicit definition of token classes, which diminishes the reusability of the model on other tasks.

To overcome this challenge, we follow the span extraction approach, also known as Extractive Question Answering (QA) in NLP [2]. Different from sequence labeling which assigns a label for every token based on their embedding, QA models predict the start and the end positions of the corresponding answer for each query. For the IE problem, a query represents a required field/tag that we want to extract from the original document context. Each query is represented as a learnable embedding vector and kept separately from the main language model encoder.

Figure 1 describes our span extraction model. Let $D = \{w_1, w_2, ..., w_n\}$ is the input sequence and $H = [h^1, ..., h^j, ..., h^n] = f(D|\Theta)$ is the hidden representation

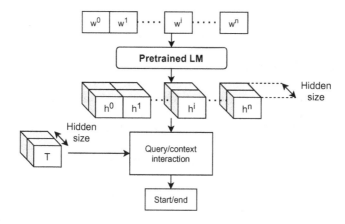

**Fig. 1.** The span extraction model.

of the input sequence produced by the pretrained encoder $f()$ with $h^i \in \mathbf{R}^c$ is the representation vector of the token $i^{th}$, $T \in \mathbf{R}^c$ is the vector representation of the query (field embedding) and $c$ is the hidden size of the encoder, the start and end position of the corresponding answer is calculated as:

$$start, end = g(T, H) \tag{1}$$

with $g()$ is represented by an attention layer [7] that captures the interaction between the query $T$ and the context $H$.

To calculate the span extraction loss, the $softmax$ operation is applied over all tokens in the context instead of over the token classes for each token as in sequence tagging. The softmax loss on all sequence tokens eliminates the class imbalance problem for shorter answer entities. The separation of question (query) embedding and context embedding also opens the potential of accumulating and reusing information on different datasets.

## 3  Pre-training Objectives with Recursive Span Extraction

### 3.1  LayoutLM for Low-Resource Languages

This section describes some effective methods for transferring the LayoutLM to low-resource languages, e.g. Japanese. Pre-training a language model from scratch with the MLM objective normally requires millions of data and can take a long time for training. While such an amount of data can be acquired in plain text, visually rich data does not naturally exist in large quantities due to the lack of word-level OCR annotations. Since pre-trained weights are available for English only, applying LayoutLM to other languages is still an open question. To overcome this, we propose a simple transferred pre-training procedure for LayoutLM that first takes advantage of the available pre-trained weights of BERT and then transfers its contextual representation to LayoutLM.

**Overcoming Data Shortage.** To transfer the LayoutLM model from English (the source language) to another language (the target language), we follow the following procedure. First, the model's word embedding layer was initialized from the pre-trained BERT of the target language and the rest layers (including positional embedding layers, encoder layers and the MVLM head) were initialized from the publicly available LayoutLM for English. After that, the model was trained with the MVLM objective using a much smaller dataset (about 17,000 samples) with 100 epochs. The final model can be fine-tuned to downstream tasks such as sequence labeling, document classification or question answering.

The word embedding layer maps plain text tokens into a semantically meaningful latent space, while the latent space can be similar, the mapping process itself is strictly distinct from language to language. We argue that spatial information is language-independent and can be useful among languages with alike reading orders. Similarly, the encoder layers are taken from the LayoutLM weights to leverage its ability to capture attention from both semantic and positional inputs. Our experiments on internal datasets show that this transferred pre-training setting yields significantly better results than without using the word embedding weights from BERT. Additionally, our dataset with 17,000 samples is not sufficient to pre-train the LayoutLM from scratch, which would give an unfair comparison between using our proposed transferred pre-training procedure and training from scratch. Thus, we plan to verify our approach on the recently introduced XFUN dataset [14].

**Overcoming Inadequate Annotations.** VRDs usually exist in image formats without any OCR annotations. Thus, to address the lack of word-level OCR annotations, we used an in-house OCR reader to extract line-level annotations. Then, word-level annotations were approximated by dividing the bounding box of each line proportionally to the length of the words in that line.

### 3.2 Span Extraction Pre-training

To address the limitation of training data requirement when pre-training the LayoutLM model, we present a new pre-training objective that re-uses existing entities' annotations from downstream IE tasks. Our pre-train task bases on a recursive span extraction formulation that extends the traditional QA method to extract multiple answer spans from a single query.

**Recursive Span Extraction.** As opposed to sequence labeling, our span extraction formulation allows a flexible independent query (field) with the context, which enables knowledge accumulation from multiple datasets by keeping updating a collection of question embeddings. However, one limitation of our span extraction model described in Fig. 1 is that the model can not extract more than one answer (entity) for each query, which is arguably crucial in many IE use cases. We, therefore, propose a strategy to extend the capability of the model by using a recursive link decoder mechanism which is described in Fig. 2.

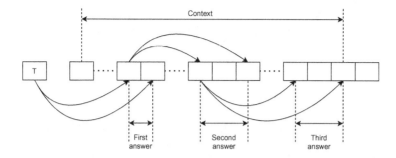

**Fig. 2.** The recursive span extraction mechanism.

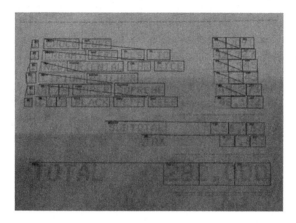

**Fig. 3.** Recursive link prediction demonstrated on a receipt example [8]. (*green* and *blue* arrows denote the target start and end positions of answer span corresponding to the query token (green bounding-boxes). The query tokens are the first token that starts each text segments. (Color figure online)

Using the same notation in Sect. 2, our recursive mechanism is described as follows :

$$s_1, e_1 = g(T, H) \tag{2}$$
$$s_{i+1}, e_{i+1} = g(h^{s_i}, H) \tag{3}$$

The first answer span is extracted by the vector representation $T$ of the query and the hidden representation $H$ of the context (Eq. 2). After that, the hidden representation corresponding to the start token of the $i$-th answer is used as the query vector to extract the $(i+1)$-th answer recursively (Eq. 3). This recursive decoding procedure stops if $s_i = e_i = 0$ or when $s_i = s_j, e_i = e_j$ with $j < i$. By using this method, we can decode all answer spans that belong to the same query/entity tag in a document context.

**Pre-training with Semantic Entities.** The recursive span extraction task forms a generic relation structure among entities, which reinforce the language model capability to represent context and spatial information in visually-rich documents. The model has to infer the continuous links between same-field items that often represented by vertical or horizontal links following the table structure. A sample visualization of a document and the corresponding continuous linking formulation can be viewed in Fig. 3.

By forcing the model to reconstruct the relation between continuous entities which belong to the same query, we hypothesize that the model can produce better performance when being fine-tuned on downstream IE tasks. Experiment results in Sect. 4 validate our hypothesis.

The proposed query-based architecture and recursive span decoding enable a generic IE formulation that is independent from the specific query tags of each document type. Recall that the query embedding was built separately from the main pre-trained model, the LayoutLM encoder will focus on modelling document context and interactions between its entities. We can pre-train the LayoutLM model seamlessly on multiple IE datasets with different key-value fields. To this end, we named this pre-training objective as *Span extraction semantic pre-training*, which allows us to pretrain LayoutLM with several small-sized IE-annotated datasets in low-resource languages, instead of using large scale document data such as [12].

# 4 Experiments

## 4.1 Experiment Setup

**Dataset.** We use two internal self-collected datasets: one for pre-training the language model and one for evaluating IE performance as a downstream task. We also evaluate the performance of our proposed method on a public dataset consists of shop receipts [8], which represent a typical example of the document understanding task in commercial systems. Table 1 shows the statistics of datasets used in our experiments.

**Table 1.** 3 VRD document datasets. *Italic* is internal.

| Data | #train | #test | #words/doc | #fields |
|------|--------|-------|-----------|---------|
| *Japanese Pre-training BizDocs* | 5170 | - | 503 | 123 |
| *Japanese Internal Invoice* | 700 | 300 | 421 | 25 |
| CORD | 800 | 100 | 42 | 30 |

*Japanese pre-training BizDocs* is a self-collected dataset. It consists of 5,170 document images and their OCR transcriptions. The dataset contains various types of business documents such as: financial report, invoice, medical receipt, insurance contract, . . . Each document type was annotated with an unique set of key-value fields which represent the important information that end-users want to extract. There are over 100 distinct fields and around 80,000 labeled entities. We use this dataset to pre-train our LayoutLM model with the proposed span-based objectives (Sect. 3.2).

*Japanese invoice* is another Invoice documents dataset that disjoints with the aforementioned pre-training data. The data consists of 1,200 document images of shipping invoice and is annotated with 25 type of key-value fields. We divide the data into 700/200/300-sample sets that correspond the training/validation/testing data respectively.

*CORD* [8] consists of Indonesian receipts collected from shops and restaurants. The dataset includes 800 receipts for training, 100 for validation, and 100 for testing. An image and a list of OCR annotations are provided alongside with each receipt. There are 30 fields which define with most important classes such as: store information, payment information, menu, total. The objective of the model is to label each semantic entity in the document and extract their values (Information Extraction) from input OCR annotations. The dataset is available publicly at https://github.com/clovaai/cord.

**Training and Inference.** The English-based LayoutLM model starts from the pre-trained weight of [13] which was provided with the paper, dubbed as *EN-LayoutLM-base*. We used the *LayoutLM-base* version from the original paper (113M parameters) due to hardware limitations. Another Japanese-based model (*JP-LayoutLM-base*) was pre-trained by using the procedure described in Sect. 3.1 with the same configuration as *LayoutLM-base*, which uses the MVLM task on our internal data. This model is intended to be used with Japanese datasets.

To conduct our experiments, we firstly pre-train the *JP-LayoutLM-base* model on the *Japanese Pre-training BizDocs* dataset with the proposed span-extraction objectives. Then, we perform fine-tuning on two IE datasets: *Japanese invoice* and *CORD* with the pre-trained weight *JP-LayoutLM-base* and *EN-LayoutLM-base* respectively. We compare the results of span-extraction with sequence labeling on these downstream IE tasks. Also, we measure the fine-tuning result with LayoutLM-base model without span-based pre-training to demonstrate the effect of the proposed method.

As for hyperparameters detail, the two models *JP-LayoutLM-base* and *EN-LayoutLM-base* share the same backbone with the hidden size of 768. In the Japanese model, we adopt the embedding layer and the vocabulary from the Japanese pre-trained BERT cl-tohoku[1] which contains 32,000 sub-words.

---

[1] https://github.com/cl-tohoku/bert-japanese.

We also set the input sequence length to 512 in all tested models. The learning rate and the optimization method were kept $5e-5$ and Adam optimizer respectively [13]. In term of loss function, with QA format, the standard cross entropy (CE) loss is used for start and end indices from all document tokens. With sequence tagging model, CE loss was used for the whole sequence classification output of the document.

**Evaluation Metrics.** We adopt the popular *entity-level F1-score (micro)* for evaluation, which is commonly used in previous studies [3,4,12,13]. The *Entity F1-score (macro)* is used as the second metric to represent the overall result on all key-value fields, since it emphasizes model performance on some rare fields/tags. This situation is commonly appeared in industrial settings where data collection is costly for some infrequent fields.

## 4.2   Results and Discussion

**CORD.** We report the results on the *test* and *dev* sub-sets of CORD in Table 3 and Table 2 respectively. The span extraction method consistently improves the performance of IE in compare to existing sequence labeling method. We can observe 1%–3% improvement in both metrics which is significant given competitive base result of default LayoutLM model. Interestingly, compared to the result from LayoutLMv2 [12], our model performs better than both *LayoutLM-large* and *LayoutLM-base-v2*, even though we only use a much lighter version of the model. This demonstrates the effectiveness of our span extraction formulation for downstream IE tasks. Note that the span extraction formulation is independent from the model architecture, thus we can expect the same improvement when starting from more complex pre-trained models. We leave this experiment as our future work.

Another finding from the results is that *F1-score macro* is increased by an significant amount alongside with entity-level score (*micro*) in both subsets. This illustrates that our method span-extraction helps to improve performance on all fields/tags, not only common tags that have more entities' annotation.

**Table 2.** Results on the *CORD (test) dataset (\*: result from original paper [12]).*

| Methods | F1 (macro) | F1 (micro) | #Params |
|---|---|---|---|
| EN-LayoutLM-base *(seq labeling)* | 80.08 | 94.86 | 113M |
| EN-LayoutLM-base *(span extraction)* | **83.46** | **95.71** | 113M |
| EN-LayoutLM-large *(seq labeling)* \* | - | 94.93 | 343M |
| EN-LayoutLM-base-v2 *(seq labeling)* \* | - | 94.95 | 200M |

**Table 3.** Results on the *CORD (dev) dataset.*

| Methods | F1-score (macro) | F1-score (micro) |
|---|---|---|
| EN-LayoutLM-base *(seq labeling)* | 80.74 | 96.16 |
| EN-LayoutLM-base *(span extraction)* | **82.13** | **97.35** |

**Japanese Invoice.** Table 4 presents the result on *Japanese invoice* data. We compare our method to sequence labeling, which shows similar improvements in term of both *F1-score* metrics. Additionally, when adding the span-based pre-training task to the model, the performance of the downstream IE task increases by a substantial amount (+1.2% in *F1-score (micro)*). The improvement illustrates that our pre-training strategy can effectively improve the IE performance, given only semantic-annotated entities from other document types. This means that we can accumulate future IE datasets when fine-tuning as a continual learning scheme to further improve the base pre-train LayoutLM model.

The *F1-score (macro)* metric is also significantly improved when using span pre-training objectives compared to starting from the raw LayoutLM model. The observed phenomenon can be explained as the knowledge between different fields is effectively exploited to improve performance of less frequent tags, which often cause the *macro* F1 metric to be lower than *micro* in both datasets. The result aligns with our initial hypothesis in Sect. 3.1.

**Table 4.** Ablation study on the *Japanese invoice dataset*

| Methods | F1-score (macro) | F1-score (micro) |
|---|---|---|
| JP-LayoutLM-base *(seq labeling)* | 85.67 | 90.15 |
| JP-LayoutLM-base *(span extraction)* | 87.55 | 91.34 |
| JP-LayoutLM-base + **span pre-training** *(seq labeling)* | 88.02 | 91.84 |
| JP-LayoutLM-base + **span pre-training** *(span extraction)* | **89.76** | **92.55** |

**Visualization.** The output of the model on the CORD (test) dataset can be visualized at Fig. 4. The model can successfully infers spatial relations between various document elements to form the final IE output. We can observe some irregular structures of the table (due to camera-capture condition) which can be processed correctly. It demonstrates model capability to understand general layout structure of input documents.

**Fig. 4.** Visualization of model output on the *CORD (test)* dataset [8]. (*red* and *blue* boxes denote the output start and end positions of answer spans corresponding to each target field. Arrows represent the links between tokens. (Color figure online)

## 5   Conclusion

This paper introduces a new span extraction formulation for the information extraction task on visually-rich documents. A pre-training scheme based on semantic entities annotation from IE data is also presented. The proposed method shows promising results on two IE datasets of business documents. Our method consistently achieves 1%–3% improvement compared to the sequence labeling approach, even with a smaller LayoutLM backbone. Furthermore, it can be applied to most existing pre-trained models for IE and opens a new direction to continuously improve the performance of pre-trained models through downstream tasks fine-tuning. For future work, we would like to consider the integration of our span extraction mechanism with the latest LayoutLMv2 model. We also plan to extend the model pre-training task with more diverse document relations such as: table structures, header-paragraph relations to further push the performance of pre-trained LM on VRDs.

## References

1. Dang, T.A.N., Thanh, D.N.: End-to-end information extraction by character-level embedding and multi-stage attentional u-net. In: BMVC, p. 96 (2019)
2. Devlin, J., Chang, M.W., Lee, K., Toutanova, K.: Bert: pre-training of deep bidirectional transformers for language understanding. arXiv preprint arXiv:1810.04805 (2018)
3. Hong, T., Kim, D., Ji, M., Hwang, W., Nam, D., Park, S.: BROS: a pre-trained language model for understanding texts in document (2021). https://openreview.net/forum?id=punMXQEsPr0
4. Hwang, W., Yim, J., Park, S., Yang, S., Seo, M.: Spatial dependency parsing for semi-structured document information extraction (2020)
5. Li, X., Sun, X., Meng, Y., Liang, J., Wu, F., Li, J.: Dice loss for data-imbalanced nlp tasks. arXiv preprint arXiv:1911.02855 (2019)

6. Liu, X., Gao, F., Zhang, Q., Zhao, H.: Graph convolution for multimodal information extraction from visually rich documents (2019). https://arxiv.org/abs/1903.11279

7. Luong, T., Pham, H., Manning, C.D.: Effective approaches to attention-based neural machine translation. ArXiv abs/1508.04025 (2015)

8. Park, S., et al.: Cord: a consolidated receipt dataset for post-OCR parsing. In: 33rd Conference on Neural Information Processing Systems (NeurIPS 2019), Vancouver (2019)

9. Qian, Y., Santus, E., Jin, Z., Guo, J., Barzilay, R.: GraphIE: a graph-based framework for information extraction (2018). http://arxiv.org/abs/1810.13083

10. Tomanek, K., Hahn, U.: Reducing class imbalance during active learning for named entity annotation. In: Proceedings of the Fifth International Conference on Knowledge Capture. pp. 105–112. K-CAP '09, Association for Computing Machinery, New York (2009). https://doi.org/10.1145/1597735.1597754

11. Vedova, L.D., Yang, H., Orchard, G.: An Invoice Reading System Using a Graph Convolutional Network, vol. 2, pp. 434–449 (2019). https://doi.org/10.1007/978-3-030-21074-8

12. Xu, Y., et al.: Layoutlmv2: multi-modal pre-training for visually-rich document understanding. In: Proceedings of the 59th Annual Meeting of the Association for Computational Linguistics (ACL) 2021, August 2021

13. Xu, Y., Li, M., Cui, L., Huang, S., Wei, F., Zhou, M.: Layoutlm: pre-training of text and layout for document image understanding. In: Proceedings of the 26th ACM SIGKDD International Conference on Knowledge Discovery & Data Mining, pp. 1192–1200 (2020)

14. Xu, Y., et al.: Layoutxlm: Multimodal pre-training for multilingual visually-rich document understanding (2021)

15. Yu, W., Lu, N., Qi, X., Gong, P., Xiao, R.: PICK: processing key information extraction from documents using improved graph learning-convolutional networks. In: 2020 25th International Conference on Pattern Recognition (ICPR) (2020)

# Recurrent Neural Network Transducer for Japanese and Chinese Offline Handwritten Text Recognition

Trung Tan Ngo, Hung Tuan Nguyen$^{(\boxtimes)}$ ⓘ, Nam Tuan Ly ⓘ, and Masaki Nakagawa ⓘ

Tokyo University of Agriculture and Technology, Tokyo, Japan
ntuanhung@gmail.com, nakagawa@cc.tuat.ac.jp

**Abstract.** In this paper, we propose an RNN-Transducer model for recognizing Japanese and Chinese offline handwritten text line images. As far as we know, it is the first approach that adopts the RNN-Transducer model for offline handwritten text recognition. The proposed model consists of three main components: a visual feature encoder that extracts visual features from an input image by CNN and then encodes the visual features by BLSTM; a linguistic context encoder that extracts and encodes linguistic features from the input image by embedded layers and LSTM; and a joint decoder that combines and then decodes the visual features and the linguistic features into the final label sequence by fully connected and softmax layers. The proposed model takes advantage of both visual and linguistic information from the input image. In the experiments, we evaluated the performance of the proposed model on the two datasets: Kuzushiji and SCUT-EPT. Experimental results show that the proposed model achieves state-of-the-art performance on all datasets.

**Keywords:** Offline handwriting recognition · Japanese handwriting · Chinese handwriting · RNN transducer · End-to-end recognition

## 1 Introduction

There is an increasing demand for automatic handwriting recognition for different purposes of document processing, educating, and digitizing historical documents in recent years. Over the past decade, numerous fundamental studies have provided large handwriting sample datasets such as handwritten mathematic expressions [1], handwritten answers [2], and historical documents [3, 4]. Moreover, many deep neural network (DNN)-based approaches are proposed to extract visual features for recognizing handwritten text [5–9]. Thanks to the acceleration of computational software and hardware, DNN has become an essential component of any handwriting recognition system.

One of the essential characteristics of handwriting is that there are many groups of categories having similar shapes [10, 11]. It is especially true in languages as Chinese or Japanese, where the number of categories is more than 7000, so that many groups of categories cannot be distinguished alone. Therefore, current handwriting recognition systems need to rely on grammar constraints or dictionary constraints as post-processing

© Springer Nature Switzerland AG 2021
E. H. Barney Smith and U. Pal (Eds.): ICDAR 2021 Workshops, LNCS 12917, pp. 364–376, 2021.
https://doi.org/10.1007/978-3-030-86159-9_26

to eliminate inaccurate predictions. These post-processing methods have been studied for a long time with several popular methods such as the n-gram model [12], dictionary-driven model [13], finite-state machine-based model [14], recurrent neural network (RNN)-based language model [15].

These linguistic post-processing processes are combined into the DNN in most practical handwriting recognition systems. Generally, the handwriting predictions are combined with the linguistic probabilities in a log-linear way and beam search to get the n-best predictions at the inference process [15]. These methods are usually named as the explicit language models, which need to be pre-trained using large corpora. On the other hand, the language models could also be implicitly embedded into DNN during the training process. In this case, however, the handwriting recognizer might overfit a text corpus if the training dataset is not diverse with many different sentence samples [7].

The limitation of the linguistic post-processing methods is the discreteness of the linguistic post-processing processes from the DNN character/text recognition models. Previous studies have shown that end-to-end unified DNN systems achieve better recognition results than discrete systems [16–18]. This is because every module in an end-to-end system is optimized simultaneously as other DNN modules, while individual systems might not achieve global optimization. Therefore, we propose to directly integrate a linguistic context module into a well-known DNN for the handwriting recognition problem.

The main contribution of this paper is to demonstrate the way and the effectiveness of RNN-Transducer for Chinese and Japanese handwriting recognition, which combines both visual and linguistic features in a single DNN model. As far as we know, this is the first attempt to combine these two different types of features using the RNN transducer model. It might open a new approach for improving handwriting recognition. The rest of the paper is organized as follows: Sect. 2 introduces the related works while Sect. 3 describes our methods. Section 4 and 5 present the experiment settings and results. Finally, Sect. 6 draws our conclusion.

## 2 Related Work

Most traditional text recognition methods are based on the segmentation of a page image into text lines and each text line into characters. The latter stage has been studied intensively [19, 20]. However, this segmentation-based approach is error-prone and affects the performance of the whole recognizer. In recent years, based on deep learning techniques, many segmentation-free methods are designed to surpass the problem and proven to be state-of-the-art for many text recognition datasets. These methods are grouped into two main approaches: convolutional recurrent neural network (CRNN) [21] and attention-based sequence-to-sequence.

Graves et al. [21] introduced Connectionist Temporal Classification (CTC) objective function and combined it with convolutional network and recurrent neural network to avoid explicit alignment between input images and their labels. The convolutional network helps extract features from images, while the recurrent network deals with a sequence of labels. This combination achieved high performance in offline handwriting recognition [9].

Besides, the attention-based sequence-to-sequence methods have been successfully adopted in many tasks such as machine translation [22] and speech recognition [23]. These methods are also applied to text recognition tasks and achieve high accuracy [11]. The attention mechanism helps to select the relevant regions of an input image to recognize text sequentially. Furthermore, the methods are also applied for lines by implicit learning the reading order and predict recognized text in any order. However, the attention-based methods require predicting exactly the same as the target result [2]. In addition, these models do not achieve high results in the Chinese and Japanese datasets compared to CRNN and CTC methods [2, 24].

Recently, some new CRNN models combined with the attention mechanism achieved state-of-the-art results in speech recognition [25]. In the Kuzushiji Japanese cursive script dataset, Nam et al. [24] also achieved the best result with the same approach. However, these modifications cannot surpass the result of the CRNN model in the SCUT-EPT Chinese examination dataset [2]. In Chinese and Japanese, a character pattern consists of multiple radicals, and the radical set is much smaller than the character set. Some researchers [26, 27] use the relationship between radicals and characters to enhance recognition. They reported the best results in both CASIA [4] and SCUT-EPT datasets.

The CRNN model has some limitations, such as requiring the output sequence not longer than the input sequence and does not model the interdependencies between the outputs. Therefore, Graves [28] introduced the RNN-Transducer as an extension of CTC to address these mentioned issues.

The RNN-Transducer defines a joint model of both input-output and output-output. Therefore, the RNN-Transducer mixes the visual features from input and context information of predicted result to find later prediction. It can be seen as an integration of a language model into the CRNN model without external data. In addition, the model also defines a distribution over the output sequence without the length constraint. However, there is no prior research about RNN-Transducer in the offline handwritten text recognition field. We will describe this model architecture and its application to Chinese and Japanese handwriting datasets in the next sections.

## 3    RNN-Transducer Architecture

Similar to RNN-Transducer in [28], its network architecture in handwriting recognition topic also has three main parts: a visual feature encoder, a linguistic context encoder, and a joint decoder. This structure is depicted in Fig. 1. At first, a convolutional neural network (CNN) extracts local features $x_t$ from an image. Then these features are encoded by a recurrent neural network (RNN) as in the CRNN model. During training, the linguistic encoder extracts context features from previous characters in ground truth text. Meanwhile, it takes input from predicted characters in the inference process. These two feature vectors are aggregated into a joint decoder. Then, this decoder predicts a density over the output sequence distribution. In this experiment, we apply a greedy search for the inference process.

Let $x = (x_1, x_2, \ldots, x_T), x_t \in X$ be an input feature sequence of length $T$ over the visual feature space $X$. Meanwhile, let $y = (y_1, y_2, \ldots, y_U), y_u \in Y$ be an output sequence of length $U$ over the output space $Y$. In addition, let $K$ is the number of categories, thus output vectors in $y$ have the size $K$.

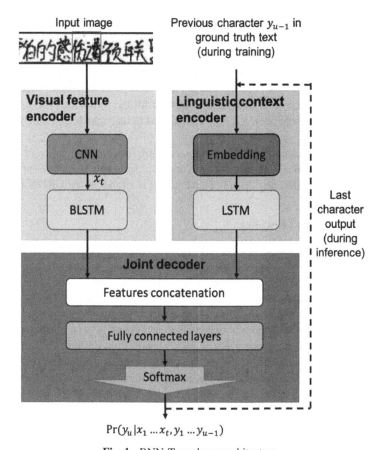

**Fig. 1.** RNN-Transducer architecture.

The extended output space $\overline{Y}$ is defined as $Y \cup \{\varnothing\}$, where $\varnothing$ denotes a blank symbol. The meaning of the blank symbol is "output nothing". Thus, it is used to represent alignments of an output text on an input sample.

Let $a = (a_{1,1}, a_{1,2}, \ldots, a_{T,U})$, $a_{i,j} \in \overline{Y}$ as alignments of $y$. For example, in case $y = (B, E, E)$ and $x = (x_1, x_2, x_3, x_4)$, some valid alignments are $(\varnothing, \varnothing, B, \varnothing, E, E, \varnothing)$, $(\varnothing, B, E, E, \varnothing, \varnothing, \varnothing)$, $(\varnothing, B, \varnothing, E, \varnothing, E, \varnothing)$, or $(\varnothing, \varnothing, B, E, \varnothing, E, \varnothing)$. Given $x$, the RNN-Transducer defines the conditional distribution $\Pr(y|x)$ as a sum over all possible alignments $a$ as Eq. (1).

$$Pr(y|x) = \sum_{a \in B^{-1}(y)} Pr(a|x) \qquad (1)$$

where $B$ is a function that removes $\varnothing$ symbols from $a$, which is defined by $B(a) = y$.

### 3.1 Visual Feature Encoder

In this work, the visual feature encoder is composed of a CNN and an RNN. As mentioned above, the CNN effectively extracts features from handwritten text images [15, 16]. This

CNN module is constructed by stacking convolutional, batch norm, and max pool layers. It extracts a feature sequence from a text image by sliding a sub-window through the input image along a defined direction.

This feature sequence is then passed to the recurrent neural network. The RNN allows information of nearby inputs to remain in the network's internal states. Therefore, the RNN can use visual contextual features and support for the following processes. In our proposed visual feature encoder, the BLSTM is used to learn the long-range context in both input directions.

Given a text line image, the CNN extracts a feature sequence $(x_1, x_2, \ldots, x_T)$. Then the recurrent layers in BLSTM scan forward and backward this sequence and compute a feature sequence $f = (f_1, f_2, \ldots, f_T)$. This encoder is similar to the CRNN model [9].

### 3.2 Linguistic Context Encoder

The linguistic context encoder consists of an input embedded layer and an RNN. The inputs are sparsely encoded as one-hot vectors of size $K + 1$. At first, the embedded layer maps one-hot variables into dense representations. It reduces the dimensionality of variables and creates meaningful vectors in the transformed space. Then, we use the LSTM for this encoder to process embedded vectors.

This network gets input from a sequence $\mathbf{y} = (y_1, y_2, \ldots, y_U)$ of length $U$ and outputs a context vector sequence $\mathbf{g} = (g_1, g_2, \ldots, g_U)$. It is similar to a deep learning language model, which extracts linguistic information from the context. These vectors are then aggregated with vectors from the visual feature encoder. Therefore, a dense vector with a smaller size than the character set is more appropriate for the aggregation purpose.

### 3.3 Joint Decoder

The joint decoder is constructed by several fully connected layers and a softmax layer. It combines features from the transcription vector $f_t$, where $1 \leq t \leq T$, and the prediction vector $g_u$, where $1 \leq u \leq U$. Then it defines the output distribution over $\overline{Y}$ as Eq. (2).

$$Pr\big(a_{t,u}|f_t, g_u\big) = softmax\Big(W^{joint}\tanh\Big(W^{visual}f_t + W^{linguistic}g_u + b\Big)\Big) \qquad (2)$$

where $W^{visual}$, $W^{linguistic}$ are the weights of linear projection, in turn, for visual feature vectors and context feature vectors to a joint subspace; $b$ is a bias for this combination; $W^{joint}$ is a mapping to the extended output space $\overline{Y}$. Furthermore, the function $Pr(a|x)$ in Eq. (1) is factorized as Eq. (3).

$$Pr(a|x) = \prod_{t=1}^{T} \prod_{u=1}^{U} Pr\big(a_{t,u}|f_t, g_u\big) \qquad (3)$$

### 3.4   Training and Inference Process

To train the RNN-Transducer model, the log-loss $L = -\log(Pr(y|x))$ is minimized. However, a naive calculation of $Pr(y|x)$ in Eq. (1) by finding all alignments $a$ is intractable. Therefore, based on Eq. (3), Graves et al. used the forward-backward algorithm [28] to calculate both loss and gradients.

In the inference process, we seek the predicted vector $a$ with the highest probability of the input image. However, the linguistic feature vector $g_u$ and hence $Pr(a_{t,u}|f_t, g_u)$ depends on all previous predictions from the model. Therefore, the calculation for all possible sequences $a$ is also intractable. In this paper, we apply the greedy search for a fast inference process through output sequences as shown in Algorithm 1.

```
Algorithm 1: Output Sequence Greedy Search
Initalise: Y=[]
g, h = LinguisticEncoder(∅)
for t = 1 to T do
  out = JointDecoder(f[t], g)
  y_pred = argmax(out)
  if y_pred != ∅:
    add y_pred to Y
    g, h = LinguisticEncoder(y_pred, h)
  end if
end for
Return: Y
```

## 4   Datasets

To evaluate the proposed RNN-Transducer model, we conducted experiments on the Kuzushiji_v1 Japanese text line and SCUT-EPT Chinese examination paper text line datasets. Our model is evaluated with Character Error Rate (CER) in each dataset. The dataset details are described in the following subsections.

### 4.1   Kuzushiji Dataset

Kuzushiji is a dataset of the pre-modern Japanese in cursive writing style. It is collected and created by the National Institute of Japanese Literature (NIJL). The Kuzushiji_v1 line dataset is a collection of text line images from the first version of the Kuzushiji dataset. The line dataset consists of 25,874 text line images from 2,222 pages of 15 pre-modern Japanese books. There are a total of 4,818 classes. To compare with the state-of-the-art result, we apply the same dataset separation in [24] as shown in Table 1. The testing set is collected from the 15th book. The text line images of the 1st to 14th books are divided randomly into training and validation sets with a ratio of 9:1.

**Table 1.** The detail of the Kuzushiji_v1 line dataset.

|  | Training set | Validation set | Testing set |
|---|---|---|---|
| Text line images | 19,797 | 2,200 | 3,878 |
| Books | $1^{st}$–$14^{th}$ |  | $15^{th}$ |

## 4.2 SCUT-EPT Chinese Examination Dataset

SCUT-EPT is a new and large-scale offline handwriting dataset extracted from Chinese examination papers [2]. The dataset contains 50,000 text line images from the examination of 2,986 writers, as shown in Table 2. They are randomly divided into 40,000 samples in the training set and 10,000 ones in the test set. We also split the training set into 36,000 samples for the sub-training set and 4,000 for the validation set. There are a total of 4,250 classes, but the number of classes in the training set is just 4,058. Therefore, some classes in the testing set do not appear in the training set. In [2], augmentation from other Chinese character datasets: CASIA-HWDB1.0-1.2 was applied. While the CRNN model covers the entire 7,356 classes, we use only the SCUT-EPT dataset for the training and testing processes.

The character set size of the SCUT-EPT dataset is smaller than that of the Kuzushiji_v1 dataset. However, the SCUT-EPT dataset has some additional problems such as character erasure, text line supplement, character/phrase switching, and noised background.

**Table 2.** The detail of the SCUT-EPT dataset.

|  | Training set | Validation set | Testing set |
|---|---|---|---|
| Text lines images | 36,000 | 4,000 | 10,000 |
| Classes | 4,058 | 4,058 | 3,236 |
| Writers |  | 2,986 |  |

## 5 Experiments

The configuration of each experiment is declared in Sect. 5.1. We compare different network configurations on the number of LSTM (BLSTM) layers belonging to the visual extractor and linguistic networks. These configurations are compared with the state-of-the-art methods for the Kuzushiji dataset in Sect. 5.2 and the SCUT-EPT dataset in Sect. 5.3. Furthermore, the correctly and wrongly recognized samples are illustrated in Sect. 5.4.

### 5.1 RNN-Transducer Configurations

We use ResNet32 pre-trained on the ImageNet dataset for the visual feature extractor, which allows our model to have good initialization for training. In detail, we use all layers of ResNet32 except the final fully connected layer and replace the final max-pooling layer of ResNet32 with our own max-pooling layer.

With the reading direction from top to bottom in the Kuzushiji Japanese dataset, the width of feature maps is normalized to 1 pixel, and the height is kept unchanged. Meanwhile, for the SCUT-EPT Chinese dataset, which has horizontal text lines, the height of feature maps is normalized to 1 pixel and their width is kept unchanged. At the recurrent layers, we adopt a number of BLSTM layers with 512 hidden nodes. A projection is applied to the sequence by a fully connected layer. Each experiment has a different number of layers and output sizes of the projection. The architecture of our CNN model in the visual extractor is shown in Table 3.

**Table 3.** Network configuration of our CNN model.

| Component | Description |
|-----------|-------------|
| Input_1 | $H \times W$ image |
| ResNet32 | All layers except the final max-pooling layer and fully connected layers |
| MaxPooling_1 | #window: (depend on the dataset) |

An embedding layer is employed at the linguistic context encoder before the recurrent network with the output vector of size 512. The following part is a number of LSTM layers with 512 hidden nodes and a projection layer. These have a similar structure with the recurrent network and the projection layer in the visual extractor network. The joint network projects a concatenated encoded vector from the visual extractor network and linguistic network to 512 dimensions via a fully connected (FC) layer. This concatenated vector is fed to a hyperbolic tangent function and then fed to another FC layer and a softmax layer. The last FC layer has the node size equal to the character set size $Y^*$.

For training the RNN-Transducer model, each training batch and validation batch have 16 and 12 samples. We preprocess and resize the input images to 128-pixel width for the Kuzushiji dataset and 128-pixel height for the SCUT-EPT dataset. We apply augmentations for every iteration, such as random affine transforms and Gaussian noise. Moreover, layer normalization and recurrent variation dropout are employed to prevent over-fitting. We utilize the greedy search in the evaluation.

For updating the RNN-Transducer model parameters, we use an Adam optimizer with an initial learning rate of 2e–4. This learning rate is applied with a linear warmup scheduler in the first epoch and a linear decreasing scheduler in the remaining epochs. The model of each setting is trained with 100 epochs. During training, the best models are chosen according to the best accuracy on the validation set.

To evaluate the effect of the number of LSTM layers and the size of the visual and linguistic encoded vectors, we conduct experiments with 3 different settings. For instance, the number of LSTM layers is from one to three, and the encoded vector sizes are 512 or 1024. All the configurations are described in Table 4.

**Table 4.** Configurations of RNN-Transducer for SCUT-EPT and Kuzushiji_v1 dataset.

| Name | Configuration |
|------|---------------|
| S1 | 1 LSTM layer + 512-dimensional encoded vector |
| S2 | 2 LSTM layer + 512-dimensional encoded vector |
| S3 | 3 LSTM layer + 1024-dimensional encoded vector |

## 5.2 Performance on Kuzushiji Dataset

Table 4 shows the evaluation of all configurations and the other methods. The lowest CER of 20.33% is from the S1 setting with a single LSTM layer and 512-dimensional encoded vectors. Our best result is lower than the state-of-the-art CER of the AACRN method by 1.15 percentage points. As mentioned before, the RNN-Transducer model is an extension of the CRNN model with the support of a linguistic network. Our model is better than the baseline CRNN model [18] by 8.01 percentage points. Therefore, it proves the efficiency of the linguistic module for the RNN-Transducer model. Furthermore, all other settings of the RNN-Transducer model also have lower CERs than the other previous methods (Table 5).

**Table 5.** Character error rates (%) on the test set of the Kuzushiji_v1 dataset.

| Method | CER (%) |
|--------|---------|
| CRNN [18] | 28.34 |
| Attention-based model [11] | 31.38 |
| AACRN [24] | 21.48 |
| S1 | **20.33** |
| S2 | 20.78 |
| S3 | 21.39 |

## 5.3 Performance on SCUT-EPT Dataset

Table 6 shows the results of the RNN-Transducer model in comparison with the other state-of-the-art methods. All our models are better than the other methods without external data and radical knowledge. The lowest CER of 23.15% is from the S3 setting with 3 LSTM layers and 1024 dimensional encoded vectors. As mentioned above, the SCUT-EPT dataset has some additional challenges so that the most complex model-S3 achieves the best performance.

Our best result is better than the state-of-the-art CER of 23.39% by the LODENet method by 0.24 percentage point, while our proposed method does not require prior knowledge of Chinese radicals for the training process. The state-of-the-art CRNN model [2] has a CER of 24.03%, which is slightly inferior to our model. Since the

LODENet model achieves a CER of 22.39% when text from Wiki corpus and prior radical knowledge are used, our models are expected to be improved further when they are utilized.

**Table 6.** Character error rate (CER) (%) on the test set of the SCUT-EPT dataset.

| Method | CER without external data | CER with external data | Required radical knowledge |
|---|---|---|---|
| CRNN [2] | 24.63 | 24.03 | |
| Attention [2] | 35.22 | 32.96 | |
| Cascaded attention [2] | 51.02 | 44.36 | |
| CTC and MCTC:WE [27] | 23.43 | - | ✓ |
| LODENet [26] | 23.39 | **22.39** | ✓ |
| S1 | 24.34 | - | |
| S2 | 24.44 | - | |
| S3 | **23.15** | - | |

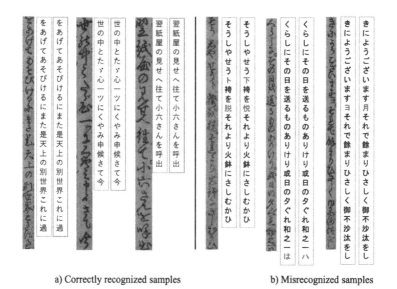

a) Correctly recognized samples            b) Misrecognized samples

**Fig. 2.** Correctly recognized and misrecognized samples in the Kuzushiji_v1 dataset.

### 5.4 Sample Visualization

Figure 2 and Fig. 3 show some correctly recognized and misrecognized samples by the best RNN-Transducer model for the Kuzushiji_v1 and that for SCUT-EPT datasets.

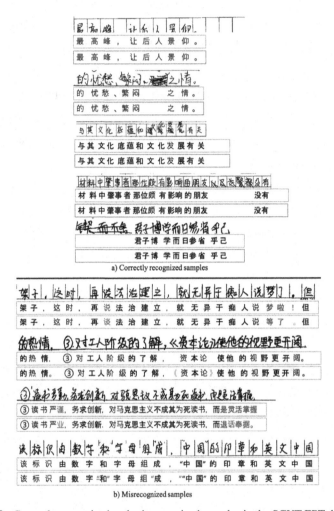

a) Correctly recognized samples

b) Misrecognized samples

**Fig. 3.** Correctly recognized and misrecognized samples in the SCUT-EPT dataset.

For each sample, the blue rectangle below or right shows the ground-truth, and the red one shows the recognition result. As illustrated in Fig. 2b, most of the errors are wrong recognitions of characters in the ground-truth rather than insertion or deletion errors. In Fig. 3a, the results prove that our model can handle challenges in the SCUT-EPT dataset, such as character erasure, text line supplement, and noised background. From the misrecognized samples as shown in Fig. 3b, our model correctly recognizes green-colored characters while the ground-truth text was mislabeled.

## 6   Conclusion

In this paper, we have presented an RNN-Transducer model to recognize Japanese and Chinese offline handwritten text line images. The proposed model is effective in modeling both visual features and linguistic context information from the input image. In the experiments, the proposed model archived 20.33% and 23.15% of character error rates on the test set of Kuzushiji and SCUT-EPT datasets, respectively. These results outperform state-of-the-art accuracies on both datasets. In future works, we will conduct experiments of the proposed model with handwritten text datasets in other languages. We also plan to incorporate an external language model into the proposed model.

**Acknowledgement.** The authors would like to thank Dr. Cuong Tuan Nguyen for his valuable comments. This research is being partially supported by A-STEP JPMJTM20ML, the grant-in-aid for scientific research (S) 18H05221 and (A) 18H03597.

# References

1. Mouchère, H., Zanibbi, R., Garain, U., Viard-Gaudin, C.: Advancing the state of the art for handwritten math recognition: the CROHME competitions, 2011–2014. International Journal on Document Analysis and Recognition (IJDAR) 19(2), 173–189 (2016). https://doi.org/10.1007/s10032-016-0263-5
2. Zhu, Y., Xie, Z., Jin, L., Chen, X., Huang, Y., Zhang, M.: SCUT-EPT: new dataset and benchmark for offline Chinese text recognition in examination paper. IEEE Access 7, 370–382 (2019)
3. Clanuwat, T., Bober-Irizar, M., Kitamoto, A., Lamb, A., Yamamoto, K., Ha, D.: Deep learning for classical Japanese literature. In: The Neural Information Processing Systems - Workshop on Machine Learning for Creativity and Design (2018)
4. Xu, Y., Yin, F., Wang, D.H., Zhang, X.Y., Zhang, Z., Liu, C.L.: CASIA-AHCDB: a large-scale chinese ancient handwritten characters database. In: Proceedings of the International Conference on Document Analysis and Recognition, ICDAR, pp. 793–798 (2019)
5. Graves, A., Liwicki, M., Fernández, S., Bertolami, R., Bunke, H., Schmidhuber, J.: A novel connectionist system for unconstrained handwriting recognition. IEEE Trans. Pattern Anal. Mach. Intell. 31, 855–868 (2009)
6. Bluche, T., Louradour, J., Knibbe, M., Moysset, B., Benzeghiba, M.F., Kermorvant, C.: The a2ia Arabic handwritten text recognition system at the openhart2013 evaluation. In: International Workshop on Document Analysis Systems, pp. 161–165 (2014)
7. Nguyen, H.T., Ly, N.T., Nguyen, K.C., Nguyen, C.T., Nakagawa, M.: Attempts to recognize anomalously deformed Kana in Japanese historical documents. In: Proceedings of the 4th International Workshop on Historical Document Imaging and Processing - HIP 2017, pp. 31–36 (2017)
8. Du, J., et al.: Watch, attend and parse: An end-to-end neural network based approach to handwritten mathematical expression recognition. Pattern Recognit. 71, 196–206 (2017)
9. Ly, N.T., Nguyen, C.T., Nguyen, K.C., Nakagawa, M.: Deep convolutional recurrent network for segmentation-free offline handwritten Japanese text recognition. In: Proceedings of the 14th International Conference on Document Analysis and Recognition, ICDAR, pp. 5–9 (2018)
10. Yin, F., Wang, Q.F., Zhang, X.Y., Liu, C.L.: ICDAR 2013 Chinese handwriting recognition competition. In: Proceedings of the International Conference on Document Analysis and Recognition, ICDAR, pp. 1464–1470 (2013)
11. Ly, N.T., Nguyen, C.T., Nakagawa, M.: An attention-based end-to-end model for multiple text lines recognition in Japanese historical documents. In: Proceedings of the 15th International Conference on Document Analysis and Recognition, pp. 629–634 (2019)

12. Hamdani, M., Doetsch, P., Kozielski, M., Mousa, A.E.D., Ney, H.: The RWTH large vocabulary arabic handwriting recognition system. In: International Workshop on Document Analysis Systems (DAS), DAS 2014, pp. 111–115 (2014)
13. Koerich, A.L., Sabourin, R., Suen, C.Y.: Large vocabulary off-line handwriting recognition: a survey (2003)
14. Nguyen, C.T., Nakagawa, M.: Finite state machine based decoding of handwritten text using recurrent neural networks. In: Proceedings of International Conference on Frontiers in Handwriting Recognition, ICFHR, pp. 246–251 (2016)
15. Ingle, R., Fujii, Y., Deselaers, T., Baccash, J.M., Popat, A.: A scalable handwritten text recognition system. In: Proceedings of the 15th International Conference on Document Analysis and Recognition, pp. 17–24 (2019)
16. Bluche, T., Louradour, J., Messina, R.: Scan, attend and read: end-to-end handwritten paragraph recognition with MDLSTM attention. In: Proceedings of the International Conference on Document Analysis and Recognition, ICDAR, pp. 1050–1055 (2017)
17. Puigcerver, J.: Are multidimensional recurrent layers really necessary for handwritten text recognition? In: Proceedings of the International Conference on Document Analysis and Recognition, ICDAR, pp. 67–72 (2017)
18. Ly, N.T., Nguyen, C.T., Nakagawa, M.: Training an end-to-end model for offline handwritten Japanese text recognition by generated synthetic patterns. In: Proceedings of International Conference on Frontiers in Handwriting Recognition, ICFHR, pp. 74–79 (2018)
19. Wang, Q.F., Yin, F., Liu, C.L.: Handwritten Chinese text recognition by integrating multiple contexts. IEEE Trans. Pattern Anal. Mach. Intell. **34**, 1469–1481 (2012)
20. Srihari, S.N., Yang, X., Ball, G.R.: Offline Chinese handwriting recognition: an assessment of current technology (2007). https://link.springer.com/article/10.1007/s11704-007-0015-2
21. Graves, A., Fernández, S., Gomez, F., Schmidhuber, J.: Connectionist temporal classification: labelling unsegmented sequence data with recurrent neural networks. In: Proceedings of the 23rd International Conference on Machine Learning - ICML 2006, pp. 369–376 (2006)
22. Bahdanau, D., Cho, K.H., Bengio, Y.: Neural machine translation by jointly learning to align and translate. In: Proceedings of the 3rd International Conference on Learning Representations (2015)
23. Bahdanau, D., Chorowski, J., Serdyuk, D., Brakel, P., Bengio, Y.: End-to-end attention-based large vocabulary speech recognition. In: the 41thinternational Conference on Acoustics, Speech and Signal Processing, pp. 4945–4949 (2016)
24. Ly, N.T., Nguyen, C.T., Nakagawa, M.: Attention augmented convolutional recurrent network for handwritten Japanese text recognition. In: Proceedings of the 17th International Conference on Frontiers in Handwriting Recognition, ICFHR, pp. 163–168 (2020)
25. Kim, S., Hori, T., Watanabe, S.: Joint CTC-attention based end-to-end speech recognition using multi-task learning. In: the 42th International Conference on Acoustics, Speech and Signal Processing, pp. 4835–4839 (2017)
26. Hoang, H.-T., Peng, C.-J., Tran, H.V., Le, H., Nguyen, H.H.: lodenet: a holistic approach to offline handwritten Chinese and Japanese text line recognition. In: Proceedings of the 25th International Conference on Pattern Recognition (ICPR), pp. 4813–4820 (2021)
27. Wigington, C., Price, B., Cohen, S.: Multi-label connectionist temporal classification. In: Proceedings of the 15th International Conference on Document Analysis and Recognition, ICDAR, pp. 979–986 (2019)
28. Graves, A.: Sequence transduction with recurrent neural networks (2012)

# MTL-FoUn: A Multi-Task Learning Approach to Form Understanding

Nishant Prabhu, Hiteshi Jain$^{(\boxtimes)}$, and Abhishek Tripathi

Perfios Software Solutions Private Limited, Mumbai, India

**Abstract.** Form layout understanding is a task of extracting and structuring information from scanned documents, and consists of primarily three tasks: *(i) word grouping, (ii) entity labeling* and *(iii) entity linking.* While the three tasks are dependent on each other, current approaches have solved each of these problems independently. In this work, we propose a multi-task learning approach to jointly learn all the three tasks simultaneously. Since the three tasks are related, the idea is to learn a shared embedding that can perform better on all three tasks. Further, the publicly available form understanding datasets are too small, and not ideal to train complex deep learning models. Multi-task learning is an effective method to provide some degree of regularization to the model for such small sized datasets. The proposed model, *MTL-FoUn*, outperforms existing approaches of learning the individual form understanding tasks on the publicly available data.

**Keywords:** Multi-task learning · Form understanding

## 1 Introduction

Document AI is an emerging field with many commercial applications. Document AI typically refers to reading, understanding and analysing of business documents [16]. Extracting structured information from business documents such as invoices, bank statements, financial documents, purchase orders and medical bills is an important activity in various industries. These documents can be of digital origin or scanned/photographed copies of physical documents. The sheer variety in the quality, templates, and formats of business documents makes document image analysis an extremely difficult problem to solve.

Document AI models are usually built to solve specific sub-tasks that are involved in document image analysis. Some of the early models tried to solve the problem of extracting text from these document images by using Optical Character Recognition (OCR) [5,11,13]. While OCR models extract text efficiently, they fail to extract the more complicated structures that are common in business documents like tables, graphs, images and so on. Table recognition involves identifying tables in a document image, extracting the structure of the

© Springer Nature Switzerland AG 2021
E. H. Barney Smith and U. Pal (Eds.): ICDAR 2021 Workshops, LNCS 12917, pp. 377–388, 2021.
https://doi.org/10.1007/978-3-030-86159-9_27

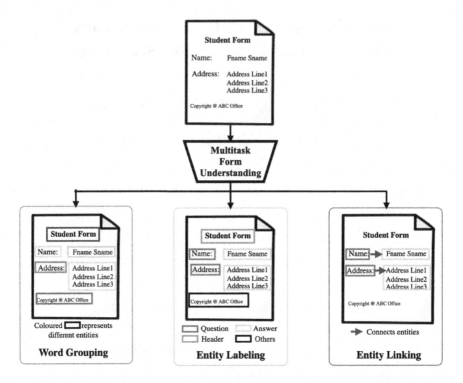

**Fig. 1.** End-to-end pipeline for key-value pair extraction illustrated with an example image.

table by identifying the individual rows and columns, and extracting the content of the identified cells. Hao *et al.* [2], Schreiber *et al.* [10] and others build more sophisticated CNN models specifically for table recognition. Another related problem is document layout analysis, which involves identifying different blocks of data in a document image. Soto *et al.* [12] and Zhong *et al.* [17] use Faster R-CNN [9] and Mask R-CNN [3] models to perform document layout analysis. Liu *et al.* [7] use a Graph Convolutional Network (GCN) to solve the task of receipt understanding (detecting entities of specific types in document images).

In this work, we focus on the form understanding task (FoUn) [6]. Forms are common methods of collecting data and are used across various domains - hospitals, industries, banks, etc. Form understanding is a relatively less explored field and involves three tasks [6] - i) *Word grouping:* is the task of grouping words of the same entity together. For example, grouping together the first and last names of a person as one entity, or grouping together all the words in an address. ii) *Semantic Entity Labeling:* is the analysis of text and spatial layout of the document to classify entities into questions, answers, headers and other categories present in the form. iii) *Entity Linking:* Assuming the existence of question-answer pairs in the document, entity linking is the task of identifying which question entity is linked to which answer entity. For example, if "Name" is a question entity, and "Mr. John Doe" is its corresponding answer entity, then "Name" is linked to "Mr. John Doe".

Recently, many approaches to form understanding has been published. However, the existing approaches consider each of the three sub tasks separately. Jaume et al. [6] formulate the problem of *word grouping* as a textline extraction task, and solve it using Tesseract [11] and Google Vision OCR engines. This naive approach performs poorly as it does not consider spatial layout and the textual content. The authors postulate the need of a learning-based algorithm to improve performance.

Jaume et al. [6] perform the task of *semantic labeling* at the entity level and Yiheng et al. [16] perform it at the word level. In [6], a multi-layer perceptron is trained over input features of the entities to classify them into 4 classes - questions, answers, header and others. The input features include BERT [1] features, spatial features *i.e.* bounding box coordinates, and length of sequences. In contrast, Yiheng et al. [16] propose LayoutLM, a pre-trained transformer network that uses both text and layout features to get the word-level embeddings. This joint pre-training with text and layout features performs better than the individual features for document layout analysis. Yiheng et al. [16] use these LayoutLM embeddings to classify each word in the document into one of 13 classes (Question, Answer, and Header in the BIES format and Other).

To solve the *Entity linking task*, Jaume et al. [6] use the same entity features as the labeling task. For every pair of entities, these features are concatenated and passed through a two-layer neural network binary classifier to predict the existence of a link between the two entities. Wang et al. [14] propose a model called *Docstruct* which uses features of 3 different modalities - semantic (BERT) features, layout features and visual features to solve the entity linking task. They model key-value pair linking as parent-child hierarchical relation, and the probability of existence of a parent-child relation between two entities is solved using an asymmetric algorithm with negative sampling [8].

All of the existing methods solve each of the three form understanding tasks (word grouping, entity labeling, and entity linking) independently and fail to exploit the inter-dependence between the three tasks. We pose the following question: *can learning these three tasks together using a single model help improve the performance of all the three tasks individually?* We hypothesize that by forcing the network to learn to perform all the three tasks simultaneously should allow the model to learn a better representation for FoUn. Thus, we propose a multi-task learning (MTL) approach to FoUn. Specifically, we use task-specific heads along with LayoutLM as a backbone to extract document features, and optimize the entire model end-to-end using loss functions that account for word grouping, entity labeling and entity linking. An overview of our multi-task approach is shown in Fig. 1.

The primary contribution of this work is the problem formulation: While existing approaches use different features (word embeddings, visual features, layout features), they fail to leverage the inter-dependence between the three tasks. To the best of our knowledge, we are the first ones to formulate the form layout understanding in a multitask setting to jointly learn entity embeddings for all three tasks. We also propose a novel embedding-learning based approach for word grouping task. The proposed model, *MTL-FoUn*, is simple, yet effective, and is based on an intuitively understandable idea. Empirical evaluations on the FUNSD dataset show that the performance of our multi-task learning setup outperforms single-task learning across all three tasks.

The paper is organized as follows: Sect. 2 explains our proposed model. Section 3 includes the experimental results using the proposed approach and discusses the performance of the proposed model. Finally the paper is concluded in Sect. 4.

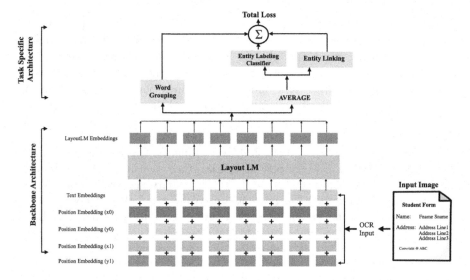

**Fig. 2.** Multi-Task Model with LayoutLM backbone: The input to the Grouping model are word vectors, whereas the input to the Labelling and Linking models are entity vectors.

## 2    Multitask Approach to Form Understanding

As discussed in Sect. 1, the task of key-value pair extraction consists of 3 distinct sub-tasks: Word Grouping, Entity Labeling, and Entity Linking. Unlike previous approaches which try to learn separate models for each of these tasks, we propose a single model trained jointly on all three tasks in a multi-task setting. The details of our proposed model, *MTL-FoUn*, are shown in Fig. 1.

Multitask learning is a learning paradigm where a single model serves to learn more than one task. The multitask model consists of a common backbone and task specific heads. The network is optimized using a weighted sum of the individual losses of all the tasks. When we optimize such a model end-to-end, the model is able to learn richer representations in the common backbone which helps solve all the sub-tasks. Thus, the performance of main task can be improved while learning the auxiliary tasks.

We consider entity linking to be our main task, and entity labelling and word grouping to be auxiliary tasks. The problem formulation of the 3 tasks and the choice of loss functions used are specifically designed such that learning the two auxiliary tasks helps the model perform better on the main task. The entity labelling task forces the model to learn to identify which entity is a key and which is a value. This auxiliary task can be seen as an inductive bias that restricts the hypothesis space for the main task and hence helps improve performance. The word grouping task is formulated as a embedding-learning problem, such that the embeddings of words in the same entity are close together and those of words in different entities are far apart. Learning this task helps the backbone model generate richer word embeddings, which in turn improves entity linking by generating better entity embeddings.

Furthermore, since there is very little training data available for FoUn, multi-task learning is an effective method to provide some degree of regularization to our model. The features learned by our model have to be suitable for all three tasks, and this prevents the model from overfitting to any one of them.

As discussed earlier, networks trained in multitask learning settings usually have two parts - common network backbone and task specific heads. The common backbone learns a representation of the input document which is further processed through task specific heads to obtain task specific outputs. We describe both these parts below.

## 2.1  Common Backbone

We use the LayoutLM model [16] as the backbone network to extract word-level features from the document. In order to keep the model simple, we use the version of LayoutLM that uses semantic (text) and layout ($2 - d$ positions) features of a given document to provide word-level features. Our LayoutLM model does not make use of visual (image) features. The features extracted by the LayoutLM backbone are used for all three downstream tasks.

## 2.2  Task Specific Heads

**Word Grouping.** Word Grouping is the task of grouping together words to form entities. An entity is a group of one or more words that together form a logical unit. For example, if the key-value pair "Name" $\rightarrow$ "Mr. John Doe" is to be extracted from the document image, then the key "Name" is one entity, and the value "Mr. John Doe" is another. We obtain vector representations for entities by simply taking the average of the word vectors belonging to that entity.

We propose a novel method of *Word Grouping* where our model learns to generate task-specific vectors for every word in the document, and then performs clustering on the learned word vectors to group words into entities. During training, we use Hinge Loss to minimize the distance between vectors of words belonging to the same entity, and maximize the distance between the vectors representing different entities. During inference, we run Agglomerative Clustering on the learned word vectors and group them into entities (each cluster represents one entity).

We pass the word-vectors obtained from LayoutLM through a fully connected layer and then compute the grouping loss $Loss_{Group}$ on its activations. Considering the following: (1) A given document is made up of a set of entities $E$, (2) Each entity is made up of a set of words, and (3) We have access to a distance function $d$ that takes two vectors as input and returns the distance between them, we can define the grouping loss for a single document as follows:

$$Loss_{Group} = \delta \sum_{e \in E} 0.5 \, mm \sum_{x,y \in e} d(x, y) + (1 - \delta) \sum_{a,b \in E} max(0, \Delta - d(a, b)) \quad (1)$$

where $\delta$ is the weighting factor for the two terms of the loss, and $\Delta$ is the margin we define to ensure that the distance between any two entities is at least $\Delta$.

**Entity Labelling.** Entity Labelling is the task of assigning a label to each entity from the set of labels $\{question, answer, header, other\}$. The labels can be described as follows:

- *question*: The entities which are the keys in the key-value pairs. These entities usually have outbound links (to answers).
- *answer*: The entities which are the answers in the key-value pairs. These entities usually have inbound links (from questions).
- *header*: Entities which appear as titles or headings in documents.
- *other*: The catch-all category that is assigned to any entity that is not a question, answer or header.

We obtain a vector representation for each entity in a given document by averaging the word vectors (as obtained from the LayoutLM backbone) belonging to that entity. These entity representations are then passed through a four-class classifier consisting of two fully connected layers. Since Entity Labelling is a classification task, we use the standard Cross-Entropy loss as the Labelling Loss.

$$Loss_{Label} = \sum_{e \in E} CrossEntropy(y_e, y'_e) \quad (2)$$

where $y$ is the ground-truth label and $y'$ is the prediction of the labelling model.

**Entity Linking.** Entity Linking is the task of predicting the probability of a link between a question entity and an answer entity. We follow an approach similar to the one proposed by Wang et al. [14], where given two vectors $x_i$ and $x_j$ representing entities $i$ and $j$, we compute the probability of a link existing from $i$ to $j$ as

$$P_{i \to j} = x_i M x_j^T \in \mathbb{R} \quad (3)$$

where $M$ is a parameter matrix.

The difference between our approach to Entity Linking and that of Wang *et al.* [14] is twofold:

- We replace the complicated feature extraction modules used by Wang *et al.* [14] (they use 4 separate models: a transformer network, a fully connected network, a CNN + LSTM network, and an attention-based influence gate) with a single simplified feature extractor (LayoutLM). Unlike the DocStruct model proposed by Wang *et al.* [14] which extracts features of each entity separately, LayoutLM extracts features by taking global context of the document into account.

- Wang *et al.* [14] skip the entity labelling step and hence are forced to mitigate the problem of class imbalance (most entity pairs do not have a link connecting them) by using negative sampling while training. Since we are training the model in a multitask setting, and entity labeling is an auxiliary task for us, we utilize the entity labels and only consider the possibility of a link existing from a *Question* to an *Answer*. We do not consider other possibilities like *Question* → *Header*, *Other* → *Answer*, *Question* → *Question* and so on. Hence, we are able to train our model without using negative-sampling.

For a given *Answer* entity, our entity linking model predicts a likelihood-score of an inbound link to it from each *Question* entity in the document. We then pass the predicted scores through a *softmax* function to obtain a probability distribution over the set of all *Question* entities. This probability distribution is then passed to a negative-log-likelihood (NLL) loss function. This procedure is repeated for all the *Answer* entities in the document, and the total loss is the sum of the loss for each *Answer* entity. The final loss function is shown in Eq. 4. The equation depicts *softmax + NLL* as *CrossEntropy*.

$$Loss_{Link} = \sum_{a \in A} CrossEntropy(y_{q \to a}, y'_{q \to a}) \tag{4}$$

where $A$ is the set of all *Answer* entities, $y_{q \to a}$ are the ground-truth links from Question $q$ to Answer $a$, and $y'_{q \to a}$ is the predicted probabilities of links from Question $q$ to Answer $a$.

We train our Multi-Task model by combining the 3 loss functions shown in Eqs. 1, 2 and 4 as

$$Loss = \alpha \cdot Loss_{Group} + \beta \cdot Loss_{Label} + \gamma \cdot Loss_{Link} \tag{5}$$

where $\alpha$, $\beta$, and $\gamma$ are hyperparameters used to weight the three loss functions.

## 3   Experiments

In this section, we describe the form understanding dataset, the baselines we use for word grouping, entity labeling and entity linking tasks and the evaluation metrics which we use to compare our models against the baselines.

### 3.1  Dataset

We evaluate the models on the publicly available *FUNSD dataset* [6] which is the current benchmark dataset for the form understanding task. It is composed of 199 annotated scanned form images. These images are noisy and vary widely in appearance hence making the form understanding task difficult. It consists of positions of single words and the links between the text fragments. The authors have provided the train/test split with 149 images considered under the train split and 50 images under the test split. The train split consists of 7411 text fragments and 4236 key value pairs, whereas the test set consists of 2332 text fragments and 1076 key value pairs.

### 3.2  Evaluation Metric

We discuss the various evaluation metrics used for model comparison for different form understanding tasks.

- Word Grouping - Word grouping can be considered as a clustering task where words are the data points and the clusters are the semantic entities. Jaume *et al.* [6] proposed using *Adjusted Rank Index (ARI)* [4] as a metric to evaluate the grouping task. ARI is used to measure the similarity between two clusterings. We use it to compare the clusters predicted by our model to the ground-truth groups of words in the dataset. ARI ranges from $-1$ to $1$ where a $0$ represents a random assignment of clusters, and $1$ represents a perfect clustering. A negative value indicates that the clustering is worse than random.
- Entity Labeling - Entity labeling is a multi-class classification problem. Thus we evaluate the models using the F1 score.
- Entity Linking - Entity linking is evaluated using in two different ways: ranking and classification. We follow the work of Wang *et al.* [14] and evaluate our linking models using the following ranking metrics: mean Average Precision (mAP) and mean Rank (mRank). These metrics measure how likely is a given answer to be correctly linked to its corresponding question (as opposed to the any other entity in the document). A higher mAP indicates a better model (higher is better), and a lower mRank indicates a better model (lower is better). To evaluate our entity linking models as classifiers, we follow Jaume *et al.* [6] and use F1 score. Reporting both types of metrics helps us directly compare our proposed models to entity linking models proposed by both Jaume *et al.* [6] and Wang *et al.* [14].

Following the evaluation procedures of Jaume *et al.* [6] and Wang *et al.* [14], the entity labelling task is evaluated assuming that the optimal word groups are available, and the entity linking task is evaluated assuming that both the optimal word groups and the optimal entity labels are available. This allows us to evaluate only one task at a time, and allows for fair comparisons with the baselines which have reported results in this format.

## 3.3 Baseline Comparisons

As discussed in Sect. 1, several different methods have been proposed to solve the three form understanding tasks - word grouping, entity labeling, and entity linking. These methods form the baselines for our model.

**Word Grouping**: Jaume *et al.* [6] propose a naive approach for the *Word grouping task*. They extract lines of text using OCR models like Tesseract and Google vision, and consider each line to be one entity. We compare the performance of our *MTL-FoUn* model with this baseline word grouping model [6] in Table 1. We see that our proposed multi-task model outperforms the naive line detection approach proposed in [6].

**Table 1.** Table showing word grouping results.

|  | Adjusted Rand Index (ARI) |
| --- | --- |
| FUNSD Paper | 0.41 |
| MTL-FoUn | 0.69 |

**Entity Labelling**: Jaume *et al.* [6] perform entity labelling by obtaining a feature vector for each entity independent of other entities in the document. Each entity is represented by a combination of the number of words in it, the BERT representation of its text, and its bounding box position. The feature representation of any given entity does not depend on any other entity. These feature representations are then passed on to a simple classifier. This is the only baseline available which is directly comparable to our approach.

Yiheng *et al.* [16] (LayoutLM) propose a classifier to classify words into questions, answers, headers and others. The difference between their approach and other approaches like ours and Jaume *et al.* [6], is that they run the classification at the word level instead of at the entity level. Since they classify words instead of entities, they use the BIESO format (13 classes) for classification.

In order to make a fair comparison, we also report the results of training their LayoutLM model for classification at the word-level with only 4 classes, and our *MTL-FoUn* model trained on the same task (word-level classification). We report the results of the Labelling task in Table 2. Our model outperforms both baselines as shown.

**Entity Linking**: Jaume *et al.* [6] propose an approach to Entity Linking that is very similar to the approach they propose for Entity Labelling. They obtain feature vectors for entities independent of each other, and then concatenate pairs of entity vectors and use them for binary classification (link exists or not). Wang *et al.* [14] on the other hand, use a more sophisticated method involving a 3 different feature extractors and an attention based influence gate. Wang *et al.* [14] also report a baseline result where they replace their feature extractors with LayoutLM and run linking at the word-level.

**Table 2.** Table showing entity labelling results.

|  | F1 | Remarks |
|---|---|---|
| FUNSD Paper | 0.57 | Entity Level (4 classes) |
| MTL-FoUn | **0.85** | |
| LayoutLM Paper | 0.79 | Word Level (BIESO - 13 classes) |
| LayoutLM | 0.84 | Word Level (4 classes) |
| MTL-FoUn | 0.83 | |

We use all 3 above mentioned models as baselines, and also report results for two variants of our own model. We train entity labelling and then finetune the trained LayoutLM backbone of our labelling model for the entity linking task. We call this model *Sequential Model*. The other model which we report results for is *MTL-FoUn*, which is the multi-task model trained simultaneously on all 3 tasks as explained in Sect. 2. The *Sequential Model* outperforms the baselines with a lower mRank and higher F1 score. The multi-task model performs better than all the models in terms of all three evaluation metric- mRank, mAP and F1.

Thus, the multi-task learning helps in all individual tasks related to form understanding.

**Table 3.** Table showing Entity Linking results. F1 and mAP - higher is better. mRank - lower is' better.

|  | mAP | mRank | F1 |
|---|---|---|---|
| FUNSD Paper | 0.23 | 11.68 | 0.04 |
| LayoutLM Word Level | 0.47 | 7.11 | - |
| DocStruct Model | 0.71 | 2.89 | - |
| Sequential Model | 0.65 | 1.45 | 0.61 |
| MTL-FoUn | **0.71** | **1.32** | **0.65** |

### 3.4 Hyperparameters and Model Details

This section describes the hyperparameters used and other details about our proposed models. As explained in Sect. 2.1, we use the text+layout version of the LayoutLM model provided by Wolf *et al.* [15]. Each of three task-specific heads consist of 2 fully connected layers that map the 768-dimensional outputs of the LayoutLM backbone to lower dimensions. We map the vectors to 64-dimensions for the word grouping task, and 256-dimensions for the other two tasks. The $\delta$ and $\Delta$ values shown in Eq. 1 for grouping loss are both set to 0.1. The distance function $d$ (as shown in Eq. 1) used to train the word grouping model is cosine-distance. The $\alpha$, $\beta$, and $\gamma$ values used to weight the three individual losses in

Eq. 5 are set to 10.0, 1.0, and 0.1 respectively. These specific values for the hyperparameters were manually chosen by running multiple experiments and observing the loss curves. Since the FUNSD dataset is relatively small with only 149 training examples, we did not rely any automated hyperparameter-learning approach.

## 4   Conclusion and Future Work

Form understanding involves three important sub tasks - word grouping to form entities, entity labeling, and entity linking tasks. In this work, we introduce a multi-task learning model, MTL-FoUn, which learns all three tasks simultaneously. We show that our multi-task approach is able to capture richer representations for words and entities, and performs better than the single task approaches used previously. We show empirically that our MTL-FoUn model performs better than all baseline works for *word grouping, entity labeling and entity linking.*

The biggest bottleneck for performance for Form Understanding is the lack of annotated data. Having more annotated data would not only help the models train better, but would open up other avenues for performance improvement such as learning hyperparameters on the validation set. Another avenue for improving performance is the inclusion of visual cues from the document image. Using CNN or Transformer based visual encoders to encode the document image, and using these visual features along with the text and layout features can help the model learn better overall representations and improve performance of all three subtasks. We intend to extend this work to incorporate some of these ideas in the future.

## References

1. Devlin, J., Chang, M.W., Lee, K., Toutanova, K.: Bert: Pre-training of deep bidirectional transformers for language understanding. arXiv preprint arXiv:1810.04805 (2018)
2. Hao, L., Gao, L., Yi, X., Tang, Z.: A table detection method for pdf documents based on convolutional neural networks. In: 2016 12th IAPR Workshop on Document Analysis Systems (DAS), pp. 287–292. IEEE (2016)
3. He, K., Gkioxari, G., Dollár, P., Girshick, R.: Mask R-CNN. In: Proceedings of the IEEE international Conference on Computer Vision, pp. 2961–2969 (2017)
4. Hubert, L., Arabie, P.: Comparing partitions. J. Classifi. **2**(1), 193–218 (1985)
5. Islam, N., Islam, Z., Noor, N.: A survey on optical character recognition system. arXiv preprint arXiv:1710.05703 (2017)
6. Jaume, G., Ekenel, H.K., Thiran, J.P.: Funsd: a dataset for form understanding in noisy scanned documents. In: 2019 International Conference on Document Analysis and Recognition Workshops (ICDARW). vol. 2, pp. 1–6. IEEE (2019)
7. Liu, X., Gao, F., Zhang, Q., Zhao, H.: Graph convolution for multimodal information extraction from visually rich documents. arXiv preprint arXiv:1903.11279 (2019)

8. Mikolov, T., Sutskever, I., Chen, K., Corrado, G., Dean, J.: Distributed representations of words and phrases and their compositionality. arXiv preprint arXiv:1310.4546 (2013)
9. Ren, S., He, K., Girshick, R., Sun, J.: Faster R-CNN: Towards real-time object detection with region proposal networks. arXiv preprint arXiv:1506.01497 (2015)
10. Schreiber, S., Agne, S., Wolf, I., Dengel, A., Ahmed, S.: Deepdesrt: deep learning for detection and structure recognition of tables in document images. In: 2017 14th IAPR International Conference on Document Analysis and Recognition (ICDAR). vol. 1, pp. 1162–1167. IEEE (2017)
11. Smith, R.: An overview of the tesseract OCR engine. In: Ninth International Conference on Document Analysis and Recognition (ICDAR 2007). vol. 2, pp. 629–633. IEEE (2007)
12. Soto, C.X., Soto, C.X.: Visual detection with context for document layout analysis. Tech. rep., Brookhaven National Lab. (BNL), Upton, NY (United States) (2019)
13. Tafti, A.P., Baghaie, A., Assefi, M., Arabnia, H.R., Yu, Z., Peissig, P.: OCR as a service: an experimental evaluation of google Docs OCR, Tesseract, ABBYY FineReader, and Transym. In: Bebis, G., et al. (eds.) ISVC 2016. LNCS, vol. 10072, pp. 735–746. Springer, Cham (2016). https://doi.org/10.1007/978-3-319-50835-1_66
14. Wang, Z., Zhan, M., Liu, X., Liang, D.: Docstruct: a multimodal method to extract hierarchy structure in document for general form understanding. arXiv preprint arXiv:2010.11685 (2020)
15. Wolf, T., et al.: Transformers: State-of-the-art natural language processing. In: Proceedings of the 2020 Conference on Empirical Methods in Natural Language Processing: System Demonstrations, pp. 38–45. Association for Computational Linguistics, October 2020. https://www.aclweb.org/anthology/2020.emnlp-demos.6
16. Xu, Y., Li, M., Cui, L., Huang, S., Wei, F., Zhou, M.: Layoutlm: pre-training of text and layout for document image understanding. In: Proceedings of the 26th ACM SIGKDD International Conference on Knowledge Discovery & Data Mining, pp. 1192–1200 (2020)
17. Zhong, X., Tang, J., Yepes, A.J.: Publaynet: largest dataset ever for document layout analysis. In: 2019 International Conference on Document Analysis and Recognition (ICDAR), pp. 1015–1022. IEEE (2019)

# VisualWordGrid: Information Extraction from Scanned Documents Using a Multimodal Approach

Mohamed Kerroumi$^{(\boxtimes)}$ ⓘ, Othmane Sayem$^{(\boxtimes)}$ ⓘ, and Aymen Shabou$^{(\boxtimes)}$ ⓘ

DataLab Groupe, Credit Agricole S.A, Montrouge, France
mohamed.kerroumi@student.ecp.fr,
{othmane.sayem,aymen.shabou}@credit-agricole-sa.fr

**Abstract.** We introduce a novel approach for scanned document representation to perform field extraction. It allows the simultaneous encoding of the textual, visual and layout information in a 3-axis tensor used as an input to a segmentation model. We improve the recent Chargrid and Wordgrid [10] models in several ways, first by taking into account the visual modality, then by boosting its robustness in regards to small datasets while keeping the inference time low. Our approach is tested on public and private document-image datasets, showing higher performances compared to the recent state-of-the-art methods.

**Keywords:** Information extraction · Multimodal · Scanned document analysis · WordGrid · Chargrid

## 1 Introduction

In a vast majority of business workflows, information extraction from templatic documents such as invoices, receipts, tax notices, etc. is largely a manual task. Automating this process has become a necessity as the number of client documents increases exponentially. Most industrial automated systems today have a rule-based approach to documents with a certain structure, and can be associated with a finite number of templates. However, documents often have a variety of layouts and structures. In order to understand the semantic content of these documents, the human brain uses the document's layout, as well as the textual and visual information available in its contents.

The challenge is to overcome rule-based systems, and to design end-to-end models that automatically understand both the visual structure of the document and the textual information it contains. For instance, in a document like an invoice, the *total amount to pay* is associated with a numerical value that appears frequently near terms such as *total*, *total to pay* and *net to pay*, and also after fields like *total before taxes*, *taxes*, *cost*, etc. Thus, as Katti et al. showed with Chargrid [10], combining both positional and textual information, was proven to be efficient for this task.

© Springer Nature Switzerland AG 2021
E. H. Barney Smith and U. Pal (Eds.): ICDAR 2021 Workshops, LNCS 12917, pp. 389–402, 2021.
https://doi.org/10.1007/978-3-030-86159-9_28

On the other hand, the visual content of a document was proven to improve model accuracy for document classification when combined with textual information [1].

In this article, we prove that adding visual information to textual and positional features improves the performance of the information extraction task. The improvement is more significant when dealing with documents with rich visual characteristics such as tables, logos, signatures, etc. We extend the work of Katti et al. [6,10] with a new approach (called **VisualWordGrid**) that combines the aforementioned modalities with two different strategies to achieve the best results in the task of information extraction from image documents.

The present paper is organized as follows: Sect. 2 presents related work for information extraction. Section 3 describes the datasets we used for evaluation. Section 4 introduces the proposed approach. Section 5 discusses the obtained results. Finally, Sect. 6 provides our conclusions regarding the new method.

## 2   Related Work

Interest in solving the information extraction task has grown in fields where machine learning is used, from Natural Language Processing (NLP) to Computer Vision (CV) domains. Depending on the representation of the document, different methods are applied to achieve the best possible performance.

For instance, NLP methods transform each document to a 1D sequence of tokens, before applying named entity recognition models to recognize the class of each word [11]. These methods can be successful when applied to documents with simple layout, such as books or articles. However, for documents like invoices, receipts or tax notices, where visual objects such as tables and grids are more common, these textual methods are less efficient. In such cases, structural and visual information are essential to achieve good performance.

Alternatively, computer vision methods can also be very efficient for this task, specifically for documents like Identity Cards which are very rich with visual features. In these approaches, only the image of the scanned document is given as an input. Object detection and semantic segmentation are among the most used techniques for field extraction [8] from these documents. The OCR engine is applied at the end of the pipeline on image crops to extract the text of detected fields. These approaches can be very useful when dealing with documents with normalized templates. For documents with various templates and layouts, these models do not perform well.

Most recent studies try to exploit the textual and the layout aspects of the document by combining both NLP and CV methods in the extraction task. In the Chargrid [10] or BertGrid [6] papers, a document is presented as a 2D grid of characters (or words) embeddings. The idea behind this representation is to preserve structural and positional information, while exploiting textual information contained in the document. Both papers reported significant increase in the performance of information extraction task compared to purely textual approaches. In a more general approach, Zhang et al. [20] recently proposed TRIE, an end-to-end text reading and information extraction approach, where a module of text reading is introduced. The model mixes textual and visual features of text reading and information extraction to reinforce both tasks mutually, in a multitask approach.

More recently, Yiheng et al. proposed the LayoutLM [16], a new method to leverage the visual information of a document in the learning process. Instead of having the text embedding of each token as the sole input, relative position of tokens in the image and the corresponding feature map of the image crop within the document were added too. Inspired by the BERT model [7], Yiheng et al. used scanned document classification task as a supervised pre-training step for LayoutLM to learn the interactions between text and layout information. Then they enforced this learning by a semi-supervised pre-training using Masked Visual-Lanquage Model (MVLM) as a multi-task learning. The dataset used for pre-training contains 11M documents and the pre-training took 170 h on 8 GPUs. Hence this approach needs large computational resources.

Other works, focused on solving the document semantic segmentation task, introduced the idea of simultaneously encoding visual, textual and structural modalities into a tensor representation. For instance, Yang et al. [19] proposed a multimodal approach to extract the semantic structure of documents using a fully convolutional neural network. Barman et al. [2] proposed a multimodal segmentation model that combines both visual and textual features to extract semantic structure of historical documents.

Compared to these related works, we propose in this paper two multimodal document representation strategies suited to the information extraction task. The first one is simpler, yet highly effective compared to state-of-the-art multimodal approaches (while improving the extraction scores). The second one is similar to the related works [2, 19] but slightly improved and adapted to the field extraction task (i.e. extracting small text regions).

## 3   Data

In order to evaluate our work, we will show our experiments on two datasets showing interesting visual structures that help improving the information extraction task using multimodal strategies.

**RVL-CDIP Dataset** [13]. It is a public dataset that was released to help improve and evaluate layout analysis techniques on scanned invoice documents. It contains 520 invoice images (Fig. 1) with their corresponding OCR files containing the extracted text, along with XML files containing the ground-truth bounding boxes of all the semantic fields. Each word in a given document of the dataset is classified into a semantic region described by a box. Among the 6 available fields, we will focus on extracting 4 of them: *Receiver, Supplier, Invoice_info, Total.*

**Tax Notice Dataset.** It is an in-house **private** dataset. It contains 3455 tax notices since 2015 (Fig. 2). The documents are in French and their templates

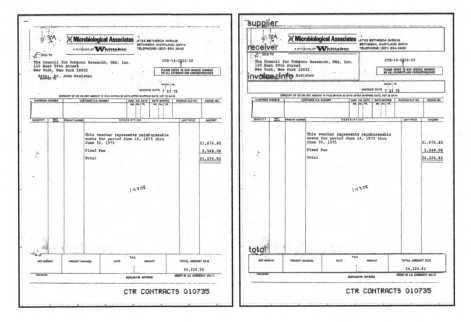

**Fig. 1.** Invoice example from RVL-CDIP with fields of interest.

changed over the years. Hence template matching could not be used as an approach for information extraction. The dataset was annotated by manually putting bounding boxes around fields of interest. There are mainly 6 entities to extract from each document: *Year, Name, Address, Type_of_Notice, Reference_Tax_Income, Family_Quotient*. The dataset contains first and second pages from tax notices, as some fields can appear on both pages, depending on the issue date. Moreover, a single page doesn't necessarily contain all fields.

## 4   Method

In this section, we introduce the **VisualWordGrid** approach, a new 2D representation of documents that extends the Chargrid philosophy by adding the visual aspect of the document to the textual and layout ones. We define two main models that differ on document representation and model architecture : **VisualWordGrid-pad** and **VisualWordGrid-2encoders**.

### 4.1   Document Representation

Our main idea is adding the visual information of the image to the textual and structural data used in the *WordGrid* representation. The most direct way for doing so is by adding the corresponding RGB channels to each pixel embedding. While this concatenation has no impact on background pixels, it adds a large amount of noise to pre-trained word embeddings. The main challenge here is to

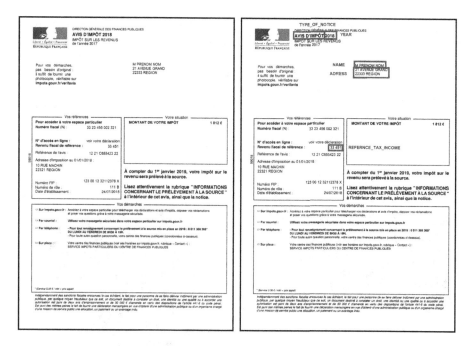

**Fig. 2.** Tax Notice *fake* example with fields of interest.

adapt the concatenation method to preserve textual embeddings, while adding the background visual information. Our representations of documents extend [10] using two strategies as follows.

Using an OCR, each document can be represented as a set of words and their corresponding bounding boxes. The textual and layout information of each document can be represented in $\mathcal{D} = \{(t_k, b_k) | k = 1, ..., n\}$, with $t_k$ the k-th token in the text of the document and $b_k = (x_k, y_k, w_k, h_k)$ its corresponding bounding box in the image.

**VisualWordGrid-pad**

Our first model representation of the document is defined as follows :

$$W_{ij} = \begin{cases} (e_d(t_k), 0, 0, 0) & if \quad \exists k \quad such \quad as \quad (i, j) \prec b_k \\ (0_d, R_{ij}, G_{ij}, B_{ij}) & otherwise \end{cases} \tag{1}$$

$$(i, j) \prec b_k \iff x_k \leq i \leq x_k + w_k \quad \wedge \quad y_k \leq j \leq y_k + h_k$$

where $d$ is the embedding dimension, $e_d$ is the word's embedding function, $0_d$ denotes an all-zero vector of size $d$, and $(R_{ij}, G_{ij}, B_{ij})$ the RGB channels of the $(i, j)^{th}$ pixel in the raw document's image.

In other words, for each point in the document's image, if this point is included in a word's bounding box $b_k = (x_k, y_k, w_k, h_k)$, the vector representing this point is the word's embedding padded by $0_3$. Thus, by setting the RGB channels to $0_3$, we drop the visual information related to this point. However if the point is not included in any word's bounding box, the vector representing this point is the concatenation of $0_d$ and the RGB channels of this point. In this case, we keep the visual information. Hence, the visual, textual and layout information of the document are encoded simultaneously in a 3-axis tensor of shape $(H, W, d + 3)$ as shown in Fig. 3, while preserving their original information.

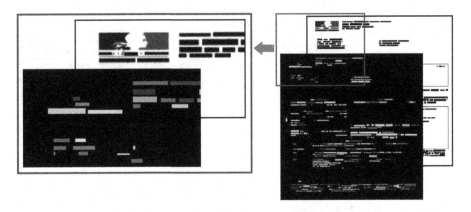

**Fig. 3.** VisualWordGrid encoding of an invoice sample. On the right, the proposed concatenation of a Wordgrid representation and the image of the document. On the left, a zoom of the previous figure.

**VisualWordGrid-2encoders**

Our second model representation is similar to the CharGrid-Hybrid approach presented in [10]. Instead of encoding the document on the character level using a one-hot encoding, we encode the document on the word level using *Word2Vec* [12] or *Fasttext* [4] embeddings. Hence, for each document we have two inputs:

- *WordGrid encoding*: This input encodes the textual and layout information of the document. This approach of encoding is similar to WordGrid presented in [10]. For words encoding, we use *Word2Vec* or *Fasttext* embeddings.

$$W_{ij} = \begin{cases} e_d(r_k) & if \quad \exists k \quad \text{such as} \quad (i,j) \prec b_k \\ 0_d & \text{otherwise} \end{cases} \tag{2}$$

- *Image*: The raw image of the document resized to match the WordGrid encoding dimensions.

## 4.2  Model Architectures

In this section, we discuss model architectures related to both strategies.

**VisualWordGrid-pad**

Once the 2D representation of the document is encoded, we use it to train a neural segmentation model. Unlike *chargrid* and *wordgrid* papers, we dropped the bounding box regression block to keep the semantic segmentation block only, since there can be at most one instance of each class in the datasets.

We use the *Unet* [14] as a segmentation model and the *ResNet34* [9] as a backbone for the encoder. The weights of the backbone are initialized using transfer learning from a model pre-trained on the ImageNet classification task. These weights are available in the open-source package *Segmentation Models* [17]. The UNet component extracts and encodes advanced features of the input grid in a small feature map, and the decoder expands this feature map to recover segmentation maps of the same size as the input grid, and thus generates the predicted label masks. We used a *softmax* activation function for the final layer of the decoder. The shape of the decoder's output is $(H, W, K + 1)$, where $K$ is the number of fields of interest, and 1 is the background class.

In the inference step, we iterate over the bounding box of each word in the OCR output, then attribute a single class to the most dominant category pixel-wise, to get the final prediction value for the corresponding field (see Fig. 4).

**VisualWordGrid-2encoders**

This model is composed of 2 encoders. The first one is the classic WordGrid encoder (2), and the second one is the raw image encoder. We keep one decoder, and for each block in the decoder, we concatenate on the skip connections from both encoders (see Fig. 4).

## 4.3  Implementation Details

In this section, we provide implementation details.

**Word Embedding Function**

We use different word embedding functions depending on the dataset. For RVL-CDIP, we propose Word2Vec pretrained embeddings on the Wikipedia corpus, publicly available thanks to *Wiki2Vec* [18]. The choice of *Wiki2Vec* is due to the good quality of the OCR files, since words are correctly recognized and most of them have their related embeddings. Unlike RVL-CDIP dataset, OCR outputs of our Tax Notice dataset are noisy, due to the quality of customer documents scans. We observed frequent misspelling errors in the Tesseract 4.1 [15] ouputs. Our experiments show that a custom *FastText* embedding trained on the corpus of the Tax Notice dataset is the best words embedding function to handle the noise. As explained in [4], Word embedding using this approach is the sum of n-grams subword embeddings. Hence, even in case of a misspelled

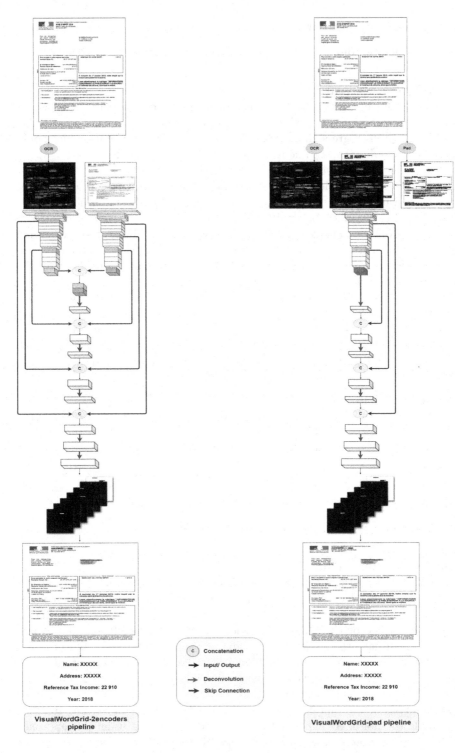

**Fig. 4.** VisualWordGrid pipelines.

or dropped character in the token, its embedding wouldn't differ too much from the embedding of the original word.

## Loss Function

The loss function we use for training is the sum of the cross entropy loss for segmentation $(L_{seg})$ and the Intersection over Union loss $(L_{IoU})$.

$$Loss = L_{seg} + L_{IoU} \tag{3}$$

– **Cross Entropy Loss**: The cross entropy loss penalizes pixel misclassification. The goal is to assign each pixel to its ground truth field.

$$L_{seg} = \sum_{x \in \Omega} -log\big(p_{\hat{l}(x)}(x)\big) \tag{4}$$

with $\hat{l}(x)$ the ground truth label of the point $x$.

– **Intersection over Union Loss**: This loss function is often used to train neural networks for a segmentation task. It's a differentiable approximation of the IoU metric and is the most indicative of success for segmentation tasks as explained in [3]. In our case, it significantly increases performances of the model compared to a model trained only with the cross entropy. The IoU metric is defined as :

$$IoU = \frac{I}{U} = \frac{|T \cap P|}{|T \cup P|} = \frac{|T \odot P|}{|T + P - (T \odot P)|} \tag{5}$$

$$L_{IoU} = 1 - IoU \tag{6}$$

where $T$ is the true labels of the image pixels and $P$ is their prediction labels. We also use the $IoU$ metric to monitor the training of our models.

## Metrics

To evaluate the performance of the different models, we used two metrics:

– **Word Accuracy Rate (WAR)**: It's the same metric as the one used in [10]. It's similar to the *Levenshtein distance* computed on the token level instead of the character level. It counts the number of substitutions, insertions and deletions between the ground-truth and the predicted instances. This metric is usually used to evaluate speech-to-text models. The WAR formula is as follows:

$$WAR = 1 - \frac{\#[insertions] + \#[deletions] + \#[substitutions]}{N} \tag{7}$$

where, $N$ is the total number of tokens in the ground truth instance for a specific field. The WAR of a document is the average on all fields.

- **Field Accuracy Rate (FAR)**: This metric evaluates the performance of the model in extracting complete and exact field information. A field is the set of words of a same entity. This metric counts the number of exact match between the ground-truth and the predicted instances. It is useful in industrial applications, as we need to evaluate the number of cases where the model succeeds to extract the whole field correctly, for control purposes for example. The FAR formula is as follows:

$$FAR = \frac{\#[\text{Fields exact Match}]}{N_{fields}} \tag{8}$$

where, $\#[\text{Fields exact Match}]$ is the number of fields correctly extracted from the document with an exact match between the ground-truth and the predicted words values, and $N_{fields}$ is the total number of fields. We note that for any processed document in the evaluation set, with no target field in the ground truth, we attribute an empty string to the value of each field, so false positives are penalized too.

In the next section, we will report for each model the average WAR and FAR metrics on the documents in the test set.

## 5   Experiments

In this section, we compare our approaches (VisualWordGrid-pad, VisualWordGrid-2encoders) to two others, on both datasets. We report their average scores ($\overline{FAR}, \overline{WAR}$) and inference time ($\overline{InferenceTime}$) on CPU for a single document, and their number of trainable parameters.

The two competing approaches are the following ones:

- **Layout Approach**: This approach is a layout encoding only. Instead of using word embedding or pixel RGB channels to encode a specific document as in (1), it uses a simpler 2D encoding suited to a segmentation task, i.e.:

$$W_{ij} = \left\{ \begin{array}{l} (1,1,1) \; if \quad \exists k \quad \text{such} \quad \text{as} \quad (i,j) \prec b_k \\ (0,0,0) \; \text{otherwise} \end{array} \right. \tag{9}$$

Then, we use this type of document encoding (Fig. 5) as input to train an information extraction model using the proposed architecture, loss and model hyper-parameters.
- **WordGrid**: This approach is very similar to BertGrid [6]. Instead of using a Bert [7] model to generate contextual embeddings, we use a *Word2Vec* pre-trained embedding for RVL-CDIP dataset, and a custom *Fasttext* embedding for the Tax Notice dataset. Equation (2) introduces the document encoding formula.

We keep the same model architecture, loss function and model hyper-parameters as proposed in the VisualWordGrid model.

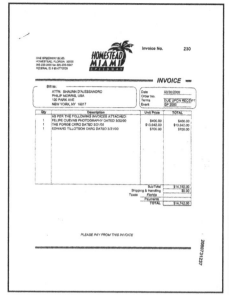

**Fig. 5.** Invoice sample and its encoding using the Layout approach

## 5.1 Datasets

Since the RVL-CDIP dataset volume is very small, we don't use a classic split of the dataset into training set and validation sets. Instead, we use a *k-fold* split of the dataset with $k = 5$. For each experiment, we do 5 tests, each one with a training on 80% of the dataset and the remaining 20% is split equally into validation and test sets. We report the average of the metrics on the 5 tests. This way, the values of the metrics don't depend on the seed of the split, and metrics are a more reliable representation of real model performance.

For the Tax Notice dataset, we assign 80% of the dataset to training, 15% to validation and 5% to test, on which we report our results. The OCR task to extract textual information was performed using the open source OCR engine Tesseract 4.1.

## 5.2 Results

For all experiments, we use *Adam* optimizer with $lr = 0.001$ and *batch_size* = 8. We use a GPU NVIDIA Quadro RTX 6000 with 24 GB GPU memory and the *Keras* framework [5] to build and train models. The inference time is measured on Intel Xeon W-2133 CPU (3.60 GHz). The Table 1 shows the scores ($\overline{WAR}$, $\overline{FAR}$) of the different approaches on the RVL-CDIP dataset.

We clearly see in the Table 1 that **VisualWordGrid-pad** gives the best FAR and WAR scores. Our proposed encoding system improves the WordGrid FAR and WAR by 1 and **7.9** respectively. Moreover, it exploits all the visual,

Table 1. Models performances on the RVL-CDIP dataset.

| Approach | $FAR$ | $WAR$ | $InferenceTime$ | #Parameters |
|---|---|---|---|---|
| Layout Only | 23.0% | 5.4% | 2.14 s | 24 439 384 |
| WordGrid | 27.7% | 10.8% | 2.22 s | 24 743 673 |
| **VisualWordGrid-pad** | **28.7%** | **18.7%** | 3.77 s | 24 753 084 |
| VisualWordGrid-2encoders | 26.9% | 17.0% | 6.08 s | 48 003 004 |

textual and structural content of documents while keeping the inference time and the number of parameters close to the WordGrid ones.

Unlike Katti et al. [10], we notice an increase in the WAR score when using the two encoders approach (VisualWordGrid-2encoders) to capture the visual and textual aspect of document. It boosts the WordGrid performance, since the WAR goes up by **6.2**. The reasons for this improvement are the modifications we added to make the model more robust in the information extraction task. We used a ResNet34 backbone for the encoder and took advantage of transfer learning to speed up the training of the model. We also changed the cross entropy loss used in [10] by adding the IoU loss to it. Notice that we used the $IoU$ as a metric for the callback.

Similarly, we tested the different approaches on the Tax Notice dataset. We reported the results in Table 2.

Table 2. Models performances on the Tax Notice dataset.

| Approach | $FAR$ | $WAR$ | $InferenceTime$ | #Parameters |
|---|---|---|---|---|
| Layout Only | 83.3% | 92.3% | 5.29 s | 24 439 674 |
| WordGrid | 83.6% | 92.4% | 5.70 s | 24 743 963 |
| VisualWordGrid-pad | 83.9% | 92.9% | 5.92 s | 24 753 374 |
| **VisualWordGrid-2encoders** | **85.8%** | **93.6%** | 6.19 s | 48 003 294 |

The VisualWordGrid-padding approach slightly improves the WordGrid scores, while the VisualWordGrid-2encoders gives the best performance but at the expense of a slightly higher inference time.

As in several industrial applications, using information extraction requires the smallest inference time. VisualWordGrid-pad would be the best choice. It leverages the visual/textual/layout information of a document while keeping the number of trainable parameters roughly the same as WordGrid.

## 6    Conclusion

VisualWordGrid is a simple, yet effective 2D representation of documents that encodes the textual, layout and visual information simultaneously. The grid-based representation includes token embeddings and the image's RGB channels.

We can take advantage of these multimodal inputs to perform several document understanding tasks. For the information extraction task, VisualWordGrid shows better results than those of state of the art models on two datasets (the public RVL-CDIP dataset and the private Tax Notice dataset), while keeping model parameters and inference time roughly the same (especially when using the padding strategy). In many fields, this approach is suitable for production.

# References

1. Audebert, N., Herold, C., Slimani, K., Vidal, C.: Multimodal deep networks for text and image-based document classification. In: Cellier, P., Driessens, K. (eds.) ECML PKDD 2019. CCIS, vol. 1167, pp. 427–443. Springer, Cham (2020). https://doi.org/10.1007/978-3-030-43823-4_35

2. Barman, R., Ehrmann, M., Clematide, S., Oliveira, S.A., Kaplan, F.: Combining visual and textual features for semantic segmentation of historical newspapers. J. Data Min. Digit. Hum. HistoInf. (2020)

3. van Beers, F., Lindström, A., Okafor, E., Wiering, M.: Deep neural networks with intersection over union loss for binary image segmentation. In: ICPRAM (2019)

4. Bojanowski, P., Grave, E., Joulin, A., Mikolov, T.: Enriching word vectors with subword information. Trans. Assoc. Comput. Linguisti. 5, 135–146 (2017)

5. Chollet, F., et al.: Keras (2015). https://github.com/fchollet/keras

6. Denk, T., Reisswig, C.: Bertgrid: Contextualized embedding for 2d document representation and understanding. CoRR abs/1909.04948, September 2019. http://arxiv.org/abs/1909.04948

7. Devlin, J., Chang, M.W., Lee, K., Toutanova, K.: BERT: Pre-training of deep bidirectional transformers for language understanding. In: NAACL-HLT (2019)

8. Hao, L., Gao, L., Yi, X., Tang, Z.: A table detection method for pdf documents based on convolutional neural networks. In: 2016 12th IAPR Workshop on Document Analysis Systems (DAS), pp. 287–292, April 2016. https://doi.org/10.1109/DAS.2016.23

9. He, K., Zhang, X., Ren, S., Sun, J.: Deep residual learning for image recognition. In: 2016 IEEE Conference on Computer Vision and Pattern Recognition (CVPR), pp. 770–778 (2016)

10. Katti, A.R., et al.: Chargrid: towards understanding 2d documents. In: Proceedings of the 2018 Conference on Empirical Methods in Natural Language Processing, Brussels, Belgium, October 31–November 4, 2018, pp. 4459–4469. Association for Computational Linguistics (2018). https://www.aclweb.org/anthology/D18-1476/

11. Lample, G., Ballesteros, M., Subramanian, S., Kawakami, K., Dyer, C.: Neural architectures for named entity recognition. In: Proceedings of the 2016 Conference of the North American Chapter of the Association for Computational Linguistics: Human Language Technologies, pp. 260–270. Association for Computational Linguistics, San Diego, June 2016. https://doi.org/10.18653/v1/N16-1030, https://www.aclweb.org/anthology/N16-1030

12. Mikolov, T., Chen, K., Corrado, G.S., Dean, J.: Efficient estimation of word representations in vector space. CoRR abs/1301.3781 (2013)

13. Riba, P., Dutta, A., Goldmann, L., Fornés, A., Ramos, O., Lladós, J.: Table detection in invoice documents by graph neural networks. In: 2019 International Conference on Document Analysis and Recognition (ICDAR). pp. 122–127 (2019)

14. Ronneberger, O., Fischer, P., Brox, T.: U-net: convolutional networks for biomedical image segmentation. In: MICCAI (2015)
15. Smith, R.: An overview of the tesseract ocr engine. In: Proceedings of the Ninth International Conference on Document Analysis and Recognition (ICDAR), pp. 629–633 (2007)
16. Xu, Y., Li, M., Cui, L., Huang, S., Wei, F., Zhou, M.: Layoutlm: pre-training of text and layout for document image understanding. In: Proceedings of the 26th ACM SIGKDD International Conference on Knowledge Discovery & Data Mining, August 2020. https://doi.org/10.1145/3394486.3403172
17. Yakubovskiy, P.: Segmentation models (2019)
18. Yamada, I., et al.: Wikipedia2vec: an efficient toolkit for learning and visualizing the embeddings of words and entities from wikipedia. arXiv preprint 1812.06280v3 (2020)
19. Yang, X., Yumer, E., Asente, P., Kraley, M., Kifer, D., Giles, C.L.: Learning to extract semantic structure from documents using multimodal fully convolutional neural network. In: Proceedings - 30th IEEE Conference on Computer Vision and Pattern Recognition (CVPR 2017), pp. 4342-4351 (2017)
20. Zhang, P., et al.: TRIE: end-to-end text reading and information extraction for document understanding (2020)

# A Transformer-Based Math Language Model for Handwritten Math Expression Recognition

Huy Quang Ung$^{(\boxtimes)}$ ⓘ, Cuong Tuan Nguyen ⓘ, Hung Tuan Nguyen ⓘ,
Thanh-Nghia Truong ⓘ, and Masaki Nakagawa ⓘ

Tokyo University of Agriculture and Technology, Tokyo, Japan
fx4102@go.tuat.ac.jp, nakagawa@cc.tuat.ac.jp

**Abstract.** Handwritten mathematical expressions (HMEs) contain ambiguities in their interpretations, even for humans sometimes. Several math symbols are very similar in the writing style, such as dot and comma or "0", "O", and "o", which is a challenge for HME recognition systems to handle without using contextual information. To address this problem, this paper presents a Transformer-based Math Language Model (TMLM). Based on the self-attention mechanism, the high-level representation of an input token in a sequence of tokens is computed by how it is related to the previous tokens. Thus, TMLM can capture long dependencies and correlations among symbols and relations in a mathematical expression (ME). We trained the proposed language model using a corpus of approximately 70,000 LaTeX sequences provided in CROHME 2016. TMLM achieved the perplexity of 4.42, which outperformed the previous math language models, i.e., the $N$-gram and recurrent neural network-based language models. In addition, we combine TMLM into a stochastic context-free grammar-based HME recognition system using a weighting parameter to re-rank the top-10 best candidates. The expression rates on the testing sets of CROHME 2016 and CROHME 2019 were improved by 2.97 and 0.83 percentage points, respectively.

**Keywords:** Language model · Mathematical expressions · Handwritten · Transformer · Self-attention · Recognition

## 1 Introduction

Nowadays, devices such as pen-based or touch-based tablets and electronic whiteboards are becoming more and more popular for users as educational media. Learners can use them to learn courses and do exercises. Especially, educational units can use those devices to support their online learning platform in the context of the SARS-CoV-2 (COVID-19) widely spreading worldwide. These devices provide a user-friendly interface for learners to input handwritten mathematical expressions, which is more natural and quicker than common editors such as Microsoft Equation Editor, MathType, and LaTeX. Due to the demands of real applications, the research on handwritten mathematical expression (HME) recognition has been conceived as an important role in document analysis since

© Springer Nature Switzerland AG 2021
E. H. Barney Smith and U. Pal (Eds.): ICDAR 2021 Workshops, LNCS 12917, pp. 403–415, 2021.
https://doi.org/10.1007/978-3-030-86159-9_29

the 1960s and very active during the last two decades. The performance of HME recognition systems has been significantly improved according to the series of competitions on recognition of handwritten mathematical expressions (CROHME) [1].

However, there remain challenging problems in HME recognition. One problem is that there are lots of ambiguities in the interpretation of HMEs. For instance, there exist math symbols that are very similar in the writing style, such as "0", "o", and "O" or dot and comma. These ambiguities challenge HME recognition without utilizing contextual information. In addition, recognition systems without using predefined grammar rules such as the encoder-decoder model [2, 3] might result in syntactically unacceptable misrecognitions. One promising solution for these problems is to combine an HME recognition system with a math language model. Employing language models for handwritten text recognition has shown effectiveness in previous research [4–6].

A mathematical expression (ME) has a 2D structure represented by several formats such as MathML, one-dimensional LaTeX sequences, and two-dimensional symbol layout trees [7]. Almost all recent HME recognition systems output their predictions as the LaTeX sequences since LaTeX is commonly used in real applications. Thus, we focus on an ME language model for the LaTeX sequences in this paper.

There are some common challenges in modeling MEs similar to natural language processing. First, there is a lack of corpora of MEs as MEs rarely appear in daily documents. Secondly, there are infinite combinations of symbols and spatial relationships in MEs. Thirdly, there are long-term dependencies and correlations among symbols and relations in an ME. For example, "(" and ")" are often used to contain a sub-expression, and if they contain a long sub-expression, it is challenging to learn the dependency between them.

There are several methods to modeling MEs. The statistical $N$-gram model was used in [8]. It assigns a probability for the $n$-th tokens given $(n-1)$ previous tokens based on the maximum likelihood estimation. However, the $N$-gram model might not represent the long dependencies due to the limitation of the context length. Increasing this length might lead to the problem of estimating a high-dimensional distribution, and it requires a sufficient amount of training corpus. In practical applications, the trigram model is usually used, and the 5-g model is more effective when the training data is sufficient. The recurrent neural network-based language model (RNNLM) proposed by [9] was utilized in HME recognition systems [1, 2]. RNNLM predicts the $n$-th token given $(n-1)$ previous tokens in previous time steps. However, they still face the problem of the long-term dependencies.

In recent years, the transformer-based network using a self-attention mechanism has achieved impressive results in natural language processing (NLP). In the language modeling task, Al-Rfou et al. [10] presented a deep transformer-based language model (TLM) for character-level modeling and showed the effectiveness against RNNLM.

In this paper, we present the first transformer-based math language model (TMLM). Based on the self-attention mechanism, the high-level representation of an input token in a sequence of tokens is computed by how it is related to the previous tokens so that TMLM can capture long dependencies and correlations in MEs. Then, we propose a method to combine TMLM into a stochastic context-free grammar-based HME recognizer. In our experiments, we show that our TMLM outperforms the $N$-gram model and RNNLM in the task of modeling MEs.

The rest of the paper is organized as follows. Section 2 briefly presents related research. Section 3 describes our method in detail. Section 4 presents our experiments for evaluating the proposed method. Finally, Sect. 5 concludes our work and discusses future works.

## 2   Related Works

Language models are well-known as generative models and autoregressive models since they predict the next state of a variable given its previous states. In NLP, Radford et al. [11, 12] proposed Generative Pre-Training models (GPT and GPT-2) with high achievements on NLP benchmarks based on the vanilla transformer-based network in [13]. Their models are trained by the casual language modeling loss, then fine-tuned for multitask learning such as text classification, question answering, and similarity. Dai et al. [14] presented a Transformer-XL for capturing extended length of context using a recurrent architecture for context segments. Transformer-XL can learn dependency that is 80% longer than RNNs, 450% longer than TLM. The inference speed is 1,800 times faster than TLM by caching and reusing previous computations. XLNet presented by Yang et al. [15], is the first model utilizing bidirectional contexts for transformer-based language models. This model significantly outperformed the conventional BERT model [16] in 20 tasks of NLP benchmarks.

There are several studies combining HME recognition systems with pre-trained language models. Wu et al. [8] combined their encoder-decoder HME recognizer with a pre-trained 4-g model to get the $N$ best paths. Zhang et al. [2] utilized a Gated Recurrent Unit-based language model (GRULM) for their HME recognizer that is an encoder-decoder model with temporal attention. This attention is to help the decoder determine the reliability of spatial attention and that of the language model per time step. The language models improved the expression rate by around 1 percentage point. Hence, the approach for combining language models into recognition systems is essential to study.

In CROHME 2019 [1], the Samsung R&D team used a probabilistic context-free grammar-based recognizer combined with two bigram language models, i.e., a language sequence model and a language model for spatial relationships. Besides, theMyScript team used LSTM-based language models for their grammar-based recognition system.

## 3   Proposed Method

Given a sequence of tokens $X = (x_1, x_2 \ldots, x_N)$, constructing a language model is to estimate the joint probability $P(X)$, which is often auto-regressively factorized as $P(X) = \prod_t P(x_t | X_{<t})$ where $X_{<t} = (x_1, \ldots, x_{t-1})$. According to this factorization, the problem reduces to estimating each conditional factor $P(x_t | X_{<t})$. In this paper, our proposed model with a self-attention mechanism encodes the context $X_{<t}$ to produce the categorical probability of the token $x_t$.

In this section, we first describe our proposed TMLM, which is mainly based on [10]. Then, we present a method for combining our model with an HME recognizer.

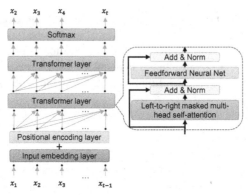

**Fig. 1.** Overview of the proposed transformer-based language model with two transformer layers.

### 3.1 Transformer-Based Math Language Model

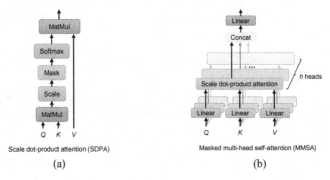

**Fig. 2.** Illustration of scale dot-product attention and masked multi-head attention

TMLM consists of three main parts: an input embedding layer, a positional encoding layer (PE), and a stack of transformer layers, as shown in Fig. 1. First, sequential input tokens $\{x_1, x_2, \ldots, x_N\}$ are fed into the input embedding to embed the categories of discrete tokens into a continuous space for better representation. Secondly, each embedded vector according to each input token is added by a PE vector to present the token's position in the sequence. The detail of the PE is presented later in this section. Thirdly, the outputs of the input embedding and positional encoding are passed into stacked transformer layers to learn high-level representation based on the self-attention mechanism. Finally, the output of the top transformer layer is input to a softmax layer to obtain the categorical probability for the token $x_t$ given $\{x_1, \ldots, x_{t-1}\}$. Although all input tokens are fed into our model at the same time, the model is restricted to attend only tokens on the left side of $x_t$ to produce $P(x_t|x_1, \ldots, x_{t-1})$ by a mask in the transformer layer.

The architecture of the transformer layer is based on the decoder of the conventional transformer-based model [16]. It consists of a masked multi-head self-attention (MMSA), layer normalization [17], and a feedforward neural network, as shown in

Fig. 1. In addition, residual connections are added for the model to learn better. Here, we present MMSA and PE, which play important roles in our model.

**Masked Multi-head Self-attention.** This layer receives the representation of input tokens and outputs the higher representation for the tokens based on how each token is related to others. MMSA includes multiple attention functions, which allow the model to attend information from different representation subspace. We firstly present a masked single-head self-attention.

A traditional attention function can be described as the mapping of a query and a set of key-value pairs to produce an output. Note that the query, the keys, and the values are all vectors. The output is a weighted sum of the values, where the weight assigned to each value is computed by a compatibility function of the query with the corresponding key of the value.

The masked single-head self-attention function, called scaled dot-product attention (SDPA), are also based on the queries ($Q$), the keys ($K$) of dimension $d_k$, and the values ($V$) of dimension $d_v$ as shown in Fig. 2(a). We compute the dot products of the query with all keys, then scale them by $\sqrt{d_k}$. Next, we apply a mask to restrict the model to attend only the left side of the current predicted token. We then apply a softmax function to obtain the weights on the values. The output of this attention function is formulated as follows:

$$Att(Q, K, V) = \text{softmax}\left(\frac{QK^T}{\sqrt{d_k}}\right)V \tag{1}$$

SDPA is called a "head" in MMSA. The architecture of MMSA including $h$ heads is shown in Fig. 2(b). With multiple heads, we project the queries, keys, and values $h$ times with three different learnable linear projections. On each of these projected versions of queries, keys, and values, we then perform SDPAs in parallel. Then, we concatenate their outputs and once again project to obtain the final output of MMSA.

**Positional Encoding.** Since tokens $(x_1, x_2, \ldots, x_N)$ are input to our model at the same time and there is no convolutional/recurrent layer, the model cannot exploit the positional information of tokens. It is a serious problem for the task of language modeling. To address it, we utilize PE having the same dimensionality as the input embedded vector, $R^{N \times d_{embed}}$ ($d_{embed}$ is the dimension of the input embedded vector). Then, we add PE to the input embedded vector to provide the positional information for our model. PE of the $p$-th token and the $i$-th dimension is computed by the sine and cosine function as follows:

$$PE(p, i) = \begin{cases} \sin\left(\frac{p}{10000^{i/d_{embed}}}\right) & if \ i \ is \ even \\ \cos\left(\frac{p}{10000^{(i-1)/d_{embed}}}\right) & otherwise \end{cases} \tag{2}$$

### 3.2 Combining Language Model with HME Recognizer

In this study, we use a language model to sort the top-$M$ best candidates outputted from the stochastic context-free grammar-based HME recognizer. Given $M$ candidates

$\{c_1, c_2, \ldots, c_M\}$ of LaTeX sequences and their corresponding scores, the combined scores are computed as follows:

$$Score_{comb}(c_i) = Score_{recog}(c_i) + \alpha \times Score_{LM}(c_i) \tag{3}$$

where $Score_{recog}(c_i)$ and $Score_{LM}(c_i)$ are the scores of the $i$-th candidate, $c_i$, from the HME recognizer and the language model, respectively. $Score_{comb}(c_i)$ is the combined score of $c_i$. $\alpha$ is a weighting parameter to balance between recognition and language scores. Note that $Score_{LM}(c_i)$ is the sum of logarithms of conditional probabilities output from the language model and normalized by the length of the candidate, $c_i$. For this combination method, we refer to the HME recognizer producing $Score_{recog}(c_i)$ based on the sum of logarithms of probability terms. The candidate having the highest combined score is the final recognition result.

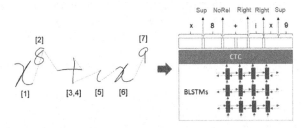

**Fig. 3.** Illustration for symbol-relation temporal classifier

## 4    Experiments

This section presents evaluations of our proposed TMLM using a corpus of LaTeX sequences. We also evaluate TMLM when combining with the HME recognizer proposed in [18] and present error analyses.

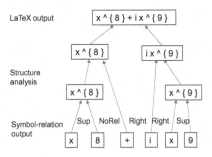

**Fig. 4.** Symbol-level parser.

## 4.1  Dataset

We uses a corpus of 68,862 LaTeX sequences provided in CROHME 2016 [19]. For preprocessing steps, we first filtered invalid syntax LaTeX sequences and removed style-related characters such as "\mathrm", "\textrm", and so on. Then, we normalized the LaTeX sequences into the same format as the output of our HME recognizer. For example, "{a}^{2}" is normalized as "a^{2}". The corpus is partitioned into a training set, a validation set, and a testing set according to the ratio of 8:1:1. The number of symbols in the dictionary is 108, including the padding "<pad>" and the end-of-sequence symbol "<eos>".

## 4.2  HME Recognizer

In this section, we present the online HME recognizer [18] used in our experiments. The recognizer receives a sequence of point-based features extracted from an input HME and outputs recognition result as a LaTeX sequence. It consists of two main stages: (1) A symbol-relation temporal classifier (SRTC) for segmenting and classifying symbols and spatial relationships in an HME and (2) A symbol-level parser (SLP). We denoted this HME recognizer as SRTC_SLP.

**Symbol-Relation Temporal Classifier.** SRTC consists of three stacked Bidirectional Long-Short Term Memory (BLSTM) layers and a Connectionist Temporal Classification (CTC) layer at the top, as shown in Fig. 3. Its input is a sequence of point-based features, including the representation of off-strokes (pen movements between strokes). Here, we ideally assume that there are no delayed strokes in the input HMEs. The stacked multiple BLSTM layers encode bidirectional context from the input and learn high-level representation. Then, the CTC layer generates a sequence of symbols and spatial relationships. There are 7 types of spatial relationships: superscript, subscript, upper, lower, horizontal, inside, and no relation (denoted as "NoRel").

**Symbol-Level Parser.** Given the output of SRTC, SLP based on the Cocke–Younger–Kasami (CYK) algorithm [20] is applied to merge recognized symbols and spatial relationships along with predefined grammar rules, as shown in Fig. 4. This bottom-up method considers many possible combinations of hypotheses at the intermediate levels. Hence, it produces several candidates at the top of the combination tree even if the less promising candidates are pruned. Each candidate has a corresponding score computed based on the classification probabilities of symbols and spatial relationships.

**Training and Testing.** SRTC_SLP was trained on the CROHME 2016 and CROHME 2019 training sets and tested on the CROHME 2016 and CROHME 2019 testing sets, respectively. Without using a language model, it achieved the expression rate of 53.44% and 52.38% on the CROHME 2016 and CROHME 2019 testing sets, respectively. This expression rate is higher than that of the state-of-the-art TAP recognizer (without using language models and/or ensemble methods) [2] by 3.22 percentage points in the expression rate on the CROHME 2016 testing set.

## 4.3  Experimental Settings

In this section, we present settings for training the proposed TMLM. We evaluated TMLM with different numbers of transformer layers. The number of heads is fixed to 4 heads. The dimension of each head is set to 16. The context length of TMLM is fixed to 256, which covers the maximum sequence length in our LaTeX corpus. The dimension of vectors of input embedding and the dimension of hidden states are set to 256 and 512, respectively. The number of hidden nodes in the feedforward neural net layer is set to 1024. The dropout rate is set to 0.1. We applied an adaptive log-softmax function proposed in [21]. Our model is trained by the AdamW optimizer [22] with a learning rate of $10^{-5}$. The model is implemented based on the "Hugging Faces" library [23]. For combining language models with SRTC_SLP, we determined the parameter $\alpha \in R^{+}$ in Eq. 3 by applying the enumeration method on $\{0, 0.1, 0.2, \ldots, 2.0\}$. The chosen $\alpha$ parameters achieved the best expression rates on the CROHME 2014 testing set.

To evaluate language models, we utilized the perplexity measurement. Given a sequence of tokens $X = (x_1, x_2, \ldots, x_N)$, the perplexity of $X$ is the exponentiated average negative log-likelihood formulated as follows:

$$ppl(X) = \exp\left\{ -\frac{1}{N} \sum_{t=1}^{N} \log p(x_t|X_{<t}) \right\} \tag{4}$$

where $p(x_t|X_{<t})$ is the conditional probability outputted from the language model.

## 4.4  Evaluation

In this section, we compare the proposed language model with the previous methods. Then, we conduct experiments to compare the performance of those models when combined with the SRTC_SLP recognizer.

**Comparisons with Other Language Modeling Methods.** We compared our TMLM with the traditional N-gram model and GRULM. We increased the context length N in the N-gram model to 11 since TMLM can attend all the past contexts for a fair comparison. For GRULM, we increased the number of GRU layers up to 3 layers for evaluating the performance as well as comparing with TMLM in the condition of a similar number of trainable parameters. The dimension of an input embedded vector and hidden states in GRULM are set as the same as in TMLM.

Table 1 presents the perplexity of the models on the testing set extracted from our LaTeX corpus, as mentioned in Sect. 4.3. The results show that our proposed TMLM models outperform all N-gram models and all GRULMs, even using fewer trainable parameters. For the N-gram models, increasing the context length can improve the perplexity, but it seems to converge when N reaches 11. GRULMs perform better than the N-gram models. Among GRULMs, the perplexity of GRULM_2L achieves the best, which implies that increasing the number of GRU layers is not adequate. On the other hand, TMLMs can learn better when increasing the number of transformer layers.

With nearly the same number of trainable parameters, TMLM_2L performs significantly better than GRULM _2L. It implies that the architecture of TMLM is much more effective than the traditional GRULM on modeling MEs.

**Table 1.** Comparisons with other language modeling methods.

| Model | #Layers in model | #Parameters | Perplexity |
|---|---|---|---|
| 3-grams | - | - | 9.603 |
| 5-grams | - | - | 7.557 |
| 9-grams | - | - | 6.550 |
| 11-grams | - | - | 6.500 |
| GRULM_1L | 1 | 1.3M | 6.050 |
| GRULM _2L | 2 | 2.8M | 6.049 |
| GRULM _3L | 3 | 4.4M | 6.377 |
| Ours: TMLM_2L | 2 | 2.7M | 4.598 |
| Ours: TMLM_5L | 5 | 6.3M | 4.509 |
| Ours: TMLM_8L | **8** | **10M** | **4.420** |

**Evaluation on Combining Language Models into the HME Recognizer.** We combined the SRTC_SLP recognizer with the language models that achieved the best performance in the previous experiment (i.e., 11-grams, GRULM_2L, and TMLM_8L). In detail, the combined score in Eq. 3 is computed for the top-10 best candidates from SRTC_SLP.

**Table 2.** Expression rates on combining the HME recognizers with language models.

| Recognition system | Expression rate (%) | |
|---|---|---|
| | CROHME 2016 | CROHME 2019 |
| SRTC_SLP | 53.44 | 52.38 |
| SRTC_SLP + 11-grams | 56.15 | 52.54 |
| SRTC_SLP + GRULM_2L | 55.36 | 52.88 |
| **(Ours) SRTC_SLP+ TMLM_8L** | **56.41** | **53.21** |
| (Zhang et al. [2]) TAP | 49.29 | - |
| (Zhang et al. [2]) TAP + GRUs | 50.41 | - |
| (Wu et al. [8]) PAL-v2 | 49.00 | - |
| (Wu et al. [8]) PAL-v2 + 4-grams | 49.35 | - |

LM: language model

Table 2 presents the expression rates of the combined recognizers on the CROHME 2016 and CROHME 2019 testing sets. The combination of SRTC_SLP and TMLM_8L achieves the best expression rates in both testing sets. TMLM_8L improves 2.97 and 0.83 percentage points of the expression rates on the CROHME 2016 and CROHME

2019 testing set, respectively. The SRTC_SLP + 11-grams is better than SRTC_SLP + GRULM_2L in the CROHME 2016 testing set, but that result is opposed in the CROHME 2019 testing set.

Table 2 also presents the expression rates of the state-of-the-art HME recognizers that utilized math language models, i.e., TAP [2] and PAL_v2 [8]. Compared to those models, SRTC_SLP combined with our TMLM_8L yields the best expression rate on the CROHME 2016 testing set. Combining the language models only improved around 1 percentage point in the case of TAP and PAL_v2 while TMLM_8L improves 2.97 percentage points. Here, we cannot conclude that our method for utilizing a math language model is better than their methods since they utilized different types of HME recognizers as well as different LaTeX corpora to train their language models. We consider conducting more experiments on the combination method as a remaining work.

**Table 3.** Percentages of corrected, miscorrected, and unchanged recognition results when combining the SRTC_SLP recognizer with language models.

| Dataset | Method | Corrected (%) | Miscorrected (%) | Unchanged (%) |
|---------|--------|---------------|------------------|----------------|
| CROHME 2016 | 11-grams | 4.01 | 1.31 | 94.68 |
| | GRULM_2L | 2.96 | **1.05** | 95.99 |
| | **TMLM_8L** | **4.62** | 1.66 | 93.72 |
| CROHME 2019 | 11-grams | 1.83 | 1.67 | 96.50 |
| | GRULM_2L | 1.92 | **1.42** | 96.66 |
| | **TMLM_8L** | **2.50** | 1.67 | 95.83 |

Table 3 shows the recognition results in more detail about the percentage of corrected cases, miscorrected cases, and unchanged cases when combining SRTC_SLP with three different language models on the CROHME 2016 and CROHME 2019 testing sets. The percentages of the corrected cases by TMLM_8L are the highest compared to 11-grams and GRULM_2L on both testing sets. However, that of the miscorrected cases by TMLM_8L is the worst compared to others. GRULM_2L caused the least miscorrections compared to others, but it could not correct many cases. 11-grams and TMLM_8L have comparable percentages of miscorrected cases, but TMLM_8L corrected more cases than 11-grams did.

### 4.5 Error Analysis

In this section, we present some samples which are corrected or miscorrected when applying TMLM_8L with the SRTC_SLP recognizer.

Figure 5(a) and Fig. 5(b) show two corrected cases. In Fig. 5(a), "$\alpha$" in the HME sample are recognized as "2" without using TMLM_8L since it seems to be similarly written as "2". However, the language score of the candidate with "$\alpha$" is significantly higher than the one with "2". Therefore, the recognizer combined TMLM_8L results in the correct prediction. TMLM_8L performs well since "$\alpha$" seems more likely to appear next to the trigonometry function (e.g., sine, cosine, and tangent) than a number. Similarly, "9" in the HME sample of Fig. 5(b) are correctly recognized by TMLM_8L.

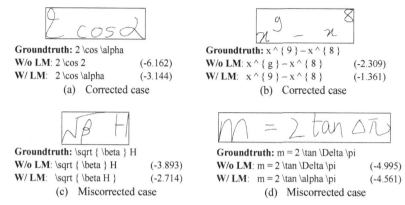

**Groundtruth:** 2 \cos \alpha
**W/o LM**: 2 \cos 2                                    (-6.162)
**W/ LM**:  2 \cos \alpha                               (-3.144)
          (a)   Corrected case

**Groundtruth:** x ^ { 9 } – x ^ { 8 }
**W/o LM**: x ^ { g } – x ^ { 8 }                       (-2.309)
**W/ LM**:  x ^ { 9 } – x ^ { 8 }                       (-1.361)
          (b)   Corrected case

**Groundtruth:** \sqrt { \beta } H
**W/o LM**: \sqrt { \beta } H                           (-3.893)
**W/ LM**:  \sqrt { \beta H }                           (-2.714)
          (c)   Miscorrected case

**Groundtruth:** m = 2 \tan \Delta \pi
**W/o LM**: m = 2 \tan \Delta \pi                       (-4.995)
**W/ LM**:  m = 2 \tan \alpha \pi                       (-4.561)
          (d)   Miscorrected case

**Fig. 5.** Examples of corrected and miscorrected cases when combining the SRTC_SLP recognizer and TMLM_8L (LM: language model). Each case shows an HME image, its ground truth, and its recognition candidates with/without TMLM_8L and their corresponding scores from TMLM_8L.

Figure 5(c) and Fig. 5(d) show two miscorrected cases. The case in Fig. 5(c) is miscorrected since the language model score of the incorrect result is higher than that of the correct result. We can realize that the context, in this case, is not clear. The case in Fig. 5(d) is miscorrected since "$\alpha$" seems more likely to appear next to the tangent symbol than "$\Delta$".

According to those examples, we can see that modeling MEs is still challenging since the context in an ME is not clear and our corpus of MEs might not be enough to estimate the distribution of MEs.

## 5   Conclusion and Future Works

This paper presented a transformer-based math language model (TMLM) for improving the recognition rate of HME recognition systems. We showed that our TMLMs perform better than the traditional language models for MEs, i.e., the $N$-gram and GRULM. The best perplexity achieved is 4.42, resulted from TMLM_8L of 8 transformer layers. Combining TMLM_8L with the online HME recognizer in [18] improved the expression rate by 2.97 and 0.83 percentage points on the CROHME 2016 and CROHME 2019 testing set, respectively.

There are several remaining works. Firstly, we should enrich the source of ME LaTeX by collecting open sources on the internet. Secondly, we should modify our TMLM to exploit the bidirectional context in MEs. Thirdly, the method for jointly training an HME recognizer and a math language model should be studied for better optimization.

**Acknowledgement.** This research is being partially supported by the grant-in-aid for scientific research (A) 19H01117 and that for Early Career Research 21K17761.

# References

1. Mahdavi, M., Zanibbi, R., Mouchere, H., Viard-Gaudin, C., Garain, U.: CROHME + TFD: competition on recognition of handwritten mathematical expressions and typeset formula detection. In: Proceedings of International Conference on Document Analysis and Recognition, pp. 1533–1538 (2019)
2. Zhang, J., Du, J., Dai, L.: Track, attend, and parse (TAP): an end-to-end framework for online handwritten mathematical expression recognition. IEEE Trans. Multimed. **21**, 221–233 (2019)
3. Zhang, J., et al.: Watch, attend and parse: an end-to-end neural network based approach to handwritten mathematical expression recognition. Pattern Recognit. **71**, 196–206 (2017). https://doi.org/10.1016/j.patcog.2017.06.017
4. Poznanski, A., Wolf, L.: CNN-N-gram for handwritingword recognition. In: Proceedings of the IEEE Computer Society Conference on Computer Vision and Pattern Recognition, pp. 2305–2314 (2016)
5. Zhu, B., Zhou, X.D., Liu, C.L., Nakagawa, M.: A robust model for on-line handwritten japanese text recognition. Int. J. Doc. Anal. Recognit. **13**, 121–131 (2010). https://doi.org/10.1007/s10032-009-0111-y
6. Reeve Ingle, R., Fujii, Y., Deselaers, T., Baccash, J., Popat, A.C.: A scalable handwritten text recognition system. In: Proceedings of the International Conference on Document Analysis and Recognition (ICDAR), pp. 17–24 (2019)
7. Zanibbi, R., Blostein, D.: Recognition and retrieval of mathematical expressions. Int. J. Doc. Anal. Recognit. **15**, 331–357 (2012)
8. Wu, J.-W., Yin, F., Zhang, Y.-M., Zhang, X.-Y., Liu, C.-L.: Handwritten mathematical expression recognition via paired adversarial learning. Int. J. Comput. Vision **128**(10–11), 2386–2401 (2020). https://doi.org/10.1007/s11263-020-01291-5
9. Mikolov, T., Karafiát, M., Burget, L., Cernocky, J.: Recurrent neural network based language model. In: 11th Annual Conference of the International Speech Communication Association, pp. 1045–1048 (2010)
10. Al-Rfou, R., Choe, D., Constant, N., Guo, M., Jones, L.: Character-level language modeling with deeper self-attention. In: 33rd AAAI Conference on Artificial Intelligence, pp. 3159–3166 (2019)
11. Radford, A., Narasimhan, K., Salimans, T., Sutskever, L.: Improving Language understanding by generative pre-training (2018). https://s3-uswest-2.amazonaws.com/openai-assets/research-covers/language-unsupervised/language_understanding_paper.pdf
12. Radford, A., Wu, J., Child, R., Luan, D., Amodei, D., Sutskever, I.: Language models are unsupervised multitask learners (2019). https://d4mucfpksywv.cloudfront.net/better-language-models/language_models_are_unsupervised_multitask_learners.pdf
13. Vaswani, A., et al. Attention is all you need. In: Advances in Neural Information Processing Systems (2017)
14. Dai, Z., Yang, Z., Yang, Y., Carbonell, J., Le, Q.V., Salakhutdinov, R.: Transformer-XL: attentive language models beyond a fixed-length context. In: ACL 2019 - 57th Annual Meeting of the Association for Computational Linguistics, pp. 2978–2988 (2019)
15. Yang, Z., Dai, Z., Yang, Y., Carbonell, J., Salakhutdinov, R., Le, Q.V.: XLNet: generalized autoregressive pretraining for language understanding. In: Advances in Neural Information Processing Systems (2019)
16. Devlin, J., Chang, M.-W., Lee, K., Toutanova, K.: BERT: pre-training of deep bidirectional transformers for language understanding - ACL anthology. In: Proceedings of the 2019 Conference of the North American Chapter of the Association for Computational Linguistics: Human Language Technologies, pp. 4171–4186 (2019)

17. Ba, J.L., Kiros, J.R., Hinton, G.E.: Layer normalization. http://arxiv.org/abs/1607.06450 (2016)
18. Nguyen, C.T., Truong, N.-T., Nguyen, H.T., Nakagawa, M.: Global context for improving recognition of online handwritten mathematical expressions. In: Proceedings of the International Conference on Document Analysis and Recognition (ICDAR) (2021)
19. Mouchere, H., Viard-Gaudin, C., Zanibbi, R., Garain, U.: ICFHR2016 CROHME: competition on recognition of online handwritten mathematical expressions. In: 2016 15th International Conference on Frontiers in Handwriting Recognition (ICFHR), pp. 607–612 (2016)
20. Cocke, J., Schwartz, J.T.: Programming Languages and Their Compilers: Preliminary Notes. Courant Institute of Mathematical Sciences, New York University (1970)
21. Grave, E., Joulin, A., Cisse, M., Grangier, D., Jegou, H.: Efficient softmax approximation for GPUsÉdouard. In: 34th International Conference on Machine Learning, pp. 1302–1310 (2017)
22. Loshchilov, I., Hutter, F.: Decoupled weight decay regularization. In: International Conference on Learning Representations (2019)
23. Wolf, T., et al.: HuggingFace's transformers: state-of-the-art natural language processing (2019). http://arxiv.org/abs/1910.03771

# Exploring Out-of-Distribution Generalization in Text Classifiers Trained on Tobacco-3482 and RVL-CDIP

Stefan Larson[1][(✉)], Navtej Singh[1], Saarthak Maheshwari[2], Shanti Stewart[3], and Uma Krishnaswamy[2]

[1] SkySync, Ann Arbor, MI, USA
{slarson,nsingh}@skysync.com
[2] University of California, Berkeley, CA, USA
[3] Oregon State University, Corvallis, OR, USA

**Abstract.** To be robust enough for widespread adoption, document analysis systems involving machine learning models must be able to respond correctly to inputs that fall outside of the data distribution that was used to generate the data on which the models were trained. This paper explores the ability of text classifiers trained on standard document classification datasets to generalize to out-of-distribution documents at inference time. We take the Tobacco-3482 and RVL-CDIP datasets as a starting point and generate new out-of-distribution evaluation datasets in order to analyze the generalization performance of models trained on these standard datasets. We find that models trained on the smaller Tobacco-3482 dataset perform poorly on our new out-of-distribution data, while text classification models trained on the larger RVL-CDIP exhibit smaller performance drops.

**Keywords:** Document classification · Text classification · Out-of-distribution generalization

## 1 Introduction

Recent years have seen great improvements in the document analysis and recognition field. These advancements have typically stemmed from higher capacity deep learning models and larger training datasets. Two representative datasets in the field of document classification are the Tobacco-3482 [1] and the RVL-CDIP [2] datasets, the latter consisting of over 400,000 training samples across 16 document categories. Indeed, the RVL-CDIP dataset has emerged as the premier benchmark for evaluating document classification algorithms.

However, there are some potential weaknesses with both of these datasets. For one, documents from both datasets come from the same domain: the tobacco

S. Maheshari, S. Stewart and U. Krishnaswamy—Work performed while author was an intern at SkySync.

E. H. Barney Smith and U. Pal (Eds.): ICDAR 2021 Workshops, LNCS 12917, pp. 416–423, 2021.
https://doi.org/10.1007/978-3-030-86159-9_30

industry. Second, the documents in both corpora are from the 1950s to 2002, and are older than—for lack of a better word—contemporary documents. These two potential weaknesses lead us to question whether models trained on RVL-CDIP and Tobacco-3482 have the capacity to generalize to *out-of-distribution* inputs. That is to say, can models trained on RVL-CDIP and Tobacco-3482 perform well on documents that come from different industries than the tobacco industry? and can such models perform well on more recent documents?

We attempt to answer these questions in this paper. We train text classifiers on the RVL-CDIP and Tobacco-3482 datasets and evaluate performance on new test datasets. These test datasets are drawn from "distributions" that are quite different from the datasets' original distributions. We find that while models trained on Tobacco-3482 and RVL-CDIP generalize well on portions of our new test datasets, they perform substantially worse on others. Moreover, models trained on Tobacco-3482 exhibit far less capacity for generalizability than those trained on RVL-CDIP.

## 2   Related Work

Measuring out-of-distribution performance is an important task that has seen increased interest in the past half-decade, especially as systems powered by machine learning make their way into more and more industries and applications. Such systems must not only perform well on in-distribution data, but must also perform well on data that is "unseen" at training time.

Out-of-distribution performance can be measured on two different types of inputs for supervised learning tasks: first, on data that does not belong to any of the target categories of the label set; and second, on data that *does* belong to the target categories but was generated or collected by a different mechanism than was used to create the original training data.

Examples of the first category include so-called *out-of-scope* [5] and *out-of-domain* [6] data in the dialog system field. Prior work has also analyzed out-of-distribution performance in image classification tasks where models trained on CIFAR datasets are tested on a wide range of data that do not belong to any of CIFAR's target categories [7].

Prior work from the second out-of-distribution category include [9] and [8], who evaluate transformer models on the NLP tasks of natural language inference, sentiment analysis, and news article classification. Other related work on *out-of-domain* sentiment analysis includes [11] and [10], who train sentiment analysis models on product reviews for a specific product category, then analyze model performance on reviews from a different category. In this way, the analysis of the second type of out-of-distribution phenomenon is the same as what is known as *distribution shift*, where the "training distribution differs from the test distribution" [12]. In our current paper, we focus only on this second category of out-of-distribution inference.

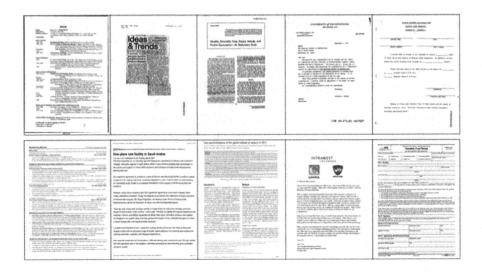

**Fig. 1.** Samples from the Tobacco-3482 corpus (top row) and our out-of-distribution evaluation corpus (bottom row). The labels for the samples are (from left to right): *resume*, *news*, *scientific*, *letter*, and *form*.

# 3    In- and Out-of-Distribution Data

## 3.1    In-Distribution Data

We use the Tobacco-3482 [1] and RVL-CDIP [2] datasets as *in-distribution* training data for our text-based document classifiers. Documents from the Tobacco-3482 dataset belong to one of 10 document categories, while the RVL-CDIP dataset contains 16 different document categories. Both the Tobacco-3482 and RVL-CDIP datasets contain documents from the IIT-CDIP test collection [3], which in turn contains documents from a corpus of publicly available documents from litigation against several tobacco-related companies [4] (the majority of these documents are from the 1950s through 2002). Examples from the Tobacco-3482 dataset can be seen in Fig. 1 (top row).

## 3.2    Out-of-Distribution Data

We selected five categories from the Tobacco-3482 dataset and five categories from the RVL-CDIP dataset and gathered new out-of-distribution testing data for these categories. The list of selected categories is shown in Table 1. We collected data from two different "distributions" for the *resume* data: first, we use internal company resumes; second, we used a search engine to scrape curriculum vitaes (CVs). (Curriculum vitaes can be thought of as long-form resumes, and alternatively 1–2 page resumes can be thought of as short-form CVs.) Importantly, we explicitly searched for CVs that did not contain the phrase "curriculum vitae".

**Table 1.** Categories from Tobacco-3482 and RVL-CDIP covered by our out-of-distribution evaluation data sets.

| Tobacco-3482 | RVL-CDIP | Num. of OOD test samples |
|---|---|---|
| form | form | 215 |
| letter | letter | 80 |
| news | news article | 30 |
| scientific | — | 133 |
| resume | resume | 267 |
| — | invoice | 26 |

Our out-of-distribution *forms* data consists of tax forms, application forms, request forms, and donation forms that we scraped via web search. All of these forms are unfilled. Our new *letter* data consists of cover letters, recommendation letters, and various other letters, all scraped via web search, as does the out-of-distribution *news* and *invoice* data. The *scientific* out-of-distribution data consist of academic research papers found via web search, the majority of which were sampled from several undergraduate research journals. Examples of our new out-of-distribution documents can be seen in Fig. 1 (bottom row). Table 1 shows the number of out-of-distribution test samples that we collected for each category. Importantly, all out-of-distribution documents come from domains/areas other than the tobacco industry, and were created relatively recently (i.e., in years 2010–2021).

As seen in Table 1, most of the chosen categories overlap between the Tobacco-3482 and RVL-CDIP datasets, with the exclusion of *invoice* (which is not a category in Tobacco-3482) and *scientific*. The *scientific* data partially consists of academic research papers (an example of which is shown in the first row and third column of Fig. 1), which is why we use academic research papers in our out-of-distribution test set. While the RVL-CDIP corpus contains categories called *scientific report* and *scientific publication*, we could not determine which of these categories was more appropriate for academic research papers, and hence we do not include this in our evaluation of RVL-CDIP.

## 4 Experiments

### 4.1 Methodology

We conduct several experiments to evaluate the out-of-distribution generalization performance of text classifiers trained on the Tobacco-3482 and RVL-CDIP datasets. First, we train models on both the Tobacco-3482 and RVL-CDIP datasets and evaluate in-distribution performance on the five categories specified in Table 1. Unlike the RVL-CDIP corpus, the Tobacco-3482 dataset does not have a specified test set, so we randomly partitioned the dataset in to train and test sets with a 90-10 split. Second, we evaluate the models by testing on our newly-collected data introduced in Sect. 3.2.

Third, we modify our out-of-distribution evaluation datasets by removing tokens from documents that are indicative of the name of the document's category. For instance, we remove all instances of the word "form" and "invoice" from the *form* and *invoice* categories, respectively. (All taboo words are listed in Table 2.) Following [13], we call this modified evaluation data the *taboo* data. The purpose of the taboo evaluation data is to test whether the model can generalize further to data that does not contain features (in this case, tokens) that we suspect may be overrepresented in the training data.

**Table 2.** Taboo words for each test category.

| Category | Taboo words |
| --- | --- |
| *form* | "form" |
| *letter* | "letter" |
| *resume* | "resume", "curriculum", "vitae" |
| *news* | "news", "article" |
| *scientific* | "article" |
| *invoice* | "invoice" |

We use Google Tesseract as our optical character recognition (OCR) engine for extracting the in-distribution training and test text data from the RVL-CDIP and Tobacco-3482 document images. Since our out-of-distribution evaluation data is entirely from Microsoft Word and PDF files with text already embedded in the files, we use an extraction tool that simply reads and processes this text data without need for OCR.

Our text classifier of choice is MobileBERT [15], which is a pre-trained transformer model that was originally trained via knowledge distillation from BERT [14]. Since it is a product of knowledge distillation, MobileBERT consists of far fewer parameters than the original BERT model, and thus ought to be less prone to overparameterization. Nevertheless we also use label smoothing with a factor of 0.1. We train and test on the first 512 tokens of each document.

## 4.2   Results

Accuracy scores on the five in-distribution categories are shown in Table 3. We note that the Tobacco-3482 dataset appears to be easier than the RVL-CDIP dataset, as the classifier's accuracy scores on the former are higher in all four categories in which there is a category overlap.

Table 4 displays the accuracy scores for the text classifiers when evaluated on the out-of-distribution data. The results of the taboo out-of-distribution tests are shown in the second rows of each main row of Table 4. Compared to the

**Table 3.** In-distribution accuracy scores.

| Train dataset | Test data | | | | | |
|---|---|---|---|---|---|---|
| | *form* | *letter* | *resume* | *news* | *scientific* | *invoice* |
| Tobacco-3482 | 93.2 | 89.5 | 100 | 89.5 | 85.2 | — |
| RVL-CDIP | 70.8 | 82.0 | 96.0 | 76.8 | — | 81.2 |

in-distribution accuracy scores in Table 3, the out-of-distribution scores are typically lower. This is expected, as the out-of-distribution data ought to be harder for the classifier to recognize. For instance, the decline in performance on the Tobacco-3482 data tends to be quite severe (e.g. from 100 to 44.3 on *resume* and 89.5 to 30.0 on *news*). Surprisingly though, the out-of-distribution performance is better than the in-distribution performance in a few cases for the classifier trained on the RVL-CDIP dataset. With the exception of *invoice*, the declines on the RVL-CDIP dataset from in- to out-of-distribution tend to be quite mild.

**Table 4.** Performance (accuracy) of document classifiers when trained on an in-distribution training dataset and evaluated on an out-of-distribution test dataset. The second row of each main row displays the results on the taboo out-of-distribution evaluation data.

| Train dataset | Test data | | | | | | |
|---|---|---|---|---|---|---|---|
| | *form* | *letter* | *resume* | *CVs* | *news* | *scientific* | *invoice* |
| Tobacco-3482 | 89.3 | 89.5 | 44.3 | 94.0 | 30.0 | 53.4 | — |
| | 87.0 | 72.5 | 44.3 | 94.0 | 26.7 | 53.4 | — |
| RVL-CDIP | 90.7 | 83.8 | 88.0 | 100 | 73.3 | — | 69.2 |
| | 86.5 | 82.5 | 88.0 | 100 | 73.3 | — | 46.2 |

In all cases, we either see a decline or no change in performance when comparing the out-of-distribution tests with the taboo tests. However, only the *invoice* category from RVL-CDIP seems to be substantially impacted, dropping 23% points from 69.2 to 46.2.

In general, these results indicate that the MobileBERT model trained on RVL-CDIP appears to be more robust and exhibits better generalization to out-of-distribution data than the MobileBERT model trained on Tobacco-3482. This is likely attributed to the fact that RVL-CDIP consists of more training data. Our biggest takeaway is that models trained on RVL-CDIP do not inherently lack the capacity to generalize to documents outside of the tobacco industry or outside of documents from a fixed time period. Indeed, all of the out-of-distribution data that we collected and on which we tested comes from non-tobacco related sources and was generated relatively recently (i.e. 2010–2021).

# 5 Future Work

Text analysis is only part of the picture in document classification, and future work will extend our analysis of out-of-distribution performance to image classification models. Future work will also evaluate text- and image-based document classifiers on out-of-distribution inputs that *do not* belong to any of the categories seen at training time.

# 6 Conclusion

At the outset of this paper we were concerned with the ability of a document classification model trained on Tobacco-3482 or RVL-CDIP to generalize well to out-of-distribution documents. By training a text-based document classifier on the Tobacco-3482 and RVL-CDIP datasets and evaluating on newly-collected out-of-distribution test data, we found that while the model trained on Tobacco-3482 tends to perform poorly on out-of-distribution data, the RVL-CDIP dataset typically endows the model with the ability to generalize to this type of data.

# References

1. Kumar, J., Doermann, D.: Unsupervised classification of structurally similar document images. In: Proceedings of the International Conference on Document Analysis and Recognition (ICDAR) (2013)
2. Harley, A.W., Ufkes, A., Derpanis, K.G.: Evaluation of deep convolutional nets for document image classification and retrieval. In: Proceedings of the International Conference on Document Analysis and Recognition (ICDAR) (2015)
3. Lewis, D., Agam, G., Argamon, S., Frieder, O., Grossman, D., Heard, J.: Building a test collection for complex document information processing. In: Proceedings of SIGIR (2006)
4. University of California, San Francisco: The Legacy Tobacco Document Library (LTDL) (2007)
5. Larson, S., et al.: An evaluation dataset for intent classification and out-of-scope prediction. In: Proceedings of the 2019 Conference on Empirical Methods in Natural Language Processing and the 9th International Joint Conference on Natural Language Processing (EMNLP-IJCNLP) (2019)
6. Xu, H., He, K., Yan, Y., Liu, S., Liu, Z., Xu, W.: A deep generative distance-based classifier for out-of-domain detection with Mahalanobis space. In: Proceedings of the 28th International Conference on Computational Linguistics (COLING) (2020)
7. Liang, S., Li, Y., Srikant, R.: Enhancing the reliability of out-of-distribution image detection in neural networks. In: Proceedings of the International Conference on Learning Representations (ICLR) (2018)
8. Hendrycks, D., Liu, X., Wallace, E., Dziedzic, A., Krishnan, R., Song, D.: Pre-trained transformers improve out-of-distribution robustness. In: Proceedings of the 58th Annual Meeting of the Association for Computational Linguistics (ACL) (2020)
9. Desai, S., Durrett, G.: Calibration of pre-trained transformers. In: Proceedings of the 2020 Conference on Emperical Methods in Natural Language Processing (EMNLP) (2020)

10. Desai, S., Zhan, H., Aly, A.: Evaluating lottery tickets under distributional shifts. In: Proceedings of the 2nd Workshop on Deep Learning Approaches for Low-Resource NLP (DeepLo) (2019)
11. Peng, M., Zhang, Q., Jiang, Y., Huang, X.: Cross-domain sentiment classification with target domain specific information. In: Proceedings of the 56th Annual Meeting of the Association for Computational Linguistics (ACL) (2018)
12. Koh, P.W., et al.: WILDS: A Benchmark of in-the-Wild Distribution Shifts. arXiv preprint arXiv:2012.07421 (2021)
13. Larson, S., et al.: Iterative feature mining for constraint-based data collection to increase data diversity and model robustness. In: Proceedings of the 2020 Conference on Empirical Methods in Natural Language Processing (EMNLP) (2020)
14. Devlin, J., Chang, M.W., Lee, K., Toutanova, K.: BERT: pre-training of deep bidirectional transformers for language understanding. In: Proceedings of the 2019 Conference of the North American Chapter of the Association for Computational Linguistics: Human Language Technologies (NAACL) (2019)
15. Sun, Z., Yu, H., Song, X., Liu, R., Yang, Y., Zhou, D.: MobileBERT: a compact task-agnostic BERT for resource-limited devices. In: Proceedings of the 58th Annual Meeting of the Association for Computational Linguistics (ACL) (2020)

# Labeling Document Images for E-Commence Products with Tree-Based Segment Re-organizing and Hierarchical Transformer

Peng Li$^{(\boxtimes)}$, Pingguang Yuan, Yong Li, Yongjun Bao, and Weipeng Yan

JD.com, Beijing, China
{lipeng464,yuanpingguang,liyong5,baoyongjun,Paul.yan}@jd.com

**Abstract.** Document images of products have been widely used in E-commence. As a kind of special data, the contents in document images are quite diverse: texts can be scattered anywhere with pictures, and both short text snippets and long text chunks exist. To predict text labels in document images, we propose a two stage approach. The first stage, named as tree-based segment re-organizing, is to resume text order and text connection through hierarchical clustering, segment reordering and segment merging. The second stage, named as hierarchical transformer, is to generate segment embeddings and predict segment labels, where segment level and document level encoder are applied. We empirically study the effects of incorporating different features and compare two kinds of attention to aggregate context, where distance and direction are measured in 1D and 2D respectively. Experiments based on a real-world dataset show that our proposed segment re-organizing method can reduce about 40% input size to the labeling model while bring negligible impact to performance. For hierarchical transformer, we empirically show that document encoder using 1D attention is more effective than 2D attention.

**Keywords:** Document image of products · Semantic labeling · Attention for post-OCR texts

## 1 Introduction

Document images such as merchandise posters (Fig. 1) have been widely used in E-commence to describe product. Compared to the text format contents such as product titles, the document images contain additional product details such as functions, co-use products, usage etc. Automatic labeling document images, i.e. detecting connected text regions and predicting their semantic types, is important for understanding products, which is the basis for customer-product match in search and recommendation.

Although text labeling has been well studied in NLP [7], document image labeling has brought up new challenge: the input is not a token sequence, but many

© Springer Nature Switzerland AG 2021
E. H. Barney Smith and U. Pal (Eds.): ICDAR 2021 Workshops, LNCS 12917, pp. 424–439, 2021.
https://doi.org/10.1007/978-3-030-86159-9_31

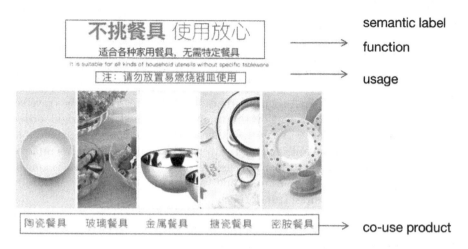

**Fig. 1.** An example document image for product "智能变频电磁炉" (Intelligent frequency conversion induction cooker). It contains three kinds of semantic contents.

separated token sequences, named as text segment, each with a bounding box to indicate position. The text order and text connection information is missing, which makes text understanding difficult. There have been several studies proposed for labeling document images. In these studies, the most commonly used datasets are about scanned images of business documents, scientific papers and examination papers [19]. The existing methods can be categorized into two types: the first one is a pipeline paradigm which does text clustering and text labeling sequentially [2], while the second one is an end-to-end paradigm, which takes raw texts of OCR as input and generate final labels directly [11]. The pipeline paradigm can optimize each component separately while end-to-end paradigm can have better generalization ability, especially when the pre-training technique is applied [12].

In this paper, we study the problem of labeling document images for products. Specifically, we first collect document images from real-world E-commerce platform and define semantic labels according to document content. Compared to the previous datasets [19], our dataset has its unique characteristics: 1) The content is quite diverse, where texts are mixed up with product pictures, and both short texts and long texts exist in document images. 2) The label granularity is coarser. The labels such as functions or sales note generally correspond to large chunks of texts while previous datasets focus on finer entity level labels such as address or date.

To predict labels, we can take a similar pipeline approach as [2]. However, due to the data specificity, three important factors need to be accounted: 1) The text segments in product document image are usually grouped into different visual regions, simple top-to-bottom and left-to-right heuristics can distort the truth segment order. 2) The number of all the segment tokens in a document can exceed the maximum sequence length of pre-train models, which needs to be tackled. 3) Some text segments are quite short and it is necessary to combine all the possible contexts such as product name and category to predict segment labels. By accounting these factors, we propose a novel two-stage labeling approach using **tree-based segment re-organizing** and **hierarchical transformer**. Specifically, the **tree-based segment re-organizing** aims to resume text order and text connection by visual information: first we group text segments into a tree structure by hierarchical segment dilation ; then we recursively traverse the tree in depth-first order and by the top-to-bottom and left-to-right order to output segment list; after that we further merge neighbor text segments if their text font sizes are similar enough. **Hierarchical transformer** employs two Transformers to encode segment by accounting segment tokens and segment context respectively. To be able to encode various product texts, we utilize a pre-trained model as the first Transformer.

Our contributions are threefold. First, to our best knowledge, it is the first attempt to label document images for products to identify coarse-grained semantic regions. Second, we propose a tree-based segment re-organizing and hierarchical Transformer solution to tackle the problem. Last, we empirically compare two kinds of distance-and-direction aware attention for aggregating context and validate the effects of different features.

## 2   Related Works

In recent years, deep learning based OCR becomes practical[1] and extracting semantic information from document images has attracted much attention.

There are several representative datasets each having its own extraction target. Specifically, SROIE [19] is a dataset composed of scanned receipts in printed English aiming to extract company, date, address, total. [6] proposed a Chinese document image dataset composed of ValueAdded Tax Invoices (VATI) and International Purchase Receipts (IPR) to extract buyer/seller, date and tax amount etc. [11] proposed an Examination Paper Head Dataset to extract key and value for person information. All the above datasets are about extracting fine-grained entities. In this paper, we focus on the document images of products and aim to label semantic regions at a coarse granularity. [9] also studied the document images of products, but their goal is to infer reading order of text segments.

Similar to sequential labeling in traditional NLP [7,14], the core task in document image labeling is also to model segment dependency. There are two main streams of methods addressing this issue. The first is to transform 2D segments to 1D segment list and model segment dependency in 1D space using sequential models such as Bi-LSTM [2]. The second is to model segment dependency directly in 2D space, where 2D position bias affects information aggregation. [13] proposed to use Transformer [17] to encode segments where the attention scores based on 2D position bias are added to the content attention scores. [4–6] proposed to use Graph Neural Networks (GNNs) to encode segments. Actually Transformer and GNN model are similar from the perspective of information

---

[1] https://github.com/paddlepaddle/paddleocr/.

flow, where each pair of inputs are modeled. Besides, [15] proposed a CNN based framework for 2D layout modeling.

To account context for segments, [2,4,5] aggregate segment embeddings while [12,13] aggregate token embeddings, and the token dependency across segments are modeled directly. As Transformer scales quadratically with input length, aggregating token embedding may be limited by the sequence length of tokens. To address this limitation, Longformer [16] is proposed for which some direct attention between tokens are skipped. Actually, aggregating segment embeddings can also be seen as a kind of simplification for aggregating token embeddings.

In this paper, we first merge possible coherent texts as new segment to make dependent tokens be well captured by segment encoder. Then we use document encoder to only aggregate high level segment embeddings as context.

## 3   Product Document Image Dataset

In this section, we introduce our new product document image dataset and the semantic labels.

We build our dataset by sampling products from Jingdong[2], one of Chinese largest E-commerce platform. The products in E-commerce are generally organized by tree structured taxonomy: each leaf node is a basic product unit called SKU and each inner node is a category containing similar SKUs. For each SKU, it has a product name and a long document image. Note that the long document image is a concatenation of multiple independent sub images, which needs to be segmented.

Specifically, we take the following steps to build the dataset: 1) We chose 2 categories to sample SKUs: "Kitchenware" and "Cookware", where "Kitchenware" is the parent node of "Cookware". For each category, we take a stratified sampling which first sample product name and then sample related SKUs. The number of sampled SKUs for "Kitchenware" is about 3 times the number for "Cookware", since the parent node "Kitchenware" contain about 10 times SKUs than "Cookware". 2) We ask annotators to split long document image to sub document images. Each sub document image is the actual input for labeling. 3) We ask annotators to label document image regions by drawing rectangles. Each rectangle corresponds to a semantic coherent region which is consistent with people understanding. Multiple neighbor regions with the same semantic can be annotated with one rectangle, as we focus on coarse-grained semantic. 4) We did OCR on the document image and collect all the bounding boxes and texts. During the above process, we also annotate table region in document images and exclude them for modeling, since table content is a kind of special texts. The final dataset contains two sub dataset: Kitchenware dataset and Cookware dataset. The Statistics about the dataset is given in Table 1.

As for the semantic labels, we ask domain expert and summarize 20 content labels. These labels can be divided int to 3 groups: the labels describing product

---

[2] https://www.jd.com/.

**Table 1.** Dataset statistics

| Category | No. of doc. images | No. of SKUs | No. of segments | No. of prod. names |
|---|---|---|---|---|
| Cookware | 19,030 | 1,866 | 85,673 | 158 |
| Kitchenware | 46,861 | 5,260 | 207,127 | 627 |

**Table 2.** Dataset labels

| Label group | Labels |
|---|---|
| Product itself | component, function, advertisement, brand name, brand detail, product name, product model, usage, sales note, product QA (question & answer), other |
| Product application | demand, demand problem, competitive product label, competitive product problem, co-use product, substitution, applicable scene, other |
| Document organization | navigation |

itself, the labels describing product application and the labels for document organization such as navigation bar. All the labels are presented in Table 2.

## 4   Our Proposed Approach

Let $D = \{(S_1, P_1), (S_2, P_2), ..., (S_{|D|}, P_{|D|}\}$ denote a document image, where $S_i$ is a text segment and $P_i$ is the corresponding segment bounding box. $P_i$ and $S_i$ are generated by OCR detection and recognition processes. The segment $S_i$ can be further represented as $S_i = (w_1^i, w_2^i, .., w_{|S_i|}^i)$, where each $w^i$ is a token in the segment. The bounding box $P_i$ can be represented as $P_i = (x_1^i, y_1^i, x_2^i, y_2^i)$, where $(x_1^i, y_1^i)$ and $(x_2^i, y_2^i)$ correspond to the coordinates of top left point and bottom right point respectively. These coordinate values are further normalized by diving the image width and height, so $x_1^i, x_2^i, y_1^i, y_2^i \in (0, 1)$.

The problem of labeling document images can be formulated as follows: given $D = \{(S_1, P_1), (S_2, P_2), ..., (S_{|D|}, P_{|D|}\}$, predict a list of semantic labels $\{t_1, t_2, ..., t_{|D|}\}$, where $t_i$ is the corresponding label for $(S_i, P_i)$. For product document image, we focus on coarse-grained semantics, so $t_i$ is in the range of 20 labels in Sect. 3. We do not incorporate BIO tag conversion to extract fine-grained entities.

To predict labels, we propose a pipeline framework consisting of two sequential steps: the first step is to re-organize segments aiming to merge coherent text segments by visual information, called tree-based segment re-organizing; the second step is to combine both text and position information for label prediction.

## 4.1    Tree-Based Segment Re-organizing

Product sellers usually take several visual tricks to present information: the coherent text segments tend to be close in distance, or have consistent font size or font color, or are aligned vertically or horizontally. It is intuitive to use visual information to merge similar and distance close segments together as a new segment for labeling. The problem is to what extent visually similar and distance close segments can be merged: on the one hand, the merge should be sensitive to find coherent segments, on the other hand, the merge should be prudent so as not to incorporate unrelated segments. Another problem is how to arrange segments in reading order. For business documents or scientific papers, the text segments are generally line by line, and it is natural to arrange segments according to top to down and left to right order. For document image of product, the texts are mixed with pictures and they are clustered into different regions. It is necessary to first group the segments as clusters and then sort segments by cluster.

Specifically, our proposed segment re-organizing method are as follows:

First, we do hierarchical clustering on segments by recursively grouping neighbors. The clustering output is a tree structure, as shown in Fig. 2, where each leaf node is a segment and each inner node is a cluster.

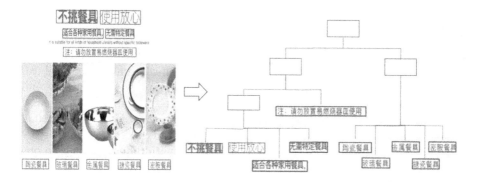

**Fig. 2.** An illustration of clustering the OCR segments

To define neighbors, we incorporate a hyper parameter $\epsilon$ to restrict the maximum allowed distance to neighbors. For each segment $S_i$, its neighbor segments should be in the range of $(0, \epsilon f_i)$, where $f_i$ is the average font size of $S_i$. The clustering can be done efficiently using morphological operation, i.e., segment dilation: the original document is first transformed to a binary image where text segment pixel is 1 and the others is 0 ; then for each segment, its boundaries is expanded by $\epsilon f_i$; after dilation, the connected segments form a cluster. Besides, we observe that if two segments are aligned vertically or horizontally, they are more probable to be coherent, so the expansion width should be enlarged for the segments with alignment. We incorporate 2 hyper parameters $(\epsilon^{av}, \epsilon^{ah}) > \epsilon$.

For each segment $S_i$, if it is vertically (horizontally) aligned with others, its vertical (horizontal) boundaries is expanded by $\epsilon^{av} f_i$ ($\epsilon^{ah} f_i$). To judge whether a segment is with alignment, we incorporate another 2 hyper parameter $\omega^v, \omega^h$. We first sort all the segments according to their vertical positions and horizontal positions to get two lists. For segment $S_i$, if the vertical (or horizontal) distance to its next segment $S_{i+1}$ in list is smaller than $\omega^v f_i$ ($\omega^h f_i$), then both $S_i$ and $S_{i+1}$ are taken as segments with vertical (or horizontal) alignment. The vertical distance and horizontal distance are given by Eq. (1) and Eq. (2) respectively.

$$\text{vdist}(S_i, S_{i+1}) = \min \left( |y_1^i - y_1^{i+1}|, |y_2^i - y_2^{i+1}|, |\frac{y_1^i + y_2^2}{2} - \frac{y_1^{i+1} + y_2^{i+1}}{2}| \right) \quad (1)$$

$$\text{hdist}(S_i, S_{i+1}) = \min \left( |x_1^i - x_1^{i+1}|, |x_2^i - x_2^{i+1}|, |\frac{x_1^i + x_2^i}{2} - \frac{x_1^{i+1} + x_2^{i+1}}{2}| \right) \quad (2)$$

Second, once we finish clustering, we arrange the segments based on the clustered tree. We recursively traverse the tree in depth-first order. For the lowest inner node, its child segments are output using the top-to-bottom and left-to-right order. In this step we actually generate a segment permutation $D^\tau$ of original document $D$.

Third, we carefully merge the consecutive segments of $D^\tau$ if they meet the merging standard: under the same lowest inner node of the clustered tree and having similar enough font sizes. We incorporate 2 hyper parameters $\psi$ and $\psi^a$ ($\psi^a > \psi$) to define font similarity. For segment $S_i$, if the font size of the next segment $S_{i+1}$ satisfies $f_{i+1} \in [(1 - \psi)f_i, (1 + \psi)f_i]$, we take $S_i$ and $S_{i+1}$ as font similar segments. If $S_i$ and $S_{i+1}$ are aligned, then the similar standard can be relaxed to $f_{i+1} \in [(1 - \psi^a)f_i, (1 + \psi^a)f_i]$. We iterate $D^\tau$ sequentially and greedily replace $S_i$ with $S_i \cup S_{i+1}$ if $S_i$ and $S_{i+1}$ meet the merging standard. After merging, we get a new segment list $D^M$ which is the input to the labeling model. Note that the length of $D^M$ is less than that of raw segment list $D$ and $D^\tau$, and the merge may combine the segments with different labels to one segment. In the training phase, we use the most common label as the merged segment label.

### 4.2 Hierarchical Transformer

To obtain segment representation, we use two Transformer based encoders: segment encoder and document encoder. The model architecture is given in Fig. 3. Specifically, segment encoder is a pre-trained model while document encoder is a labeling oriented Transformer where the relative position of segments is important. By using two transformers, we split direct dependency of words across difference segments. We argue that for document image of products, there exists several local coherent regions and the direct dependency of words for different segments is relatively weak.

Actually, there are two advantages of using hierarchical transformers: 1) Compared to using one transformer on segment concatenation [14], the computation overhead for attention is reduced, as the computation complexity for attention is usually quadratic to input length. 2) Theoretically, segment encoder based on pre-trained model can encode complete semantic for segment, as the number of segment tokens is usually less than the maximum allowed length of pre-trained models such as BERT [20].

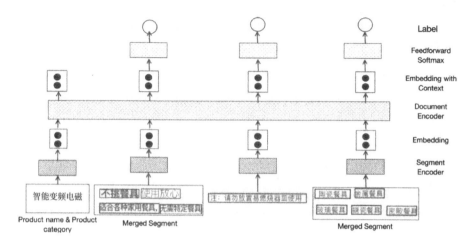

**Fig. 3.** Our proposed hierarchical transformer: segment encoder is to encode merged segment in $D^M$ and document encoder is to generate context aware segment representations. Product name and product category are combined and added to the head of $D^M$ as an additional context segment.

**Segment Encoder.** The segment encoder takes each $S_i \in D^M$ as input and generate a corresponding vector $\mathbf{e}(S_i)$. Specifically, the raw input $S_i = (w_1^i, w_2^i, .., w_{|S_i|}^i)$ is firstly converted to $S_i = ([\text{CLS}], w_1^i, w_2^i, .., w_{|S_i|}^i, [\text{SEP}])$, where [CLS] and [SEP] are two special tokens. The hidden state of the first token [CLS] from pre-trained model is taken as segment representation. Note that there are many pre-trained models available and they can replace each other [20–22] as segment encoder.

In our implementation, the token tensors from different segments are first transformed to equal size, i.e., the maximum sequence length $maxlen$. For those segments with length greater than $maxlen$, we only keep the first $maxlen - 2$ tokens for computing segment vectors.

**Document Encoder.** The document encoder is used to aggregate context for segment representation. Accounting that some text segments are ambiguous, especially short segments, it is necessary to take all the contexts into consideration. Here we incorporate product name and product category into modeling, which are two kinds of additional information available during dataset construction. Specifically, we concatenate product name and category as one segment $S_0$, which is added to the head of document $D^M$. $S_0$ is encoded with segment encoder as other segments. All the segment embeddings are further encoded with document encoder. We use a labeling oriented Transformer as our document encoder where two important components need to be defined: the input embedding and the attention.

*Input Embedding.* For document image, visual information such as position is an additional information of segments. It is natural to assume visual features can help predict segment labels. We combine both text and visual features to calculate input embedding. Specifically, for the $i$-th segment, the input embedding is defined as follows:

$$\mathbf{e}_i = \mathbf{e}(S_i) + \mathbf{e}(l_i) + \mathbf{e}(P_i) \tag{3}$$

Here, $\mathbf{e}(S_i)$ is the text embedding generated by segment encoder with dimension $d^{seg}$. $l_i$ and $P_i$ are the segment length and segment position. $\mathbf{e}(\cdot)$ represents embedding function for each feature.

Note that $l_i$ is a categorical feature having discrete values while $P_i = (x_1^i, y_1^i, x_2^i, y_2^i)$ is a group of numeric features with continuous values. For $l_i$, we use the following function called Position Embedding Conversion (PEC), to calculate embeddings.

$$\mathbf{e}(l_i) = PEC(l_i) \tag{4}$$
$$PEC(v) = \text{Dropout}(\text{ReLU}\left(\text{Linear}(PE(v))\right)) \tag{5}$$

where $PE(\cdot) \in \mathbb{R}^{d^{seg}}$ is the position embedding function in Transformer [17], each dimension of $PE(\cdot)$ corresponds to a sinusoid. The use of $PE(\cdot)$ is to set a good initial value suitable to Transformer initialization.

As for the numeric features $P_i = \{x_1^i, y_1^i, x_2^i, y_2^i\} \in (0,1)$, we first convert them to categorical features using bucketing: a hyper parameter $C^P$ is used to split $(0,1)$ to $C^P$ equal spaced intervals and the feature conversion function can be represented by Eq. (6)

$$B(v) = k \quad \text{if } v \in \left(k/C^P, (k+1)/C^P\right) \tag{6}$$

The final embedding function $\mathbf{e}(P_i)$ is given by Eq. (7)

$$\mathbf{e}(P_i) = \sum_{v \in \{x_1^i, y_1^i, x_2^i, y_2^i\}} PEC\left(B(v)\right) \tag{7}$$

*Attention.* The attention in Transformer is used to aggregate context. Previous study [18] shows that dropping scale factor of attention and using distance-and-direction aware attention are important for sequential labeling. Similarly, these two factors are also important for document image labeling. Specifically, a single attention head transforms the input vector $\mathbf{e}_i$ to three vectors: queries $\mathbf{q}_i \in \mathbb{R}^{d^{seg}}$, keys $\mathbf{k}_i \in \mathbb{R}^{d^{seg}}$ and values $\mathbf{v}_i \in \mathbb{R}^{d^{seg}}$. Then the attention after dropping scale factor can be represented as $\text{Attn}(Q, K, V) = \text{softmax}(A)V$, where $A_{ij} = \mathbf{q}_i^T \mathbf{k}_j$. To define distance-and-direction aware attention, we explore two attention formula which calculates attention from the perspective of 1 dimension (1D attention) and 2 dimension (2D attention) respectively.

For **1D Attention**, we use a similar equation as in [18] to calculate attention scores: the distance and direction are measured according to the positions in input list $D^M$ and the relative position embedding $\mathbf{R}_{i-j}$ is multiplied to $\mathbf{q}_i$, $\mathbf{k}_j$ :

$$A_{ij} = \mathbf{q}_i^T \mathbf{k}_j + \mathbf{q}_i^T \mathbf{R}_{i-j} + \mathbf{k}_j^T \mathbf{R}_{i-j} + \mathbf{u}^T \mathbf{k}_i + \mathbf{v}^T \mathbf{R}_{i-j} \tag{8}$$

where $\mathbf{u} \in \mathbb{R}^{d^{seg}}$, $\mathbf{v} \in \mathbb{R}^{d^{seg}}$ are learnable parameters, $\mathbf{R}_{i-j} \in \mathbb{R}^{d^{seg}}$ is the relative positional embedding. The equation for $\mathbf{R}_{i-j} \in \mathbb{R}^{d^{seg}}$ is given in Eq. (9), where $i$ is in the range $[0, d/2]$.

$$\mathbf{R}_{i-j} = [..., \sin(\frac{i-j}{10000^{2t/d}}), \cos(\frac{i-j}{10000^{2t/d}}), ...] \tag{9}$$

For **2D Attention**, we use a similar equation as in [13] to calculate attention scores: the relative position bias are measured in 2D plane of image, and the relative position embedding are added to $\mathbf{q}_i^T \mathbf{k}_j$ directly.

$$A_{ij} = \mathbf{q}_i^T \mathbf{k}_j + \mathbf{E}^1 \left( B(x_1^i) - B(x_1^j) \right) + \mathbf{E}^2 \left( B(x_2^i) - B(x_2^j) \right)$$
$$+ \mathbf{E}^3 \left( B(y_1^i) - B(y_1^j) \right) + \mathbf{E}^4 \left( B(y_2^i) - B(y_2^j) \right) \tag{10}$$

where $\mathbf{E}^1, \mathbf{E}^2, \mathbf{E}^3, \mathbf{E}^4 \in \mathbb{R}^{C^P \times d^{seg}}$ are four learnable embedding matrices.

## 5 Experiments

### 5.1 Implementation Details

To build the model, we split the data set to training set, development set and test set with 8:1:1 ratio. We use precision, recall and F1 as evaluation measures and the performance is reported based on the raw segments in $D^\tau$.

For tree-based segment re-organizing, there are three groups of hyper parameters: the parameters for judging vertical and horizontal alignment $\omega^v, \omega^h$, the parameters for dilation $\epsilon, \epsilon^{av}, \epsilon^{ah}$ and the parameters for measuring font similarity $\psi, \psi^a$ between segments. We set $\omega^v = \omega^h = 0.3$, $\epsilon = 0.2$, $\epsilon^{av} = 0.6$, $\epsilon^{ah} = 0.6$, $\psi = 0.06$ and $\psi^a = 0.3$ respectively. For hierarchical transformer, we use albert_tiny[3] as our segment encoder. albert_tiny is pre-trained on Chinese corpus, it has 3 layers and the dimension of output embedding $d^{seg} = 312$. We set the maximum sequence length for segment encoder to 32. The number of layers in document encoder Transformer is set to 1. To convert the numeric feature $P_i$, the number of buckets $C^P$ is set to 10.

---

[3] https://huggingface.co/voidful/albert_chinese_tiny.

To train the hierarchical transformer, we take a two-phase training strategy: the first phase is to freeze the parameters of segment encoder and learn the parameters of document encoder with a large learning rate. The second phase is to fine-tune the learned model from the first phase and update the parameters for both segment encoder and document encoder. For each phase, we train for 20 epochs and keep the model with the best performance on the development set. For the first phase, we set the learning rate and batch size to 1e−2 and 128 respectively. For the second phase, we set the learning rate and batch size to 1e−5 and 96 respectively.

## 5.2   Experimental Results

**Effects of Using Tree-Based Segment Re-organizing.** Recall that segment re-organizing will merge raw segments of document image, and the merged segment is the basic prediction unit of the second stage model, so each raw segment in a merged segment will have the same label. The segment re-organizing is fully unsupervised, and it is possible that raw segments with different labels to be merged. We study the impact of this merging to the final labeling performance. The maximum accuracy after segment re-organizing is given in Table 3. From the table, we can see that there is little impact, about 1%, on the upper bound performance 1. Besides, we do the statistics on the average number of segments in document image before and after segment re-organizing. Note that the number of segments can be reduced by almost 40%, which greatly reduces the size of model input. The above two aspects show the effectiveness of our proposed segment re-organizing method.

**Table 3.** Effects of using tree-based segment re-organizing

|  | Cookware dataset | | | Kitchenware dataset | | |
|---|---|---|---|---|---|---|
|  | train | dev | test | train | dev | test |
| Max accuracy | 0.9899 | 0.9920 | 0.9904 | 0.9875 | 0.9865 | 0.9881 |
| Avg. no. of raw segments | 7.572 | 7.252 | 7.535 | 7.125 | 7.278 | 7.252 |
| Avg. no. of merged segments | 4.535 | 4.349 | 4.394 | 4.407 | 4.493 | 4.450 |

**Effects of Using 1D and 2D Attention.** We examine the performances of two distance-and-direction aware attention formula for document encoder. The results are given in Table 4. From the table, we can see that the performance using 1D attention is superior than the performance using 2D attention. There are two possible reasons to explain this result: 1) The 1D attention measures distance and direction by segment offsets in 1D list, which follows the top to bottom and left to right order and is usually consistent with reading order. Considering that segment positions in document image are quite diverse, it is difficult to model segment dependency using original 2D coordinate. 2) The 1D attention calculates the interaction between relative position bias and content (queries and keys), but 2D attention only adds a relative position based attention score to content based attention score, for which the interaction between position and content is not accounted. The above analysis show that for coarse-grained segment labeling, modeling context using reading order in 1D is probable a better choice than modeling dependency in 2D, although 2D attention using 2D coordinate is shown effective on word level [13]. In the follow-up sections, We only report the performance measures using 1D attention.

**Table 4.** Performance comparison using 1D attention and 2D attention

|              | Cookware dataset | | Kitchenware dataset | |
|--------------|--------|--------|--------|--------|
|              | dev    | test   | dev    | test   |
| 1D attention | **0.7686** | **0.7834** | **0.7805** | **0.7856** |
| 2D attention | 0.6949 | 0.7221 | 0.7671 | 0.7789 |

**Effects of Using Different Features.** We present the effects of using different features in Table 5. From the table, we can see that segment length, i.e., the number of tokens, can improve the labeling performance. We surprisingly found that adding position feature does not bring further improvement as shown in previous studies [12,13]. This may be caused by that absolute segment position does not correlate with segment types, for which the segments of same type can scatter everywhere in a document image. This is also consistent with our actual observation.

**Table 5.** Performance comparison using different features

|                            | Cookware dataset | | Kitchenware dataset | |
|----------------------------|--------|--------|--------|--------|
|                            | dev    | test   | dev    | test   |
| Text                       | 0.7681 | 0.7815 | 0.7737 | 0.7855 |
| Text + length              | **0.7686** | **0.7834** | **0.7805** | **0.7856** |
| Text + length + position   | 0.7676 | 0.7812 | 0.7769 | 0.7855 |

**Effects of Using Product Name and Category.** In this part, we validate the effects of incorporating additional product name and category information for segment context modeling. As shown in Table 6, the labeling results using product name and category for Kitchenware dataset is better, but the advantage of using product name and category is not obvious for Cookware dataset. We think this is because the contents of Kitchenware are more diverse than those of Cookware, and product name and category name can provide more information gain for broader categories.

**Table 6.** Performance comparison of using context with or without product name & category

|  | Cookware dataset | | Kitchenware dataset | |
|---|---|---|---|---|
|  | dev | test | dev | test |
| Context with product name & category | **0.7686** | 0.7834 | **0.7805** | **0.7856** |
| Context without product name & category | 0.7669 | **0.7858** | 0.7731 | 0.7817 |

**Performance Comparison for Different Labels.** The labeling performance for each semantic label are given in Table 7 and Table 8.

**Table 7.** The performance of each label on Cookware dataset

|  | Label | Precision | Recall | F1 | support |
|---|---|---|---|---|---|
| Product itself | component | 0.7647 | 0.1646 | 0.2708 | 79 |
|  | function | 0.8354 | 0.8968 | 0.8650 | 6,540 |
|  | advertisement | 0.7345 | 0.6897 | 0.7114 | 2,314 |
|  | brand name | 0.1071 | 0.1154 | 0.1111 | 26 |
|  | brand detail | 0.5909 | 0.4483 | 0.5098 | 29 |
|  | product name | 0.2921 | 0.2031 | 0.2396 | 128 |
|  | product model | 0.7788 | 0.7232 | 0.7500 | 448 |
|  | usage | 0.8281 | 0.7873 | 0.8072 | 1,199 |
|  | sales note | 0.7797 | 0.8119 | 0.7955 | 1,600 |
|  | product QA | 0.5605 | 0.7316 | 0.6347 | 190 |
|  | other | 0.4906 | 0.4370 | 0.4623 | 540 |
| Product application | demand | 0.0000 | 0.0000 | 0.0000 | 6 |
|  | demand problem | 0.4773 | 0.5122 | 0.4941 | 41 |
|  | competitive product label | 0.4000 | 0.3077 | 0.3478 | 13 |
|  | competitive product problem | 0.6939 | 0.3579 | 0.4722 | 95 |
|  | co-usage product | 0.7992 | 0.8263 | 0.8125 | 236 |
|  | substitution | – | – | – | – |
|  | applicable scene | 0.5214 | 0.4966 | 0.5087 | 147 |
|  | other | 0.1154 | 0.0370 | 0.0561 | 81 |
| Document organization | navigation | 0.8577 | 0.7108 | 0.7774 | 619 |

**Table 8.** The performance of each label on Kitchenware dataset

|  | Label | Precision | Recall | F1 | Support |
|---|---|---|---|---|---|
| Product itself | component | 0.7067 | 0.3487 | 0.4670 | 456 |
|  | function | 0.8239 | 0.8963 | 0.8586 | 13,642 |
|  | advertisement | 0.7003 | 0.7233 | 0.7116 | 5,215 |
|  | brand name | 0.0000 | 0.0000 | 0.0000 | 25 |
|  | brand detail | 0.0000 | 0.0000 | 0.0000 | 97 |
|  | product name | 0.5055 | 0.3660 | 0.4246 | 377 |
|  | product model | 0.8207 | 0.8388 | 0.8296 | 2,717 |
|  | usage | 0.7171 | 0.7836 | 0.7489 | 1,511 |
|  | sales note | 0.8160 | 0.8023 | 0.8091 | 5,185 |
|  | product QA | 0.7148 | 0.7765 | 0.7444 | 255 |
|  | Other | 0.6711 | 0.4470 | 0.5366 | 1,926 |
| Product application | demand | 0.6667 | 0.0769 | 0.1379 | 26 |
|  | demand problem | 0.6871 | 0.6914 | 0.6892 | 162 |
|  | competitive product label | 0.4000 | 0.1333 | 0.2000 | 15 |
|  | competitive product problem | 0.2373 | 0.1333 | 0.1707 | 105 |
|  | co-usage product | 0.4048 | 0.1619 | 0.2313 | 164 |
|  | substitution | – | – | – | – |
|  | applicable scenes | 0.6805 | 0.4530 | 0.5439 | 362 |
|  | other | 0.6471 | 0.4425 | 0.5256 | 174 |
| Document organization | navigation | 0.8548 | 0.8064 | 0.8299 | 1,467 |

From the table, we can see that the distribution of labels is quite uneven. Function, advertisement, sales note and product model are the top 4 labels with the largest quantities and their labeling performances are around 0.8. Product QA and usage label also have a passable performance around 0.6–0.7. Note that the segments of these labels are usually long texts and their statistical features are well reflected in the dataset. The labeling performance for brand name, competitive product label, competitive product problem and applicable scene are not very good. These segments are usually short texts and judging their labels need additional information such as common sense knowledge.

# 6   Conclusion

In this paper, we study the problem of labeling document images in E-commence. We build a dataset through sampling products from a real-world E-commence website and denote the corresponding semantic labels. Accounting that the characteristics of document images, we propose a two stage labeling approach: the first stage is to resume coherent segments by visual information, and the second stage is to make prediction using hierarchical transformer. The hierarchical transformer utilizes a pre-trained model as segment encoder and a labeling oriented Transformer as document encoder. The document encoder is to aggregate segment context, where additional product name and category are included. We compare two attention formula for document encoder, where distance and direction are measured in 1D and 2D respectively. Experiments show that the performance of using 1D attention is superior than that of using 2D attention, and product name and product category are helpful for the datasets with diverse products. Furthermore, we find that for segment level modeling, using the absolute 2D position as input embedding may not be that effective, which is in contrary to the previous methods using 2D position for token level modeling.

# References

1. Esser, D., Schuster, D., Muthmann, K., Berger, M., Schill, A.: Automatic indexing of scanned documents: a layout-based approach. Doc. Recognit. Retrieval XIX **8297**, 118–125 (2012)
2. Hwang, W., et al.: Post-OCR parsing: building simple and robust parser via BIO tagging. In: Workshop on Document Intelligence at NeurIPS 2019 (2019)
3. Guo, H., Qin, X., Liu, J., Han, J., Liu, J., Ding, E: EATEN: entity-aware attention for single shot visual text extraction. In: ICDAR, pp. 254–259 (2019)
4. Qian, Y., Santus, E., Jin, Z., Guo, J., Barzilay, R.: GraphIE: a graph-based framework for information extraction. In: NAACL-HLT, pp. 751–761 (2019)
5. Yu, W., Lu, N., Qi, X., Gong, P., Xiao, R.: PICK: processing key information extraction from documents using improved graph learning-convolutional networks. In: ICPR 2020 (2020D)
6. Liu, X., Gao, F., Zhang, Q., Zhao, H.: Graph convolution for multimodal information extraction from visually rich documents. In: NAACL-HLT, pp. 32–39 (2019)
7. Ma, X., Hovy, E.: End-to-end sequence labeling via bi-directional LSTM-CNNs-CRF. In: ACL, pp. 1064–1074 (2016)
8. Zhang, P., et al.: TRIE: end-to-end text reading and information extraction for document understanding. In: ACMMM, pp. 1413–1422 (2020)
9. Li, L., Gao, F., Bu, J., Wang, Y., Yu, Z., Zheng, Q.: An end-to-end OCR text re-organization sequence learning for rich-text detail image comprehension. In: Vedaldi, A., Bischof, H., Brox, T., Frahm, J.-M. (eds.) ECCV 2020. LNCS, vol. 12370, pp. 85–100. Springer, Cham (2020). https://doi.org/10.1007/978-3-030-58595-2_6
10. Zhang, X., Wei, F., Zhou, M.: HIBERT: document level pre-training of hierarchical bidirectional transformers for document summarization. In: ACL, pp. 5059–5069 (2019)
11. Wang, J., et al.: Towards robust visual information extraction in real world: new dataset and novel solution. In: CoRR 2021 (2021)
12. Xu, Y., Li, M., Cui, L., Huang, S., Wei, F., Zhou, M.: LayoutLM: pre-training of text and layout for document image understanding. In: ACM-SIGKDD, pp. 1192–1200 (2020)
13. Garncarek, U., Powalski, R., Stanisawek, T., Topolski, B., Graliński, F.: LAMBERT: layout-aware language modeling using BERT for information extraction. In: CoRR 2020 (2020)
14. Cohan, A., Beltagy, I., King, D., Dalvi, B., Weld, D.S.: Pretrained language models for sequential sentence classification. In: CoRR 2019 (2019)
15. Katti, A.R., et al.: Chargrid: towards understanding 2D documents. In: EMNLP, pp. 4459–4469 (2018)

16. Iz, B., Matthew, E.P., Arman, C.: Longformer: the long-document transformer. In: CoRR 2020 (2020)
17. Ashish, V., et al.: Attention is all you need. In: NIPS 2017 (2017)
18. Yan, H., Deng, B., Li, X., Qiu, X: TENER: adapting transformer encoder for named entity recognition. In: CoRR 2019 (2019)
19. Huang, Z., et al.: ICDAR 2019 competition on scanned receipt OCR and information extraction. In: ICDAR, pp. 1516–1520 (2019)
20. Devlin, J., Chang, M., Lee, K., Toutanova, K.: BERT: pre-training of deep bidirectional transformers for language understanding. In: ACL, pp. 4171–4186 (2019)
21. Lan, Z., Chen, M., Goodman, S., Gimpel, K., Sharma, P., Soricut, R.: ALBERT: a lite BERT for self-supervised learning of language representations. In: ICLR 2020 (2020)
22. Sun, Y., Wang, S., Li, Y., Feng, S., Wu, H.: ERNIE: enhanced representation through knowledge integration. In: CoRR 2019 (2019)

# Multi-task Learning for Newspaper Image Segmentation and Baseline Detection Using Attention-Based U-Net Architecture

Anukriti Bansal[1], Prerana Mukherjee[2]($\boxtimes$), Divyansh Joshi[2],
Devashish Tripathi[2], and Arun Pratap Singh[2]

[1] The LNM Institute of Information Technology, Jaipur, Rajasthan, India
anukriti.bansal@lnmiit.ac.in
[2] School of Engineering, Jawaharlal Nehru University, Delhi, India
{prerana,divyan95_soe,devash76_soe,arun70_soe}@jnu.ac.in

**Abstract.** In this work, we propose an end-to-end language agnostic multi-task learning based U-Net framework for performing text block segmentation and baseline detection in document images. We leverage the performance of U-Net by augmenting attention layers between the contracting and expansive path via skip connections. The generalization ability of the model is validated on handwritten images as well. We perform exhaustive experiments on ICPR2020 challenge dataset and obtain a test accuracy of 96.09% and 99.44% for simple track baseline detection and text block segmentation respectively, 97.47% and 98.51% complex track baseline and text block segmentation respectively. The source code is made publicly available at https://github.com/divyanshjoshi/Attention-U-Net-Newspaper-Text-Block-Segmentation.

**Keywords:** Multi-task learning · Newspaper document images · Attention · Text block segmentation · Text baseline detection

## 1 Introduction

With the increase in information dissemination, the usage of e-newspapers, digital content has become more prolific. Particularly, this further surges the demand of more research in the domain of document image processing. Thus, it becomes also imperative to maintain and archive such different kinds of digital documents with minimal cost and in most efficient manner. Out of these, the most complex layout is present in newspaper images as it has an overload of both text and graphics. There are two main challenges while dealing with newspaper document images, i) the structure of old newspapers are more complex than the recent e-newspapers as pages are subjected to more degeneration due to issues

---

A. Bansal and P. Mukherjee—The authors have contributed equally.

E. H. Barney Smith and U. Pal (Eds.): ICDAR 2021 Workshops, LNCS 12917, pp. 440–454, 2021.
https://doi.org/10.1007/978-3-030-86159-9_32

such as poor scan quality, wear and tear of pages etc., and ii) huge variations in the layout structure of different publishing houses and variants within the same edition in different issues.

In order to exploit the information present in such documents, it becomes necessary to segment these articles in a way that they become more decipherable. There could be many interesting applications such as indexing and retrieval tasks [26,27] once this complex layout is extracted. In this work, we address to solve the two relevant problems in the page segmentation domain: i) text block segmentation-it enables to separate the text and graphical components and treat them independently, and ii) text based baseline detection-it helps in identifying the lines belonging to each block components. Motivated by the above mentioned challenges and advantages, we develop a novel method of logical labelling of such documents particularly newspaper images in our context. We leverage the use of multi-task learning paradigm to jointly learn the shareable parameters for performing two complementary tasks. Inspired by the much celebrated U-Net [24,25] architecture prominently utilized in medical imagery, we augment an attention block in the pipeline of the Modified U-Net architecture catering multi-task learning. The core idea is to provide successive layers as opposed to a contracting network, here pooling operations are replaced by up-sampling operations, thus increasing the newspaper page layout resolution. High resolution features from contracting path when combined with the up-sampled out, followed by a convolution layer can give more precise segmentation result. The large number of feature channels allow the propagation of contextual information across successive layers. The expansive path is symmetric to the contractive path connected by skip connections and the attention block. The attention block is a self attention grid based gating module which help to concentrate the attention coefficients on the localized regions. Finally, the output is bifurcated into two outputs: text block output and baseline output.

## 1.1  Related Work

Page segmentation and baselines detection has been an important and active area of research since long back [23]. Here we review some of the recent work using both hand-crafted features and advanced machine learning and deep learning algorithms.

## 1.2  Hand Crafted Features Methods

Document layout analysis is very important research direction and has several applications in optical character recognition for geometric and logical analysis. Geometric layout analysis [7] or typically named as page segmentation algorithms require splitting the page into homogeneous regions consisting of text or graphics. Most of the techniques attempt to solve it using ruling lines (horizontal or vertical) detection. Logical layout analysis [8,13] requires the segregation into logical units such as headlines or paragraphs and then form a consistent relationships amongst them. Article segmentation tasks are highly dependent on the task's complexity [5]. In [22], authors solve the problem of article detection in digitised newspaper images. As most of the prior works assume that the text segmentation is ordered whereas in this work, the authors proposed 2D Markovian process to encode the appropriate reading order inside the geometric text blocks. Bansal et al. [4] utilized fixed point models to solve the task of article segmentation. Other works include text line segmentation in unconstrained printed text documents [14], straight line based segmentation [2] which include handcrafted features for performing the requisite tasks.

### 1.3   Deep Learning Feature Based Methods

In [12], the authors utilized cascaded instance aware segmentation technique based on multi-scale fully convolutional neural network (FCN). It consists of two major components: i) text block region segmentation framework, ii) rotation invariant instance aware segmentation which further disintegrates the text block regions into requisite text or word lines. In [11], authors provide an end-to-end framework for page segmentation for performing 3 types of instance segmentation which includes: text block, tables and figures. They propose a multi-scale multi-task FCN learning framework which enables page segmentation and element wise contour detection. On one hand the semantic segmentation task helps in pixel wise prediction the various elements whereas on the other hand the contour detection pipeline identifies the nearby edges around each element. A conditional random field network is trained on the output from the semantic segmentation and contour detection branches which further improvises the segmentation output. They also utilize some heuristic rules based post-processing to identify the individual table elements. Lee et al. [17], authors proposed trainable multiplication layers (TMLs) in the standard U-Net convolutional neural network. These TMLs extract the co-occurrence features across the layers to detect the presence of any periodic textual element re-occurring (such as tables, text line structures etc.) or textural similarities in various elements in the text. In [19], authors proposed a machine learning approach for page segmentation, the first step includes classification score generation for various page components and second step involves the connected component analysis to group the semantically and spatially close components in the page layout.

### 1.4   Major Contributions

In view of the above discussions, our contributions can be summarized in the following aspects:

1. We propose an end-to-end novel learning based deep neural network having two complementary branches, which is utilized to solve the problem of text block segmentation and baseline detection in document images. To the best of the authors' knowledge, this is the first work that segment text blocks as well as detect baselines in arbitrary type of documents in a unified multi-task learning based framework.

2. We augment an attention block in the U-Net pipeline which consists of convolutional neural network with shared feature learning for text block segmentation and baseline detection to focus on the correct regions of interest.
3. We evaluate the proposed approach for both tasks to demonstrate superior performance on Text Block Segmentation Competition ICPR2020-NewsEye dataset as compared to the existing frameworks. The generalization ability of the model to detect baselines on various types of document images is also shown on the real handwritten documents from the ICDAR 2017 handwritten baseline detection dataset.

The organization of the paper is as follows. In Sect. 2, we explain the proposed multi-task learning framework with attention gates for the text-block segmentation and base line detection. Details of the experimental evaluations on various standard datasets are discussed in Sect. 3. Finally we conclude the paper in Sect. 4 and provide avenues for future research.

## 2    Proposed Multi-task Learning-Based Framework

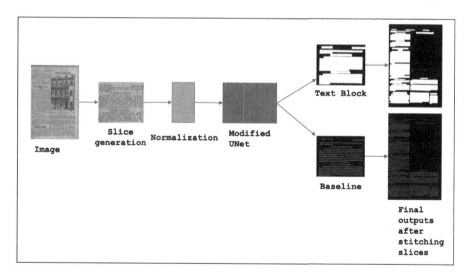

**Fig. 1.** Overview of the proposed method: Input document images are first augmented using a unique image slicing-based method. Intensity values of each slice is normalized (max-normalization) in the range of 0 to 1 before feeding into an attention-based U-Net architecture. The model output two images, one containing text-blocks and the other containing baselines.

## 2.1  Overview

In this work, we propose a multi-task learning of deep neural network for developing a single model for the segmentation of text blocks and the detection of baselines from the document images of different layouts and type. We train convolutional neural networks (CNN hereafter) that follow a U-Net architecture, with attention gates to learn to focus on the target regions more efficiently. The overall pipeline of the proposed method is shown in Fig. 1. Initially, we generate non-overlapping slices of the input image, which are then normalized in the intensity range of 0 to 1 using lambda normalization (max-normalization) layer. They are then fed into the modified U-Net architecture, which outputs two type of images: i) The first image is the segmented image with text-blocks. It contains labels of all the pixels that belongs to text-blocks, and ii) the second detects baselines by labeling all the pixels that belongs to document text lines. All the output image slices are then resized to the original input image size by the combining different slices together.

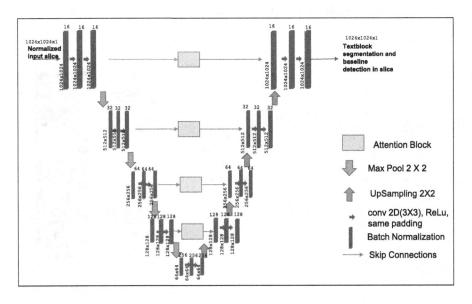

**Fig. 2.** Modified U-Net module.

CNNs have performed well in document layout classification and page segmentation [16,20]. The U-Net architecture is a type of Fully Convolutional Network (FCN) that is composed of two paths. The first path is the contraction path (also known as down-sampling or encoder), which is a stack of the convolutional and max-pooling layers used for capturing the context in the image. The second path is the expanding path (also known as up-sampling or decoder), which is used for precise localization using transposed convolutions. There are several advantages of using CNNs in U-Net-based architecture: (1) one of the major advantages is that U-Net does not require a large volume of images during the training phase as compared to other data hungry deep-learning and CNN-based algorithms, (2) the images of varying sizes can be given as input to the U-Net architecture, and (3) they are computationally very efficient.

We further improves the accuracy and the computational efficiency of the proposed model for text-block segmentation and baseline detection by incorporating attention gates. Attention gates can automatically learn to focus on the important regions of interest in the image without any additional regulation while training the model. These gates generate on-the-fly suggestions for the important regions by highlighting some salient features with respect to a particular task. It improves the model accuracy by suppressing the feature activation in irrelevant regions. In addition to this, the attention gates do not require large number of model parameters and therefore do not introduce a significant computational overhead.

For the development and evaluation of our model, we used images from the page segmentation and baseline detection-based competitions' datasets of ICPR2020[1] [21] and ICDAR2017 [9]. To evaluate the performance of the proposed model, we use region overlapping-based statistical metric, namely dice score. Our model gives promising results when tested on the simple and complex tracks of the ICPR2020 NewsEye challenge dataset. The generalization ability of our model is established by showing the encouraging results for an entirely different set of images from ICDAR2017 baseline detection dataset on handwritten document images.

The following paragraphs explain the proposed framework in detail.

## 2.2   Modified U-Net Architecture

**U-Net Architecture.** Architecture of the modified U-Net architecture used for the purpose of segmentation of text blocks and baseline detection is shown in Fig. 2. The network consists of a contracting path, that comprises of the repeated application of two $3 \times 3$ convolutions, each followed by a rectified linear unit (ReLU) and a $2 \times 2$ max pooling operations with stride 2 for downsampling. Number of feature channels are doubled at each downsampling step. Each step in the expansive path, on the other hand, consists of an upsampling of the feature map followed by a $2 \times 2$ convolution (that halves the number of feature channels), thereafter a concatenation with the corresponding cropped feature map from the contracting path and two $3 \times 3$ convolutions via skip connects, each followed by a ReLU. At the final layer a $1 \times 1$ convolution is used to map each 16-component feature vector to two classes: one for text-block/baselines and the other one for non-text. Intermediate batch-normalization layers are added to avoid over-fitting and to develop a stable model.

---

[1] https://www.mathematik.uni-rostock.de/forschung/projekte/citlab/projects/text-block-segmentation-competition-icpr2020/.

**Attention Gates.** This paper uses grid-based gating, which allows attention gates to focus on specific regions of an image and thus helps in better localization and subsequent segmentation. We use additive attention gates (Bahdanau et al.) [3], which are computationally little expensive but have observed to have better accuracy over multiplicative attention gates [18]. These attention gates are integrated in the U-Net architecture to highlight the important features that are transferred through the skip connections (refer Fig. 2). Activation gates make use of the contextual information to focus on the region of interest. The gates separate the noise and other irrelevant details from the coarse level features before passing them through the skip connections. This is done just before the concatenation operation such that only appropriate activations gets merged. Additionally, activation gates filter the neuron activations both during the forward and the backward pass. Gradients originating from non-text regions are down weighted during the backward pass. This enables model parameters in shallower layers to be updated mostly based on the textual regions [3].

## 3 Experimental Results and Discussion

The experimental program is coded using Keras [6] API of TensorFlow framework [1,10]. We use Adam optimizer [15] for training with Dice loss as the loss function (Eq. 1). Model is trained for 100 epochs with batch size equals to 2. Initial learning rate is 1e−4 with decay = learning rate/100. The computer processor is Intel Core i5-7200U with 8 GB RAM. The following paragraphs describe important aspects related to the experimental evaluation and the obtained results.

### 3.1 Dataset Description and Pre-processing

We used historical newspaper images from the page segmentation and baseline detection based competition from ICPR2020. It primarily consists of two tracks: simple images (text only) and complex images (text+graphics with varied layouts). The ground truth of these scanned pages consists of baseline coordinates and text regions. It consists of 80 images (40 from simple track and 40 from complex track), out of which 60 images, 30 images from each track, were used for training and 20 images, 10 images from each track, are used for validation to form separate models for simple and complex tracks each. The test set consists of test images from ICPR2020, and randomly picked handwritten images from ICDAR2017 and newspaper images from internet. Since data for training the model is very limited, we augment the data by performing image slicing. We divided the images into different non-overlapping slices, which contain both text and non-text part. In order to determine the size of these slices, we experimented with multiple dimensions and empirically chose two slice dimensions: $1024 \times 1024$ and $512 \times 512$. These dimensions captures the page-layout information efficiently and were giving good results. After image slicing, training set contains 441 sliced patch images of dimension $1024 \times 1024$ and 1914 sliced patch images of size $512 \times 512$.

We have not performed any other pre-processing operation such as binarization or removal of ruling lines, which may cause some segmentation errors. The proposed model learns to identify pixels that belong to text-blocks and baselines and helps in segmenting-out the non-text regions like graphics, images and ruling lines.

## 3.2  Evaluation Metrics

To evaluate the overall accuracy text-block and baseline, we utilize Dice coefficient and F-value as the basic evaluation metrics.

**Dice Coefficient.** Dice loss helps in determining the overlap between predicted and the actual mask and is calculated as follows,

$$D = \frac{2 \sum_i^N pred_i gt_i}{\sum_i^N pred_i^2 + \sum_i^N gt_i^2} \tag{1}$$

Here, $N$ is the total number of pixels, $pred_i$ and $gt_i$ are the pairs of corresponding pixel values in the predicted image and the ground truth image respectively. We also compute the accuracy of the proposed model by comparing the overall percentage of the correctly classified text and non-text pixels.

**F-Value.** This measure helps in finding the correctness of the found text blocks and associates with how well the baselines are detected. It requires the computation of precision and recall matrices. The precision $(P)$ suggests how well a set of ground truth baselines are covered by a set of hypotheses baselines. It doesn't account for the segmentation errors. The recall value $(R)$ suggests how well a set of hypotheses baselines is covered by a set of ground truth baselines, while penalizing segmentation errors. F-value $(F)$ is computed as follows,

$$F = \frac{2.R.P}{R + P} \tag{2}$$

where $P, R, F \epsilon [0, 1]$.

## 3.3  Results of Text-Block Segmentation

We provide explainability of the proposed model's performance by visualizing the class activation maps at different hierarchical levels of the convolutional block. In Fig. 3, we have demonstrated the heatmaps generated on addition of the attention block in the U-Net architecture across the various layers. We observe that in higher levels in the upsampling, the text blocks and baselines are more concentrated in the heatmaps demonstrating the localization efficiency and accuracy of the model. In Fig. 4, we show the text block segmentation results obtained using slices of size $512 \times 512$ and $1024 \times 1024$. It can be observed that the combination result in case of slices of size $512 \times 512$ contains holes and leads to under-segmentation error. When we tried to increase the input dimensions of the proposed model from $512 \times 512$ to $1024 \times 1024$, we observe that the loss got reduced from 0.09 to 0.0678 and perform better labeling. This can be attributed to the fact that there is heavy compression of information of images of size (4000, 2500) to (512, 512), moreover they do not capture the vertical ruling lines as can be seen in Fig. 4(d).

### 3.4 Results of Baseline Detection

In Fig. 5, we provide the qualitative analysis of multi-task learning results using proposed methodology on images from simple and complex tracks of ICPR 2020 challenge dataset. It can be seen the proposed methodology is also able to deal with newspaper images, which contains both text and graphic regions. In Fig. 6, we provide the slice based processing ($1024 \times 1024$) results. We also show the effective generalization ability of the method to perform text block as well as baseline detection results on the images from the different dataset as can be seen in the case of Fig. 6d)–i). It is important to note that the Fig. 6f)–i) results are on handwritten ICDAR 2017 documents whereas the model is trained on only newspaper images from the ICPR2020 images, thus it validates the good generalizability of the model. Table 1 shows the performance analysis of individual tasks with the proposed multi-task learning framework. Table 3 shows the comparative analysis of the proposed method with the top 3 challenge results on baseline detection using the F-measure as used in the ICPR 2020 NewsEye dataset. The compared methods utilize rule based method followed by split and merging baselines, textual post processing, deep learning based instance segmentation approach (MaskRCNN) or graph attention networks. We did not resort to any kind of pre or post-processing and have adapted a single network to perform both the tasks simultaneously with shared weights.

**Experimental Analysis:** Results of the deep learning-based algorithms are very sensitive to the choice of hyperparameters, which may vary every time they are executed. Therefore, we have provided results of U-Net and modified U-Net (with attention) with different regularization parameters in Table 2. The results suggest that the modified U-Net gives best results in both simple and complex track of ICPR2020 dataset.

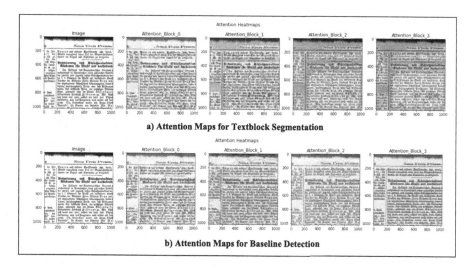

Fig. 3. Attention mechanism in the proposed framework at different hierarchical levels (a) Attention heatmaps for text block segmentation (b) Attention heatmaps for baseline detection.

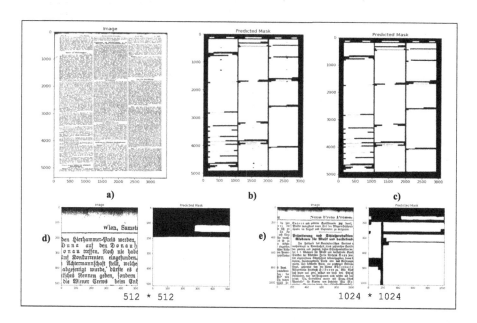

Fig. 4. Text block prediction results with the proposed method a) Input image b) Prediction results using slices (512 * 512) c) Prediction results using slices (1024 * 1024) d) Text block segmentation in Slices (512 * 512) e) Text block segmentation in Slices (1024*1024).

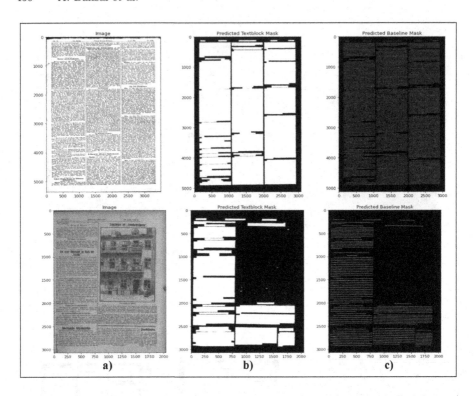

**Fig. 5.** Visual results of multi-task learning results using proposed methodology. a) Input image b) text block segmentation and c) Baseline detection on ICPR 2020-NewsEye Dataset (Simple Track-top row) and Complex Track-bottom row).

**Table 1.** Performance analysis of the proposed model on ICPR2020 competition-NewsEye dataset

| Task | Training accuracy | Training dice loss | Validation accuracy | Validation dice loss | Test accuracy | Test dice loss |
|---|---|---|---|---|---|---|
| Simple track baseline detection | 95.94 | 0.2880 | 94.57 | 0.3737 | 96.09 | 0.2681 |
| Simple track text block segmentation | 99.32 | 0.0050 | 99.10 | 0.0061 | 99.44 | 0.0041 |
| Complex track baseline detection | 96 | 0.3329 | 94.77 | 0.4372 | 97.47 | 0.4088 |
| Complex track text block segmentation | 98.57 | 0.0115 | 93.43 | 0.0508 | 98.51 | 0.0172 |

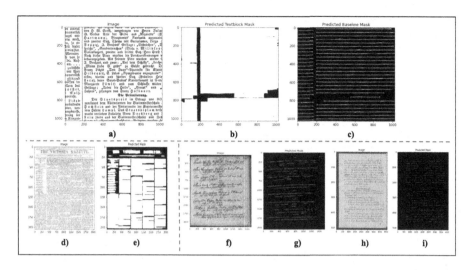

**Fig. 6.** Slice (1024 * 1024) based processing with the proposed method a) Input image b) text block segmentation and c) Baseline detection. d)–g) shows results on random Internet images. d)–e): Input image and Test block segmentation; f)–g): Input image and Baseline Detection results (ICDAR 2017 Simple Image); h)–i): Input image and Baseline Detection results (ICDAR 2017 Complex Image).

**Table 2.** Ablation studies on variants on the U-Net architecture

| Trained on | Model | Training accuracy | Training dice loss | Validation accuracy | Validation dice loss | Test accuracy | Test dice loss |
|---|---|---|---|---|---|---|---|
| Simple Track ICPR 2020 | U-Net+Dropout (downsampled to 1024 * 1024) | 97.68 | 0.0161 | 98.54 | 0.0092 | 98.51 | 0.0094 |
| | U-Net+BN (downsampled to 1024 * 1024) | 99.32 | 0.0050 | 99.10 | 0.0061 | 99.44 | 0.0041 |
| | U-Net+BN (downsampled to 2048 * 2048) | 97.78 | 0.0222 | 97.72 | 0.0228 | 99.09 | 0.0510 |
| | U-Net+BN+dropout (downsampled to 2048 * 2048) | 99.16 | 0.0225 | 99.05 | 0.0199 | 99.05 | 0.0221 |
| | U-Net+BN+dropout+ Slicing (1024 * 1024) | 98.94 | 0.0101 | 99.10 | 0.0082 | 99.24 | 0.0074 |
| | U-Net+BN+dropout+ Slicing (512 * 512) | 98.27 | 0.0235 | 98.96 | 0.0136 | 99.10 | 0.0083 |
| | U-Net+BN+Attention (downsampled to 1024 * 1024) | 99.48 | 0.0246 | 99.09 | 0.0247 | 99.18 | 0.0244 |
| | U-Net+BN+Attention+ Slicing (1024 * 1024) | 99.54 | 0.0241 | 99.12 | 0.0241 | 99.24 | 0.0240 |
| Complex Track ICPR 2020 | U-Net+BN +Slicing (512 * 512) | 97.98 | 0.0158 | 94.47 | 0.0469 | 94.88 | 0.0345 |
| | U-Net+BN+Slicing (1024 * 1024) | 97.23 | 0.0307 | 94.89 | 0.0479 | 94.11 | 0.0396 |
| | U-Net+BN+Attention+ Slicing (1024 * 1024) | 98.51 | 0.0259 | 96.14 | 0.0391 | 95.11 | 0.0296 |

**Table 3.** Comparative analysis on ICPR 2020 NewsEye dataset challenge baseline results using F-measure metric.

| Methods | F-value simple track | F-value somplex track |
|---|---|---|
| CIT lab | 0.934 | 0.768 |
| Cinnamon AI and HCMC | 0.999 | - |
| HEC | 0.995 | 0.887 |
| Lenovo research and SCUT | 0.997 | 0.954 |
| Proposed method | 0.995 | 0.671 |

## 4   Conclusion

This paper presented an effective multi-task learning based method for text block segmentation and baseline detection. The proposed method is flexible with respect to several aspects. Firstly, it is free of any pre-processing blocks such as ruling line detection, superpixel, connected component detection etc. and doesn't require any tuning parameter to perform the segmentation tasks. Secondly, we utilized attention based U-Net architecture which is adaptive to different types of documents and page components. In future, we will attempt to include the geometric constraints such as contour edges belonging to the distinct elements to localize the baselines with its representative text block. An amalgamation of local and global text block features can be explored as a future research direction.

## References

1. Abadi, M., et al.: TensorFlow: a system for large-scale machine learning. In: 12th {USENIX} Symposium on Operating Systems Design and Implementation ({OSDI} 2016), pp. 265–283 (2016)
2. Alhéritière, H., Cloppet, F., Kurtz, C., Ogier, J.M., Vincent, N.: A document straight line based segmentation for complex layout extraction. In: 2017 14th IAPR International Conference on Document Analysis and Recognition (ICDAR), vol. 1, pp. 1126–1131. IEEE (2017)
3. Bahdanau, D., Cho, K., Bengio, Y.: Neural machine translation by jointly learning to align and translate. arXiv preprint arXiv:1409.0473 (2014)
4. Bansal, A., Chaudhury, S., Roy, S.D., Srivastava, J.: Newspaper article extraction using hierarchical fixed point model. In: 2014 11th IAPR International Workshop on Document Analysis Systems, pp. 257–261. IEEE (2014)
5. Beretta, R., Laura, L.: Performance evaluation of algorithms for newspaper article identification. In: 2011 International Conference on Document Analysis and Recognition, pp. 394–398. IEEE (2011)
6. Chollet, F., et al.: Keras (2015). https://github.com/fchollet/keras
7. Clausner, C., Antonacopoulos, A., Derrick, T., Pletschacher, S.: ICDAR 2019 competition on recognition of early Indian printed documents-REID 2019. In: 2019 International Conference on Document Analysis and Recognition (ICDAR), pp. 1527–1532. IEEE (2019)

8. Coquenet, D., Chatelain, C., Paquet, T.: Span: a simple predict and align network for handwritten paragraph recognition. arXiv preprint arXiv:2102.08742 (2021)
9. Diem, M., Kleber, F., Fiel, S., Grüning, T., Gatos, B.: CBAD: ICDAR 2017 competition on baseline detection. In: 2017 14th IAPR International Conference on Document Analysis and Recognition (ICDAR), vol. 1, pp. 1355–1360. IEEE (2017)
10. Gulli, A., Pal, S.: Deep Learning with Keras. Packt Publishing Ltd., Birmingham (2017)
11. He, D., Cohen, S., Price, B., Kifer, D., Giles, C.L.: Multi-scale multi-task FCN for semantic page segmentation and table detection. In: 2017 14th IAPR International Conference on Document Analysis and Recognition (ICDAR), vol. 1, pp. 254–261. IEEE (2017)
12. He, D., et al.: Multi-scale FCN with cascaded instance aware segmentation for arbitrary oriented word spotting in the wild. In: Proceedings of the IEEE Conference on Computer Vision and Pattern Recognition, pp. 3519–3528 (2017)
13. Iikura, R., Okada, M., Mori, N.: Improving BERT with focal loss for paragraph segmentation of novels. In: Dong, Y., Herrera-Viedma, E., Matsui, K., Omatsu, S., González Briones, A., Rodríguez González, S. (eds.) DCAI 2020. AISC, vol. 1237, pp. 21–30. Springer, Cham (2021). https://doi.org/10.1007/978-3-030-53036-5_3
14. Kaur, R.P., Jindal, M.K., Kumar, M.: TxtLineSeg: text line segmentation of unconstrained printed text in Devanagari script. In: Singh, V., Asari, V.K., Kumar, S., Patel, R.B. (eds.) Computational Methods and Data Engineering. AISC, vol. 1257, pp. 85–100. Springer, Singapore (2021). https://doi.org/10.1007/978-981-15-7907-3_7
15. Kingma, D.P., Ba, J.: Adam: a method for stochastic optimization. In: Proceedings of the 3rd International Conference for Learning Representations (ICLR) (2015)
16. Kosaraju, S.C., et al.: Dot-net: document layout classification using texture-based cnn. In: 2019 International Conference on Document Analysis and Recognition (ICDAR), pp. 1029–1034. IEEE (2019)
17. Lee, J., Hayashi, H., Ohyama, W., Uchida, S.: Page segmentation using a convolutional neural network with trainable co-occurrence features. In: 2019 International Conference on Document Analysis and Recognition (ICDAR), pp. 1023–1028. IEEE (2019)
18. Luong, M.T., Pham, H., Manning, C.D.: Effective approaches to attention-based neural machine translation. arXiv preprint arXiv:1508.04025 (2015)
19. Maia, A.L., Julca-Aguilar, F.D., Hirata, N.S.: A machine learning approach for graph-based page segmentation. In: 2018 31st SIBGRAPI Conference on Graphics, Patterns and Images (SIBGRAPI), pp. 424–431. IEEE (2018)
20. Mechi, O., Mehri, M., Ingold, R., Amara, N.E.B.: Text line segmentation in historical document images using an adaptive U-Net architecture. In: 2019 International Conference on Document Analysis and Recognition (ICDAR), pp. 369–374. IEEE (2019)
21. Michael, J., Weidemann, M., Laasch, B., Labahn, R.: ICPR 2020 competition on text block segmentation on a newseye dataset. In: Del Bimbo, A., et al. (eds.) ICPR 2021. LNCS, vol. 12668, pp. 405–418. Springer, Cham (2021). https://doi.org/10.1007/978-3-030-68793-9_30
22. Naoum, A., Nothman, J., Curran, J.: Article segmentation in digitised newspapers with a 2D Markov model. In: 2019 International Conference on Document Analysis and Recognition (ICDAR), pp. 1007–1014. IEEE (2019)
23. O'Gorman, L.: The document spectrum for page layout analysis. IEEE Trans. Pattern Anal. Mach. Intell. **15**(11), 1162–1173 (1993)

24. Oktay, O., et al.: Attention U-Net: Learning where to look for the pancreas. arXiv preprint arXiv:1804.03999 (2018)

25. Ronneberger, O., Fischer, P., Brox, T.: U-Net: convolutional networks for biomedical image segmentation. In: Navab, N., Hornegger, J., Wells, W.M., Frangi, A.F. (eds.) MICCAI 2015. LNCS, vol. 9351, pp. 234–241. Springer, Cham (2015). https://doi.org/10.1007/978-3-319-24574-4_28

26. Shiah, C.Y.: Content-based document image retrieval based on document modeling. J. Intell. Inf. Syst. **55**, 287–306 (2020)

27. Vidal, E., et al.: The Carabela project and manuscript collection: large-scale probabilistic indexing and content-based classification. In: 2020 17th International Conference on Frontiers in Handwriting Recognition (ICFHR), pp. 85–90. IEEE (2020)

# Data-Efficient Information Extraction from Documents with Pre-trained Language Models

Clément Sage[1,2(✉)], Thibault Douzon[1,2], Alex Aussem[1], Véronique Eglin[1], Haytham Elghazel[1], Stefan Duffner[1], Christophe Garcia[1], and Jérémy Espinas[2]

[1] Univ Lyon, CNRS, LIRIS, Villeurbanne, France
{alexandre.aussem,veronique.eglin,haytham.elghazel,stefan.duffner,
christophe.garcia}@liris.cnrs.fr
[2] Esker, Villeurbanne, France
{clement.sage,thibault.douzon,jeremy.espinas}@esker.com

**Abstract.** Like for many text understanding and generation tasks, pre-trained languages models have emerged as a powerful approach for extracting information from business documents. However, their performance has not been properly studied in data-constrained settings which are often encountered in industrial applications. In this paper, we show that LayoutLM, a pre-trained model recently proposed for encoding 2D documents, reveals a high sample-efficiency when fine-tuned on public and real-world Information Extraction (IE) datasets. Indeed, LayoutLM reaches more than 80% of its full performance with as few as 32 documents for fine-tuning. When compared with a strong baseline learning IE from scratch, the pre-trained model needs between 4 to 30 times fewer annotated documents in the toughest data conditions. Finally, LayoutLM performs better on the real-world dataset when having been beforehand fine-tuned on the full public dataset, thus indicating valuable knowledge transfer abilities. We therefore advocate the use of pre-trained language models for tackling practical extraction problems.

**Keywords:** Pre-training · Language models · Business documents · Information extraction · Document understanding · Document intelligence · Few-shot learning · Intermediate learning

## 1 Introduction

Business documents are files that describe all the internal and external transactions occurring in a company. Such documents cover a wide variety of types, including invoices, purchase orders, receipts, vendor contracts, financial reports and employment agreements. To cope with the increasing volume of business documents to process, academic and industrial practitioners have leveraged AI techniques to automatically read, understand and interpret them [24]. This research

© Springer Nature Switzerland AG 2021
E. H. Barney Smith and U. Pal (Eds.): ICDAR 2021 Workshops, LNCS 12917, pp. 455–469, 2021.
https://doi.org/10.1007/978-3-030-86159-9_33

topic, recently referred to as Document Intelligence (DI), comprises multiple disciplines ranging from Natural Language Processing, Computer Vision over Information Retrieval to Knowledge Representation and Reasoning among others.

Nowadays, business documents are still often distributed in non-machine-readable formats such as images of scanned documents or PDFs filled with unstructured data [6]. One crucial task in Document Intelligence is thus to parse the text of these documents to retrieve valuable semantic information. It may be extracting the value of fields that repeatedly appear in the documents, e.g. the total amount in restaurant receipts [16] or analyzing the structure of forms by identifying all their key-value pairs [17]. To tackle the diversity and complexity of document structure and content, current Information Extraction (IE) approaches employ deep neural networks that learns from annotated documents. Yet, as for many tasks in DI, labeling documents is a challenge in IE since it involves significant human expertise in the targeted application domain [24]. Besides, the extraction objectives are highly specific to the type of documents to process, hindering the reusability of a trained IE model. In [26,32], the authors obtain high-quality annotations from the end users of commercialized document automation software but those users expect to rapidly leverage the benefits of automated IE. Therefore, DI practitioners usually seek to minimize the amount of supervision required to design performing automation tools, especially knowing the wide spectrum of document types that a company may receive or emit.

Following the current trend in the NLP field, a number of works [14,28,35,36] have proposed language models that are pre-trained on large collections of documents and then fine-tuned and evaluated on several document analysis tasks such as information extraction but also document-level classification and visual question answering. Their pre-trained models have considerably outperformed the previous state-of-the art models that were trained from scratch, whether they are evaluated on benchmarks with large-scale [13] or relatively restrained [16,17,27] annotated sets for training. However, this comparison has not been conducted in even more data-constrained settings that are encountered in practical applications of IE models. In this paper, we aim to quantify to what extent the pre-trained models are sample-efficient for IE tasks by comparing LayoutLM [36]—a pre-trained language model recently proposed for encoding 2D documents—with two models without pre-training. We present three main findings that we experimentally validated using the public SROIE benchmark [16] as well as a private real-world dataset:

- The pre-trained LayoutLM exhibits remarkable few-shot learning capabilities for IE, reaching more than 80% of its full performance with as few as 32 documents for fine-tuning.
- This model is significantly more data-efficient than a strong non-pretrained baseline in the lowest data regimes, hitting the same levels of extraction performance with around 30 times fewer samples for the real-world dataset.
- Finally, the pre-trained model displays helpful knowledge transfer between IE tasks since learning beforehand to extract information on the full SROIE

dataset improves the performance of up to 10% when fine-tuning the model on the private dataset.

Corroborating the data efficiency of such models already observed in other NLP tasks [2,4,15], our results show that using pre-trained models dramatically reduces the amount of annotations required for achieving satisfying performance which is appreciable for industrial IE systems.

## 2 Related Works on Information Extraction (IE)

**Fully Supervised Models.** Historically tackled by rule-based approaches [3,21], the IE task has lately been dominated by machine learning based solutions [5]. Most ML approaches first employ an encoder, usually a few neural network layers, to obtain contextualized high-level representations of all the tokens of the document. Then, a decoder module composed of a couple of dense layers is immediately applied to these representations to classify each token according to the type of information that it carries. Most works adopting this sequence labeling approach for extracting information have focused on constituting more powerful representations of the document tokens. The first encoders to appear were recurrent neural networks [26,33] that operate on an uni-dimensional arrangement of tokens. Later, encoders that explicitly consider the two dimensional structure of business documents have been proposed, thus leveraging physical layout information. These methods either represent a document as a graph of tokens [10,22,29,37] or a regularly shaped grid on which the tokens are embedded [7,8,18,39]. Some convolutional layers are then applied to these models of document to obtain the token representations. In addition to better understanding the document layout, some authors [18,25] also include the pixel values of the document images in the input for capturing clues not conveyed by the text modality such as table ruling lines, logos and stamps.

In all these extraction models, the whole set of their parameters, except perhaps the token embeddings [8], are learned in a fully supervised task-specific way. Specifically, they are attributed random values at the beginning of the model training. The parameters values are then updated by directly minimizing the cross-entropy loss on the target IE dataset. While being successful for most IE tasks, this results in a costly process since a massive amount of weights need to be learned from scratch.

**Pre-trained Models.** Since the recent development of language modeling techniques [2,9], NLP models for understanding and generating text are not learned from scratch anymore [30]. Rather, the mainstream approach to reach state-of-the-art performance on many downstream tasks is to adapt the parameters of models that have already learned powerful representations of the language. Such pre-training is performed in a self-supervised way on a large quantity of text data. Starting from LayoutLM [36], pre-trained models that were originally operating on serialized text have been extended to process the spatially distributed text contained in business documents, e.g. text blocks and tables. To that end, positional embedding vectors relative to their absolute 2D coordinates are included into the token representations that are given to the Transformer encoder. Before fine-tuning the model on the downstream tasks like the fully supervised models, LayoutLM is first pre-trained on millions of document pages [20] using a self-supervised Masked Visual-Language Modeling (MVLM) task that naturally expands the main pre-training objective of BERT [9].

This work further inspires other language models dedicated to two dimensional documents. While the image modality was introduced only at the fine-tuning stage in LayoutLM, later models [14,28,35] include visual descriptors from convolutional layers directly into the token representations used for pre-training. These recent works mainly focus on adding new pre-training objectives complementing MVLM to more effectively mix the text, layout and image modalities when learning the document representations, for example the topic-modeling and document shuffling tasks of [28], the Sequence Positional Relationship Classification (SPRC) objective [34], the text-image alignment and matching tasks leveraged in [35] and the 2D area-masking strategy from [14]. Moreover, [14,35] both modify the computation of the self-attention scores to better encompass the relative positional relationships among the tokens of the document. Finally, [28] has resorted to page index embeddings and the Longformer's [1] self-attention that scales linearly with the sequence length in order to process multi-page and longer documents.

All these pre-trained models largely surpass fully supervised models and have established state-of-the-art performance on multiple document understanding benchmarks, including common information extraction datasets [16,17,27]. Yet, all the experiments have been performed with the full training set of the downstream tasks for fine-tuning, thus not studying the potential of pre-trained models to learn IE with few annotated data compared to models without such pre-training. Our contribution consists here in showing how pre-trained models can lead to a performance gain on low-resource downstream IE tasks.

## 3    Models

In our experiments, we follow the sequence labeling approach for performing IE. The evaluated models are composed of an encoder delivering contextualized representations of the tokens and a linear classifier that decodes this sequence of representations to extract information. All models only differ by their encoder.

### 3.1    Encoder

As shown in Fig. 1, we use three different networks for encoding the business documents. We compare a pre-trained encoder with two fully supervised encoders.

**Pre-trained Model.** As pre-trained model, we use LayoutLM from [36] since this is the only IE work that publicly releases their pre-trained model parameters. We use its base-uncased version[1] which consists of a 12-layer Transformer

---

[1] https://github.com/microsoft/unilm/tree/master/layoutlm.

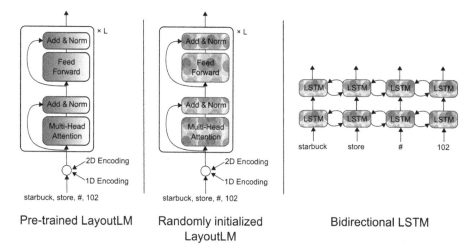

**Pre-trained LayoutLM**    **Randomly initialized LayoutLM**    **Bidirectional LSTM**

**Fig. 1.** The different architectures used in our experiments for encoding documents. From left to right: Transformer-based LayoutLM [36] with pre-trained weights, LayoutLM with random initialization and a 2-layer bidirectional LSTM also randomly initialized.

with a hidden size of 768 and 12 attention heads per layer, resulting in 113 millions weights. It is built upon the BERT base-uncased model with 4 additional embedding vectors to represent the position of each token in the document page. This 2D positional encoding, coupled with a pre-training task that strongly binds the token's semantic representation with their surrounding, allows LayoutLM to take advantage of the structure of the documents. Although proposed in their paper for the fine-tuning stage, we do not leverage the image modality since it brings marginal improvements for IE. We thus solely rely on the text and its layout for constructing token embeddings. We refer the reader to their paper for more details about its architecture and pre-training stage.

**Fully Supervised Models.** For fully supervised models, we use 2 encoders that are trained from scratch on the IE tasks. First, we reuse the LayoutLM model but we discard pre-training and randomly initialize all its parameters. However, as confirmed by our early experiments, this encoder version performs poorly in low-resource settings due to its massive amount of parameters to learn from scratch. Secondly, we propose a smaller fully supervised baseline that has shown success in past IE works [26,33]. This is a 2-layer bidirectional LSTM network (BLSTM) with a 128 hidden size. We reuse the same sub-word tokenizer as LayoutLM and employ only textual embeddings for tokens. The resulting model contains 8.5 millions parameters.

Following standard practises, Transformer and embedding layers are respectively initialized with a truncated normal and Gaussian distributions. BLSTM layers resort to Glorot initialization [12].

## 3.2  Decoder

On top of each of these 3 encoders, we add a dense softmax layer to predict the information type carried by each document token. Since the fields to extract can be spread over multiple tokens, the BIESO labeling scheme [31] is utilized to denote the beginning (B), continuation (I) and end (E) of a field value while S classes stand for single token values. This results in 4 output classes per field, with the additional class O for tokens not conveying any relevant information. At inference time, we determine the class of a token by getting its highest probability and reduce the resulting list of BIESO classes to obtain the field level predictions. If a document has more than 512 tokens, its text is split in multiple sequences that are independently processed by the extraction model.

## 4  Datasets

As illustrated in Fig. 2, we consider two IE datasets that cover different document types and extraction objectives.

(a) receipt from SROIE                    (b) purchase order from PO-51k

**Fig. 2.** A document sample for each dataset alongside their expected field values to extract. For PO-51k, we show a fictive purchase order due to privacy reasons.

## 4.1    Scanned Receipts OCR and Information Extraction (SROIE)

We train and evaluate the models on the public SROIE dataset [16] containing restaurant receipts. We only consider its information extraction task that aims to retrieve the name and address of the company issuing the receipt, the total amount and date. The dataset gathers 626 receipts for training and 347 receipts for test. We further randomly split the training set to constitute a validation set of 26 receipts. While not stated in [16], the document issuers are shared between the training and test sets.

Each receipt is given the ground-truth value for the four targeted fields. The comparison with the model predictions is made in terms of exact matching of strings, leading to precision, recall and F1 score metrics[2]. For the sake of readability, we only report the F1 scores averaged over all the targeted fields. To establish the BIESO labels, we look for the receipt words matching the ground-truth field values. For the total amount, a value may match different sets of words, e.g. the amounts without taxes or after rounding. If so, we select the bottom most occurrence having the keyword total in its line.

We use the provided OCR results containing a list of text segments and their bounding boxes. As noticed by many submissions in the leaderboard including LayoutLM's authors, they contain a number of brittle text recognition errors, e.g. a comma interpreted as a dot. This highly impacts the evaluation results based on exact matching. Therefore, following previous works, we manually fix them in the test set while we perform fuzzy matching for deriving the token labels in the training set. The order of text segments being sometimes faulty, we also re-arrange them from top-to-bottom.

## 4.2    Real-World Purchase Orders (PO-51k)

To prove the efficiency of the IE models, we also conduct experiments on a private dataset composed of 51,000 English purchase orders that were processed on a commercial document automation solution. We split the dataset in 40k, 1k and 10k documents for training, validation and test sets. Unlike SROIE, these three subsets contain different document issuers, respectively 6200, 870 and 1700 issuers. This induces that for a large portion of the test set, the layout and content organization of documents have not been seen at training time.

We aim to extract 3 different fields among these purchase orders: the document number, the date and the total amount. The ground truth for these fields is directly provided by the end-users of the automation software, ensuring high-quality annotations. We employ the same methodology as in SROIE for evaluating the models. Text of documents is retrieved thanks to a commercial OCR engine.

Since LayoutLM is not designed for handling multi-page documents, we only consider the first page of documents. Because of this limitation, there may be no

---

[2] The metric values are obtained at: https://rrc.cvc.uab.es/?ch=13&com=evaluation&task=3.

value to predict for a target field. In practice, roughly 25% of the documents miss a total amount on the first page while only 10% of the documents are affected for the two other fields.

## 5    Experiments

### 5.1    Experiment Settings

We use the following settings in all our experiments. To evaluate data efficiency, we restrict the training set to 8, 16, 32, 64, 128, 256 and 600 randomly selected documents for both datasets. For PO-51k, we additionally study the extraction performance when training with 2k, 8k and 40k samples. We repeat each experiment 5 times, each time with different random seeds and thus different selected training documents. We plot the average $\mu$ of the 5 F1 scores as well as the shaded region $[\mu - \sigma, \mu + \sigma]$ for representing the standard deviation $\sigma$. We use a log scale over the number of training documents to better visualize the lowest-resource regimes.

As in [36], we use the Adam optimizer with an initial learning rate of $5e-5$, linearly decreasing it to 0 as we reach the maximum number of training steps. For the BLSTM model, we employ a higher initial learning rate of $5e-3$ since the former value was not giving a good convergence. For each run, we set the maximum number of training steps to 1k for the pre-trained LayoutLM and 2k for models without pre-training. We proceed to early stopping on the validation set to choose the model checkpoint to evaluate or use for a further training run. We employ a batch size of 8 for all runs in SROIE. For PO-51k, we set the batch size to 16 for all runs, except for 8 and 40k training docs where we fix it to respectively 8 and 32 in order to see at least once each training document. Following the results of language models fine-tuning in low-resource settings [15], we update the entire model in all runs.

All training runs are performed on a single 12 Go TITAN XP GPU. We have released the code for reproducing the experiments on the SROIE dataset[3].

### 5.2    Few-Shot Learning

For both datasets, we first study the performance when the models independently learn the IE task from a few annotated samples. After initializing them from scratch or from pre-trained weights, we fine-tune the models for variable numbers of training documents. We report below their results on the whole test set.

**SROIE.** We show F1 scores for the SROIE dataset in the Fig. 3. We first notice that we get to an average F1 score of 0.9417 when the pre-trained LayoutLM is fine-tuned on 600 receipts. This is in accordance with the 0.9438 F1 score reported in its paper [36] when considering the 626 documents of the original

---

[3] https://github.com/clemsage/unilm

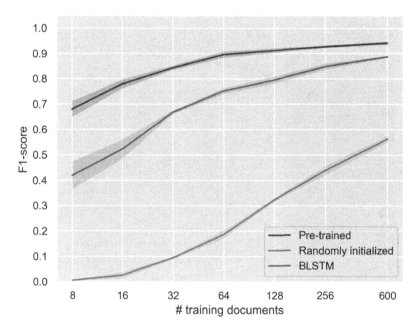

**Fig. 3.** Few-shot extraction performance on the SROIE [16] test set for the pre-trained LayoutLM [36] against its randomly initialized version and a BLSTM network.

training set. The model convergence is really fast, hitting 90% of its full performance with only 32 documents, i.e. a 18 times smaller training set.

Unsurprisingly, we observe that the pre-trained LayoutLM achieves significantly better performance than fully supervised models whatever the number of training documents. Yet, the fewer training documents we make use of, the larger is the difference of F1 score between these two classes of models. For instance, even if the BLSTM network reaches a near similar level of performance with 600 documents (0.8874 against 0.9417), it performs significantly worse than LayoutLM in more data-constrained regimes: the gap of F1 score attains 0.2612 for 8 training receipts. This is even more noticeable for the randomly initialized LayoutLM which completely fails to extract the fields when trained with 8 documents. When offered the full training set, the model does not even outperform its pre-trained counterpart that makes use of only 8 documents.

As expected [38], the performance variance is greater in the lowest data regimes. Yet, the pre-training effectively reduces the variance, making pre-trained models less dependent on the choice of fine-tuning documents.

**PO-51k.** We show F1 scores for the PO-51k dataset in the Fig. 4. We observe similar learning curves for all models, including the pre-trained model that hits 92% of its maximal performance with only 128 samples, i.e. 312 times fewer training documents. In the lowest data regimes, the gap between LayoutLM and the fully supervised baselines is even wider than for SROIE. Indeed, the difference

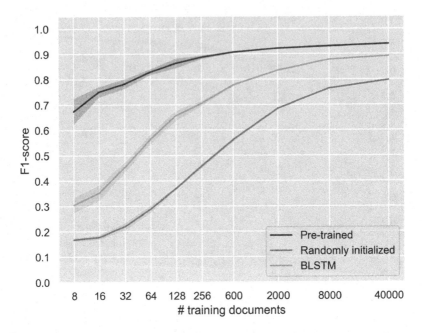

**Fig. 4.** Few-shot extraction performance on the PO-51K test set for the pre-trained LayoutLM [36] against its randomly initialized version and a BLSTM network.

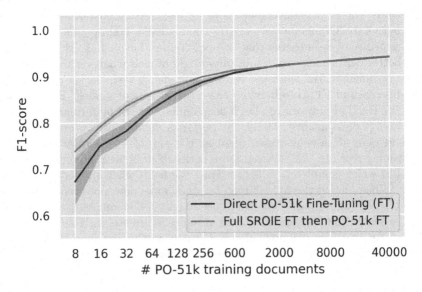

**Fig. 5.** Test F1 scores of pre-trained LayoutLM when transferring extraction knowledge from SROIE to PO-51k tasks. The IE performance is always improved by resorting to SROIE as an intermediate task, the boost being significant with few available PO-51k documents for fine-tuning.

with the BLSTM model is on average of 0.37 F1 score until 32 documents while it was on average of 0.23 points for SROIE. The BLSTM trained with 600 documents performs on par with LayoutLM fine-tuned on only 32 documents, i.e. a order of magnitude less annotations. We also note that this real-world dataset is notoriously more complex than SROIE since a few hundreds documents are not enough to achieve full convergence of the F1 scores. We finally underline the sample inefficiency of LayoutLM trained from scratch with a F1 score at 40k training documents that still lags behind both its pre-trained counterpart and the BLSTM.

On both datasets, we have confirmed that the pre-training stage extensively reduces the amount of annotations needed to reach specific performance for downstream IE tasks.

## 5.3   Intermediate Learning

In these experiments, we analyze to what extent learning to extract information from given documents decreases the annotation efforts for later performing IE on another document distribution. Specifically, we first fine-tune the pre-trained LayoutLM on the SROIE task using its full training set and then transfer the resulting model on the PO-51k dataset and study its few-shot performance. This simulates an actual use case where a practitioner leverages publicly available data to later tackle IE in more challenging industrial environments.

Since the fields to extract are not identical between the SROIE and PO-51k tasks, we remove the final classifier layer on top of LayoutLM after the fine-tuning on SROIE. We replace it with a randomly initialized layer that matches the number of fields in PO-51k. Even if this imposes to learn the decoder parameters from scratch between the two IE tasks, there are only a few thousands compared to the million weights of the encoder. We therefore hope that LayoutLM can still transfer some knowledge from SROIE to PO-51k tasks.

**SROIE to PO-51k.** We compare the few-shot performance on PO-51k when having firstly fine-tuned on SROIE with the results obtained when directly employing the pre-trained LayoutLM weights. We show results of these intermediate learning experiences in Fig. 5.

We note that the fine-tuning on SROIE considerably improves the extraction for few PO-51k examples with a boost of 0.065 (+10%) F1 score for 8 documents. For 600 examples or more, the effect of intermediate learning disappears with a performance indistinguishable from directly fine-tuning on PO-51k. Fine-tuning beforehand on the SROIE dataset also helps to reduce the variance when it is significant: between 8 to 32 PO-51k documents, the mean standard deviation decreases from 0.031 to 0.017 (−45%) when resorting to intermediate learning.

Therefore, if the amount of annotated documents at their disposal is limited, we encourage IE practitioners not to directly fine-tune the pre-trained models on their task but first use publicly available IE datasets to enhance performance.

# 6 Conclusion

In this paper, we showed that pre-trained language models are highly beneficial for extracting information from few annotated documents. On a public dataset as well as on a more demanding industrial application, such a pre-trained app-roach consistently outperformed two fully supervised models that learn from scratch the IE task. We finally demonstrated that pre-training brings additional improvements when transferring knowledge from an IE task to another.

In the future, we will further investigate the potential of pre-trained models for intermediate learning. Under the current sequence labeling paradigm, the decoder still needs to be learned from scratch for each IE task, presumably hindering the transferability of extraction knowledge between downstream tasks. We hypothesize that resorting to decoders with reusable weights may help to better leverage the knowledge learned from the intermediate IE task. We have particularly in mind the question answering format [11] which has already shown success for zero-shot relation extraction [19]. We also plan to confirm that the sample efficiency of pre-trained models is observed for other document analysis tasks such as document level classification [13] or visual question answering [23].

**Acknowledgment.** The work presented in this paper was supported by Esker. We thank them for providing the PO-51k dataset and for insightful discussions about these researches.

# References

1. Beltagy, I., Peters, M.E., Cohan, A.: Longformer: the long-document transformer. arXiv preprint arXiv:2004.05150 (2020)
2. Brown, T.B., et al.: Language models are few-shot learners. In: Advances in Neural Information Processing Systems 33: Annual Conference on Neural Information Processing Systems 2020, NeurIPS 2020, Virtual, 6–12 December 2020 (2020). https://proceedings.neurips.cc/paper/2020/hash/1457c0d6bfcb4967418bfb8ac142f64a-Abstract.html
3. Cesarini, F., Gori, M., Marinai, S., Soda, G.: INFORMys: a flexible invoice-like form-reader system. IEEE Trans. Pattern Anal. Mach. Intell. **20**(7), 730–745 (1998)
4. Chen, Z., Eavani, H., Chen, W., Liu, Y., Wang, W.Y.: Few-shot NLG with pre-trained language model. In: Proceedings of the 58th Annual Meeting of the Association for Computational Linguistics, Online, pp. 183–190. Association for Computational Linguistics, July 2020. https://doi.org/10.18653/v1/2020.acl-main.18. https://www.aclweb.org/anthology/2020.acl-main.18
5. Chiticariu, L., Li, Y., Reiss, F.: Rule-based information extraction is dead! Long live rule-based information extraction systems! In: Proceedings of the 2013 Conference on Empirical Methods in Natural Language Processing, pp. 827–832 (2013)
6. Cohen, B., York, M.: Ardent partners' accounts payable metrics that matter in 2020. Technical report, Ardent Partners (2020). http://ardentpartners.com/2020/ArdentPartners-AP-MTM2020-FINAL.pdf

7. Dang, T.A.N., Thanh, D.N.: End-to-end information extraction by character-level embedding and multi-stage attentional U-Net. In: 30th British Machine Vision Conference 2019, BMVC 2019, Cardiff, UK, 9–12 September 2019, p. 96. BMVA Press (2019). https://bmvc2019.org/wp-content/uploads/papers/0870-paper.pdf

8. Denk, T.I., Reisswig, C.: BERTgrid: contextualized embedding for 2D document representation and understanding. In: Workshop on Document Intelligence at NeurIPS 2019 (2019). https://openreview.net/forum?id=H1gsGaq9US

9. Devlin, J., Chang, M.W., Lee, K., Toutanova, K.: BERT: pre-training of deep bidirectional transformers for language understanding. In: Proceedings of the 2019 Conference of the North American Chapter of the Association for Computational Linguistics: Human Language Technologies, Volume 1 (Long and Short Papers), pp. 4171–4186 (2019)

10. Gal, R., Ardazi, S., Shilkrot, R.: Cardinal graph convolution framework for document information extraction. In: Proceedings of the ACM Symposium on Document Engineering 2020, pp. 1–11 (2020)

11. Gardner, M., Berant, J., Hajishirzi, H., Talmor, A., Min, S.: Question answering is a format; when is it useful? arXiv preprint arXiv:1909.11291 (2019)

12. Glorot, X., Bengio, Y.: Understanding the difficulty of training deep feedforward neural networks. In: Proceedings of the Thirteenth International Conference on Artificial Intelligence and Statistics, pp. 249–256. JMLR Workshop and Conference Proceedings (2010)

13. Harley, A.W., Ufkes, A., Derpanis, K.G.: Evaluation of deep convolutional nets for document image classification and retrieval. In: 2015 13th International Conference on Document Analysis and Recognition (ICDAR), pp. 991–995. IEEE (2015)

14. Hong, T., Kim, D., Ji, M., Hwang, W., Nam, D., Park, S.: BROS: a pre-trained language model for understanding texts in document (2021). https://openreview.net/forum?id=punMXQEsPr0

15. Howard, J., Ruder, S.: Universal language model fine-tuning for text classification. In: Proceedings of the 56th Annual Meeting of the Association for Computational Linguistics (Volume 1: Long Papers), Melbourne, Australia, pp. 328–339. Association for Computational Linguistics, July 2018. https://doi.org/10.18653/v1/P18-1031. https://www.aclweb.org/anthology/P18-1031

16. Huang, Z., et al.: ICDAR 2019 competition on scanned receipt OCR and information extraction. In: 2019 International Conference on Document Analysis and Recognition (ICDAR), pp. 1516–1520. IEEE (2019)

17. Jaume, G., Ekenel, H.K., Thiran, J.: FUNSD: a dataset for form understanding in noisy scanned documents. In: 2nd International Workshop on Open Services and Tools for Document Analysis, OST@ICDAR 2019, Sydney, Australia, 22–25 September 2019. pp. 1–6. IEEE (2019). https://doi.org/10.1109/ICDARW.2019.10029

18. Katti, A.R., et al.: Chargrid: towards understanding 2D documents. In: Proceedings of the 2018 Conference on Empirical Methods in Natural Language Processing, pp. 4459–4469 (2018)

19. Levy, O., Seo, M., Choi, E., Zettlemoyer, L.: Zero-shot relation extraction via reading comprehension. In: Proceedings of the 21st Conference on Computational Natural Language Learning (CoNLL 2017), pp. 333–342 (2017)

20. Lewis, D., Agam, G., Argamon, S., Frieder, O., Grossman, D., Heard, J.: Building a test collection for complex document information processing. In: Proceedings of the 29th Annual International ACM SIGIR Conference on Research and Development in Information Retrieval, pp. 665–666 (2006)

21. Li, Y., Krishnamurthy, R., Raghavan, S., Vaithyanathan, S., Jagadish, H.V.: Regular expression learning for information extraction. In: Proceedings of the 2008 Conference on Empirical Methods in Natural Language Processing, Honolulu, Hawaii, pp. 21–30. Association for Computational Linguistics, October 2008. https://www.aclweb.org/anthology/D08-1003

22. Liu, X., Gao, F., Zhang, Q., Zhao, H.: Graph convolution for multimodal information extraction from visually rich documents. In: Proceedings of the 2019 Conference of the North American Chapter of the Association for Computational Linguistics: Human Language Technologies, Volume 2 (Industry Papers), Minneapolis, Minnesota, pp. 32–39. Association for Computational Linguistics, June 2019. https://doi.org/10.18653/v1/N19-2005. https://www.aclweb.org/anthology/N19-2005

23. Mathew, M., Karatzas, D., Jawahar, C.: DocVQA: a dataset for VQA on document images. In: Proceedings of the IEEE/CVF Winter Conference on Applications of Computer Vision (WACV), pp. 2200–2209, January 2021

24. Motahari, H., Duffy, N., Bennett, P., Bedrax-Weiss, T.: A report on the first workshop on document intelligence (DI) at NeurIPS 2019. ACM SIGKDD Explor. Newslett. **22**(2), 8–11 (2021)

25. Palm, R.B., Laws, F., Winther, O.: Attend, copy, parse end-to-end information extraction from documents. In: 2019 International Conference on Document Analysis and Recognition (ICDAR), pp. 329–336. IEEE (2019)

26. Palm, R.B., Winther, O., Laws, F.: CloudScan - a configuration-free invoice analysis system using recurrent neural networks. In: 2017 14th IAPR International Conference on Document Analysis and Recognition (ICDAR), pp. 406–413. IEEE (2017)

27. Park, S., et al.: CORD: a consolidated receipt dataset for post-OCR parsing. In: Workshop on Document Intelligence at NeurIPS 2019 (2019). https://openreview.net/forum?id=SJl3z659UH

28. Pramanik, S., Mujumdar, S., Patel, H.: Towards a multi-modal, multi-task learning based pre-training framework for document representation learning. arXiv preprint arXiv:2009.14457 (2020)

29. Qian, Y., Santus, E., Jin, Z., Guo, J., Barzilay, R.: GraphIE: a graph-based framework for information extraction. In: Proceedings of the 2019 Conference of the North American Chapter of the Association for Computational Linguistics: Human Language Technologies, Volume 1 (Long and Short Papers), pp. 751–761 (2019)

30. Qiu, X., Sun, T., Xu, Y., Shao, Y., Dai, N., Huang, X.: Pre-trained models for natural language processing: a survey. Sci. China Technol. Sci. 1–26 (2020)

31. Ramshaw, L.A., Marcus, M.P.: Text chunking using transformation-based learning. In: Armstrong, S., Church, K., Isabelle, P., Manzi, S., Tzoukermann, E., Yarowsky, D. (eds.) Natural Language Processing Using very Large Corpora, pp. 157–176. Springer, Dordrecht (1999). https://doi.org/10.1007/978-94-017-2390-9_10

32. Sage, C., Aussem, A., Eglin, V., Elghazel, H., Espinas, J.: End-to-end extraction of structured information from business documents with pointer-generator networks. In: Proceedings of the Fourth Workshop on Structured Prediction for NLP, pp. 43–52. Association for Computational Linguistics, Online, November 2020. https://www.aclweb.org/anthology/2020.spnlp-1.6

33. Sage, C., Aussem, A., Elghazel, H., Eglin, V., Espinas, J.: Recurrent neural network approach for table field extraction in business documents. In: 2019 International Conference on Document Analysis and Recognition (ICDAR), pp. 1308–1313. IEEE (2019)

34. Wei, M., He, Y., Zhang, Q.: Robust layout-aware IE for visually rich documents with pre-trained language models. In: Proceedings of the 43rd International ACM SIGIR conference on research and development in Information Retrieval, SIGIR 2020, Virtual Event, China, 25–30 July 2020, pp. 2367–2376. ACM (2020). https:// doi.org/10.1145/3397271.3401442
35. Xu, Y., et al.: LayoutLMv2: multi-modal pre-training for visually-rich document understanding. arXiv preprint arXiv:2012.14740 (2020)
36. Xu, Y., Li, M., Cui, L., Huang, S., Wei, F., Zhou, M.: LayoutLM: pre-training of text and layout for document image understanding. In: Proceedings of the 26th ACM SIGKDD International Conference on Knowledge Discovery & Data Mining, pp. 1192–1200 (2020)
37. Yu, W., Lu, N., Qi, X., Gong, P., Xiao, R.: PICK: processing key information extraction from documents using improved graph learning-convolutional networks. arXiv preprint arXiv:2004.07464 (2020)
38. Zhang, T., Wu, F., Katiyar, A., Weinberger, K.Q., Artzi, Y.: Revisiting few-sample BERT fine-tuning. In: International Conference on Learning Representations (2021). https://openreview.net/forum?id=cO1IH43yUF
39. Zhao, X., Wu, Z., Wang, X.: CUTIE: learning to understand documents with convolutional universal text information extractor. arXiv preprint arXiv:1903.12363 (2019)

# ICDAR 2021 Workshop on Graph Representation Learning for Scanned Document Analysis (GLESDO)

# GLESDO 2021 Preface

Robust reading, also known as automatic document image processing, is an essential task in various applications areas such as data invoice extraction, subject review, medical prescription analysis, etc. and holds significant commercial potential. Several approaches are proposed in the literature, but dataset availability and data privacy challenge it.

Considering the problem of information extraction from documents, different aspects must be taken into account, such as (1) document classification, (2) text localization, (3) OCR (optical character recognition), (4) table extraction, and (5) key information detection. In this context, graph-based approaches are attractive methods for document processing. In fact, graphs are a natural way to represent the connections among objects (text, blocks, images, etc.) and to discover novel and hidden knowledge from data. The extracted text from scanned documents can be represented in the shape of a graph to exploit the best features of their characteristics. On the other hand, understanding spatial relationships is critical for text document extraction results for some applications such as invoice analysis. The aim is to capture the structural connections between keywords (invoice number, date, amounts) and the main value (the desired information). An effective approach requires a combination of spatial and textual information.

In this workshop, we aimed to bring together experts from industry, science, and academia to exchange ideas and discuss ongoing research in graph representation learning for scanned document analysis.

The first edition of GLESDO was held in virtual form on September 5, 2021, for halfday, in conjunction with ICDAR 2021. We had one invited speaker: Sandeep Tata from Google Research, San Francisco, USA. The Program Committee was selected to reflect the interdisciplinary nature of the field. For this First edition, we welcomed two kinds of contributions: short and long papers. We received a total of five submissions. Each paper was reviewed by two members of the Program Committee via EasyChair. A double-blind review was used for these short paper submissions, and the authors were welcome to anonymize their submissions. The five submissions were evaluated and accepted for presentation at the workshop, and are published in these proceedings.

September 2021

Rim Hantach
Rafika Boutalbi
Philippe Calvez
Balsam Ajib Trinov
Thibault Defourneau

# Organization

## General Chairs

Rim Hantach                Engie, France
Rafika Boutalbi            Universität Stuttgart, Germany

## Program Committee Chairs

Philippe Calvez            Engie, France
Balsam Ajib                Trinov, France
Thibault Defourneau        Trinov, France

## Program Committee

Muhammad Muzzamil Luqman    La Rochelle University, France
Pau Riba                   CVC, Spain
Abdel Belaid               Loria, France
Anastasiia Iurshina        University of Stuttgart, Germany
Balsam Ajib                Trinov, France
Thibault Defourneau        Trinov, France
Gisela Lechuga             Engie, France

# Representing Standard Text Formulations as Directed Graphs

Frieda Josi[1]([⊠])(iD), Christian Wartena[1](iD), and Ulrich Heid[2]

[1] University of Applied Sciences and Arts Hanover, Expo Plaza 12,
30539 Hanover, Germany
{frieda.josi,christian.wartena}@hs-hannover.de
[2] University of Hildesheim, Lübecker Straße 3, 31141 Hildesheim, Germany
heidul@uni-hildesheim.de
https://im.f3.hs-hannover.de/en/
https://www.uni-hildesheim.de/fb3/institute/iwist/

**Abstract.** In order to ensure validity in legal texts like contracts and case law, lawyers rely on standardised formulations that are written carefully but also represent a kind of code with a meaning and function known to all legal experts. Using directed (acyclic) graphs to represent standardized text fragments, we are able to capture variations concerning time specifications, slight rephrasings, names, places and also OCR errors. We show how we can find such text fragments by sentence clustering, pattern detection and clustering patterns. To test the proposed methods, we use two corpora of German contracts and court decisions, specially compiled for this purpose. However, the entire process for representing standardised text fragments is language-agnostic. We analyze and compare both corpora and give an quantitative and qualitative analysis of the text fragments found and present a number of examples from both corpora.

**Keywords:** Graph-based text representations · Legal writings · Standardised formulation

## 1 Introduction

In legal writings, like contracts or court decisions parts of the text are frequently reused. This type of text reuse is different from plagiarism since this is completely legal; there is no single source of a reused text, but most fragments are used ubiquitous since many years. Reusing smaller or larger text fragments is not only done for efficiency reasons, but the use of standardized phrases and passages is essential for the proper understanding and interpretation of legal writing.

Standardized expressions can vary from short phrases to long passages. In the present work we focus on standardized formulations that consist of several sentences. Surprisingly, hardly any attempt has been undertaken to try to identify such passages automatically. Though the existence and importance of such

E. H. Barney Smith and U. Pal (Eds.): ICDAR 2021 Workshops, LNCS 12917, pp. 475–487, 2021.
https://doi.org/10.1007/978-3-030-86159-9_34

passages has been noted by various authors there is also no clear definition of a standardized passage. When we try to identify standardized passages we are faced with two problems: (1) when is a passage just a repetition of a few sentences used by one author in similar documents and from which point on can we call it a standardized passage? And (2) when is a passage a variation of a commonly used passage and how much has it to deviate to become an independent formulation? Related to the second issue is also the question which variant among all variants found is the most representative one.

Since we often find many similar passages or variants of the same passage, we propose not to choose one, but to represent standardized passages as directed acyclic graphs (DAG). In an empirical study on two large legal corpora, we show that it is possible to cluster all frequent sequences of sentences in small clusters that are mutually almost disjunct and can be represented in uncluttered directed graphs. We see that many, but not all of these *DAGs* are good candidates for standardized passages. We are still far away from a general definition of a standardized passage, but we think that the investigations on the large legal corpora presented here can contribute to a better understanding of the nature of standardized passages and their role in legal documents.

## 2    Phrasemes and Standardized Passages in Legal Writing

In the legal domain texts play an important role. The primary function of a legal text is not to inform or to convince the reader, nor to describe something. Rather a legal text or more precisely a *constitutive* text is a declaration in the sense of Searle's illocutionary speech acts [17] and thus creates a reality and shapes its environment [4,11]. A contract or a court decision is not a model of the reality, but actors are bound to do what the text prescribes. Thus, though these texts are produced en masse, they need to be formulated carefully and their interpretation should be clear and without ambiguities. Thus, often standardized formulations are used to ensure the correct formulation and interpretation (see e.g. [15, p.251]) and the reuse of such passages is an essential trait of legal and normative texts (see e.g. [21]). The standardized formulation often has a specific interpretation and legal consequences both known to the (legally educated) reader and author. Engberg [6,7] stresses that in many text types, like e.g. court decisions, it is important to use exactly the conventional formulation, since this works as a signal to the (expert) reader, who will be disturbed and might misunderstand the text if the convention is not followed.

Standardized formulations are quite flexible and show a considerable degree of variation. Names of persons, dates etc. can be exchanged but also words or phrases can be added or deleted. A standardized formulation can consist of just a few words or of several sentences. Phrases or short pieces of text that are frequently used and have a fixed meaning are well known from other disciplines and from general language and are called idiomatic expressions, phrasemes, phraseologisms or routine formulae or routine expressions [2,16]. All these terms are used as synonyms or with slightly different meanings but always for short phrases.

For longer passages Lindroos [13] found the following terms: text patterns (*Textmuster*), formulaic or schematic texts (*formelhafte oder schematisch gestaltete Texte*), stereotyped texts (*stereotypisierte Texte*) or preformed structures (*vorgeformten Strukturen*). Wozniak [21] uses the term pragmatic phrasealogisms for all types of recurrent phrases and text and calls longer recurrent passages *Kleintexte* (Small texts), formulaic (short) texts or textual phrasealogisms (*textwertige Phraseologismen*). Płomińska [15] uses the terms micro and macro routine expressions for recurrent phrases and recurrent units that consist of one or more sentences, respectively. Płomińska further classifies routine expressions found in court decisions according to their function.

According to Wozniak [21] there is no consensus whether standardized passages can have the status of phrasealogisms. However, she notes that such passages frequently occur in certain text types, like court decisions or contracts. Often these textual phrasealogisms have a fixed syntactic structure with a variable lexical filling. The number of slots that need to be filled differs for each text type and the part of the text the passage belongs to. E.g. the final provisions in contract show less lexical variable slots than the legal instructions in a notification.

## 3   Related Work

There is abundant research on the possibilities to automatically find collocations and short phrases with an idiomatic meaning or that are typical for a discipline or text type. However, not much work was done on the automatic recognition of longer routine expressions. Wahl and Gries [19] is an exception, but they still focus on the phrase level and units shorter than complete sentences. Finding reused text passages and sentences is often important for the analysis of documents. Kliche et al. [12] developed a tool suite where users, can upload the texts to be analyzed and then define patterns to find reused texts in the newsletter corpus. Another research project on text reuse in newspaper articles is described by Clough et al. [5]. The similarity of the documents is calculated using the n-gram overlap, with n-gram lengths of 1 to 10 words. For the overlaps between the sentences of these similar documents the substring matching algorithm Greedy-String-Tiling [20] is used. Recently, a number of papers have described the reuse of texts in legal documents. Burgess et al. [3] present a tool to discover the reuse of text passages from laws for new bills in order to make it more difficult for lobbyists to influence legislation. They assume that lobbyists include similar text formulations in the bills. To detect this text reuse, they use Elasticsearch to find relevant documents on the Internet. Then, they use the Smith-Waterman algorithm for aligning text passages to identify related text passages in the law and in the bill. Finally, they calculate the similarity of these two text passages using the Jaccard coefficient. Graph-based representations have been used before to capture variation in texts. Filippova [8] identifies word changes in a set of similar sentences by transforming the words in the sentences into a graph. Based on the shortest sentence in the set, the paths in the sentence are verified. In this way,

**Table 1.** Data overview of both corpora.

|  | Case law corpus | Contract corpus |
|---|---|---|
| Documents | 4,250 | 2,167 |
| Sentences | 308,832 | 751,281 |
| Tokens | 7,202,106 | 10,433,943 |

longer and more complex sentences can be succinctly reduced to the shortest variant (multi-sentence compression). Ma et al. [14] propose paraphrasing sentences using a graph-based method. The sentences used are from one domain, so the authors could ensure that the thematic scope of the sentences is quite similar. In the first step, they use the Word aligner by Sultan et al. [18] to align pairs of sentences. They then generate *DAGs* for each group of sentences to identify paraphrases.

## 4   Legal Corpora

We have compiled two corpora with two different types of legal texts. The first corpus consists of court decisions and the second of contract texts.[1]

**Case Law Corpus.** The first corpus contains anonymized court decisions from German criminal procedures from the years 2015 to the beginning of 2020. The court decisions are published by the Federal Court of Justice (*BGH*). The court decisions were crawled directly from the website of the case law database and are available in HTML format for further work. The source for the documents in the case law corpus is given under Appendix A.1. An overview of the number of documents, tokens, sentences and sentence clusters for all used corpora is shown in Table 1.

**Contract Corpus.** The second corpus contains contracts of the Hamburg City Administration and the Bremen City Administration. Some cooperation agreements between universities are also included. Among these contracts are several contracts that universities have concluded with external service providers. All contracts are available in PDF format. The contract texts had to be extracted, cleaned and prepared for further processing. The quality of the scanned contracts from the city administrations is not so good. There are documents with a lower scanning resolution, pages have been scanned at different angles, and so on. Moreover, all information that represents personal data has been blacked out. The contracts in this corpus are from the years 2014 to 2019 and are publicly

---

[1] The sources for the documents compiled for both corpora will be published on our website: http://textmining.wp.hs-hannover.de/juver.html. Likewise, we publish the developed methods and also the document collections on our project page.

available under the *Data License Germany Attribution 2.0* or *Data License Germany Null Version 2.0* license. The corpus is available under http://textmining. wp.hs-hannover.de/juver.html. The sources used in this corpus are listed under Appendix A.2.

# 5 Method

## 5.1 Recurrent Sentences

In order to abstract away from these small variations, we cluster all sentences into clusters with very similar sentences. We use the minimum link (agglomerative) clustering algorithm and the Jaccard coefficient between the sets of character trigrams extracted from each sentence as a similarity measure. In the experiments described below, we require a minimal Jaccard index of 0.75 between two sentences in order to be considered for merging their clusters. Given the large number of sentences we cannot compute the similarity between each pair of sentences. Using a word index, for each sentence we retrieve for each sentence the first 100 sentences with the highest number of common words (excluding stop words). Only for these sentences we compute the trigram similarity.

In order to put sentences in which a number or date is changes into the same cluster, we remove stop words and replace numbers and dates by a general token.

As we will see below, in almost all cases the results are quite intuitive. There are, however, a few cases in which the result is not as we would like it to be. In the first place errors in sentence segmentation obviously cannot be undone by the clustering. In the second place, if there are too many (small) OCR-errors, the trigram overlap between two sentences is often large enough, but sometimes the number of common tokens is too small and the sentence pair is not considered as a candidate for computing the exact similarity.

## 5.2 Standardized Passages

Standardized legal formulations often go beyond the sentence level and can extend to long paragraphs. Not all frequent sequences of sentences that we find in our corpora are routine formulations. We also have cases where almost the same contract is concluded with various partners and only a few names and amounts and dates are changed in a standard contract. For the moment, we will not try to draw a line between standardized contracts and standardized passages as the distinction is not clear cut and we find all kind of intermediate cases.

**Frequent Sequences of Sentences.** We start our search for standardized passages by selecting all sequences that exceed a given minimum frequency. We start counting all pairs of two sentences and then proceed step wise to longer sequences. For efficiency we use the fact that a sequence $\langle s_1, s_2 \ldots s_n \rangle$ only can be frequent if both $\langle s_2 \ldots s_n \rangle$ and $\langle s_2 \ldots s_{n-1} \rangle$ are, very similar to the procedure followed in the Apriori algorithm for pattern detection in databases [1].

**Clustering of Sequences and Representation as Directed Graphs.** Once we collected all frequent patterns, there are many overlapping ones. Patterns can either be a proper subpattern of another pattern, or we have partially overlapping patterns.

We now consider each pattern as a directed graph in which each sentence is a vertex (node) and in which there is an edge from $s_1$ to $s_2$ if $s_2$ follows $s_1$ in the detected pattern. Now we add start and end nodes to each graph and cluster them. To do so, we need a similarity measure between graphs. Various similarity measures based on the number of common edges or common can be used. We decided to simply use the number of common edges as a similarity measure and to use again the minimum link (agglomerative) clustering algorithm. We do not use any stopping criterion and thus two graphs that share an edge are guaranteed to be in the same cluster. This has the advantage that, when analyzing the corpus, there is no ambiguity and there are no conflicting overlapping sequences in the corpus. Finally, we merge all graphs from a cluster into one new graph.

The graphs build in this way can be cyclic. For some purposes it is advantageous if we have acyclic graphs. For this purpose, we use a number of heuristics to remove edges. First, we find a number of cycles by searching the shortest path from each vertex to itself. Now we remove the edge that is on the largest number of cycles. If there is a tie we remove the edge going to the vertex with the highest in-degree. If there is still a tie, we remove the edge starting in the vertex with the highest out-degree. If there are still several possibilities we remove the edge with the lowest count (number of occurrences, see below). This process is repeated until the graph is acyclic. In the examples below, removed edges will be displayed in red.

# 6   Corpus Analysis

The graphs found are based on the patterns counted in the corpus. Here we used some minimum frequency. When merging the graphs, new possibilities arise to traverse the graph. Some of them might occur in the corpus but with a frequency below the initial threshold. E.g. consider the situation in which we have a threshold of 10 and sequences $\langle s_1 s_2 s_3 \rangle$ and $\langle s_2 s_3 s_4 \rangle$ that both occur 15 times. These sequences will be merged into one graph that has a path $\langle s_1 s_2 s_3 s_4 \rangle$ corresponding to a sequence that eventually occurs in the corpus bit less than 10 times. Thus, after building our graphs we count all instances of the possible patterns again and add weights to the edges of the graph, indicating how often each pair of two sentences connected by the edges occurs in the corpus.

Both corpora consist of a number of documents. We divide each document into sections, defined as a heading or a series of headings followed by normal text. For the case law corpus, we can extract the required information for this easily from the HTML structure. For the contract corpus, we rely on our classification of text elements (see [9]). For the case law corpus we use a threshold of 10 occurrences for a pattern to be considered. For the contract corpus we take 50 occurrences as a lower boundary.

**Table 2.** Sentences, patterns and graphs

|                           | Case law corpus | Contract corpus |
|---------------------------|-----------------|-----------------|
| Sentences                 | 308,832         | 751,281         |
| Unique sentences          | 197,811         | 448,288         |
| Sentence clusters         | 178,579         | 381,895         |
| Patterns                  | 161             | 1605            |
| Pattern graphs            | 94              | 227             |
| Pattern/graph occurrences | 7,569           | 26,954          |

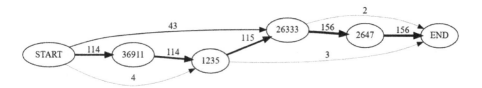

**Fig. 1.** Example of a typical linear sequence of sentences represented as a DAG with various options to start and end from the contract corpus. Sentences are represented by their IDs.

## 6.1  Statistics

In the case law corpus, we find 161 patterns that can be clustered into 94 graphs that have in total 7,569 instances in the corpus. For the contract corpus we find 227 graphs with 26,954 instances. Details are given in Table 2.

Let us have a closer look at the graphs. The majority of the graphs consists just of two nodes. In case law corpus 79 out of 94 graphs consist just of two sentences while the largest graph comprises 7 sentences. For the contract corpus 124 graphs have only two sentences. Here, the largest graph has 17 sentence nodes. In most graphs all sentences are on the longest path. This means that there are various options where to start and stop, but there are no options where one sentence or another sentence can be chosen. In most cases there seem to be optional parts at the start and the end of the standardized passage. In some cases, also the beginning or the continuation was not properly recognized, because of too much variation, OCR errors or segmentation errors. Figure 1 shows an example of such a "linear" graph. For the case law corpus only 3 graphs do not have such a linear structure. In the contract corpus 12 graphs have a nonlinear structure.

Since the clustering is based on the number of common edges, it is possible that a sentence is part of several graphs. For the case law corpus, we find 142 sentences that are part of at least one graph; 29 of them show up in at least two graphs, the most frequent one in 16 graphs. This most frequent sentence turns out to be the heading *Gründe* (Reasons). The most volatile real sentence is part of 8 graphs and reads *Gegen dieses Urteil wendet sich der Angeklagte mit seiner auf die Verletzung formellen und materiellen Rechts gestützten Revision.* (The defendant opposes this judgment with his appeal based on the violation of formal and substantive law).

In the contract corpus, we found 715 sentences that are part of a longer frequent sequence, 78 of which occur in at least two graphs. The most frequent one, however, is only a part of two graphs, and again is just a single word: *Einzelpreis [EUR]* (Unit price [EUR]). The sentence showing up in the largest number of different graphs, three to be precise, reads: *Die einzelnen Aufgaben und die Verteilung der Zuständigkeiten sind wie folgt geregelt:* (The individual tasks and the distribution of responsibilities are settled as follows:).

## 6.2   Examples of Sentence Clusters

We have discussed sentence clustering in detail elsewhere, see [10] (to appear) and briefly described in Sect. 5.1. Here we will just give an impression of the results.

It turns out that the sentence clustering is quite essential in the whole approach. If we do not cluster at all, we miss many interesting sequences, since several occurrences of a sentence have changed names, dates or other small variations. If too many similar sentences end up in a cluster, we find different continuations that in fact correspond to different sentences in the same cluster. We found that too low trigram similarity requirement in combination with single link clustering in some cases leads to long chains of sentences that are increasingly different.

A typical example of a cluster is given by the following four sentences:

– *Die Gefährlichkeitsprognose begegnet ebenfalls durchgreifenden rechtlichen Bedenken.*
– *Auch die Gefährlichkeitsprognose begegnet durchgreifenden rechtlichen Bedenken.*
– *7 c) Auch die Gefährlichkeitsprognose begegnet durchgreifenden rechtlichen Bedenken.*
– *11 c) Zuletzt begegnet auch die Gefährlichkeitsprognose durchgreifenden rechtlichen Bedenken.*
– Translation: "The danger prognosis also encounters sweeping legal concerns."

## 6.3   Examples and Analysis of the Found Passages

Since it is unclear what a standardized passage exactly is and what standardized passages occur in our corpora, it is impossible to give a numeric evaluation of our approach. Instead, let us have a look at the sequences that are found.

The results found are of course highly dependent on the minimum support required for each sequence. If we lower that requirement much more sequences would be found, but we would expect them to be less general.

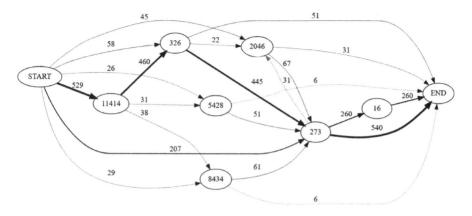

**Fig. 2.** Example of a complex graph from the case law corpus.

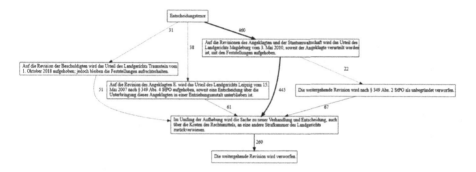

**Fig. 3.** Alternative representation of the graph from Fig. 2

**Case Law Corpus.** As mentioned above most graphs are very simple. Let us nevertheless have a look at one of the most complex graphs found, as it shows the potential of this type of representation. Figure 2 shows a graph with some sentences on the revision of a decision in its full complexity. If we remove the start and end node (end thus the information on possible subsequences) and also remove the cycle in the graph, the structures become quite clear and it is even possible to add the text to the nodes (Fig. 3). The structure now becomes quite clear. First we have a heading ('Decision Tenor') followed by three variants of the statement that the decision of a regional court is overturned on the appeal of the accused. In the first case, the findings remain, in the third variant, the findings are also overturned, the second variant refers to a more special case. Then a sentence about the further procedure and the costs follow. Finally, there are two variants of the sentence stating that a further revision is not possible that can appear at different positions in the passage.

If we take a closer look at the large number of simple graphs, we see many cases, that consist of a heading (like 'Decision' or 'Reasons') followed by a typical opening sentence for that section. Furthermore, we have a lot of sequences in which the first part gives the decision of the court, especially on an appeal, followed by one or two sentences on the court costs. A typical example is the following sequence:

1. *Die weitergehende Revision wird als unbegründet verworfen.* (The further revision is rejected as unfounded.)
2. *Im Umfang der Aufhebung wird die Sache zu neuer Verhandlung und Entscheidung, auch über die Kosten des Rechtsmittels, an eine andere Strafkammer des Landgerichts zurückverwiesen.* (To the extent of the annulment, the matter will be referred back to another criminal division of the regional court for a new hearing and decision, including the costs of the appeal.)

Two further frequent types of sequences are constituted by passages about the role and power of the court and by definitions of certain facts, e.g.:

1. *Bedingten Tötungsvorsatz hat, wer den Eintritt des Todes als mögliche Folge seines Handelns erkennt (Wissenselement) und billigend in Kauf nimmt (Willenselement).* (Those who recognize the occurrence of death as a possible consequence of their actions (element of knowledge) and approve of it (element of will) have a conditional killing intention. )
2. *Beide Elemente müssen durch tatsächliche Feststellungen belegt werden.* (Both elements must be substantiated by factual findings.)

**Contract Corpus.** In the contract corpus, we find a different situation. This is partially due to the bad PDF-quality of the downloaded contracts and the OCR errors made. Here we often find long sequences extending over a section heading. We only search sequences within sections, but sometimes the headings are not recognized.

As an example, consider the following sentence cluster. This cluster has 26 variants, but if we ignore the variants caused by OCR-errors, 2 versions remain:

1. *3.1 Infrastruktur Die Leistung des Auftragnehmers erfolgt ausschließlich auf unterstützten Plattformen, die durch Hersteller freigegeben sind.*
2. *Die Leistung des Auftragnehmers erfolgt ausschließlich auf unterstützten Plattformen, die durch Hersteller freigegeben sind.*
3. Translation: [3.1 Infrastructure] The service of the contractor is carried out exclusively on supported platforms that have been approved by the manufacturer.

All variants together are found 154 times in the corpus, 88 times preceded by the sentence *3.1 Infrastructure* and 52 time preceded by each time exactly the same sentence with system configuration. Similar as in this example variants in the graphs are often due to segmentation errors and so to say repaired by the clustering of the sequences.

The last example also shows a further characteristic of the contract corpus. This sentence does not look like a routine formulation but seems to be part of a long section (the above example was from a graph with 11 sentences and 3 headings) with general conditions for IT-systems that is copied or appended to many contracts.

Again, we also find many section headings followed by a typical first sentence, but now also we find many headings followed by a subheading.

In our corpus there are many contracts on IT-services. Since we only consider sequences occurring at least 50 times, we have a lot of very specific passages on the availability of systems, back-ups, etc.

Nevertheless, also typical general contract formulations show up:

1. *Mit diesem Vertrag wird eine etwaige Vorvereinbarung abgelöst.* (This contract replaces any preliminary agreement.)
2. *Rechte und Pflichten der Vertragsparteien bestimmen sich ab dem Zeitpunkt seines Wirksamwerdens ausschließlich nach diesem Vertrag.* (Rights and obligations of the contracting parties are exclusively determined by this contract from the time it becomes effective. )

**Comparison of the Corpora.** The contract corpus poses much more challenges to extract proper sequences, as the texts are more structured with headings, subheadings, tables, lists, appendices, etc. Moreover, the contracts are only available as a scanned PDF. This combination makes it hard to extract proper sequences of sentences.

If we compare the sequences found a general trend seems to be that in court decisions, we find many short sequences that can be seen as instances of what are called routine expressions in the literature discussed above. In the contracts we also find this type of formulations, but not as many. Here, the page-wise copying of terms and conditions from one contract to another seems to be the main source for recurring sentence sequences.

# 7   Conclusion and Future Work

Routine expressions consisting of several sentences have been observed and discussed in the literature. We are not aware of any previous attempt to detect such schematic text fragments automatically. One of the challenges is the great flexibility and many variations these formulations have. To overcome this, we comprise similar sentences in clusters and represent routine expressions as directed graphs of the sentence clusters. We have shown that we effectively can find many such longer formulations by a common pattern detection algorithm and subsequent clustering of the patterns found.

The present work is a first exploration of the topic of automatic detecting standardized formulations and can be extended in many directions. One of the following steps we want to do is to get more insight in the variations that are possible and the aspects of a formulation that have to be constant. This also can lead to a first application that readers can point to remarkable deviations from a standard text.

# A    Appendices

## A.1    Sources for *Case Law Corpus*

1. Bundesgerichtshof (BGH) – Decisions from criminal law:
   https://www.hrr-strafrecht.de/hrr/db/abfrage.php?type=erweitert&
   sortieren=relevanz&sortrichtung=ab&gericht=BGH&aktenzeichen=&
   datvon=&datbis=&volltext=&kurzbeschreibung=&norm=StGB&
   medium=-&verknuepfung=und&sz=2.

## A.2    Sources for *Contract Corpus*

1. Stadtverwaltung Hansestadt Hamburg – City administration of Hamburg:
   http://suche.transparenz.hamburg.de/dataset?q=vertrag&esq_title=&
   check_all_
2. Stadtverwaltung Bremen – City administration of Bremen:
   https://www.transparenz.bremen.de, Keyword: *Vertrag*
3. Cooperation contracts between universities and also between universities and
   service providers: We searched specifically for contract files on university web-
   sites and added them to *Contract corpus*.

# References

1. Agrawal, R., Srikant, R.: Fast algorithms for mining association rules in large
   databases. In: Proceedings of the 20th International Conference on Very Large
   Data Bases, VLDB 1994, pp. 487–499. Morgan Kaufmann Publishers Inc. (1994)
2. Burger, H., Dobrovol'skij, D., Kühn, P., Norrick, N.R.: Phraseologie: Objektbere-
   ich, Terminologie und Forschungsschwerpunkte. In: Burger, H., Dobrovol'skij, D.,
   Kühn, P., Norrick, N.R. (eds.) Phraseologie. Ein internationales Handbuch zeit-
   genössischer Forschung, pp. 1–10. Mouton de Gruyter, Berlin (2007)
3. Burgess, M., et al.: The legislative influence detector: finding text reuse in state
   legislation. In: Proceedings of the 22nd ACM SIGKDD International Conference
   on Knowledge Discovery and Data Mining - KDD 2016. pp. 57–66. ACM Press
   (2016). https://doi.org/10.1145/2939672.2939697
4. Busse, D.: Sprache und Recht, pp. 383–393. J.B. Metzler, Stuttgart (2018). https://
   doi.org/10.1007/978-3-476-04624-6_37
5. Clough, P., Gaizauskas, R., Piao, S.S.L., Wilks, Y.: METER: MEasuring TExt
   reuse. In: Proceedings of the 40th Annual Meeting of the Association for Computational Linguistics (2002). http://dx.doi.org/10.3115/1073083.1073110. Conference
   Name: ACL-02 Library Catalog: eprints.whiterose.ac.uk Meeting Name: ACL-02
   Pages: 152–159 Place: Philadelphia Publisher: ACL
6. Engberg, J.: Signalfunktion und Kodierungsgrad von sprachlichen Merk-
   malen in Gerichtsurteilen. HERMES J. Lang. Commun. Bus. 65–82 (1992).
   https://doi.org/10.7146/hjlcb.v5i9.21506
7. Engberg, J.: Does routine formulation change meaning? - The impact of genre on
   word semantics in the legal domain, pp. 31–48. De Gruyter Mouton (2000). https://
   www.degruyter.com/view/book/9783110826005/10.1515/9783110826005.31.xml

8. Filippova, K.: Multi-sentence compression: finding shortest paths in word graphs. In: Proceedings of the 23rd Int. Conference on Computational Linguistics, COLING 2010, pp. 322–330. Association for Computational Linguistics (2010)

9. Josi, F., Wartena, C.: Structural analysis of contract renewals. In: Proceedings of the ACM CIKM 2018 Workshops, Turin (2018)

10. Josi, F., Wartena, C., Ulrich, H.: Identifizierung von häufig vorkommenden Textabschnitten in juristischen Korpora. In: 56th Linguistics Colloquium, vol. 56. Peter Lang (2021, to appear)

11. Kjær, A.L.: On the structure of legal knowledge: the importance of knowing legal rules for understanding legal texts. In: Language, Text, and Knowledge. Mental Models of Expert Communication, pp. 127–161 (2000)

12. Kliche, F., Blessing, A., Heid, U., Sonntag, J.: The eIdentity text ExplorationWorkbench. In: Calzolari, N., et al. (eds.) Proceedings of the Ninth International Conference on Language Resources and Evaluation (LREC 2014). European Language Resources Association (ELRA) (2014)

13. Lindroos, E.: Dissertation: Im Namen des Gesetzes. Eine vergleichende rechtslinguistische Untersuchung zur Formelhaftigkeit in deutschen und finnischen Strafurteilen. Fachsprache **37**(3), 218–222 (2015). https://doi.org/10.24989/fs.v37i3-4.1293

14. Ma, D., Chen, C., Golshan, B., Tan, W.C.: Essentia: mining domain-specific paraphrases with word-alignment graphs. In: Proceedings of the Thirteenth Workshop on Graph-Based Methods for Natural Language Processing (TextGraphs-13), pp. 52–57. Association for Computational Linguistics (2019). https://doi.org/10.18653/v1/D19-5307

15. Płomińska, M.: Routine expressions in German legal texts - an attempt at typology. Colloquia Germanica Stetinensia **29**, 239–253 (2020). https://doi.org/10.18276/cgs.2020.29-13

16. Sailer, M.: Idiom and phraseology. In: Aronoff, M. (ed.) Oxford Bibliographies in Linguistics. Oxford University Press, New York (2013). https://doi.org/10.1093/obo/9780199772810-0137

17. Searle, J.R.: A taxonomy of illocutionary acts. Language, mind, and knowledge 07 (1975). http://conservancy.umn.edu/handle/11299/185220. Accepted 2017-03-16T18:32:14Z Publisher: University of Minnesota Press, Minneapolis

18. Sultan, M.A., Bethard, S., Sumner, T.: Back to basics for monolingual alignment: exploiting word similarity and contextual evidence. Trans. Assoc. Comput. Linguist. **2**, 219–230 (2014). https://doi.org/10.1162/tacl_a_00178

19. Wahl, A., Gries, S.T.: Computational extraction of formulaic sequences from corpora. Comput. Phraseol. **24**, 83 (2020)

20. Wise, M.J.: Neweyes: a system for comparing biological sequences using the running Karp-Rabin greedy string-tiling algorithm. In: Proceedings. International Conference on Intelligent Systems for Molecular Biology, vol. 3, pp. 393–401 (1995)

21. Woźniak, J.: Pragmatische Phraseologismen in ausgewählten Rechtstexten-ein Systematisierungsversuch. Lingwistyka Stosowana/Applied Linguistics/Angewandte Linguistik, pp. 149–162 (2017)

# Multivalent Graph Matching for Symbol Recognition

D. K. Ho[✉], J. Y. Ramel, and N. Monmarché

Laboratoire d'Informatique Fondamentale et Appliquée de Tours,
Université de Tours, Tours, France
{kieu.ho,jean-yves.ramel,nicolas.monmarche}@univ-tours.fr

**Abstract.** This paper deals with a symbol recognition problem where symbols are deformed in different ways. To address this difficulty, the multivalent graph matching problem is considered to allow more matching possibilities between the symbols. More precisely, a solution to the multivalent graph matching problem is formulated as an extended graph edit distance minimum with additional splitting and merging operations. This minimization problem is tackled with a variant of ant colony optimization, the max-min ant system. Besides, a neighborhood search strategy added to the solution building process of the max-min ant system aims to accelerate the computational time. The efficiency of the proposed approach is illustrated in a symbol dataset in several aspects, covering both quality and quantity analyses. The result shows the interest in using multivalent graph matching to deal with noisy symbols and their meaningful interpretability in the context of sub-part correspondence against other bijective graph matching-based approaches.

**Keywords:** Multivalent graph matching · Extended graph edit distance · Max-min ant system · Neighbor search · Symbol recognition · Classification

## 1 Introduction

Several applications in real life demand the determination of an explainable similarity measure between the objects rather than a numerical value. This is the case in biometric identification, symbol recognition, medical diagnostics or handwritten document analysis [5,6,11,13,20,23]. Among the existing approaches, the graph-based approach is promising because it provides the sub-part correspondence as well as the similarity measure. In particular, a first step called graph representation is done to transform the compared objects into the corresponding graphs based on the features and topologies. Then, the problem of defining the similarity of objects turns into the problem of matching graphs regarding the features of vertices and edges of the graphs. Within some domains such as 2D and 3D image analysis, biological and biomedical applications, we can find applications of graph-based approach [8,18,20,22]. Those strengthen the power of graph-based techniques in digital society nowadays.

© Springer Nature Switzerland AG 2021
E. H. Barney Smith and U. Pal (Eds.): ICDAR 2021 Workshops, LNCS 12917, pp. 488–503, 2021.
https://doi.org/10.1007/978-3-030-86159-9_35

A Graph Matching problem (GM) belongs to either exact GM or inexact GM. For exact GM, the matching between vertices and edges must respect both the attributes and the structure of the graph [7]. Consequently, it is well adapted to symbolic attributes and in a non-noisy context. Inexact GM is more flexible since it can violate the previous constraints about the attributes or the structure. In the family of inexact GM, Graph Edit Distance (GED) is a way to produce inexact GM. Solving the GED problem leads to minimize the distance or dissimilarity measure between two graphs through a series of edit operations. Regularly, these edit operations are substitution, deletion, insertion of vertices or edges, with a specific cost associated to each operation. Two main categories of methods, exact and heuristic, are proposed to tackle the GED. Unfortunately, the exact methods like A* [9] have expensive computation time. Meanwhile, the heuristic methods like A*-Beam Search [12] have lower computational time with acceptable solution quality. Furthermore, the exact methods seem to be infeasible for solving the GM problem with large graph size from more than 100 nodes in an acceptable time. Therefore, more and more heuristic approaches are developed for GED problem.

According to [19], the multivalent GM problem is more general than the GED because it permits one vertex in one graph matching with more than one correspondence in the other graph. In some cases, one feature in one object can correspond to multiple ones in the other object [1,4]. Furthermore, in pattern recognition problems, distortions of the graph can occur, and error-tolerant graph matching techniques should be used to allow node (edge) association even if they are not absolute similar [7]. All these matching situations are special cases of multivalent matching. Therefore, multivalent GM can be seen as the most general GM problem, and we will contribute to bring the solution to solve it in this work.

While reviewing the literature, we found some interesting works dealing with the multivalent GM problem. In [1], the authors apply the GED-based GM technique for recognizing diatoms. However, only merging and splitting nodes are completed in addition to the classic operations to accommodate the problem. In [3], the authors also suggest a GED-based approach with node merging to evaluate the similarity between images. Likewise in the previous works, the edge relations related to the fusion of nodes are still unclear. In contrast, in works [2,3], the authors specify the edge merging when it is necessary. However, the experimental results are not compared to other methods. In [4], a non-bijective GM approach is utilized for seeking the correspondence between an original image and its over-segmented ones. Due to the problem context, it is a bit confused between node/edge substitution and node/edge merging. In [16,17], the authors propose a general similarity measure for the multivalent GM and employ Ant Colony Optimization (ACO), reactive tabu search to tackle the problem. However, the measure, which is intended for graphs with symbolic attributes, restricts the scope of applications. In [10], the authors address the multivalent GM problem as an extension of GED called Extended Graph Edit Distance (ExGED). A formal formulation of ExGED based on the concept of a cost matrix is given, and they provide a way to calculate the costs of merging/splitting edges in extended cases. Nonetheless, the computational time is a drawback of the approach.

From the mentioned works, we see that the GED-based approach is popular for a multivalent GM problem. Following the direction of the work [10], we formalize the multivalent GM problem as an ExGED one and specifying the costs for edge merging and splitting induced by the edit operations done on nodes. These costs are integrated into a cost matrix that considers both semantic and topological features. We also apply the ACO-based methods, which is Max-Min Ant System (MMAS), to resolve the problem. But, we enhance the performance of MMAS by reducing its computational time. Particularly, a neighbor search strategy based on the previously selected nodes is applied. Hence, the number of candidates for the next steps will be shrunked and accelerating the solution construction process of MMAS. The effectiveness of the proposed approach is presented through numerical experiments for the symbol recognition problem. The results show a significant improvement in MMAS's execution time while reserving a good solution quality.

The paper is organized as follows. Firstly, a problem formulation from GED to ExGED is presented in Sect. 2. Secondly, the cost matrix construction for ExGED is detailed in Sect. 3. Thirdly, the MMAS algorithm and the strategy to reduce its computational time are specified in Sect. 4. Next, the efficiency of the proposed approach is shown in Sect. 5. Finally, the conclusion sums up the principal points and some suggestions in the future.

## 2    From GED to ExGED Problem Formulation

### 2.1    Graph Edit Distance

**Definition 1.** *An attributed graph contains 4-tuple $G = (V, E, \mu, \xi)$, where*

$V, E$ *are sets of vertices and edges, respectively,*
$l_V, l_E$ *are sets of vertex and edge labels, respectively*
$\mu : V \mapsto l_V$: *function that assigns labels to vertices*
$\xi : E \mapsto l_E$: *function that assigns labels to edges.*

**Definition 2.** *An edit path is a sequence of edit operations $(ed_i)$ to transform one graph to another graph, denoted $\lambda(G_1, G_2) = \{ed_i\}$. A valid edit path should follow these conditions: 1) deleting a vertex implies deleting its related edges; 2) inserting an edge is only permitted if the two vertices already exist; 3) inserting an edge must not create more than one edge between two vertices (selfloops) [7].*

**Definition 3.** *Given two graphs $G_1 = (V_1, E_1), G_2 = (V_2, E_2)$, the graph edit distance (GED) is a dissimilarity measure between $G_1$ and $G_2$ and is defined by:*

$$d_{min}(G_1, G_2) = \min_{\lambda \in \Theta(G_1, G_2)} \sum_{ed_i \in \lambda} c(ed_i), \tag{1}$$

*where $\Theta(G_1, G_2)$ is the set containing all valid edit paths $\lambda$ between $G_1$ and $G_2$, $c(ed_i)$ is the cost of each edit operation $ed_i$ [7].*

Classical edit operations are given in Table 1. The cost of each operation is defined according to either the node or edge labels.

## 2.2   Extended Graph Edit Distance Problem Formulation

Extended graph edit distance (ExGED) is also a dissimilarity measure derived from the costs of the edit operations. In addition to traditional edit operations in GED, the splitting and merging operations are supplemented to consider multivalent matching as follows.

**Definition 4.** *Given two attributed graphs* $G_1 = (V_1, E_1, \mu_1, \xi_1)$ *and* $G_2 = (V_2, E_2, \mu_2, \xi_2)$, *we define:*

- *merging the set* $S_{mer} = \{u_i \in V_1, i \geq 2\}$ *to* $v \in V_2$ *is noted* $S_{mer} \rightarrow v$
- *splitting* $u \in V_1$ *into the set* $S_{spl} = \{v_j \in V_2, j \geq 2\}$ *is noted* $u \rightarrow S_{spl}$

These two operators are also mentioned in Table 1. Usually, doing node merging and splitting can lead to edge splitting and merging, this will be discussed more precisely when the cost of each operation will be detailed.

**Table 1.** Availability of edit operations for GED and ExGED (with $u \in V_1, v \in V_2, e_1 \in E_1, e_2 \in E_2, \varepsilon$ the *virtual* vertex or edge). $S_{mer}$ and $S_{spl}$ are subsets of $V_1$ and $V_2$ defined in Definition 4.

| Operation | Notation | Cost function notation | GED | ExGED |
|---|---|---|---|---|
| Vertex substitution | $u \rightarrow v$ | $c(u \rightarrow v)$ | ✓ | ✓ |
| Vertex deletion | $u \rightarrow \varepsilon$ | $c(u \rightarrow \varepsilon)$ | ✓ | ✓ |
| Vertex insertion | $\varepsilon \rightarrow v$ | $c(\varepsilon \rightarrow v)$ | ✓ | ✓ |
| Edge substitution | $e_1 \rightarrow e_2$ | $c(e_1 \rightarrow e_2)$ | ✓ | ✓ |
| Edge deletion | $e_1 \rightarrow \varepsilon$ | $c(e_1 \rightarrow \varepsilon)$ | ✓ | ✓ |
| Edge insertion | $\varepsilon \rightarrow e_2$ | $c(\varepsilon \rightarrow e_2)$ | ✓ | ✓ |
| Vertex merging | $S_{mer} \rightarrow v$ | $c(S_{mer} \rightarrow v)$ | | ✓ |
| Vertex splitting | $u \rightarrow S_{spl}$ | $c(u \rightarrow S_{spl})$ | | ✓ |

## 3   Cost Matrices for ExGED

### 3.1   Definition of the Cost Matrix for Node Operations

Inspired by the idea in [15], a cost matrix for ExGED is also built with five types of edit operations. These operations are organized in separate blocks along with their corresponding costs. Formally, given two attributed graphs $G_1, G_2$ as above, we denote that $n = |V_1|, m = |V_2|$. $P_{mer}^k = \{S_{mer}^i\}$ is a set of all possibilities for merging of nodes in $G_1$ and $h = |P_{mer}^k|$. $P_{spl}^k = \{S_{spl}^j\}$ is a set of all possibilities for splitting of nodes in $G_2$ and $l = |P_{spl}^k|$. $k$ is a parameter that describes the maximum size of sets $S_{mer}^i$ and $S_{spl}^j$, $k \geq 2$. The cost matrix is demonstrated as below.

$$
\mathbf{C} =
\begin{array}{c}
\begin{array}{cccc|c|ccc}
1 & \cdots & m & & \varepsilon & 1 & \cdots & l \\
\end{array} \\
\left(
\begin{array}{ccc|c|ccc}
c_{1,1} & \cdots & c_{1,m} & c_{1,\varepsilon} & c_{1,S_{spl}^1} & \cdots & c_{1,S_{spl}^l} \\
\vdots & \ddots & \vdots & \vdots & \vdots & \ddots & \vdots \\
c_{n,1} & \cdots & c_{n,m} & c_{n,\varepsilon} & c_{n,S_{spl}^1} & \cdots & c_{n,S_{spl}^l} \\
\hline
c_{\varepsilon,1} & \cdots & c_{\varepsilon,m} & 0 & \infty & \cdots & \infty \\
\hline
c_{S_{mer}^1,1} & \cdots & c_{S_{mer}^1,m} & \infty & \infty & \cdots & \infty \\
\vdots & \ddots & \vdots & \vdots & \vdots & \ddots & \vdots \\
c_{S_{mer}^h,1} & \cdots & c_{S_{mer}^h,m} & \infty & \infty & \cdots & \infty
\end{array}
\right)
\begin{array}{c}
1 \\ \vdots \\ n \\ \varepsilon \\ 1 \\ \vdots \\ h
\end{array}
\end{array}
$$

where $c_{i,j}$ denotes the cost of a node substitution (with $(i,j) \in \{1 \ldots n\} \times \{1 \ldots m\}$), $c_{i,\varepsilon}$ denotes the cost of a node deletion, $c_{\varepsilon,j}$ denotes the cost of a node insertion, $c_{i,S_{spl}}$ denotes the cost of a node splitting and $c_{S_{mer},j}$ denotes the cost of a node merging.

The cost matrix $\mathbf{C}$ is not a square matrix as in GED case because of dimension reductions on deletion block $((n \times n) \to (n \times 1))$ and insertion one $((m \times m) \to (1 \times m))$. This dimensional reduction aims to save more memory but still preserving the property of a GED cost matrix. Moreover, by introducing splitting and merging operations, the size of matrix $\mathbf{C}$ increases with $h$ rows and $l$ columns. $h$ and $l$ are strongly influenced by $k$. The bigger $k$ is the higher values of $h, l$ get. Consequently, the size of $\mathbf{C}$ will grow up significantly. Thus, choosing the number of $k$ would be very important, especially in a big graph. Regularly, the parameter $k$ is problem-dependent and based on expert knowledge.

### 3.2   Definition of the Costs for Edge Operations in Extended Case

To keep the computational cost reasonable, we propose to integrate the estimated costs of edge operations involved in each node operation inside the previous matrix $\mathbf{C}$. Consequently, we need to estimate a cost for the edge operations in the extended case. The detail is given below.

- For node substitution $u_i \to v_j$, two sets of incident edges of $u_i$ and $v_j$ are computed, called $E_{u_i}$ and $E_{v_j}$, respectively. Then, a square edge cost matrix $\mathbf{Ce}$ is built from $E_{u_i}$ and $E_{v_j}$ based on the cost functions of edge operations in Table 1. The size of $\mathbf{Ce}$ is $(|E_{u_i}| + |E_{v_j}|) \times (|E_{u_i}| + |E_{v_j}|)$. Finally, the Munkres's algorithm is applied on $\mathbf{Ce}$ to find the minimum sum of edge operation costs [15], or $c_{i,j} \leftarrow c(u_i \to v_j) + \text{Munkres}(\mathbf{Ce})$.
- For node deletion, deleting a node $u_i$ will remove all its adjacent edges, or $c_{i,\varepsilon} \leftarrow c(u_i \to \varepsilon) + \sum_{e \in E_{u_i}} c(e \to \varepsilon)$.
- For node insertion, inserting a node $v_j$ will insert all its adjacent edges, or $c_{\varepsilon,j} \leftarrow c(\varepsilon \to v_j) + \sum_{e \in E_{v_j}} c(\varepsilon \to e)$.
- For node merging $S_{mer} \to v$, two sets of incident edges to nodes in $S_{mer}$ and $v$ are computed first, denoted $E_{S_{mer}}$ and $E_v$, respectively. Let $E_{loop}$ be the set of edges connecting the nodes $u_i \in S_{mer}$, we introduce $E'_{S_{mer}} = E_{S_{mer}} \setminus E_{loop}$.

Then, an edge cost matrix $\mathbf{Ce}$ for two sets $E'_{S_{mer}}$ and $E_v$ is built similarly as for node substitution. The Munkres's algorithm is also used to find the minimum cost for $E'_{S_{mer}}$ and $E_v$. Finally, the total cost for node merging is $c_{S_{mer},v} \leftarrow c(S_{mer} \rightarrow v) + \text{Munkres}(\mathbf{Ce}) + \sum_{e \in E_{loop}} c(e \rightarrow \varepsilon)$.

– For node splitting $u_i \rightarrow S_{spl}$, the computational steps of edge cost are similar to node merging: $c_{u,S_{spl}} \leftarrow c(u \rightarrow S_{spl}) + \text{Munkres}(\mathbf{Ce}) + \sum_{e \in E_{loop}} c(\varepsilon \rightarrow e)$.

### 3.3   Illustrative Example

Here we give an example of the cost computation both for node and edge operations. Let $G_1$ and $G_2$ be 2 unlabelled graphs as in Fig. 1, and each node can be merged or split to maximum 2 other nodes ($k = 2$). Suppose that nodes with an edge between them can be considered as candidates for merging/splitting. Other words, $P^k_{mer} = \{ab, bc\}$, and $P^k_{spl} = \{AB, AC, BC, CD\}$ are sets of merging and splitting nodes in $G_1$ and $G_2$, respectively. We have $n = |V_1| = 3$ and $m = |V_2| = 4$. All cost functions are defined as in Table 2.

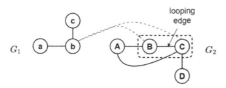

**Fig. 1.** The example graph with a partial matching $\lambda = \{b \rightarrow B, C\}$.

**Table 2.** Cost functions for edit operations in the example.

| Operation | Node operation cost | Edge operation cost |
|---|---|---|
| Substitution | $c(u \rightarrow v) = 1$ | $c(e_1 \rightarrow e_2) = 1$ |
| Insertion | $c(\epsilon \rightarrow v) = 2$ | $c(\epsilon \rightarrow e_2) = 2$ |
| Deletion | $c(u \rightarrow \epsilon) = 10$ | $c(e_1 \rightarrow \epsilon) = 10$ |
| Merging | $c(S_{mer} \rightarrow v) = 10$ | $c(e \rightarrow \epsilon) = 10, e \in E_{loop}$ |
| Spliting | $c(u \rightarrow S_{spl}) = 1$ | $c(\epsilon \rightarrow e) = 0, e \in E_{loop}$ |

With the partial matching $\lambda = \{b \rightarrow B, C\}$ in Fig. 1, we need to compute the sets of incidents edges to the involved nodes as follows.

– Incident edges to $b$: $E_b = \{(a, b), (b, c)\}$
– Looping edges: $E_{loop} = \{(B, C)\}$
– Incident edges to $B, C$: $E_{BC} = \{(A, B), (A, C), (B, C), (C, D)\}$
– Incident edges to $B, C$, except $E_{loop}$: $E'_{BC} = \{(A, B), (A, C), (C, D)\}$

Then, the edge cost matrix $\mathbf{Ce}$ for $E_b$ and $E_{BC}$ is computed in Eq. (3.3). The total cost for the partial matching $\lambda = \{b \to B, C\}$ is:

$$c_{b,BC} = c(b \to B, C) + \text{Munkres}(\mathbf{Ce}) + c(\varepsilon \to BC) = 1 + 4 + 0 = 5.$$

$$\mathbf{Ce} = \begin{array}{c} ab \\ bc \\ \varepsilon_{AB} \\ \varepsilon_{AC} \\ \varepsilon_{CD} \end{array} \begin{pmatrix} \begin{array}{ccc|cc} AB & AC & CD & \varepsilon_{ab} & \varepsilon_{bc} \\ c_{ab,AB} & c_{ab,AC} & c_{ab,CD} & c_{ab,\varepsilon} & \infty \\ c_{bc,AB} & c_{bc,AC} & c_{bc,CD} & \infty & c_{bc,\varepsilon} \\ \hline c_{\varepsilon,AB} & \infty & \infty & 0 & 0 \\ \infty & c_{\varepsilon,AC} & \infty & 0 & 0 \\ \infty & \infty & c_{\varepsilon,CD} & 0 & 0 \end{array} \end{pmatrix} = \begin{pmatrix} \begin{array}{ccc|cc} AB & AC & CD & \varepsilon_{ab} & \varepsilon_{bc} \\ 1 & 1 & 1 & 10 & \infty \\ 1 & 1 & 1 & \infty & 10 \\ \hline 2 & \infty & \infty & 0 & 0 \\ \infty & 2 & \infty & 0 & 0 \\ \infty & \infty & 2 & 0 & 0 \end{array} \end{pmatrix}$$

$$(2)$$

## 4   MMAS for ExGED

### 4.1   Algorithmic Scheme

In this work, we use the Max-Min Ant System (MMAS) which is a variant of Ant Colony Optimization (ACO) [21]. An ant colony is employed to generate solutions to the considered problem in a parallel manner. At each iteration, each ant constructs a complete matching based on the transition probabilities. A construction graph denoted $G_{ants}$, is utilized to know the possibilities for one ant. In other words, $G_{ants}$ is the search space that shows all the possibilities of node matching between 2 graphs. In this work, $G_{ants}$ is considered as a complete undirected graph where each node in $G_{ants}$ represents an edit operation between nodes in $G_1$ and $G_2$. Edges in $G_{ants}$ aim to build incrementally the best edit path. Pheromones of natural ants here correspond to probabilities that are shared among ants to build a solution. An initial pheromone value is laid on each vertex of $G_{ants}$ and the pheromone value changes according to the performance of the best matching found and the pheromone update schemes.

### 4.2   Construction Graph

A construction graph $G_{ants}$ provides all possible node matching (included extended case) between two graphs $G_1$ and $G_2$. In other word, $G_{ants}$ is built to explore all the edit paths for the ExGED problem defined in the previous section. Figure 2 presents the construction graph built from the graphs Fig. 1. The candidates for each edit operation is the Cartesian products of node sets as in Table 3. Remark that when the possibility of an ant to move on $G_{ants}$ depends on the current partial edit path. For instance, if the partial edit path is $\lambda = \{b \to B, C\}$, all candidates contain one of these nodes are pruned (gray nodes). That means the current ant cannot move on these positions until the edit path is complete.

**Table 3.** Candidates for each edit operation in the construction graph, where $V_1, V_2$ are sets of nodes in $G_1, G_2$, respectively, $\varepsilon$ is a virtual node, $P^k_{mer}, P^k_{spl}$ are sets of merging and splitting nodes, respectively.

| Operation | Substitution | Deletion | Insertion | Merging | Splitting |
|---|---|---|---|---|---|
| Candidates | $V_1 \times V_2$ | $V_1 \times \varepsilon$ | $\varepsilon \times V_2$ | $P^k_{mer} \times V_2$ | $V_1 \times P^k_{spl}$ |

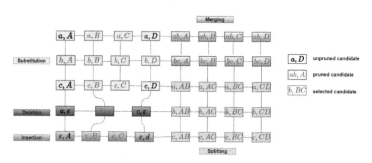

**Fig. 2.** Construction graph representing all possible node matching of two graphs in Fig. 1. Some edges are not presented for readability but all nodes are fully connected.

### 4.3 Construction of a Complete Matching and Neighborhood Search Strategy

At each iteration, every ant builds a complete edit path. It starts with an empty edit path $\lambda = \{\}$, then it adds a new candidate $v_i \in G_{ants}$ to $\lambda$ based on the transition probability, until $\lambda$ contains all nodes in both graphs. The transition probability is computed based on two factors: the heuristic and the pheromone. The heuristic is derived from the cost matrix $\mathbf{C}$ and the pheromone is laid on the construction graph $G_{ants}$. Let $\tau_{v_i}, \eta_{v_i}$ be respectively the pheromone and heuristic values of the candidate $v_i$, then the transition probability of $v_i$ is:

$$Pr_{v_i} = \frac{[\tau_{v_i}]^\alpha \times [\eta_{v_i}]^\beta}{\sum_{j=1}^{|cand|} [\tau_{v_j}]^\alpha \times [\eta_{v_j}]^\beta} \tag{3}$$

Besides, in this work, we enhance the method in [10] by considering a neighbor search strategy to reduce the computational time of MMAS. That means from a previously selected node in $G_1$, we only focus on its neighbors to find the next candidates. If no neighbor is found, a random choice is made. Due to this local strategy search, it may miss good solutions but it is less time-consuming. Admit that in the case of a complete graph, this strategy is useless because all nodes have the same number of neighbors. A demonstration for the improvement is given in Fig. 3.

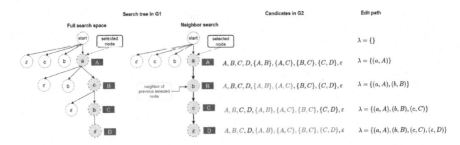

**Fig. 3.** Difference of search tree in [10] and the one with neighborhood search strategy in our work. The gray nodes indicate pruned nodes in $G_2$ according to the partial solution built.

## 4.4   Pheromone Update

The pheromone updating process happens on each node in $G_{ants}$ after all ants have built their solution at the current iteration, including reinforcement and evaporation. The reinforcement is done only on the best solution of the current iteration ($\lambda_{itbest}$). The evaporation is done for all nodes according to the evaporation rate $\rho \in [0, 1]$. Then, the pheromone values are constrained in the interval $[\tau_{min}, \tau_{max}]$ [21].

$$\tau_{v_i} = (1 - \rho) * \tau_{v_i} + \Delta_{v_i}, \quad \Delta_{v_i} = \begin{cases} \frac{1}{1+c(\lambda_{itbest})} & \text{if} \quad v_i \in \lambda_{itbest} \\ 0 & \text{otherwise} \end{cases} \tag{4}$$

## 5   Experiments

Here the MMAS algorithm is used to seek the best matching between two graphs. This is because the search space is huge and the underlying combinatorial problem can not be solved exactly in a reasonable time [2]. However, one needs to tune the parameters of MMAS to be well adapted to the problem. All experiments are implemented in Python 3.8.5 and run on Windows 10 Intel(R) Core(TM) i7-8750 CPU @ 2.20 GHz, RAM of 16.0 Go. From now on, the proposed method in this work is called im-MMAS-ExGED.

### 5.1   Data Set and Graph Representation

For the experiments, we use the models in the SESYD dataset which contains architectural and electrical symbols[1]. The deformed symbols are created by adding noises to break one line into several lines or rotating or scaling the symbols. This new dataset representing noisy symbols is publicly available as in following link.

---

[1] Can be accessed here: http://mathieu.delalandre.free.fr/projects/sesyd/.

For the graph representation, nodes are lines and edges are relations between lines. Node's attributes are the relative length ($l$) which is normalized according to the longest line in each graph. Edge's attributes are the normalized relative angle ($\theta \in [0,1]$) and type of relation ($rel$). The relation $rel$ includes T-Junction ($T$), Parallel ($P$), Successive ($S$), L-Junction ($L$) and intersection (X) [14]. We give an example in Fig. 4.

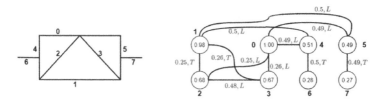

**Fig. 4.** The symbol (left) and its graph representation (right) where numerical attributes of nodes (blue) and edges (red) are normalized. (Color figure online)

From the observation, we suppose that the $S$-relation appears when there are noises. So, we set nodes that are connected by edges with $S$ relations will be candidates for merging or splitting. Also, $k = 3$ is set according to the dataset (Fig. 5).

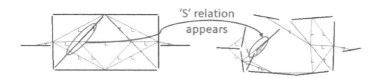

**Fig. 5.** The $S$ relation appears when lines in the model (left) are split into several lines in the deformed symbol (right).

## 5.2   Definition of Cost Functions for Edit Operations

Normally, the proposal of cost functions is based on the graph knowledge, such as the node/edge's attributes and the graph's topology. Therefore, we define the cost functions for the node and edge operations vis-à -vis the previous graph representation. For node operations, the cost is proportional to the length attribute ($l$). That means deleting/inserting a long line should cost more than a shorter one. Substitution of two lines of equivalent length should be zero. Likewise, merging lines will be cheaper if their merging generates a line of adequate length. For edge operations, the cost depends on the relation type ($rel$) and the relative angle ($\theta$). The values of edge cost functions are also in $[0, 1]$ to be compatible with the node cost functions. Details of the cost functions are in Table 4.

**Table 4.** Cost functions of edit operations for the data set.

| Node operation cost | Edge operation cost |
|---|---|
| $c(u \rightarrow v) = \|l_u - l_v\|$ | $c(e_1 \rightarrow e_2) = \begin{cases} \|\theta_{e_1} - \theta_{e_2}\| & \text{if } rel_{e_1} = rel_{e_2} \\ 1 & \text{otherwise} \end{cases}$ |
| $c(\varepsilon \rightarrow v) = l_v$ | $c(\varepsilon \rightarrow e_2) = 1$ |
| $c(u \rightarrow \varepsilon) = l_u$ | $c(e_1 \rightarrow \varepsilon) = 1$ |
| $c(S_{mer} \rightarrow v) = \|\sum_{u_i \in S_{mer}} l_{u_i} - l_v\|$ | $c(e \rightarrow \varepsilon) = 0, e \in E_{loop}$ |
| $c(u \rightarrow S_{spl}) = \|l_u - \sum_{v_i \in S_{spl}} l_{v_i}\|$ | $c(\varepsilon \rightarrow e) = 0, e \in E_{loop}$ |

### 5.3 Parameter Setting for MMAS

In this part, we study the impacts of principal parameters of MMAS and local search on the results of GM problems. Each parameter study is run 30 times and the average results are presented. Figure 6 shows the influence of parameters $\alpha, \beta, \rho$ to the final costs. Based on the results, we chose $\alpha = 1, \beta = 3, \rho = 0.01$ and 3-opt local search for later experiments. For other parameters, we also do the experiments but they do not affect the results. We set $nb_{ants} = 5, [\tau_{min}, \tau_{max}] = [0.1, 2.0]$ for later experiments.

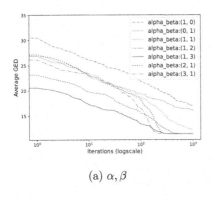

(a) $\alpha, \beta$

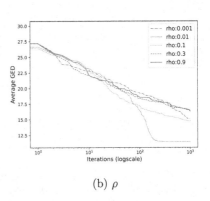

(b) $\rho$

**Fig. 6.** Influences of principal parameters to the convergence of MMAS to ExGED (with 5 ants, 300 iterations, $[\tau_{min}, \tau_{max}] = [0.1, 2.0]$).

### 5.4 Matching Quality Analysis: Interest of Using Merging and Splitting

In this section, we evaluate the matching quality regarding the sub-part correspondence or node-to-node matching. The result of the approach is compared to one of the bipartite methods for GED [15] (denoted BP-GED) and the one in [10] (denoted old-ExGED-MMAS). We regulate a matching is reasonable if it matches correctly the split lines in the noisy symbols with its origin line in the model. The node correspondence is checked manually. We take 5 models and 6 levels of distortions for each model for this experiment.

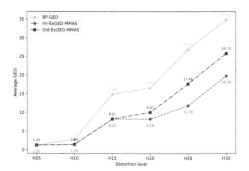

**Fig. 7.** Average costs of BP-GED, im-MMAS-ExGED and old-MMAS-ExGED in terms of distortion levels.

Figure 7 shows the average cost given by three methods in terms of distortion levels. Note that the cost is proportional to the distortion level in general, higher cost for higher distortion. In all cases, the BP-GED obtains a higher dissimilarity than the ExGED-based two methods. It confirms the appropriateness of the ExGED for this kind of distortion against the GED. Looking into the sub-part correspondence, the cost differences of methods are proportional with the achieved mappings. Specifically, ExGED-based methods can find more appropriate matching than BP-GED thanks to merging and splitting. A demonstration is given in Fig. 8 and Table 5. Also, the results given by the im-MMAS-ExGED are pretty competitive with the old-MMAS-ExGED. Their costs are not very bias but the im-MMAS-ExGED is much more time-saving than the old-MMAS-ExGED. For instance, the old-MMAS-ExGED needs 705.048 s to produce the result, meanwhile the im-MMAS-ExGED only needs 149.425 s with $maxiter = 500$ in this experiment. This consolidates that the proposed method is less time-consuming but still gives a good result compared with the old-MMAS-ExGED.

**Table 5.** Node correspondence at distorted positions between the model 032 and its noisy versions in Fig. 8 given by BP-GED and ExGED-based methods.

| Methods | Level 5 | Level 10 | Level 15 |
|---|---|---|---|
| BP-GED | $\varepsilon \to 9$ | $1 \to 5; \varepsilon \to 9$ | $2 \to 14; 3 \to 17; 0 \to 16; 6 \to 18$ |
| ExGED-based | $\varepsilon \to 8$ | $1 \to \{5,9\}$ | $2 \to \{14,17\}; 4 \to \{16,18\}$ |

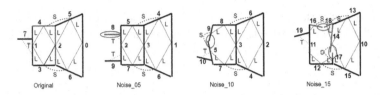

**Fig. 8.** Drawing of nodes (in bold) and edges (dashed lines) on the symbol 032 and its distortion levels. Ellipse indicates noise position, same noise position illustrated with the same color. (Color figure online)

## 5.5   Classification Problem

In the above section, we point out the interest of using merging and splitting for well recognizing the noisy symbols in terms of matching quality analysis for a small dataset. In this part, we try to indicate the efficiency of the proposed method for a bigger dataset in a classification problem. To serve this purpose, we design the experiment as follows. We select randomly 20 models in the symbol dataset as the training set (Fig. 9) and around 20 noise levels of each model as the testing set. As a consequence, the size of the training set ($\#train = 20$) is much smaller than the testing set ($\#test = 418$). We expect that the proposed method can get a good classification rate with such small training information. We use one-nearest neighbor classifier or KNN with $K = 1$ to do this task. So, the cost induced by GED-based or ExGED-based methods are utilized directly for the classification. The performance of the im-ExGED-MMAS is measured against these of BP-GED [15], Greedy algorithm for ExGED (Greedy-ExGED), old-ExGED-MMAS [10], and GED-MMAS. The parameter setting for this experiment is described in Table 6.

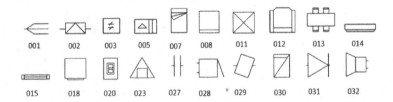

**Fig. 9.** Models for classification problem in the experiment.

**Table 6.** Parameter setting for symbol recognition problem.

| Parameter | $(\alpha, \beta)$ | $maxiter$ | $nb_{ants}$ | $\rho$ | $(\tau_{min}, \tau_{max})$ | $k$ |
|---|---|---|---|---|---|---|
| Value | (1,3) | 50,100 | 5 | 0.01 | (0.01,2.0) | 2,3 |

The results are shown in Table 7. "Accuracy" is percentage of the number of testing symbols correctly classified. $\#matchings$ is the total number of matchings done for each method. For BP-GED and Greedy-ExGED, $\#matchings = \#test \times \#train$. For the remaining methods, $\#matchings = \#test \times \#train \times nb_{ants} \times maxiter$. "Time/Matching(s)" is the ration between the total computational time and the total number of matchings. Through Table 7, we observe the following main points. Firstly, the MMAS-based approaches give a better accuracy but more time-consuming than the other approaches (BP-GED, Greedy-ExGED). It is natural because the MMAS needs to repeat the searching process several times to improve its solution quality. Nevertheless, examining the average time for constructing a matching, the MMAS-based approaches are more time-saving. Secondly, between GED-based and ExGED-based approaches, the ExGED-based ones are usually more robust with a higher classification rate, especially the im-ExGED-MMAS. Thirdly, better results are obtained with $k = 3$ rather than $k = 2$ since it provides extra matching possibilities. In contrast, it demands more computational time due to the size increase of the cost matrix. Finally, the im-ExGED-MMAS proves its effectiveness against the old-ExGED-MMAS, which supports our initial expectation of balancing between solution quality and computational time.

**Table 7.** Symbol classification results given by GED-based and ExGED-based approaches, the best result of each method is in bold.

| Solver | $(k, maxiter)$ | Accuracy (%) | Time/matching (s) | Total time (s) | $\#matchings$ |
|---|---|---|---|---|---|
| BP-GED | | **83.014** | 0.008 | 66.621 | 8360 |
| Greedy-ExGED | (2,1) | 74.402 | 0.120 | 96.347 | 8360 |
| | (3,1) | **74.880** | 0.012 | 100.561 | 8360 |
| Im-ExGED-MMAS | (2,50) | 84.450 | 0.002 | 4130.574 | 2090000 |
| | (3,50) | 85.885 | 0.002 | 4326.690 | 2090000 |
| | (2,100) | 84.211 | 0.002 | 8401.295 | 4180000 |
| | (3,100) | **86.124** | 0.002 | 8267.924 | 4180000 |
| Old-ExGED-MMAS | (2,50) | 84.211 | 0.005 | 9996.545 | 2090000 |
| | (3,50) | 85.167 | 0.005 | 10271.611 | 2090000 |
| | (2,100) | 83.732 | 0.005 | 20481.097 | 4180000 |
| | (3,100) | **85.407** | 0.005 | 20754.161 | 4180000 |
| GED-MMAS | (_,50) | **83.253** | 0.004 | 9451.112 | 2090000 |
| | (_,100) | 83.253 | 0.004 | 18683.444 | 4180000 |

# 6   Conclusion

This paper proposes an enhancement made to the performance of the proposed method MMAS-ExGED in [10] for the multivalent GM problem. More precisely, a neighborhood searching strategy is applied during the solution building process of MMAS to reduce the computational time. In other words, this strategy focuses only on the neighbor of the previously selected nodes to seek the next candidates. The efficiency of the proposal is shown in an application of symbol recognition at various levels of distortions. The numerical results demonstrate both qualitative (quality and interpretability of the generated matchings) and quantitative (classification performance) aspects. It confirms the advantage of using splitting and merging for symbols at multiple noise levels rather than the traditional GED-based methods. Also, it consolidates our hypothesis that the enhancement can keep the balance between the solution quality and the execution time. These optimist results encourage us to apply the proposed method to other datasets in the upcoming time, for instance, brain connectome comparison at different scales.

# References

1. Ambauen, R., Fischer, S., Bunke, H.: Graph edit distance with node splitting and merging, and its application to diatom identification. In: GbRPR, pp. 95–106 (2003)
2. Berretti, S., Bimbo, A.D., Pala, P.: Graph edit distance for active graph matching in content based retrieval applications. Open Artif. Intell. J. **1**(1) (2007)
3. Berretti, S., Del Bimbo, A., Pala, P.: A graph edit distance based on node merging. In: ICIVR, pp. 464–472 (2004)
4. Boeres, M.C., Ribeiro, C.C., Bloch, I.: A randomized heuristic for scene recognition by graph matching. In: International Workshop on Experimental and Efficient Algorithms, pp. 100–113 (2004)
5. Bunke, H., Riesen, K.: Recent advances in graph-based pattern recognition with applications in document analysis. Pattern Recogn. **44**(5), 1057–1067 (2011)
6. Conte, D., Foggia, P., Sansone, C., Vento, M.: Graph matching applications in pattern recognition and image processing. In: IEEE ICIP, vol. 2, pp. II-21 (2003)
7. Conte, D., Foggia, P., Sansone, C., Vento, M.: Thirty years of graph matching in pattern recognition. IJPRAI **18**(03), 265–298 (2004)
8. Fischer, S., Gilomen, K., Bunke, H.: Identification of diatoms by grid graph matching. In: S+SSPR, pp. 94–103 (2002)
9. Hart, P.E., Nilsson, N.J., Raphael, B.: A formal basis for the heuristic determination of minimum cost paths. IEEE SMC **4**(2), 100–107 (1968)
10. Ho, K.D., Ramel, J., Monmarché, N.: Multivalent graph matching problem solved by max-min ant system. In: S+SSPR, pp. 227–237 (2020)
11. Lades, M., et al.: Distortion invariant object recognition in the dynamic link architecture. IEEE Trans. Comput. **42**(3), 300–311 (1993)
12. Neuhaus, M., Riesen, K., Bunke, H.: Fast suboptimal algorithms for the computation of graph edit distance. In: S+SSPR, pp. 163–172 (2006)
13. Noma, A., Pardo, A., Cesar, R.M., Jr.: Structural matching of 2D electrophoresis gels using deformed graphs. Pattern Recogn. Lett. **32**(1), 3–11 (2011)
14. Qureshi, R.J., Ramel, J.Y., Cardot, H., Mukherji, P.: Combination of symbolic and statistical features for symbols recognition. In: ICSPCN, pp. 477–482 (2007)
15. Riesen, K., Bunke, H.: Approximate graph edit distance computation by means of bipartite graph matching. Image Vis. Comput. **27**(7), 950–959 (2009)
16. Sammoud, O., Solnon, C., Ghédira, K.: Ant algorithm for the graph matching problem. In: EvoCOP, pp. 213–223 (2005)
17. Sammoud, O., Sorlin, S., Solnon, C., Ghédira, K.: A comparative study of ant colony optimization and reactive search for graph matching problems. In: EvoCOP, pp. 234–246 (2006)

18. Shokoufandeh, A., Dickinson, S.: A unified framework for indexing and matching hierarchical shape structures. In: IWVF, pp. 67–84 (2001)

19. Sorlin, S., Solnon, C., Jolion, J.M.: A generic graph distance measure based on multivalent matchings. In: Kandel, A., Bunke, H., Last, M. (eds.) Applied Graph Theory in Computer Vision and Pattern Recognition, pp. 151–181. Springer, Heidelberg (2007). https://doi.org/10.1007/978-3-540-68020-8_6

20. Stauffer, M., Fischer, A., Riesen, K.: Filters for graph-based keyword spotting in historical handwritten documents. Pattern Recogn. Lett. **134**, 125–134 (2020)

21. Stützle, T., Hoos, H.H.: Max-min ant system. Future Gener. Comput. Syst. **16**(8), 889–914 (2000)

22. Tefas, A., Kotropoulos, C., Pitas, I.: Using support vector machines to enhance the performance of elastic graph matching for frontal face authentication. IEEE PAMI **23**(7), 735–746 (2001)

23. Zaslavskiy, M., Bach, F., Vert, J.P.: A path following algorithm for the graph matching problem. IEEE PAMI **31**(12), 2227–2242 (2008)

# Key Information Recognition from Piping and Instrumentation Diagrams: Where We Are?

Rim Hantach[(✉)], Gisela Lechuga, and Philippe Calvez

CSAI Lab ENGIE, Paris, France
rim.hantach@external.engie.com,
{gisela.lechuga,philippe.calvez1}@engie.com

**Abstract.** Nowadays, the increase of technical drawings in different industries such as construction, mechanical and the energy sector makes the task of information analysis and interpretation more complex and fastidious. In this context, the automatic digitization of these drawings is becoming important. Piping and instrumentation diagram (P&ID) is a type of engineering drawing where the flow and components are represented by lines, texts and symbols. In this paper, we propose an industrial research approach in order to detect symbols, texts and lines. We focus on the application of recent computer vision and natural language processing techniques to automatically detect and recognize the different components. First experimental results on real-world data show that the proposed pipeline can achieve competitive results.

**Keywords:** P&ID · Computer vision · NLP

## 1 Introduction

Piping and instrumentation diagrams, P&IDs, are a type of technical drawing that provide a graphical representation of a process system that include the piping, vessels, control valves, instrumentation, and other process components and equipments in a given system. Such diagrams are the primary schematic drawing used for laying out a process control system's installation and is most commonly used in the engineering field, such as during the designing of a manufacturing process in a processing plant. These facilities usually require complex chemical or mechanical configurations that are represented through the use of symbols and connections on a P&ID. They can be used to streamline a process, keep track of different pieces of equipment or guide the design of a new facility.

There are standard symbols used to represent the components in P&IDs, this representation is not made to scale, which add to the complexity of the task. Most of these symbols are also associated to words, letters or numbers to further identify and specify the components being represented. Even though standard symbols are used there can exist some variations on how they are depicted across P&IDs from different origins and industries.

© Springer Nature Switzerland AG 2021
E. H. Barney Smith and U. Pal (Eds.): ICDAR 2021 Workshops, LNCS 12917, pp. 504–508, 2021.
https://doi.org/10.1007/978-3-030-86159-9_36

In recent years, several attempts have been made at performing this task of automated reading. Given the schematic aspect of P&IDs, the goal of this work is to identify different symbols, their associated text and the different connections between components.

The rest of the paper is structured as follows: Sect. 2 describes some P&ID approaches in the state-of-the-art. Section 3 details the proposed pipeline and some experiments done so far. Section 4 concludes the paper and discusses possible future directions.

## 2   State of the Art

Digital engineering drawings have seen great attention across industries. Analyzing and processing these drawings are becoming important for decision making. In this context, several efforts have been made in different domain applications such as planning, construction, oil, gas, etc. In this section, we detail some works related to P&ID data extraction and automatic digitizing.

Dong-Yeol Yun et al. [8] propose a new object detection approach in order to recognize the different symbols in P&ID. To do that, three steps have been defined: i) region proposal using a sequential image-processing, ii) dataset annotation iii) regions classification where a convolutional network has been trained to classify the proposed regions and extract the information related to the symbols. In [7], authors introduced a new texts and symbols recognition in a P&ID based on local binary pattern (LBP) and spatial pyramid matching (SPM) for image feature extraction. Arroyo et al. [1] aim to convert the P&ID documents into an object-oriented plant description. This is by applying optical recognition and semantic analysis.

Premanand Ghadekar introduces an end-to-end GPU algorithm [3]. It starts by extracting the meta data of each instance related to the instruments, valves and equipment's present in the P&ID. Then it detects the lines and texts. The authors combine computer vision methods and deep learning models (Retina net with a Resnet-101 backbone) to classify symbols, semantically analyze texts and detect connections between equipment's. In [5], authors use template matching for symbols recognition, these have been extracted and registered automatically in the database. For line and texts recognition, the sliding window method and aspect ratio calculation has been used. A recent paper extends the previous works by applying graph search to detect the connections between symbols in P&IDs [6]. This paper is based on state-of-the-art methods to detect symbols where a Convolutional Neural Network has been used, recognize and interpret text and detect connections between symbols across lines. The proposed method can be applied in different applications such as diagram search, equipment-to-sensor mapping, and asset hierarchy creation.

## 3    Methodology

In this paper, we propose a pipeline for automatically extracting key information from P&ID's. This pipeline is based on computer vision methods to detect lines, text and symbols.

### 3.1    Line Detection Using Kernels

As the lines encountered in a P&ID are either vertical or horizontal and always straight in shape the use of convolutional kernels for their detection becomes well adapted. We tested a combination of a vertical and a horizontal kernels to make the line detection. Since thicker and thinner lines are connecting the different elements, we take this information within our kernels in order to capture as many line width variation as possible. We can see an example of detected lines in Fig. 1a.

As it can be seen in the case of the line detection, improvements are still needed. For future work we aim at removing spurious line detections, Some possible implementation could be to consider the areas with text as non line regions, as not to mix the detection of text and lines. This could also be extended to the regions with detected symbols. Besides, an improvement in the detection of dotted lines would also provide an easier to read result.

### 3.2    Text Detection Using Character Region Awareness for Text Detection

Several algorithms exists in the literature to perform this task, we chose to implement the CRAFT algorithm [2] and apply it to P&ID scanned images.

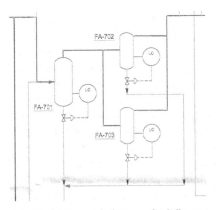

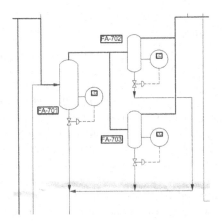

(a) Detection of lines of different widths using horizontal and vertical detection kernels.

(b) Text detection using the CRAFT algorithm.

**Fig. 1.** Preliminary line and text detection results

Without any pre-processing of the image, the results show a good detection performance. We can see an example of detected text bounding boxes in Fig. 1b.

### 3.3 Symbol Detection

For this section we consider two possible approaches when a small amount of data is available and when we have access to a bigger dataset. Having very few examples in our dataset and the lack of annotated images, we opted for a morphological approach using the circle Hough Transform. As P&ID come in different sizes and resolutions, different symbols may come in different sizes. To try this approach, we opted to focus on the detection of primary device and control symbols. In order to have a robust detection, we varied the radius of the circles in order to detect all of them. We provide an example of detected primary device and control symbols in Fig. 2a.

When datasets contain a large number of examples, other approaches can be considered, such as the use of deep neural networks. To illustrate that, annotated and trained a YOLOv5 neural network [4] to detect gate valves. We only had access to 8 P&ID's with varying amounts of gate valves. In order to have enough examples for the training we realized random crops around the object of interest. We can see the detected valves in Fig. 2b.

The use of object detection neural network also provide for an adaptable tool to detect different types of symbol, given access to a sufficiently big dataset.

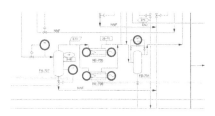

(a) Detection of Primary Device and Control Symbols.

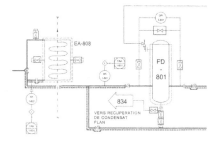

(b) Symbol detection using Yolov5.

**Fig. 2.** Preliminary symbol detection results

## 4   Conclusion

In this paper we proposed a pipeline for automatically extracting key information, such as text, symbols and connections from P&ID's. This pipeline is based on various computer vision and deep learning methods inspired from state of the art approaches. The proposed symbols detection allows for a high flexibility and can easily be trained on new classes, given large volume of annotated data or the right morphological approach is used.

Despite the recent advances in the automatic feature extraction of P&ID's using computer vision and machine learning approach, it remains a challenging task. The lack of a benchmark dataset and open access P&ID's further increase this challenge. As text and symbol detection methods improve the focus should be given to finding the links between different symbols.

As future works, our proposed pipeline can be improved by combining recent models for better performance. It is also possible to develop a new deep learning method for symbols detection and text associations. Experiments on different real-world datasets will also be tested.

# References

1. Arroyo, E., Hoernicke, M., Rodríguez, P., Fay, A.: Automatic derivation of qualitative plant simulation models from legacy piping and instrumentation diagrams. Comput. Chem. Eng. **92**(C), 112–132 (2016)
2. Baek, Y., Lee, B., Han, D., Yun, S., Lee, H.: Character region awareness for text detection. In: Proceedings of the IEEE/CVF Conference on Computer Vision and Pattern Recognition, pp. 9365–9374 (2019)
3. Ghadekar, P.: Intelligent agent for automatic engineering diagram digitization with deep learning. Biosci. Biotechnol. Res. Commun. **13**, 01–06 (2020)
4. Jocher, G., et al.: ultralytics/yolov5: v5.0 - YOLOv5-P6 1280 models, AWS, Supervise.ly and YouTube integrations (2021)
5. Kang, S.O., Lee, E., Baek, H.K.: A digitization and conversion tool for imaged drawings to intelligent piping and instrumentation diagrams (p&id). Energies **12**, 2593 (2019)
6. Mani, S., Haddad, M.A., Constantini, D., Douhard, W., Li, Q., Poirier, L.: Automatic digitization of engineering diagrams using deep learning and graph search. In: Proceedings of the IEEE/CVF Conference on Computer Vision and Pattern Recognition (CVPR) Workshops, June 2020
7. Tan, W.C., Chen, I.M., Tan, H.K.: Automated identification of components in raster piping and instrumentation diagram with minimal pre-processing. In: 2016 IEEE International Conference on Automation Science and Engineering (CASE), pp. 1301–1306 (2016)
8. Yun, D.Y., Seo, S.K., Zahid, U., Lee, C.J.: Deep neural network for automatic image recognition of engineering diagrams. Appl. Sci. **10**(11), 4005 (2020)

# Graph Representation Learning in Document Wikification

Mozhgan Saeidi$^{(\boxtimes)}$ , Evangelos Milios, and Norbert Zeh

Dalhousie University, Halifax, Canada
mozhgan.saeidi@dal.ca, {eem,nzeh}@cs.dal.ca

**Abstract.** Wikification (entity annotation) is a challenging task in Natural Language Processing (NLP). It is a method to automatically enrich a text with links to Wikipedia as a knowledge base. Wikification starts from detecting ambiguous mentions in the document, and later tries to disambiguate those mentions. In the core of the Wikification task, there is one other important NLP task: word representation. This paper proposes a new word representation for senses of a mention with Graph convolutional networks architecture. Senses are the possible meanings of one mention, based on the knowledge base. In our representation modeling, we used the context document and the first paragraph of each Wikipedia page to enhance our contextual representation. Using the nearest neighbor algorithm for disambiguating the mentions via our sense representations, we show the efficiency of our representations. The results of comparing our method with recent state-of-the-art methods show the efficiency of our solution.

**Keywords:** Representation learning · Graph convolutional networks · Wikification · Document ambiguity

## 1 Introduction

The task of "**Entity Recognition and Disambiguation**" (**ERD**) is to identify mentions[1] of entities and link them to a relevant entry in an external knowledge base, which is also known under the names of "**Entity Linking**", "**Wikification**" or more generally "**Text Annotation**" [1]. Given an input document, a Wikifier links entities of the document to the most relevant corresponding Wikipedia pages. Automated document annotation has become an important topic due to the vast growth in Natural Language Processing (NLP) applications [53]. One benefit of document annotation is enhancing text readability and its unambiguousness by inserting connections between the text and an external knowledge base, like Wikipedia, which is the most popular among online encyclopedias [41].

---

[1] A mention can be one or more tokens.

© Springer Nature Switzerland AG 2021
E. H. Barney Smith and U. Pal (Eds.): ICDAR 2021 Workshops, LNCS 12917, pp. 509–524, 2021.
https://doi.org/10.1007/978-3-030-86159-9_37

The problem of Wikification is closely related to other core NLP tasks, such as Named-Entity Recognition (NER) and Word Sense Disambiguation (WSD) [54]. Wikification, in particular, is the task of associating a word in context with the most suitable meaning from the predefined sense inventory of Wikipedia. Named-Entity Recognition involves identifying certain occurrences of noun phrases as belonging to particular categories of named-entities [34]. These expressions refer to names, including person, organization, and location names, and numeric expressions including time, date, money, and percent expressions [21]. In Word Sense Disambiguation, our knowledge base is not limited to Wikipedia, which is the difference between Wikification and WSD [23]. In WSD, the knowledge base is a treasury like WordNet and Wikipedia [18]. Details and performance of each WSD method are highly dependent on the knowledge base to link to.

The knowledge bases are different in their nature [2]; for example, WordNet is a lexical graph database of semantic relations (e.g. synonyms, hyponyms, and meronyms) between words. Synonyms are grouped into synsets with short definitions and usage examples. WordNet can thus be seen as a combination and extension of a dictionary and thesaurus [4]. Wikipedia is a hyperlink-based graph between encyclopedia entries. So, Wikification, WSD, and NER are closely related but still different because of their underlying knowledge bases. In this paper, our focus is on the Wikification problem, so we did not mention WSD and NER systems in detail [19], and we focus on providing required details of Wikification systems. We compared our proposed method with the Wikifiers by different parameter settings.

The Wikification process involves two steps: *Spotting* or *mention detection*, i.e. identifying the terms that should be wikified, and *Entity Linking* or *Disambiguation to Wikipedia*, i.e. identifying the relevant Wikipedia page among a set of candidate pages [51]. The spotter operates on the text to extract all entity ambiguous mentions and assigns all potential entity candidates for each mention. The entity linker disambiguates the candidate entities by selecting the most probable sense entity for each mention [50]. Our focus is on the second step, and we use the output of the recent spotting system [46] as the input to our algorithm.

A human reader can identify the correct meaning of each word based on the context in which the word is used. Computational methods try to mimic this approach. These methods often represent their output by linking each word occurrence to an explicit representation of the chosen sense [56]. There are two approaches to tackle this problem: The machine learning-based approach and the knowledge-based approach. In the machine learning-based approach, systems are trained to perform the task [45,52]. The knowledge-based approach requires external lexical resources such as Wikipedia, WordNet [30], a dictionary, or a thesaurus. The machine learning-based approaches mainly focus on achieving maximal precision or recall and have their drawbacks of run-time and space requirement at the time of classifier training. So, knowledge-based Wikification methods still have advantages to study. Among different knowledge-based methods, coherence-based has been more effective to explain it [11]. In the coherence-based approach, one important factor is the coherence of the whole text after disambiguation, while in other approaches, this factor might change to considering the coherence of each sentence or paragraph. It is a significant challenge to perform Wikification accurately but also fast enough to process long text documents [26]. One coherence-based approach models the relatedness between the senses and a *key-entity* in disjoint windows of the text to speed up the approach and disambiguates every word so that the total pairwise relatedness of all chosen word senses and key-entity of the same window is maximized. This method is computationally expensive, and run-time performance is considered as a secondary issue in most of the existing Wikification methods.

Embeddings have been shown to play an important role in different NLP tasks [33,45], especially in disambiguation tasks [48]. Embeddings based on pre-trained deep language models have attracted much interest recently as they have proved to be superior to classical embeddings for several NLP tasks, including WSD. These models, e.g., ELMO [37], BERT [10], XLNET [58], encode several pieces of linguistic information in their word representations. These representations differ from static neural word embeddings [36] in that they are dependent on the surrounding context of the word. This difference makes these vector representations especially interesting for disambiguation, where effective contextual representations can be highly beneficial for resolving lexical ambiguity. These representations enabled sense-annotated corpora to be exploited more efficiently [24].

Here, we face the problem of link ambiguity, meaning a phrase can be usually linked to more than one Wikipedia page in which the correct link depends on the context where it occurs. For example, the word "bar" can be linked to different articles, depending on whether it was used in a business or musical context.

In this study, we overview different current approaches for text embedding with focusing on the contextualized sense representation. We also provide an overview of the disambiguation methods, and the most used ones in the literature. Our novel contribution provides a new representation learning using the graph deep learning approach and uses the nearest neighbor heuristic algorithm for disambiguating the document. We finally compare the performance of our proposed approach with our representations with the most recent approaches in the disambiguation task.

## 2 Background

In this section, First we provide an overview for wikification systems, and second, we go through a general background of the embedding approaches. Our focus is on the pre-trained deep language embedding models.

### 2.1 Document Disambiguation with Wikipedia

A large group of NLP systems for various word disambiguation tasks rely on Lexical Knowledge Bases. These knowledge bases vary significantly in their structure, size, and subject, making them more appropriate for certain domains. For instance, WordNet was used for *synonym exploration* [13,32], and the sophisticated Unified Medical Language System (UMLS) ontology was used for *medical text dissambiguation* [17,25]. Disambiguation based on Wikipedia has been demonstrated to be comparable in terms of coverage to domain-specific ontology [55] since it has broad coverage, with documents about entities in a variety of domains [26]. Moreover, Wikipedia has unique advantages over the majority of other knowledge bases [61]. One advantage is the text in Wikipedia is primarily factual and available in a variety of languages. The other advantage is about the articles which can be directly linked to the entities they describe in other knowledge bases. Also, mentions of entities in Wikipedia articles often provide a link to the relevant Wikipedia pages, thus providing labeled examples of entity mentions and associated anchor texts in various contexts, which could be used for supervised learning in Wikification.

**Knowledge-Based Approaches.** In this type of approach, WSD is considered a graph-based problem. They use the semantic network structure, e.g., Wikipedia, WordNet, BabelNet, to find the correct meaning based on its context for each input word [33]. The latest work in this series is SensEmBERT [47] which shows the power of language models combined with a vast amount of knowledge in a semantic network to produce latent semantic representations of senses in multiple languages. ARES followed this model and created sense embeddings for the lexical meanings within a lexical knowledge base. These embeddings lie in a space that is comparable to that of contextualized word vectors [48].

As mentioned in the previous section, using any knowledge base for text disambiguation requires an "entity linker". When UMLS is used as the knowledge base, MetaMap is widely accepted as the entity linker [3,15]. In the case of Wikipedia, the entity linker is referred to as **Wikifier**. In most studies, a Wikifier uses two groups of features, local and global [57]. Local features include the context around the entity mention and some data-driven statistics regarding the mention-entity relation, such as *commonness* or *prior probabilities* [43]. The most famous example of global features is the **semantic coherence measure**. This feature is established based on the assumption that words in a given *neighbourhood* (i.e., a segment of the text) will tend to share a common topic. Examples of widely accepted Wikifiers include Wikify! [28], Wikipedia Miner [31], TagME [12], and GLOW [43].

The system of Wikify! is a wikification method that disambiguates and ranks candidates to indicate the most valuable ones to the user in terms of the meaning. Wikify! uses the Lesk algorithm [20] to choose the most appropriate sense. The Lesk algorithm identifies the most likely meaning for an ambiguous word based on the *contextual overlap* between the content of the Wikipedia pages corresponding to the candidate senses and the local context of the ambiguous word.

Wikify! has been the first widely accepted Wikifier. However, it has already been outperformed by more recent Wikifiers, such as large-scale named entity disambiguation [9]. In large-scale named entity disambiguation [9], the process is based on maximizing the agreement between the contextual information corresponding to each candidate sense and the context of the anchor text. The information for each candidate sense is a combination of various features extracted from its Wikipedia page. One of these features is the set of incoming Wikipedia links for each candidate sense, including irrelevant entities that happen to have significant overlap, in terms of the number of common words, with the anchor text. As a result, an irrelevant candidate entity with such incoming Wikipedia links could end up getting selected.

Wikipedia Miner [31] outperforms the Wikify! and the large-scale named entity methods [9,28] introduced above. The entity disambiguation approach of Wikipedia Miner relies on the graph structure of Wikipedia. This structure is used to perform disambiguation based on two concepts: *commonness* and *relatedness*. The commonness of a sense is defined by the number of times it is used as a destination in Wikipedia. Hence, commonness is sometimes referred to as prior probability. The relatedness of a candidate sense is its similarity to the context. Their approach aims to balance the commonness of a sense with its relatedness to the surrounding context. They use machine learning to combine these features so that the balance can be adjusted from document to document.

The following two methods, namely TagMe and GLOW, while still using commonness and relatedness, are very close to our solution. One of them is based on a voting system, and the other is using a ranking approach, which are two aspects of our proposed solution.

In the system of TagMe [12] a voting scheme is proposed, where the candidates for each mention can vote for all candidate entities of other mentions based on relatedness; Candidates with higher prior probabilities have stronger votes. In our proposed solution, the voting power of each candidate sense depends on its rank based on the previous voting round. Moreover, in TagMe, candidates with a prior probability below a fixed threshold are pruned. Then TagMe uses two algorithms to decide the chosen sense for each mention: disambiguation by classifier (DC), which uses a probabilistic approach based on prior probability and relatedness to select the correct candidate; and disambiguation by threshold (DT), which makes a shortlist of the top candidates with relatedness above a predefined threshold, and then chooses the candidate with the highest prior probability among them. Later, they released a more efficient version of TagMe as WAT algorithm [40].

The system of GLOW [43] uses two sub-systems; a ranker and a linker. The goal of the ranker is to select the best candidate, and the linker decides if the recommended sense by the ranker is good or not. The ranker sub-system uses two sets of features, namely local and global. Local features calculate the similarities between mentions and their candidate entities, incorporating the term frequency inverse document frequency (TF-IDF) vectors for Wikipedia pages and the anchor text [14]. The global features measure the coherence among all candidate senses in terms of the sum of pairwise normalized Google distance (NGD) [8], and Point-wise Mutual Information (PMI) [5] across all mentions in the whole disambiguation text. The linker sub-system is trained as a linear support vector machine (LSVM) to separate correct and incorrect linker outputs based on data collected from Wikipedia, which provides positive and negative examples for each mention according to Wikipedia's gold standard.

The state-of-the-art Wikifier RedW [49] is a run-time oriented Wikification solution. RedW is based on Wikipedia redirects and can wikify massive corpora with good performance. This approach is based on mapping the *longest sub-string match* between the mention and the Wikipedia entity titles. RedW assumes that a term often matches its Wikipedia title or a corresponding Wikipedia redirect page. They have this assumption because one advantage of redirects over anchor dictionaries is their dynamic nature. Redirects are updated both automatically and by Wikipedia editors. Hence they are expected to contain less noise compared to automatically created dictionaries. RedW creates a table of all Wikipedia titles, including the redirect titles. For every text, RedW tries to match its n-grams to the table, preferring longer matches. Unlike TagMe and Glow, RedW does not consider global features such as coherence. However, TagMe is designed for short text Wikification and is inferior to RedW compared to long text, and Glow has not been compared against this scheme. However, experiments show our proposed solution outperforms RedW, as we consider the coherence of the text. There is one recent work which is an application of Wikification task by using RedW method. In this work [60], authors try to navigate through information pollution by capturing the provenance of every claim. They define a provenance graph for a given natural language claim, aiming to understand where the claim may come from and how it has evolved. Specifically, to wikify a source mention for each claim, they adapt the redirect-based wikification method of RedW since RedW is efficient and context-free.

**Supervised Approaches.** The supervised approaches use sense-annotated data for their training. These types of methods have traditionally gained state-of-the-art results in terms of accuracy. Even before introducing pre-trained language models, supervised WSD methods have been shown to outperform the knowledge-based models. At the same time, their defect is the *knowledge acquisition bottleneck*, which makes it challenging to construct broad manually curated corpora. It limits the ability of these methods to scale to new words [35].

Neural sequencing models are trained for end-to-end word sense disambiguation task [42]. They re-framed WSD as a translation task that sequences of words are translated into sequences of senses. Later, some works showed the potential of contextual representation for WSD [27,37]. Sense embeddings initialization using glosses and adapted the skip-gram objective of word2vec is done by [7] to learn and improve the sense embeddings jointly with word embeddings. Later, by the appearance of NASARI vectors [6], sense embeddings were created using structural knowledge from a large multilingual semantic network. These methods represent sense embeddings in the same space as the pre-trained word embeddings, while they suffer from fixed embedding spaces. The LMMS representation considers creating sense-level embeddings with complete coverage of WordNet and shows the power of this representation for WSD by applying a simple Nearest Neighbors (k-NN) method [24]. ARES used this 1-NN method with its representations and showed improved results in disambiguating. In the following, we briefly introduce the recent state-of-the-art embeddings and then analyze their results.

## 2.2   Language Modelling Representation

Most NLP tasks now use semantic representations derived from language models. There are static word embeddings and contextual embeddings. In this section, we cover aspects of the word and contextual embeddings that are especially important to our work.

**Static Word Embeddings.** Word embeddings are distributional semantic representations usually with one of two goals: predict context words given a target word (Skip-Gram), or the inverse (CBOW) [29]. In both, the target word is at the center, and the context is considered as a fixed-length window that slides over tokenized text. These models produce dense word representations. One limit for word embeddings, as mentioned before, is meaning conflict around word types. This limitation affects the capability of these word embeddings for the ones that are sensitive to their context [44].

**Contextual Word Embeddings.** The problem mentioned as a limitation for the static word embeddings is solved in this type of embeddings. The critical difference is that the contextual embeddings are sensitive to the context. It allows the same word types to have different representations according to their context. The first work in contextual embeddings is ELMO [37], which is followed by BERT [10], as the state-of-the-art model. The critical feature of BERT, which makes it different, is the quality of its representations. Its results are task-specific fine-tuning of pre-trained neural language models. The recent representations which we analyze their effectiveness are based on these two models [38,39].

## 3   Graph Representation Modelling

This section presents our sense embedding approach, which is a novel method based on deep graph convolutional networks. This representation is a context-aware representation model of Wikipedia senses by combining the semantic and textual information derived from the document context of the mention and the first paragraph of the mention's Wikipedia page. In this representation, we used the power of neural language models, i.e., BERT [10]. We divide our approach into the following subdivisions.

### 3.1    Concept Embedding

We use concept which refers to word or mention, as well as $s$ sense. We use the long heuristic search and the table of Wikipedia titles for extracting mentions in the document. when we match them together, the mentions of the document which are corresponding with a Wikipedia page. For each mention $m$ in the document, we generate BERT representation of $m$ with $R(m)$. In our experiments, we used BERT-base-cased[2], since it has been shown the performance of the BERT-base-cased model, in comparison with uncased- is better in the task of sense disambiguation. We repeat the same procedure to produce the representations of senses for each mention, with $R(s)$. The length of each one of these representations is 300, as the dimension of BERT representations.

### 3.2    Context Representation

When we work with Wikipedia page of each mention, we consider the first paragraph of each Wikipedia page. The first paragraph is the main part of each Wikipedia page with the most noticeable information about the mention (title) of the page. Using the same pre-trained language model, we generate the representation of the first paragraphs of each mention in the Wikipedia page. In our implementations, if the length of the first paragraph is more than 512 words, the algorithm does not include the rest of the words. While our experiments show it does not happen, since the first paragraph of all the Wikipedia pages of the mentions in the used dataset are not including more than 512 words. We show this representation by $R(P)$. On the other hand, we have the input document from where we extracted our mentions. For each mention, our algorithm considers the paragraph which includes the mention. We use the representation of this document paragraph in our settings by $R(D)$. Starting from a Wikipedia page of mention $m$, we collect the set of its senses in the Wikipedia knowledge base, which are the redirect pages, i.e., all the redirects that are connected to $m$. We use these senses as the nodes of the graph that we are going to build and connect the redirect ones, which builds the graph's edges.

### 3.3    Sense Representation

In this part of building our representations, we merge the contextual information computed in the two previous steps to enrich the representation with additional information of the document and the semantic network, which is the first paragraph of the Wikipedia page. For each sense $s$ of a mention $m$, now we have four variables of $R(m)$, $R(s)$, $R(P)$, and $R(D)$, each with dimension 300.

### 3.4    Graph Convolutional Network

Graph Convolutional Networks (GCN) are a very powerful multilayer neural network architecture for machine learning on graphs [16]. GCN operates directly

---

[2] https://huggingface.co/bert-base-cased.

on a graph and induces embedding vectors of nodes based on the properties of their neighborhoods. In fact, they are so powerful that even a randomly initiated 2−layer GCN can produce useful feature representations of nodes in networks[3]. Formally, consider a graph $G = (V, E)$, where $V(|V| = n)$ and $E$ are sets of nodes and edges, respectively. Every node is assumed to be connected to itself, i.e., $(v, v) \in E$ for any $v$ which the reason for this assumption is mentioned at the end of this paragraph. Let $X \in R^{n \times m}$ be a matrix containing all $n$ nodes with their features, where $m$ is the dimension of the feature vectors, each row $x_v \in R_m$ is the feature vector for $v$. We introduce an adjacency matrix $A$ of $G$ and its degree matrix $D$, where $D_{ii} = \sum_j A_{ij}$. Because of self-loops, the diagonal elements of $A$ are all 1. We now have a graph, its adjacency matrix $A$, and a set of input feature $X$. After applying the propagation rule $f(X, A) = AX$ and $X = I$, the representation of each node (each row) is now a sum of its neighbor's features. In other words, the graph convolutional layer represents each node as an aggregate of its neighborhood. The reason for considering the self-loops in the graph is the aggregated representation of a node to include its own features.

For a one-layer GCN, the new k-dimensional node feature matrix $L^{(1)} \in R^{n \times k}$ is computed as:

$$L^{(1)} = \rho(\hat{A}XW_0) \tag{1}$$

where $\hat{A}$ is $D^{-0.5}AD^{-0.5}$, the normalized symmetric adjacency matrix and $W_0 \in R^{m \times k}$ is the weight matrix. The $\rho$ is the activation function (RELU); $\rho(x) = max(0, x)$. GCN can capture information only about immediate neighbors with one layer of convolution. When multiple GCN layers are stacked, information about larger neighborhoods are integrated;

$$L^{(j+1)} = \rho(\hat{A}L^jW_j) \tag{2}$$

which $j$ is the layer number and $L^0 = X$. In other words, the size of the second dimension of the weight matrix determines the number of features at the next layer. The feature representations can be normalized by node degree with transforming the adjacency matrix $A$ by multiplying it with the inverse degree matrix $D$. First we used the simple propagation rule $f(X, A) = D^{-1}AX$, while then improved it. The improved version is inspired by a recent work [16] that proposes a fast approximate spectral graph convolutions using a spectral propagation rule $f(X, A) = \sigma(D^{-0.5}\hat{A}D^{-0.5}XW)$. They showed this property is very useful, that connected nodes tend to be similar (e.g. have the same label).

We consider each mention of the document as one node of the graph, and a new added node (redirect link) will connect with its nearest neighbor by using cosine similarity, which makes the edges of the graph. The cosine similarity between two nodes on the edges makes the weight matrix. The number of nodes in the text graph $|V|$ is the number of mentions. For each sense $s$, we use an integrated representation of its mention $m$ with its own representation, i.e.,. $R(m, s)$. We set the feature matrix $X$ as extracted representation of BERT as

---

[3] The notation we used for GCN in this paper are the same as notations in [59].

input to GCN. The dimension of the feature matrix here is 600, as it is the representation length of two BERT embeddings, one for the mention and the other for the sense. We name our representation **msBERT**.

As mentioned, formally, the weights of edge between node $i$ and node $j$ defines as:

$$W_{ij} = \text{cosine sim}(R(i), R(j)) = \frac{R(i).R(j)}{||R(i)||||R(j)||} \tag{3}$$

which $R(i)$ is representation of node $i$.

After building the graph, we feed it into a simple 2−layers GCN as [16], the second layer node (mention,sense) embeddings are fed into a softmax classifier:

$$Z = softmax(\hat{A}RELU(\hat{A}XW_0)W_1) \tag{4}$$

where

$$\hat{A} = D^{-0.5}AD^{-0.5}$$

and

$$softmax(x_i) = \frac{1}{Z}exp(x_i)$$

with $S = \sum_i exp(x_i)$. The loss function is the one defined in [59] as:

$$L = -\sum_{d \in Y} \sum_{f=1}^{F} Y_{df} ln Z_{df} \tag{5}$$

where $Y_D$ is the set of mention indices that have labels and $F$ is the dimension of the output feature. $Y$ is the label indicator matrix. Similar to [59], the weight parameters $W_0$ and $W_1$ can be trained via gradient descent. The $\hat{A}XW_0$ contains the first layer (mention, sense) and embeddings, and $\hat{A}RELU(\hat{A}XW_0)W_1$ contains the second layer (mention, sense) and embeddings. This two-layer GCN performs message passing between nodes to two steps away, maximum. Therefore, the two-layer GCN allows the exchange of information between pairs of nodes. This GCN model on our experimental datasets (next section) shows better performance than a one-layer model and models with more than two layers. This shows the validity of our model, based on similar results in other recent works [16,22].

## 4    Disambiguation Model

We used a 1-nearest neighbor approach to test **msBERT** on the disambiguation task. For each target mention $m$ in test set, we computed its contextual embedding by means of BERT and compared it against the embeddings of msBERT associated with the senses of $m$, via training our GCN. Hence, we took as prediction for the target word the sense corresponding to its nearest neighbour. We note that the embeddings produced by msBERT are created by concatenating two BERT representations, i.e., context and sense (see Sect. 3.4), hence we repeated the BERT embedding of the target instance to match the number of dimensions. In contrast to most supervised systems, this approach does not rely on the Most Frequent Sense (MFS) backoff strategy, i.e., predicting the most frequent sense of a redirect link in Wikipedia for instances unseen at training time, as msBERT ensures full coverage for the English nominal senses. At the time of cosine similarity calculation, we include considering the distances of the $R(m, s)$ with $R(D, P)$. It integrates more context at the time of finding the nearest neighbor.

### 4.1 Evaluation Datasets

We have two main distinct categories of datasets that we used in our experiment; First, we carried out the evaluation on the English all-words WSD framework by Raganato et al. [42], comprising five standard test sets, namely, Senseval-2 (Edmonds and Cotton, 2001), Senseval-3 (Snyder and Palmer, 2004), SemEval-07 (Pradhan et al., 2007), SemEval-13 (Navigli et al., 2013), SemEval-15 (Moro and Navigli, 2015) along with ALL, i.e., the concatenation of all the test sets. Second, we used the five Wikification datasets of KORE, MSNBC, AQUAINT, Wiki5000, and Wiki30000. The first three in this category are all from news, and the last two are the Wikipedia pages.[4]

### 4.2 msBERT Configuration

We used Wikipedia as input corpus since it is the largest general-domain resource currently available. We varied the number of senses (redirect pages) to give as input to the GCN between 5 and 25 with a 5 step and selected the value n = 5 by manually assessing the quality of a sample of the disambiguation output.

**Table 1.** F-Measure performance of WSD evaluation framework on the test sets of the all-words English datasets.

| Model | Senseval-2 | Senseval-3 | Semeval-7 | Semeval-13 | Semeval-15 | All |
|---|---|---|---|---|---|---|
| BERT | 77.1 ± 0.3 | 73.2 ± 0.4 | 66.1 ± 0.3 | 71.5 ± 0.2 | 74.4 ± 0.3 | 73.8 ± 0.3 |
| LMMS | 76.1 ± 0.6 | 75.5 ± 0.2 | 68.2 ± 0.4 | 75.2 ± 0.3 | 77.1 ± 0.4 | 75.3 ± 0.2 |
| SensEmBERT | 72.4 ± 0.1 | 69.8 ± 0.2 | 60.1 ± 0.4 | 78.8 ± 0.1 | 75.1 ± 0.2 | 72.6 ± 0.3 |
| ARES | 78.2 ± 0.3 | 77.2 ± 0.1 | 71.1 ± 0.2 | 77.2 ± 0.2 | 83.1 ± 0.2 | 77.8 ± 0.1 |
| msBERT | 79.6 ± 0.5 | 79.3 ± 0.2 | 74.6 ± 0.1 | 78.1 ± 0.4 | 81.5 ± 0.4 | 76.5 ± 0.7 |

### 4.3 Comparison Systems

We compared msBERT against the best-performing knowledge-based systems evaluated on the disambiguation framework. These systems include the mentioned ones in the background section. **Wikisim** [46] which is the most recent

---

[4] We used this dataset of the second category from: https://github.com/asajadi/ wikisim/tree/master/datasets.

key-entity Wikifier, and **TagME** [12] which is available as a web service[5]. We also compare with **GLOW** [43] and **Wikipedia Miner** [31] since we have their $F1$ measures on three common datasets. Lastly, we compare our approach with **RedW** [49], which is a context-free run-time oriented Wikifier. The latest knowledge base approach in this task is SensEmBERT, which is included in the compared systems.

## 5    Experimental Results

We now report the results of the evaluation we carried out on the English disambiguation task. In Table 1 we report the results of the underlying systems as well as our proposed system. This comparison shows the effectiveness of different approaches in finding the correct sense for each ambiguous mention, based on its context. This results show our msBERT representation is as effective as the recent state-of-the-art contextualized embeddings in lexical ambiguity. Our experiments aim to explore the effectiveness of the proposed approach for the lexical ambiguity problem. The idea of engaging more context to disambiguate entity concepts more related to their context is supported by the results in Table 1. In our other experiment, we compared the results of disambiguating the text using our approach with other knowledge-based approaches. The results of this experiment are shown in Table 2. Using proposed embedding msBERT for disambiguating the text, we are more confident about considering all relevant information in the document for disambiguating the ambiguous mentions. We compare the senses of each mention with the knowledge base information and document content, which was missing in the previous methods. The improvement in our results over the baselines is statistically significant ($\chi^2$ test with $p < 0.05$). For this aim, one comparison we run is measuring precision, recall, and F1 measure of our algorithm with the baseline approaches.

**Table 2.** F-measure on the Wikification task of unsupervised knowledge-based approaches of GLOW, Wikipedia Miner, TAGME, Wikisim, and RedW in comparison with our proposed algorithm using the msBERT representation.

| Method | Kore | AQUAINT | MSNBC | Wiki5000 | Wiki30000 |
|---|---|---|---|---|---|
| GLOW | 0.68 | 0.72 | 0.73 | 0.65 | 0.66 |
| Wikipedia Miner | 0.70 | 0.73 | 0.68 | 0.67 | 0.70 |
| TAGME | 0.68 | 0.64 | 0.55 | 0.60 | 0.61 |
| Wikisim | 0.64 | 0.62 | 0.58 | 0.57 | 0.57 |
| RedW | – | 0.80 | – | 0.62 | – |
| msBERT | 0.74 | 0.79 | 0.67 | 0.73 | 0.75 |

---

[5] https://tagme.d4science.org/tagme/.

# 6   Conclusion

In this paper, we presented msBERT, an approach for producing embeddings of senses in English. The msBERT can couple the information within the knowledge base with contextual information from the document mentioned in it. This feature results in high-quality latent representations for the concepts within a lexical knowledge base. Our experiments showed that despite relying on English data only msBERT is comparable with all its alternatives on the English disambiguation task. Our graph modeling produces the integrated representation of each ambiguous mention with all its possible senses in the Wikipedia knowledge base. The other novel idea of integrating the document representation along with the first paragraph of the Wikipedia page for each mention and each sense improved the efficiency of our representations, specifically for the lexical ambiguity task. As future work, we plan to exploit the information brought by our embeddings to other related tasks, including word sense disambiguation using other knowledge bases and experiments on different datasets.

# References

1. Aghaebrahimian, A., Cieliebak, M.: Named entity disambiguation at scale. In: IAPR Workshop on Artificial Neural Networks in Pattern Recognition Proceeding, pp. 102–110 (2020)
2. Aleksandrova, D., Drouin, P., Lareau, F.C.C.O., Venant, A.: The multilingual automatic detection of 'e nonc é s bias 'e s in wikip é dia. ACL (2020)
3. Amos, L., Anderson, D., Brody, S., Ripple, A., Humphreys, B.L.: UMLS users and uses: a current overview. J. Am. Med. Inform. Assoc. **27**(10), 1606–1611 (2020)
4. Azad, H.K., Deepak, A.: A new approach for query expansion using Wikipedia and wordnet. Inf. Sci. **492**, 147–163 (2019)
5. Bouma, G.: Normalized (pointwise) mutual information in collocation extraction. In: Proceedings of GSCL, pp. 31–40 (2009)
6. Camacho-Collados, J., Pilehvar, M.T.: From word to sense embeddings: a survey on vector representations of meaning. J. Artif. Intell. Res. **63**, 743–788 (2018)
7. Chen, X., Liu, Z., Sun, M.: A unified model for word sense representation and disambiguation. In: EMNLP, pp. 1025–1035 (2014)
8. Cilibrasi, R.L., Vitanyi, P.M.: The google similarity distance. IEEE Trans. Knowl. Data Eng. **19**(3), 370–383 (2007)
9. Cucerzan, S.: Large-scale named entity disambiguation based on Wikipedia data. In: EMNLP, pp. 708–716 (2007). https://www.aclweb.org/anthology/D07-1074. pdf
10. Devlin, J., Chang, M.W., Lee, K., Toutanova, K.: BERT: pre-training of deep bidirectional transformers for language understanding. arXiv preprint arXiv:1810.04805 (2018)
11. Dixit, V., Dutta, K., Singh, P.: Word sense disambiguation and its approaches. CPUH Res. J. **1**(2), 54–58 (2015)
12. Ferragina, P., Scaiella, U.: TAGME: on-the-fly annotation of short text fragments (by Wikipedia entities). In: ACM, pp. 1625–1628 (2010)
13. Hajar, E.H., Mohammed, B.: Using synonym and definition wordnet semantic relations for implicit aspect identification in sentiment analysis. In: NISS, pp. 1–5 (2019)

14. Jones, K.S.: A statistical interpretation of term specificity and its application in retrieval. J. Documentation 53–60 (1972)
15. Kim, M.C., Nam, S., Wang, F., Zhu, Y.: Mapping scientific landscapes in UMLs research: a scientometric review. J. Am. Med. Inform. Assoc. **27**(10), 1612–1624 (2020)
16. Kipf, T.N., Welling, M.: Semi-supervised classification with graph convolutional networks. arXiv preprint arXiv:1609.02907 (2016)
17. Kraljevic, Z., et al.: MedCAT-medical concept annotation tool. arXiv preprint arXiv:1912.10166 (2019). https://arxiv.org/ftp/arxiv/papers/1912/1912.10166.pdf
18. Kwon, S., Oh, D., Ko, Y.: Word sense disambiguation based on context selection using knowledge-based word similarity. Inf. Process. Manage. **58**(4), 102551 (2021)
19. Lee, J., Fuxman, A., Zhao, B., Lv, Y.: Leveraging knowledge bases for contextual entity exploration. In: Proceedings of ACM, pp. 1949–1958 (2015)
20. Lesk, M.: Automatic sense disambiguation using machine readable dictionaries: how to tell a pine cone from an ice cream cone. In: Systems Documentation, pp. 24–26 (1986)
21. Li, B.: Named entity recognition in the style of object detection. arXiv preprint arXiv:2101.11122 (2021)
22. Li, Q., Han, Z., Wu, X.M.: Deeper insights into graph convolutional networks for semi-supervised learning. In: AAAI, Proceedings of the AAAI Conference on Artificial Intelligence, New Orleans, vol. 32, pp. 234–242 (2018)
23. Logeswaran, L., Chang, M.W., Lee, K., Toutanova, K., Devlin, J., Lee, H.: Zero-shot entity linking by reading entity descriptions. arXiv preprint arXiv:1906.07348. https://arxiv.org/pdf/1906.07348.pdf (2019)
24. Loureiro, D., Jorge, A.: Language modelling makes sense: propagating representations through wordnet for full-coverage word sense disambiguation. In Proceedings of ACM, pp. 5682–5691 (2019)
25. Mao, Y., Fung, K.W.: Use of word and graph embedding to measure semantic relatedness between unified medical language system concepts. J. Am. Med. Inform. Assoc. **27**(10), 1538–1546 (2020)
26. Martinez-Rodriguez, J.L., Hogan, A., Lopez-Arevalo, I.: Information extraction meets the semantic web: a survey. Semant. Web Preprint **11**, 255–335 (2020)
27. Melamud, O., Goldberger, J., Dagan, I.: context2vec: learning generic context embedding with bidirectional LSTM. In: SIGNL, pp. 51–61 (2016)
28. Mihalcea, R., Csomai, A.: Wikify!: linking documents to encyclopedic knowledge. In: ACM, pp. 233–242 (2007)
29. Mikolov, T., Chen, K., Corrado, G.S., Dean, J.: Efficient estimation of word representations in vector space. In: Proceedings of ICLR, vol. 4, pp. 321–329 (2013)
30. Miller, G.A., Beckwith, R., Fellbaum, C., Gross, D., Miller, K.J.: Introduction to wordnet: an on-line lexical database. Int. J. Lexicography **3**(4), 235–244 (1990)
31. Milne, D., Witten, I.H.: Learning to link with Wikipedia. In: Proceedings of the 17th ACM Conference on Information and Knowledge Management, pp. 509–518 (2008)
32. Munirsyah, M., Bijaksana, M.A., Astuti, W.: Development synonym set for the English wordnet using the method of comutative and agglomerative clustering. Jurnal Sisfokom (Sistem Informasi dan Komputer) **9**(2), 171–176 (2020). http://jurnal.atmaluhur.ac.id/index.php/sisfokom/article/download/855/633
33. Navigli, R.: Word sense disambiguation: a survey. ACM Comput. Surv. (CSUR) **41**(2), 1–69 (2009)

34. Nguyen, D.B., Hoffart, J., Theobald, M., Weikum, G.: AIDA-light: high-throughput named-entity disambiguation. In: LDOW, vol. 14, pp. 22–32 (2014)
35. Pasini, T., Elia, F.M., Navigli, R.: Huge automatically extracted training sets for multilingual word sense disambiguation. arXiv preprint arXiv:1805.04685 (2018)
36. Pennington, J., Socher, R., Manning, C.D.: Glove: global vectors for word representation. In: Proceedings of the 2014 Conference on Empirical Methods in Natural Language Processing, EMNLP, Qatar, pp. 1532–1543 (2014)
37. Peters, M., et al.: Deep contextualized word representations. Association for Computational Linguistics, pp. 2227–2237 (2018)
38. Peters, M.E., Logan IV, R.L., Schwartz, R., Joshi, V., Singh, S., Smith, N.A.: Knowledge enhanced contextual word representations. arXiv preprint arXiv:1909.04164 (2019)
39. Peters, M.E., Neumann, M., Zettlemoyer, L., Yih, W.T.: Dissecting contextual word embeddings: architecture and representation. In: EMNLP, pp. 1499–1509 (2018)
40. Piccinno, F., Ferragina, P.: From TagME to WAT: a new entity annotator. In: Proceedings of the First International Workshop on Entity Recognition & Disambiguation, pp. 55–62. ACM (2014)
41. Raganato, A., Bovi, C.D., Navigli, R.: Automatic construction and evaluation of a large semantically enriched Wikipedia. In: IJCAI, pp. 2894–2900 (2016)
42. Raganato, A., Bovi, C.D., Navigli, R.: Neural sequence learning models for word sense disambiguation. In: Proceedings of the 2017 Conference on Empirical Methods in Natural Language Processing, pp. 1156–1167 (2017)
43. Ratinov, L., Roth, D., Downey, D., Anderson, M.: Local and global algorithms for disambiguation to Wikipedia. In: Proceedings of the 49th Annual Meeting of the Association for Computational Linguistics: Human Language Technologies-Volume 1, pp. 1375–1384 (2011)
44. Reisinger, J., Mooney, R.: Multi-prototype vector-space models of word meaning. In: Human Language Technologies: The 2010 Annual Conference of the North American Chapter of the Association for Computational Linguistics, pp. 109–117 (2010)
45. Saeidi, M., Sousa, S.B.d.S., Milios, E., Zeh, N., Berton, L.: Categorizing online harassment on Twitter. In: Joint European Conference on Machine Learning and Knowledge Discovery in Databases, pp. 283–297 (2019)
46. Sajadi, A.: Semantic analysis using Wikipedia graph structure. Ph.D. thesis, Dalhousie University (2018)
47. Scarlini, B., Pasini, T., Navigli, R.: SensEmBERT: context-enhanced sense embeddings for multilingual word sense disambiguation. In: AAAI, pp. 8758–8765 (2020)
48. Scarlini, B., Pasini, T., Navigli, R.: With more contexts comes better performance: contextualized sense embeddings for all-round word sense disambiguation. In: EMNLP, pp. 3528–3539 (2020)
49. Shnayderman, I., et al.: Fast end-to-end wikification. arXiv preprint arXiv:1908.06785 (2019)
50. Singh, H., Bhattacharyya, P.: A survey on word sense disambiguation. ACM Comput. Surv. (CSUR) (2019)
51. Song, Y., Roth, D.: Machine learning with world knowledge: the position and survey. arXiv preprint arXiv:1705.02908 (2017)
52. Sysoev, A., Nikishina, I.: Smart context generation for disambiguation to Wikipedia. In: Conference on Artificial Intelligence and Natural Language, pp. 11–22 (2018)

53. Szymański, J., Naruszewicz, M.: Review on wikification methods. AI Commun. **27**(2), 97–111 (2019)
54. Wang, Y., Wang, M., Fujita, H.: Word sense disambiguation: a comprehensive knowledge exploitation framework. Knowl. Based Syst. 105–117 (2019)
55. Weikum, G., Dong, L., Razniewski, S., Suchanek, F.: Machine knowledge: creation and curation of comprehensive knowledge bases. arXiv preprint arXiv:2009.11564 (2020)
56. West, R., Paranjape, A., Leskovec, J.: Mining missing hyperlinks from human navigation traces: a case study of Wikipedia. In: Proceedings of the 24th International Conference on World Wide Web, pp. 1242–1252 (2015)
57. Xin, K., Hua, W., Liu, Y., Zhou, X.: LoG: a locally-global model for entity disambiguation. World Wide Web **24**, 1–23 (2020)
58. Yang, Z., Dai, Z., Yang, Y., Carbonell, J., Salakhutdinov, R.R., Le, Q.V.: XLNet: generalized autoregressive pretraining for language understanding. Adv. Neural Inf. Process. Syst. **32**, 221–229 (2019)
59. Yao, L., Mao, C., Luo, Y.: Graph convolutional networks for text classification. In: Proceedings of the AAAI Conference on Artificial Intelligence, AAAI, Honolulu, vol. 33, pp. 7370–7377 (2019)
60. Zhang, Y., Ives, Z., Roth, D.: "who said it, and why?" provenance for natural language claims. In: ACL, pp. 4416–4426 (2020)
61. Zhao, G., Wu, J., Wang, D., Li, T.: Entity disambiguation to Wikipedia using collective ranking. Inf. Process. Manage. **52**(6), 1247–1257 (2016)

# Graph-Based Deep Generative Modelling for Document Layout Generation

Sanket Biswas[1]([✉]) [ID], Pau Riba[1] [ID], Josep Lladós[1] [ID], and Umapada Pal[2] [ID]

[1] Computer Vision Center and Computer Science Department, Universitat Autònoma de Barcelona, Bellaterra, Spain
{sbiswas,priba,josep}@cvc.uab.es
[2] CVPR Unit, Indian Statistical Institute, Kolkata, India
umapada@isical.ac.in

**Abstract.** One of the major prerequisites for any deep learning approach is the availability of large-scale training data. When dealing with scanned document images in real world scenarios, the principal information of its content is stored in the layout itself. In this work, we have proposed an automated deep generative model using Graph Neural Networks (GNNs) to generate synthetic data with highly variable and plausible document layouts that can be used to train document interpretation systems, in this case, specially in digital mailroom applications. It is also the first graph-based approach for document layout generation task experimented on administrative document images, in this case, invoices.

**Keywords:** Document synthesis · Graph Neural Networks · Document layout generation

## 1 Introduction

The variability and diversity of complex layouts and graphical entities in digital mailroom documents prevent us from tackling document understanding problems separately, and that such specificity has been a great barrier towards deriving off-the-shelf document analysis solutions, usable by nonspecialists. Apparently, OCR-based engines are the most widely recognized products in this research community. For instance, imagine a business firm having thousands of documents to process, analyze, and transform to carry out day-to-day operations. Examples of such documents might include receipts, invoices, forms, statements, contracts, and many more pieces of data, which are highly unstructured or semi-structured, and it is essential to be able to quickly analyze and understand the information embedded within the unevenly structured data in these cases. In most of these Document Image Analysis and Recognition (DIAR) applications, the document content has been broadly classified into two structural entities: (1) physical and (2) logical structural entities. While the physical structure describes the visual aspect of the document by representing the specific objects and their

© Springer Nature Switzerland AG 2021
E. H. Barney Smith and U. Pal (Eds.): ICDAR 2021 Workshops, LNCS 12917, pp. 525–537, 2021.
https://doi.org/10.1007/978-3-030-86159-9_38

mutual positions, the logical structure assigns a definite semantic meaning to each of these objects.

In recent times, deep CNN-based methods have tried to deduct the visual differences between object classes: while the visual characteristics of certain graphical elements (e.g., plots, charts, figures) differ conspicuously from text elements, the same cannot be said for tables, where the major differences from the surrounding content lie mostly stored in the layout information and its context. Moreover, trying to train these deep CNN models from scratch may be quite impractical due to the requirement of a large amount of training examples and the need of precisely annotated document datasets, which are scarcely available in the community. The key reason may be that most of these documents (administrative documents, for example) contain sensitive information (identity name, bank details, health information and so on) and are not publicly released by government agencies or business firms to be used in cloud services. Hence, as in many other applications requiring intensive training, data augmentation through synthetic generation is a solution. In the case of document structure recognition, there is an important need to generate synthetic document layouts that can encode the structural information of the real data and can be used during training to transfer enough knowledge to the model. Patil et al. [9] formulated this task as Document Layout Generation (DLG), where they used a recursive neural network approach to map the structured representation of semi-structured documents (in the form of tree-level hierarchies) to a code representation, the space of which is approximated by a Gaussian. New hierarchies representing plausible 2D document layouts were sampled from such distributions. In this work, we tackle the problem by encoding the structured hierarchies in the form of graph representation.

Graphs possess the ability to represent two types of contextual knowledge: (1) geometric/intrinsic, spatial structure of the document with positional information of object categories like tables, and (2) semantic/extrinsic, conceptual connections between the different object categories in a document. Therefore, graphs emerge as a suitable model to represent document layouts. The revolution of deep learning has also seen considerable progress in the area of graph-based representation and learning. Graph Neural Networks (GNNs) [6,14] as deep learning approaches have extended the power of CNNs to non-Euclidean geometries to capture long distance/different levels semantics based on the relations between objects. GNN's eventually learn a state embedding that contains the neighborhood information for each node (which can represent different entities or objects). The embedding is constructed at different graph convolution layers, so it encodes the information of a subgraph centered at a node. In the scenario of document interpretation, GNNs embed a description of a local layout as a context of a given document element. A recent application example was the detection of tables in case of administrative document images [20,21].

In summary, the main contributions of this work are as follows: (1) a novel approach has been proposed for DLG task using GNN's to generate synthetic data applied to administrative invoice documents where we render data in the form of diverse graphs that can actually match the structural characteristics of the target data. (2) The proposed graph-based generative modelling for such administrative documents also helps to invoke anonymity for the sensitive information (e.g. names, addresses, billing information, total amount etc.) it might contain in the document images. As shown in Fig. 1, the nodes in the graph represent the different entities (e.g. header, table, supplier etc.) in the document, while the edges represent the visibility relations (horizontal or vertical) between the neighbouring nodes. (3) All experiments of our model have been performed on administrative invoices collection from the RVL-CDIP [11] dataset. As a result, a new synthetic invoice dataset has been created for augmenting the train data during table detection and layout analysis tasks.

The rest of this paper is organized as follows. Section 2 provides a review of the relevant state of the art. In Sect. 3 we describe the main contribution of our work. Section 4 provides experimental validation with some relevant results of our proposed approach, both qualitatively and quantitatively. Finally, Sect. 5 concludes the work throwing some light on its future scope and benefits.

## 2   Related Work

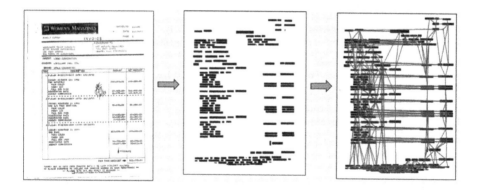

**Fig. 1.** Graph representation of the structure of an invoice image

### 2.1   Geometric Deep Learning

Geometric deep learning [6,14] has emerged as an extension of deep learning models to non-Euclidean domains, such as graphs and manifolds. To refer to neural networks applied to graph-structured data, the term Graph Neural Networks [23] was coined.

The GNN methods help to learn representations at the node, edge and graph level considering the underlying topological information. Based on the fundamental architecture, GNN methods can be aptly divided into two categories:

spatial and spectral methods. Spatial methods extend the idea of Convolutional Neural Networks (CNNs) for images and define a set of operations involving the local neighbourhood to compute a new representation [7,17]. On the other hand, spectral methods use the knowledge of spectral graph theory [24] and consider graph Laplacians for defining convolution operations in graph domain [6,14]. Gilmer et al. [10] generalized both the domains of GNN, and defined their approach in terms of a Neural Message Passing (NMP) pipeline. These fundamental architectures have been further extended to new tasks involving graphs, such as the generative variational graph autoencoder [15], learning graph edit distance between a pair of graphs [22], graph matching [28], etc.

## 2.2 Document Layout Generation

The study and analysis of the structural properties and relations between entities in documents is a fundamental challenge in the field of information retrieval. Although local tasks like the Optical Character Recognition (OCR) have been addressed with a considerably high model performance, the global and highly variable nature of document layouts has made their analysis some what more ambiguous. Previous works on structural document analysis mostly relied on the different kinds of specifically devised methods and applications [1,4,12,19]. Recent works have shown that deep learning based approaches have significantly improved the performance of these models in quality. A very standard approach in this regard was proposed by Yang et al. [26] which uses a joint visual and textual representation in a multimodal way of understanding, viewing the layout analysis as a pixel-wise segmentation task. But such modern deep learning based approaches typically require a very heavy amount of high-quality training data, that often calls for suitable methods to synthetically generate documents with real-looking layout [16] and content [18]. Our work actually focuses on the direction of research on synthetic layout generation, showing that our generated synthetic data can be extremely beneficial to augment training data for document analysis tasks.

Preserving the reliable representation of layouts has shown to be very useful in various graphical design contexts, which typically involve highly structured and content-rich objects. One such recent intuitive understanding was established by Li et al. [16] in their LayoutGAN, which aims to generate realistic document layouts using Generative Adversarial Networks (GANs) with a wireframe rendering layer. Zheng et al. [29] used a GAN-based approach to generate document layouts but their work focused mainly on content aware generation, that primarily uses the content of the document as an additional prior. Biswas et al. [2] devised a generative GAN-based model to synthesize realistic document images, guided by a spatial layout(bounding boxes with object categories) given as a reference by the user. However to use a more highly structured object generation, it is very important to focus operate on the low dimensional vectors unlike CNN's. Hence, in the most recent literature, Patil et al. [9] has exploited this highly structured positional information along with content to generate document layouts. They have used recursive neural networks which operate on the low dimensional vectors and employ two-layer perceptrons to merge any two vectors, which make them computationally cheaper and help them train with fewer samples. The recursive neural networks are coupled with Variational Autoencoders (VAEs) in their resulting model architecture and provides state-of-the-art results for generating synthetic layouts for 2D documents. They have also introduced a novel metric for measuring document similarity, called DocSim, and used this metric to show the novelty and diversity of the generated layouts.

Using geometric relations between the different entities in documents can actually help to preserve the structural information along with the content as seen in the work by Riba et al. [21] on table detection in invoice documents using GNNs. Figure 1 clearly illustrates how they have used graph modelling for document images to capture the geometrical structure of an invoice and using this knowledge can help us to generate more realistic synthetic samples for training. In this work we have used a similar kind of graph modelling for exploiting the structural information of an invoice image. Carbonell et al. [5] also used GNNs for recognition of structural components like named entities in semi-structured administrative documents. Traditional generative models for graphs [25] are usually hand-crafted to model a particular family of graphs, and thus they do not have the capacity to directly learn the generative model from observed data. To find a solution, one such graph-based generative model using GNNs was proposed by You et al. [27] on molecular data generation. They used sequential generation with Recurrent Neural Networks (RNNs) on top of graph based representations and get state-of-the-art results on molecular data generation. But there has not been any substantial work in the literature which has applied such graph-based generative models for document layout analysis tasks. In this context, it is indeed a challenging problem which we tackle in this work. As case study, we will work in the context of administrative documents, primarily focusing on invoices. Automated generation of synthetic document layouts will allow us to train document interpretation systems in a more efficient way for all kinds of document layout analysis tasks.

## 3   Method

In this work, we have explored a new research direction in the DIAR domain using the application of GNN. A hierarchical and scalable framework has been designed and implemented to exploit highly powerful graph representations in semi-structured administrative documents like invoices. Every document image has been modelled as a visibility graph which is fed to our generative model to synthesize meaningful document layouts. Figure 1 depicts the structure of an invoice document and how it has been modelled as a visibility graph. We considered a document graph whose nodes are graphical or named entities such as tables, figures, header, date, etc. while the edges represent their spatial relationships. We aim to learn a distribution $p(G)$ over an undirected graph $G = (V, E)$ defined by the node set $V = \{v_1, \ldots, v_n\}$ and edge set $E = \{(v_i, v_j) \mid v_i, v_j \in V\}$ that has a node ordering $\pi$ to map nodes to rows/columns of adjacency matrix

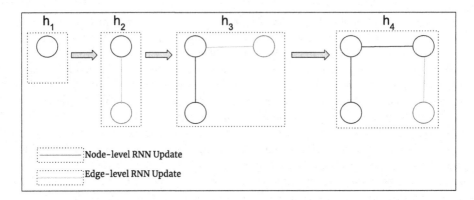

**Fig. 2. The Graph Generation Process:** Given figure illustrates the graph generation process during inference time. The Node-level RNN update encodes the graph hidden state $h$ , updated by the predicted adjacency vector $S_i^{\pi}$ for every node. The Edge-level RNN updates the sequence of edges when every new node is added.

$A^{\pi}$. This node ordering scheme has been adapted to enhance time efficiency of our training model. An adjacency matrix $A_{i,j}^{\pi}$ under a node ordering $\pi$ can be represented as $A_{i,j}^{\pi} = \mathbb{1}\left[\left(\pi\left(v_i\right), \pi\left(v_j\right)\right) \in E\right]$. So, in this work, we have proposed a graph generation framework applied in context to document images, that learns to generate realistic graphs by training on a representative set of graphs known as visibility graphs modelled from documents.

The main idea is to represent graphs of different node orderings as sequences, and then build a generative model on top of these sequences. As illustrated in the graph generation framework in Fig. 2, we decomposed the entire process into two parts: one that generates a sequence of nodes (Node-level update) and then another process that generates a sequence of edges for every new generated node (Edge-level update) which will be explained in more detail in the below subsections in a step-wise manner.

### 3.1   Graph to Sequence Mapping

We aim to learn a distribution $p_{model}(G)$ over graphs, based on a set of observed graphs $G = (G_1, ..., G_s)$, sampled from a data distribution $p(G)$, where each graph, may have a different set of nodes and edges. During training time instead of learning $p(G)$ directly, whose sample space is really complex to define, we instead sample a node ordering $\pi$ to get a set of sequences $S^{\pi}$ as our observations and learn $p(S^{\pi})$ instead. This help us to learn a model autoregressively due to the $S^{\pi}$. At inference time, we can simply sample G directly by computing $p(G)$ without this mapping. The mapping function $f_S$ from graphs to sequences, for a graph $G \sim p(G)$ with $n$ nodes under node ordering $\pi$ can be determined in Eq. 1.

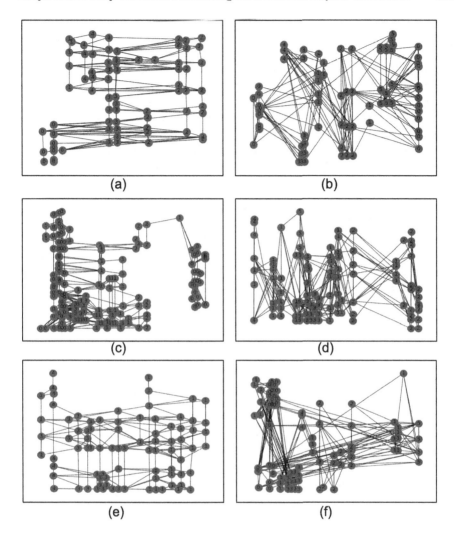

**Fig. 3.** Qualitative analysis of our generative model with figures (a), (c) and (e) representing the test graphs from real invoice data and (b), (d) and (f) representing the generated graphs from our model.

$$S^\pi = f_S(G, \pi) = (S_1^\pi, \ldots, S_n^\pi) \tag{1}$$

Every element $S_i^\pi$ is actually an adjacency vector that represents the edges between the present node $\pi(v_i)$ and previous nodes $\pi(v_j)$ already in graph. This can be further represented by the Eq. 2.

$$S_i^\pi = \left(A_{1,i}^\pi, \ldots, A_{i-1,i}^\pi\right)^T, \forall i \in \{2, 3, 4 \ldots, n\} \tag{2}$$

## 3.2   GRNN Framework

As a result of the graph to sequence mapping, our next step would be to generate the adjacency matrix of a graph G by generating these adjacency vectors $A_{i,i}^{\pi}$ of each node in a step by step sequential process. This can output different networks with variable number of nodes, while preserving the important topological properties of the generated graph. While transforming the learning distribution from $p(G)$ to $p(S^{\pi})$, we can further decompose $p(S^{\pi})$ as the product of conditional distributions over the elements due to its sequential nature as shown in Eq. 3.

$$p\left(S^{\pi}\right) = \prod_{i=1}^{n+1} p\left(S_i^{\pi} \mid S_1^{\pi}, \dots, S_{i-1}^{\pi}\right) \tag{3}$$

Now to model this still complex distribution, we used recurrent neural networks (RNNs) that consist of a state-transition function and an output-function as shown in Eqs. 4 and 5:

$$h_i = f_{\text{trans}}\left(h_{i-1}, S_{i-1}^{\pi}\right) \tag{4}$$

$$\theta_i = f_{\text{out}}\left(h_i\right) \tag{5}$$

where $h_i$ is a vector that encodes the updated generated graph information, $S_{i-1}^{\pi}$ is the adjacency vector for the last updated node $i-1$ and $\theta_i$ denotes the distribution of next node's adjacency vector i.e. $S_i^{\pi} \sim P_{\theta_i}$. So theoretically, the proposed Graph Recurrent Neural Network (GRNN) framework utilizes a hierarchical RNN as shown in Fig. 2, where the first (i.e. graph-level) RNN generates nodes and update the state of the graph. The second RNN (i.e. edge-level) generates the edges of a given node. To achieve a scalable modeling, we let these networks share their weights across all the time steps $i$ during the training phase.

## 3.3   Learning via Breadth-First Search

A great insight in our proposed method is rather than learning to generate graphs with using the Breadth-First Search (BFS) node orderings, instead of random node permutations. The BFS function takes a random permutation $\pi$ as input, picks $\pi(v_1)$ as starting node and appends the node neighbours into the BFS queue in an order defined by $\pi$. The modified equation for mapping function from graphs to sequences can rewritten as shown in Eq. 6.

$$S^{\pi} = f_S(G, \text{BFS}(G, \pi)) \tag{6}$$

This technique helps the model to be trained on all possible BFS orderings, instead of all possible node permutations. As the BFS function is deterministic in nature and many-to-one, i.e. one same ordering can eventually map multiple permutations, it reduces number of sequences we need to consider. It also makes the learning much simpler by reducing the number of edge predictions and adding possible edges only for the nodes that are considered in the BFS queue itself.

# 4    Experimental Validation

For our experimental validation, we have used the RVL-CDIP dataset [11]. In particular, the invoice subset has been split into 70 and 30% of the samples for training and evaluation respectively. Additionally, two extra datasets have been evaluated, namely, Protein [3] and Community [13], to evaluate the robustness of our method on other domains.

## 4.1    Datasets

**RVL-CDIP Invoices** [11]. The RVL-CDIP (Ryerson Vision Lab Complex Document Information Processing) is a well-known document information database containing 16 classes of documents with about 400,000 images in grayscale. For evaluating our synthetic graph generation framework, we chose the 518 images from the Invoice class, annotated with 5 different regions belonging to class header, table, address and so on. Here each invoice page is represented by a graph, with the nodes corresponding to the entities of different class in the invoice.

**Protein** [3]. 918 protein graphs have been used with every node representing an amino acid and two nodes are connected if they are at a distance threshold of 6 Angstrom.

**Community** [13]. The community dataset contains a collection of 500 two-community graphs, where each community generated by Erdos-Renyi model (E-R) [8], represents a node in the graph.

## 4.2    Data Preparation for Graph Representation

In this stage, given the image of a document, we apply physical layout techniques to detect the graphical regions. Given an invoice document, we represent each detected entity corresponding to a 7-dimensional vector containing the position of the bounding boxes and the histogram of its content (numbers, alphabets or symbol). This encoded information will be used to generate a visibility graph in order to represent the structural information of the document. We consider $G = (V, E)$ to be a visibility graph. The set of edges $E$ represent visibility relations between nodes. Two entities are said to be connected with an edge if and only if the bounding boxes are vertically or horizontally visible, i.e. a straight horizontal or vertical line can be traced between the bounding box of two entities without crossing any other. Also, long edges covering more than a quarter of the page height are discarded. An example of a visibility graph sample with corresponding node embedding is shown in Fig. 1.

### 4.3    Training Setup for Graph Recurrent Neural Network

Once the document has been processed and visibility graph has been generated, we feed them to our Graph Recurrent Neural Network (GRNN) model framework with the 7-dimensional node input space to get projected to a higher order space encoding with individual node features preserving the structural content information of the document. The graph-level RNN used in our work uses 4 layered GRU with 128 dimensional hidden state. To output the adjacency vector prediction, the edge-level RNN uses 4 layered GRU cells with 16 hidden dimensional state. To get the predicted adjacency vector in the output, the edge-level RNN maps the 16 dimensional hidden state to a 8 dimensional vector through a MLP and ReLU activation, then another MLP maps the vector to a scalar with sigmoid activation. We initialize the edge-level RNN by the output of the graph-level RNN when generating the start of sequences $S_{i-1}^{\pi}$. We use the highest layer hidden state of the graph-level RNN to initialize with a linear layer to match the dimensionality. During the training time, ground truth has been used rather than the model's own predictions. During the inference time, the model is allowed to use its own predicted graph samples at each time step to generate a graph. The Adam Optimizer has been used for minibatch size of 32. We set the learning rate to be 0.001 which is decayed by 0.2 at every 100th epoch in all experiments.

### 4.4    Evaluation Schema

The evaluation of the quality of generated graphs is quite hard to estimate. A fair comparison between the test graph and generated graph is required. By visualizing the sets of test graphs and the generated graphs, a fair qualitative comparison can be done. From Fig. 3 we can infer a qualitative comparison between the test and generated samples.

For a quantitative evaluation scheme, we have used the Maximum Mean Discrepancy (MMD) measures to calculate the distance between the two sets of graphs (in this case, the test sample and the generated sample). In our experiments, the derived MMD scores between the graphs have been calculated for degree and clustering coefficient distributions, along with the average orbit count statistics as shown in Table 1. The lower the scores, the better the real structure of the entities has been preserved.

### 4.5    Experiments on Administrative Invoice Documents

Experiments on the subset of administrative document (invoices) taken from RVL-CDIP [11] has been conducted and we report the first baseline for document layout generation using a Graph Neural Network(GNN) framework.

As illustrated in Fig. 3, we illustrate some of the qualitative results with the document graphs that we generated from our proposed GRNN model. The visualizations of the graph samples suggest that the generated graphs visually preserve the appearance of the reference one, so the model roughly learns to preserve both syntactic and semantic information for different entities. Eventually, we can create synthetic samples of invoices by generating more and more graph samples and also providing them during the inference time. Since this is the first baseline approach to use graph generative models in document datasets.

However, Table 1 depicts the quantitative results we obtained for RVL-CDIP Invoice dataset and we compared our model performance with some molecular datasets like Protein [3] and Community [13] present in the graph literature. Results clearly show that there is a huge room for improvement in the graph generative framework for documents when compared to the performance in the above mentioned benchmark molecular datasets. The 'tables' entity is a regular structured entity and our model works well for generating table classes in realistic positions. But the title, date and other entities in administrative documents do not contain uniform information about its structural relations and its quite difficult for the model to learn those semantic content.

**Table 1.** Summary of the final model results for Document Layout Generation

| Dataset | Degree ($\downarrow$) | Clustering ($\downarrow$) | Orbit ($\downarrow$) |
|---|---|---|---|
| Protein [3] | 0.014 | 0.002 | 0.039 |
| Community [13] | 0.034 | 0.102 | 0.037 |
| RVL-CDIP Invoices [11] | 0.373 | 0.166 | 0.188 |

## 5 Conclusion

In this work, we have presented a novel approach to automatically synthesize document layouts structures. The proposed method is able to understand the complex interactions among the different layout components and generate plausible layouts for 2D documents. The graph-based generative approach also explores the power of GNN's towards the learning and generation of complex structured layouts for administrative invoices as a case study.

The future scope of this work will be mainly focused on two research lines. Firstly, there is a requirement for a more efficient evaluation of synthetically generated layouts when compared to real document layout samples both quantitatively and qualitatively. Secondly, exploiting this generated layout samples for supervision purposes can enhance the performance on well-defined tasks such as table detection or document layout analysis.

**Acknowledgment.** This work has been partially supported by the Spanish projects RTI2018-095645-B-C21, and FCT-19-15244, and the Catalan projects 2017-SGR-1783, the CERCA Program/Generalitat de Catalunya and PhD Scholarship from AGAUR (2021FIB-10010). We are also indebted to Dr. Joan Mas Romeu for all the help and assistance provided during the data preparation stage for the experiments.

# References

1. Baird, H.S., Bunke, H., Yamamoto, K.: Structured Document Image Analysis. Springer Science and Business Media, Heidelberg (2012). https://doi.org/10.1007/978-3-642-77281-8
2. Biswas, S., Riba, P., Lladós, J., Pal, U.: Docsynth: a layout guided approach for controllable document image synthesis. In: International Conference on Document Analysis and Recognition (ICDAR) (2021)
3. Borgwardt, K.M., Ong, C.S., Schönauer, S., Vishwanathan, S., Smola, A.J., Kriegel, H.P.: Protein function prediction via graph kernels. Bioinformatics 21(suppl_1), i47–i56 (2005)
4. Breuel, T.M.: High performance document layout analysis. In: Proceedings of the Symposium on Document Image Understanding Technology, pp. 209–218 (2003)
5. Carbonell, M., Riba, P., Villegas, M., Fornés, A., Lladós, J.: Named entity recognition and relation extraction with graph neural networks in semi structured documents. In: 2020 25th International Conference on Pattern Recognition (ICPR), pp. 9622–9627. IEEE (2021)
6. Defferrard, M., Bresson, X., Vandergheynst, P.: Convolutional neural networks on graphs with fast localized spectral filtering. In: Advances in Neural Information Processing Systems, pp. 3844–3852 (2016)
7. Duvenaud, D.K., et al.: Convolutional networks on graphs for learning molecular fingerprints. In: Advances in Neural Information Processing Systems (2015)
8. Erdős, P., Rényi, A.: On the evolution of random graphs. Publ. Math. Inst. Hung. Acad. Sci 5(1), 17–60 (1960)
9. Gadi Patil, A., Ben-Eliezer, O., Perel, O., Averbuch-Elor, H.: Read: recursive autoencoders for document layout generation. In: Proceedings of the IEEE/CVF Conference on Computer Vision and Pattern Recognition Workshops, pp. 544–545 (2020)
10. Gilmer, J., Schoenholz, S.S., Riley, P.F., Vinyals, O., Dahl, G.E.: Neural message passing for quantum chemistry. arXiv preprint arXiv:1704.01212 (2017)
11. Harley, A.W., Ufkes, A., Derpanis, K.G.: Evaluation of deep convolutional nets for document image classification and retrieval. In: 2015 13th International Conference on Document Analysis and Recognition (ICDAR), pp. 991–995. IEEE (2015)
12. Kasturi, R., O'gorman, L., Govindaraju, V.: Document image analysis: a primer. Sadhana 27(1), 3–22 (2002)
13. Kim, J., Lee, J.G.: Community detection in multi-layer graphs: a survey. ACM SIGMOD Rec. 44(3), 37–48 (2015)
14. Kipf, T.N., Welling, M.: Semi-supervised classification with graph convolutional networks. arXiv preprint arXiv:1609.02907 (2016)
15. Kipf, T.N., Welling, M.: Variational graph auto-encoders. arXiv preprint arXiv:1611.07308 (2016)
16. Li, J., Yang, J., Hertzmann, A., Zhang, J., Xu, T.: Layoutgan: generating graphic layouts with wireframe discriminators. arXiv preprint arXiv:1901.06767 (2019)
17. Li, Y., Tarlow, D., Brockschmidt, M., Zemel, R.: Gated graph sequence neural networks. arXiv preprint arXiv:1511.05493 (2015)
18. Liu, T.F., Craft, M., Situ, J., Yumer, E., Mech, R., Kumar, R.: Learning design semantics for mobile apps. In: Proceedings of the 31st Annual ACM Symposium on User Interface Software and Technology (2018)
19. O'Gorman, L.: The document spectrum for page layout analysis. IEEE Trans. Pattern Anal. Mach. Intell. 15(11), 1162–1173 (1993)

20. Qasim, S.R., Mahmood, H., Shafait, F.: Rethinking table recognition using graph neural networks. In: 2019 International Conference on Document Analysis and Recognition (ICDAR), pp. 142–147. IEEE (2019)
21. Riba, P., Dutta, A., Goldmann, L., Fornés, A., Ramos, O., Lladós, J.: Table detection in invoice documents by graph neural networks. In: 2019 International Conference on Document Analysis and Recognition (ICDAR) (2019)
22. Riba, P., Fischer, A., Lladós, J., Fornés, A.: Learning graph distances with message passing neural networks. In: 2018 24th International Conference on Pattern Recognition (ICPR) (2018)
23. Scarselli, F., Gori, M., Tsoi, A.C., Hagenbuchner, M., Monfardini, G.: The graph neural network model. IEEE Trans. Neural Networks **20**(1), 61–80 (2008)
24. Shuman, D.I., Narang, S.K., Frossard, P., Ortega, A., Vandergheynst, P.: The emerging field of signal processing on graphs: extending high-dimensional data analysis to networks and other irregular domains. IEEE Signal Process. Mag. **30**(3), 83–98 (2013)
25. White, D., Wilson, R.C.: Spectral generative models for graphs. In: 14th International Conference on Image Analysis and Processing (ICIAP 2007), pp. 35–42. IEEE (2007)
26. Yang, X., Yumer, E., Asente, P., Kraley, M., Kifer, D., Lee Giles, C.: Learning to extract semantic structure from documents using multimodal fully convolutional neural networks. In: Proceedings of the IEEE Conference on Computer Vision and Pattern Recognition, pp. 5315–5324 (2017)
27. You, J., Ying, R., Ren, X., Hamilton, W.L., Leskovec, J.: GraphRNN: generating realistic graphs with deep auto-regressive models. arXiv preprint arXiv:1802.08773 (2018)
28. Zanfir, A., Sminchisescu, C.: Deep learning of graph matching. In: Proceedings of the IEEE Conference on Computer Vision and Pattern Recognition (2018)
29. Zheng, X., Qiao, X., Cao, Y., Lau, R.W.: Content-aware generative modeling of graphic design layouts. ACM Trans. Graph. (TOG) **38**(4), 1–15 (2019)

# Author Index

Printed in the United States
by Baker & Taylor Publisher Services